ECOTOXICOLOGY

A DERIVATIVE OF ENCYCLOPEDIA OF ECOLOGY

EDITORIAL BOARD

ECOTOXICOLOGY

A DERIVATIVE OF ENCYCLOPEDIA OF ECOLOGY

Editor-in-Chief

SVEN ERIK JØRGENSEN

Copenhagen University,
Faculty of Pharmaceutical Sciences,
Institute A, Section of Environmental Chemistry,
Toxicology and Ecotoxicology,
University Park 2,
Copenhagen,
Denmark

Associate Editor-in-Chief

BRIAN D. FATH

Department of Biological Sciences,
Towson University,
Towson, Maryland,
USA

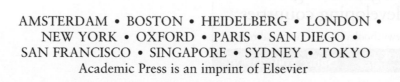

AMSTERDAM • BOSTON • HEIDELBERG • LONDON •
NEW YORK • OXFORD • PARIS • SAN DIEGO •
SAN FRANCISCO • SINGAPORE • SYDNEY • TOKYO
Academic Press is an imprint of Elsevier

ELSEVIER

ACADEMIC PRESS

Academic Press is an imprint of Elsevier
Radarweg 29, 1043 NX Amsterdam, The Netherlands
Linacre House, Jordan Hill, Oxford OX2 8DP, UK
525 B Street, Suite 1900, San Diego, CA 92101-4495, USA
30 Corporate Drive, Suite 400, Burlington, MA01803, USA

British Library Cataloguing in Publication Data
A catalogue record for this book is available from the British Library

Library of Congress Cataloging-in-Publication Data
Ecotoxicology: a derivative of Encyclopedia of ecology/editor-in-chief, Sven Erik Jørgensen;
associate editor-in-chief, Brian D. Fath.
    p. ; cm.
  Includes bibliographical references and index.
  ISBN 978-0-444-53628-0
1. Environmental toxicology–Encyclopedias. I. Jørgensen, Sven Erik, 1934-
    II. Fath, Brian D. III. Encyclopedia of ecology.
  [DNLM: 1. Environmental Pollution–adverse effects. 2. Ecosystem. 3. Hazardous Substances–adverse
    effects. 4. Hazardous Substances–pharmacokinetics. WA 670 E215 2010]
  RA1226.E255 2010
  577.03–dc22

                                                              2010008422

ISBN: 978-0-444-53628-0

For information on all Academic Press publications
visit our website at elsevierdirect.com

Printed and bound by CPI Group (UK) Ltd, Croydon, CR0 4YY

CONTENTS

PART C: CHEMICALS WITH ECOTOXICOLOGICAL EFFECTS

LIST OF CONTRIBUTORS

H J Allen
US EPA, NRMRL, Cincinnati, OH, USA

I T Aighewi
University of Maryland Eastern Shore, Princess Anne, MD, USA

O Anderbrant
Lund University, Lund, Sweden

K A Anderson
Oregon State University, Corvallis, OR, USA

M Ashmore
University of York, York, UK

S M Bard
Dalhousie University, Halifax, NS, Canada

S M Bartell
E2 Consulting Engineers, Inc., Maryville, TN, USA

V N Bashkin
VNIIGAZ/Gazprom, Moscow, Russia

J B Belden
Baylor University, Waco, TX, USA

W N Beyer
USGS Patuxent Wildlife Research Center, Beltsville, MD, USA

K Borgå
Norwegian Institute for Water Research, Oslo, Norway

B W Brooks
Baylor University, Waco, TX, USA

J P Carbone
Rohm and Haas Company, Spring House, PA, USA

D B Chambers
SENES Consultants Limited, Richmond Hill, ON, Canada

T E Chow
University of Michigan – Flint, Flint, MI, USA

P C Chrostowski
CPF Associates, Inc., Takoma Park, MD, USA

J M Conley
University of Tennessee at Chattanooga, Chattanooga, TN, USA

O A Demidova
Moscow State University, Moscow, Russia

K G Drouillard
University of Windsor, Windsor, ON, Canada

S A Dyer
Washington Savannah River Company, Aiken, SC, USA

A Fairbrother
Parametrix, Inc., Bellevue, WA, USA

S Fernandes
SENES Consultants Limited, Richmond Hill, ON, Canada

K F Gaines
Eastern Illinois University, Charleston, IL, USA

A Garva
SENES Consultants Limited, Richmond Hill, ON, Canada

F A P C Gobas
Simon Fraser University, Burnaby, BC, Canada

L J Guillette Jr.
University of Florida, Gainesville, FL, USA

R M Harper
Western Washington University, Bellingham, WA, USA

W E Hillwalker
Oregon State University, Corvallis, OR, USA

A L Iamiceli
Italian National Institute of Health, Rome, Italy

S H Imam
Western Regional Research Center (ARS-USDA), Albany, CA, USA

A B Ishaque
University of Maryland Eastern Shore, Princess Anne, MD, USA

S E Jørgensen
Copenhagen University, Copenhagen, Denmark

C Kantar
Mersin University, Mersin, Turkey

M A Q Khan
University of Illinois, Chicago, IL, USA

S F Khan
University of Denver, Denver, CO, USA

J R Kuykendall
ChemRisk, Inc., Boulder, CO, USA

W G Landis
Western Washington University, Bellingham, WA, USA

C Y Lin
University of California, Davis, CA, USA

L B Martello
ENVIRON International Corporation, San Francisco, CA, USA

S Matsui
Kyoto University, Kyoto, Japan

J D Maul
Texas Tech University, Lubbock, TX, USA

L S McCarty
LS McCarty Scientific Research & Consulting, Markham, ON, Canada

J P Meador
NOAA Fisheries, Seattle, WA, USA

R Miniero
Italian National Institute of Health, Rome, Italy

J P Myers
Environmental Health Sciences, White Hall, VA, USA

S Nagata
Kobe University, Kobe, Japan

W Naito
National Institute of Advanced Industrial Science and Technology, Tsukuba, Japan

M C Newman
College of William and Mary, Gloucester Point, VA, USA

M C Newman
Virginia Institute of Marine Science, Gloucester Point, VA, USA

H Okamura
Kobe University, Kobe, Japan

R A Pastorok
Integral Consulting, Inc., Mercer Island, WA, USA

W J G M Peijnenburg
RIVM – Laboratory for Ecological Risk Assessment, Bilthoven, The Netherlands

H Phillips
SENES Consultants Limited, Richmond Hill, ON, Canada

D Preziosi
Integral Consulting, Inc., Berlin, MD, USA

K H Reinert
AMEC Earth & Environmental, Inc., Plymouth Meeting, PA, USA

S M Richards
University of Tennessee at Chattanooga, Chattanooga, TN, USA

G F Riedel
Smithsonian Institution, Edgewater, MD, USA

D Rudnick
Integral Consulting, Inc., Mercer Island, WA, USA

A M Scheuhammer
Environment Canada, Ottawa, ON, Canada

C J Schmitt
US Geological Survey, Columbia, MO, USA

F Shattari
University of Boston, Boston, MA, USA

A J Stewart
Oak Ridge Associated Universities, Oak Ridge, TN, USA

R F Stewart
Bay Materials, LLC, Menlo Park, CA, USA

S H Swan
University of Rochester, Rochester, NY, USA

B H Sørensen
Copenhagen University, Copenhagen, Denmark

L V Tannenbaum
US Army Center for Health Promotion and Preventive Medicine, Aberdeen, MD, USA

C W Theodorakis
Southern Illinois University Edwardsville, Edwardsville, IL, USA

G O Thomas
Lancaster University, Lancaster, UK

R M Tinnacher
Colorado School of Mines, Golden, CO, USA

R S Tjeerdema
University of California, Davis, CA, USA

L van den Berg
University of York, York, UK

F S vom Saal
University of Missouri – Columbia, Columbia, MO, USA

W T Waller
University of North Texas, Denton, TX, USA

R J Wenning
ENVIRON International Corporation, San Francisco, CA, USA

B R Zaidi
University of Puerto Rico, Mayaguez, PR, USA

Y Zhao
Virginia Institute of Marine Science, Gloucester Point, VA, USA

X Zhou
Kobe University, Kobe, Japan

PREFACE

E cotoxicology is a subdiscipline of ecology that focuses on the effect of toxic substances on ecosystems and their living components. A full understanding of the ecological effects of toxic substances will require a deep knowledge of the reactions and processes of the toxic substances in nature. Biogeochemical kinetics of toxic substances is therefore an integrated part of ecotoxicology.

Today, the industrialized countries use about 100 000 chemicals, but we have a limited knowledge of only 1% of the information that is needed to make a proper environmental risk assessment of these chemicals. The need for a comprehensive treatment of ecotoxicology supplemented in the future hopefully with an increasingly wider knowledge is therefore obvious.

Part A of this book presents the basic concepts of ecotoxicology. It contains the nomenclature, the most focal topics, the history and the three very fundamental concepts of dose–response, body residues, and assimilative capacity. Further details and ecotoxicological core concepts of environmental management, ERA, are given in the last two chapters of Part A.

Part B includes 17 chapters, in which the ecotoxicological properties of the chemicals are given in details. Part C has 22 chapters and these are devoted to the presentation of ecotoxicologically most important chemicals – from the point of their effects and widespread uses. Parts B and C give two different overviews of ecotoxicology: one based on the harmful propertics and the other based on the harmful chemicals.

It is my hope that this book will be utilized by ecologists, ecotoxicologists, and environmental managers in their effort to reduce the harmful effects of toxic substances on nature and human health.

The book is a derivative of the recently published *Encyclopedia of Ecology*. Due to an excellent work by the section editor of Ecotoxicology, Steve Bartell, it has been possible to present a comprehensive overview of the ecotoxicological problems of today, including the basic concepts of ecotoxicology, the harmful ecotoxicological effects, and the most harmful chemicals.

I would like to thank Steve Bartell and all the authors of the ecotoxicological entries, who made it possible to produce this broad and up-to-date coverage of ecotoxicology.

Sven Erik Jørgensen
Copenhagen, October 2009

PART A

Ecotoxicology as an Ecological Subdiscipline, Basic Concepts

Introduction

S E Jørgensen, Copenhagen University, Copenhagen, Denmark

Ecotoxicology as an Ecological Subdiscipline	Toxic Chemicals: Overview of Their Properties and Their Ecotoxicological Effects
Effects and Risk Assessment	Further Reading

Ecotoxicology as an Ecological Subdiscipline

Ecotoxicology is a subdiscipline of ecology that focuses on the effects of toxic substances on ecosystems and their living components. A full understanding of the ecological effects of toxic substances will require an extensive knowledge of the reactions and processes of the toxic substances in nature. Biogeochemical kinetics of toxic substances is therefore an integrated part of ecotoxicology.

Contaminants can be classified into two groups:

1. nonthreshold or gradual agents, which are harmful at almost any concentration and amount, and
2. threshold agents, which have a harmful effect only above or below some concentration or threshold level.

The first group represents toxic substances that are in the focus of ecotoxicology. In principle, environmental management attempts to avoid their presence in the environment completely, as they may be harmful at even the smallest concentration. It is not feasible in practice; however, for some toxic substances such as pesticides in drinking water, it is required that the concentrations are below the detection limit, which is a very low concentration – usually measured in $ng\,l^{-1}$. The toxicological threshold effect concentration (TEC; also see the chapter 'Food-Web Bioaccumulation Models') of some toxic substances is known. TEC is the concentration below which no toxicological effect can be observed.

The second group includes various nutrients such as phosphorus, nitrogen, silica, carbon, vitamins, and biochemically important metal ions (iron, zinc, calcium, and so on). These components and their effects and reactions are covered in general ecology and in a number of ecological subdisciplines, such as limnology, forest ecology, agricultural ecology, and terrestrial ecology.

Figure 1 illustrates the difference between gradual and threshold agents.

An ecotoxicological periodic table can be found in Appendix 1. The elements are divided into four groups. The 13 elements that are important as building blocks in biological material are in green. They participate in all ecological processes and cycle in ecological networks in all ecosystems. The 10 elements that are trace elements, which means that they are present in small concentrations in biological material, are in blue. These elements often play an important biochemical role, for example as coenzymes. They may in some context have ecotoxicological effects, particularly at higher concentrations. The 21 elements with toxicological effects are in red. They are the focus of ecotoxicology and it is important to know their environmental concentration and effects. The elements Po, Rn, Ra, U, Np, Pu, and Th have radioactive effects. The remaining elements are represented in yellow and they are of minor interest to ecology and its subdisciplines due to their low concentrations in biological material and minor effects on ecological processes.

Effects and Risk Assessment

The obvious questions in ecotoxicology for each of the many compounds with ecotoxicological effects are as follows:

1. Which concentration do we find in the environment? In which ecosystems and in which parts of the ecosystems?
2. Could these concentrations cause any harm?
3. To answer the second question, we need an answer to another question: What is the relationship between concentration and effect?
4. When we characterize the relationship between probable concentrations in the different parts of the environment and the corresponding effects, what is the risk for a harmful effect?
5. Should we and could we reduce this risk?

These five questions are the pragmatic ideas behind the application of environmental risk assessments (ERAs). The chapter 'Ecological Risk Assessment' gives the basic environmental management principles behind ERA, and the following text describes in detail how to perform an ERA.

The environmental risks of all chemicals should be known to be able to phase out the most environmentally

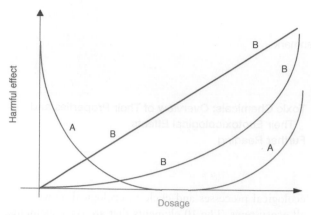

Figure 1 A is a threshold agent. A certain dosage is needed, which implies that a too little or a too high dosage is harmful. B is a non-threshold or gradual agent, that is harmful in any concentration. The harmful effect could be linear or non-linear as shown on the graph.

threatening chemicals and set standards for the use of all other chemicals. The standards should ensure that there is no serious risk in using the chemicals, provided that the standards are followed carefully. Modern abatement of pollution therefore includes ERA, which may be defined as the process of assigning magnitudes and probabilities to the adverse effects of human activities. The process involves identification of hazards such as the release of toxic chemicals to the environment by quantification of the relationship between an activity associated with an emission to the environment and its effects. The entire ecological hierarchy is considered in this context, which implies that the effects at different levels such as the cell (biochemical level), the organism, the population, the ecosystem, and the entire ecosphere should be considered.

The application of ERA is rooted in the recognition that

1. the cost of elimination of all environmental effects is impossibly high and
2. decisions in practical environmental management must always be made on the basis of incomplete information.

We use about 100 000 chemicals in such amounts that they could be a threaten to the environment, but we know only about 1% of what we need to know to be able to make a proper and complete ERA of these chemicals. This chapter provides a short introduction to available estimation methods that have been recommended to be applied in case of lack of information in the literature concerning the properties of chemical compounds. A list of relevant properties is also given in this context, and it is discussed what these properties mean with respect to environmental impact.

ERA belongs to the same 'assessment family' as environmental impact assessment (EIA), which attempts to assess the impact of a human activity. EIA is predictive and comparative and is concerned with all possible effects on the environment, including secondary and tertiary (indirect) effects, while ERA attempts to assess the probability of a given (defined) adverse effect as a result of a human activity.

Both ERA and EIA use models to find the expected environmental concentration (EEC), which is translated into impacts for EIA and into risks of specific effects for ERA.

Legislation and regulation of domestic and industrial chemicals with respect to environmental protection have been implemented in Europe and North America for decades. Both regions distinguish between existing chemicals and newly introduced chemicals.

At the UNCED meeting in Rio de Janeiro in 1992 on the Environment and Sustainable Development, it was decided to create an Intergovernmental Forum on Chemical Safety (IGFCS, Chapter 19 of Agenda 21). The primary task is to stimulate and coordinate global harmonization in the field of chemical safety, covering the following principal themes: assessment of chemical risks, global harmonization of classification and labeling, information exchange, risk reduction programs, and capacity building in chemicals management.

Uncertainty plays an important role in risk assessment. Risk is the probability that a specified harmful effect will occur or, in the case of a graded effect, the relationship between the magnitude of the effect and its probability of occurrence.

Risk assessment has emphasized risks to human health and has to a certain extent ignored ecological effects. However, it has been acknowledged that some chemicals that pose little or no risk to human health can cause severe effects on other organisms such as aquatic organisms. Examples are chlorine, ammonia, and certain pesticides. An up-to-date risk assessment therefore has to consider the entire ecological hierarchy, which is the ecologist's view of the world in terms of levels of organization. Organisms interact directly with the environment and it is organisms that are exposed to toxic chemicals. The species sensitivity distribution is therefore more ecologically credible. The reproducing population is the smallest meaningful level in an ecological sense. However, populations do not exist in vacuum, but require a community of other organisms of which the population is a part. The community occupies a physical environment with which it forms an ecosystem.

Moreover, both the various adverse effects and the ecological hierarchy have different scales in time and space, which must be included in a proper ERA (see **Figure 2**). For example, oil spills occur at a spatial scale similar to those of populations, but they are briefer

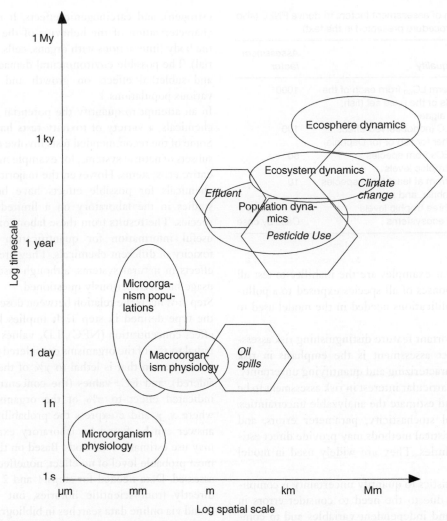

Figure 2 The spatial scale and timescale for various hazards (hexagons, italic) and for the various levels of the ecological hierarchy (circles, nonitalic).

than population processes. Therefore, the risk assessment of an oil spill requires considerations of reproduction and recolonization that occur on a longer timescale and that determine the magnitude of the population response and its significance to natural population variance.

Uncertainties in risk assessment are most commonly taken into account by application of safety factors. Uncertainties are due to three basic causes:

1. the inherent randomness of the world (stochasticity),
2. errors in execution of assessment, and
3. imperfect or incomplete knowledge.

The inherent randomness refers to uncertainty that can be described and estimated but cannot be reduced because it is characteristic of the system. Meteorological factors such as rainfall, temperature, and wind are effectively stochastic at levels of interest for risk assessment. Many biological processes such as colonization, reproduction, and mortality also need to be described stochastically.

Human errors are inevitable attributes of all human activities. This type of uncertainty includes incorrect measurements, data recording errors, computational errors, and so on.

The uncertainty is considered by use of an assessment (safety) factor that varies from 10 to 1000. The choice of assessment factor depends on the quantity and quality of toxicity data (see **Table 1**). The assessment or safety factor is used in step 3 of the ERA procedure, presented below. Relationships other than the uncertainties originating from randomness, errors, and lack of knowledge may be considered when the assessment factors are selected, for example, cost–benefit. This implies that the assessment factors for drugs and pesticides for instance may be given a lower value due to their possible benefits.

Lack of knowledge results in undefined uncertainty that cannot be described or quantified. It is a result of practical constraints on our ability to accurately describe, count, measure, or quantify everything that pertains to a

Table 1 Selection of assessment factors to derive PNEC (also see step 3 of the procedure presented in the text)

Data quantity and quality	Assessment factor
At least one short-term LC_{50} from each of the three trophic levels of the base set (fish, zooplankton, and algae)	1000
One long-term NOEC (nonobserved effect concentration, either for fish or for Daphnia)	100
Two long-term NOECs from species representing two trophic levels	50
Long-term NOECs from at least three species (normally fish, Daphnia, and algae) representing the three trophic levels	10
Field data or model ecosystems	Case by case

risk estimate. Clear examples are the inability to test all toxicological responses of all species exposed to a pollutant and the simplifications needed in the model used to predict the EEC.

The most important feature distinguishing risk assessment from impact assessment is the emphasis in risk assessment on characterizing and quantifying uncertainty. It is therefore of particular interest in risk assessment to be able to analyze and estimate the analyzable uncertainties. They are natural stochasticity, parameter errors, and model errors. Statistical methods may provide direct estimates of uncertainties. They are widely used in model development.

The use of statistics to quantify uncertainty is complicated in practice due to the need to consider errors in both dependent and independent variables and to combine errors when multiple extrapolations should be made. Monte Carlo analysis is often used to overcome these difficulties.

Model errors include inappropriate selection or aggregation of variables, incorrect functional forms, and incorrect boundaries. The uncertainty associated with model errors is usually assessed by field measurements utilized for calibration and validation of the model. The modeling uncertainty for ecotoxicological models is in principle not different from that for other ecological models.

Risk assessment of chemicals may be divided into nine steps, which are shown in **Figure 3**. The nine steps correspond to questions which the risk assessment attempts to answer to be able to quantify the risk associated with the use of a chemical. The nine steps are presented in detail below with reference to **Figure 3**.

Step 1: Which hazards are associated with the application of the chemical? This involves gathering data on the types of hazards – possible environmental damage and human health effects. The health effects include congenital, neurological, mutagenic, endocrine disruption (so-called estrogen), and carcinogenic effects. It may also include characterization of the behavior of the chemical within the body (interactions with organs, cells, or genetic material). The possible environmental damage includes lethal and sublethal effects on growth and reproduction of various populations.

In an attempt to quantify the potential danger posed by chemicals, a variety of toxicity tests have been devised. Some of the recommended tests involve experiments with subsets of natural systems, for example microcosms, or the entire ecosystems. However, the majority of tests on new chemicals for possible effects have been confined to studies in the laboratory on a limited number of test species. The results from these laboratory assays provide useful information for quantification of the relative toxicity of different chemicals. They are used to forecast effects in natural systems, although justification of their usage has been seriously questioned.

Step 2: What is the relation between dose and responses of the type defined in step 1? It implies knowledge of no effect concentration (NEC), LD_x values (the dose that is lethal to x% of the organisms considered), LC_y values (the concentration that is lethal to y% of the organisms considered), and EC_z values (the concentration giving the indicated effect to z% of the organisms considered), where x, y, and z express the probability of harm. The answer can be found by laboratory examination or we may use estimation methods. Based on these answers, the most probable level of no effect, noneffect level (NEL), is assessed. Data needed for steps 1 and 2 can be obtained directly from scientific libraries, but are increasingly found via online data searches in bibliographic and factual databases. Data gaps should be filled with estimated data. It is very difficult to gain complete knowledge about the effect of a chemical at all levels, from cells to ecosystem. Some effects are associated with very small concentrations, for example, the estrogen effect. It is therefore far from sufficient to know NEC and LD_x, LC_y, and EC_z values.

Step 3: Which uncertainty (safety) factors reflect the amount of uncertainty that must be taken into account when experimental laboratory data or empirical estimation methods are extrapolated to real situations? Usually, safety factors of 10–1000 are used. The choice is discussed above and will usually be in accordance with **Table 1**. In case of good knowledge about the chemical, a safety factor of 10 may be applied. If, on the other hand, it is estimated that the available information has a very high uncertainty, a safety factor of 10 000 may be recommended in a few cases. Most frequently, safety factors of 50–100 are applied. NEL times the safety factor is called the predicted noneffect level (PNEL). The complexity of ERA is often simplified by deriving the predicted no effect concentration (PNEC) for different environmental components (water, soil, air, biotas, and sediment).

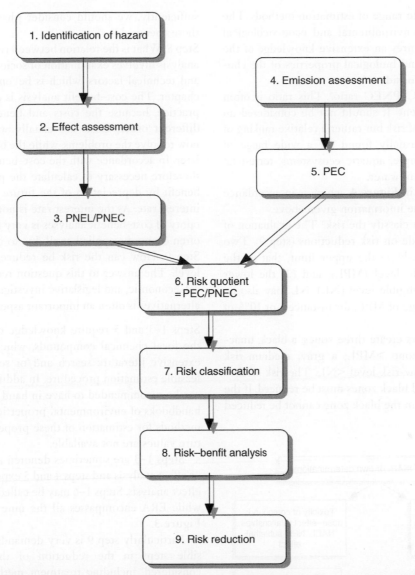

1. Identification of hazard

2. Effect assessment

3. PNEL/PNEC

4. Emission assessment

5. PEC

6. Risk quotient = PEC/PNEC

7. Risk classification

8. Risk–benfit analysis

9. Risk reduction

Figure 3 The presented procedure in nine steps to assess the risk of chemical compounds. Steps 1–3 require an extensive use of ecotoxicological handbooks and ecotoxicological estimation methods to assess the toxicological properties of the chemical compounds considered, while step 5 requires the selection of a proper ecotoxicological model. PEC, predicted environmental concentration; PNEC, predicted no effect concentration; PNEL, predicted noneffect level.

Step 4: What are the sources and quantities of emissions? The answer requires thorough knowledge of the production and use of the chemical compounds considered, including an assessment of how much of the chemical is wasted in the environment through production and use? The chemical may also be a waste product, which makes it very difficult to determine the amounts involved. For instance, the very toxic dioxins are waste products of incineration of organic waste.

Step 5: What is (are) the actual exposure concentration(s)? The answer to this question is called the predicted environmental concentration (PEC). Exposure can be assessed by measuring environmental concentrations. It may also be predicted by a model, when the emissions are known.

The use of models is necessary in most cases either because we are considering a new chemical or because the assessment of environmental concentrations without the use of models requires a very large number of measurements to determine the variations in concentrations in time and space. Furthermore, it provides an additional certainty to compare model results with measurements, which implies that it is always recommended to both develop a model and make at least a few measurements of concentrations of the ecosystem components, when and where it is expected that the highest concentration will occur. Most models will demand an input of parameters, describing the properties of the chemicals and the organisms, which will also require an extensive application of handbooks and a wide range of estimation methods. The development of an environmental and ecotoxicological model therefore requires an extensive knowledge of the physical, chemical, and biological properties of the chemical compound(s) considered.

Step 6: What is PEC/PNEC ratio? This ratio is often called the risk quotient. It should not be considered an absolute assessment of risk but rather a relative ranking of risks. The ratio is usually found for a wide range of ecosystems, for example, aquatic ecosystems, terrestrial ecosystems, and groundwater.

Steps 1–6 are shown in **Figure 4**, which is in accordance with **Figure 3** and the information given above.

Step 7: How will you classify the risk? The evaluation of risks is made to decide on risk reductions (step 9). Two risk levels are defined: (1) the upper limit, that is, the maximum permissible level (MPL), and (2) the lower limit, that is, the negligible level (NL). NL may also be defined as a percentage of MPL, for instance 1 or 10% of MPL.

The two risk limits create three zones: a black, unacceptable, high-risk zone >MPL; a gray, medium-risk level; and a white, low-risk level <NL. The risk of chemicals in the gray and black zones must be reduced. If the risk of the chemicals in the black zone cannot be reduced

sufficiently, we should consider phasing out the use of these chemicals.

Step 8: What is the relation between risk and benefit? This analysis involves examination of socioeconomic, political, and technical factors, which is beyond the scope of this chapter. The cost–benefit analysis is difficult to apply in practice, because the costs and benefits are often of a different order. Costs are usually what must be spent now to solve the problems, while the benefits are obtained later. In accordance with the cost–benefit procedure, it is therefore necessary to calculate the present value of the benefit by depreciation of the future value by use of an interest rate. As the interest rate is not known, the uncertainty of cost–benefit analyses is very high and the results often cannot be applied at all due to uncertainty.

Step 9: How can the risk be reduced to an acceptable level? The answer to this question requires deep technical, economic, and legislative investigation. Assessment of alternatives is often an important aspect in risk reduction.

Steps 1–3 and 5 require knowledge of the properties of the focal chemical compounds, which again implies an extensive literature search and/or selection of the best feasible estimation procedure. In addition to 'Beilstein', it can be recommended to have in hand a number of useful handbooks of environmental properties of chemicals and methods for estimation of these properties in case literature values are not available.

Steps 1–3 are sometimes denoted as effect assessment or effect analysis and steps 4 and 5 exposure assessment or effect analysis. Steps 1–6 may be called risk identification, while ERA encompasses all the nine steps presented in **Figure 3**.

Particularly step 9 is very demanding, as several possible steps in the reduction of the risk should be considered, including treatment methods, cleaner technology, and substitutes to replace the examined chemical.

During the last 5–6 years, in North America, Japan, and the European Union, consideration has been given to treating medicinal products similarly to other chemical products, as there is in principle no difference between medicines and other chemical products. However, this only resulted in the introduction, from 1 January 1998, of the application of ERA for new veterinary medicinal products. At present, technical directives of the European Union and the United States related to medicinal products for human use do not include any reference to ecotoxicology and the assessment of their potential risk. However, a detailed technical draft guideline issued in 1994 indicates that the approach applicable to veterinary medicine would also apply to human medicinal products. Presumably, ERA will be applied to all medicinal products in the near future, when sufficient experience with veterinary medicinal products has been achieved. Veterinary medicinal products, on the other hand, are

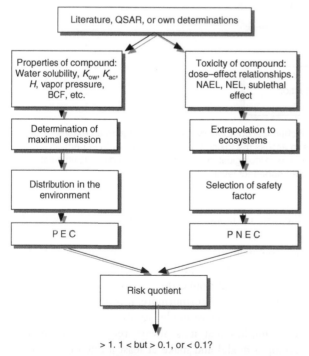

> 1. 1 < but > 0.1, or < 0.1?

Figure 4 Steps 1–6 (**Figure 3**) are shown in more detail for practical applications. The result of these steps naturally leads to assessment of the risk quotient. BCF, biological concentration factor; NAEL, nonadverse effect level; NEL, noneffect level; PEC, predicted environmental concentration; PNEC, predicted no effect concentration.

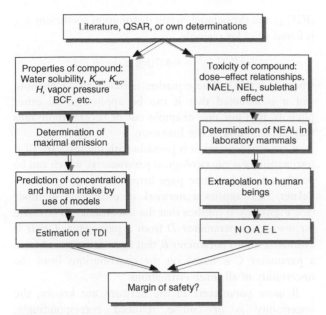

Figure 5 Environmental risk assessment for human exposure. It leads to a margin of safety, which corresponds to the risk quotient in **Figures 3** and **4**. NAEL, nonadverse effect level; NEL, noneffect level; NOAEL, nonobserved adverse effect level; TDI, tolerable daily intake.

released into the environment in larger amounts; for example, the use of manure as fertilizer on agricultural fields in spite of its possible content of veterinary medicine could be a source of contamination.

It is also possible to perform an ERA where the human population is in focus. The 10 steps corresponding to **Figure 4** are shown in **Figure 5**, which is in principle not significantly different from **Figure 4**. The principles for the two types of ERA are the same. **Figure 5** uses the nonadverse effect level (NAEL) and nonobserved adverse effect level (NOAEL) to replace the PNEC and the PEC is replaced by the tolerable daily intake (TDI).

This type of ERA is of particular interest for veterinary medicine, which may contaminate food products for human consumption. For instance, the use of antibiotics in pig feed has attracted a lot of attention, as antibiotics may be found as residues in pig meat or may contaminate the environment through the application of manure as natural fertilizer.

Toxic Chemicals: Overview of Their Properties and Their Ecotoxicological Effects

ERA requires, depending on the procedure followed, a good knowledge of the properties and the effects of the chemical in question. The basic properties that determine the distribution of the chemical in the environment are biodegradability, water solubility, solubility in fat tissue,

the vapor pressure, Henry's constant, boiling point, melting point, and biological concentration factor. Particularly, the first two properties are extremely important. A chemical with a high biodegradability will rapidly disappear and the risk will therefore be reduced correspondingly. A chemical with little biodegradability (it is often denoted as refractory) will, on the other hand, persist in the environment for a very long time, which increases the risk correspondingly. Chemicals with high water solubility are very mobile and are therefore easily distributed widely, while chemicals with low water solubility are soluble in fat tissues and will therefore biomagnify and bioaccumulate.

The vapor pressue and Henry's constant determine the distribution between water and air, while the biological concentration factor is important for the distribution between organisms and their environment (air, soil, or water).

There is a wide spectrum of effects of chemicals on organisms, and these are presented in the chapter 'Ecotoxicology Nomenclature: LC, LD, LOC, LOEC, MAC' and in Part B. A substance may be lethal, affect the growth, have various sublethal effects influencing more or less specific biochemical and physiological ractions, have reproductive toxicity, cause mutagenesis, cancer, and/or teratogenesis, and be an endocrine disruptor. In addition, the toxic effects may be chronic or acute.

The properties that determine the concentration in an organism and its various organs as a function of time are also dependent on other properties of the organism. These properties encompass uptake rate from the medium (air, soil, or water), excretion rate, biological concentration factor, and the biological half-life in the body.

The nomenclature used for the wide spectrum of crucial properties of toxic substances is presented in the next chapter.

As already mentioned, the industrialized countries use 100 000 chemicals and to be able to carry out an ERA of all these chemicals, we need to know at least 10 basic properties of each of the chemicals. In addition, we should know the effects of each chemical if not on all species at least on say 10 000 different species and families that could represent the about 5–10 million species that are on the Earth. For these 10 000 species, we also need to know the properties that determine the concentrations of the chemical in the organs of the species. It implies that we need to know $100\,000 \times 10$ basic properties or 10^6 basic properties plus about 20 properties for the interactions between 100 000 chemicals and 10 000 species or totally $100\,000 \times 20 \times 10\,000 = 2 \times 10^{10}$. At the most, we are able to perform a reliable ERA for about 500 chemicals. For about 5000 chemicals, we can make an ERA, which is based on literature values and on relatively good estimations of some of the properties we need for performance of the ERA. For the remaining 94 500 chemicals, we cannot make a proper ERA.

It is recommendable under all circumstances to get as much information about the properties of the applied chemicals as possible. Several databases on the Internet are very helpful in this context. These can be found under the section Relevant Websites.

We have adequate data for only about 1% of what we should know about toxic substances in the environment. An alternative is to estimate the properties of the 100 000 chemicals by various estimation methods. We can, for instance, use the allometric principles to estimate the properties for all species, provided that we know the properties for a few species and we know the size of the species.

Furthermore, it is possible to estimate the properties of the chemicals from their chemical structure. This is possible by the use of Quantitative Structure Activity Relationship (QSAR) methods. Two classes of Quantitative Structure Activity Relationship (QSAR) estimation methods are available: specific methods, which are valid for a specific group of chemicals, and general methods, which can be applied to all organic and inorganic compounds. The specific methods, of course, have a more limited application, but generally less uncertainty. The journal *SAR and QSAR in Environmental Research* has published numerous specific estimation methods. An example of specific methods is presented below. The following relationship between 50% growth inhibitory concentration (IGC$_{50}$) and the octanol–water distribution coefficient K_{ow} is found valid for phenols:

$$\log IGC_{50}^{-1} = 0.627(\log k_{ow}) - 0.772$$

where $r^2 = 0.783$. The equation is based on 54 phenols, and it is assumed that it can be applied to all other phenols. It is just one example out of several thousands that can be found in the literature.

A general estimation is possible with the software EEP (estimation of ecotoxicological parameters), which can be purchased on the home page http://www.ecologicalmodel.net. EEP applies a network of estimation methods (see **Figure 6**). It implies that the uncertainty to estimate for instance a parameter D from a parameter A, that is estimated from a parameter B, that again is estimated from a parameter C is based on the contributions from the uncertainty of all three estimations.

If more parameters in the network are known, the uncertainty is of course reduced correspondingly. Furthermore, a general method always has more uncertainty, because the variability of properties among chemicals in general is higher than that among chemicals from a specific group of chemicals. Therefore, the general estimation methods that are applied in a network in EEP should preferably be used mainly to give a first more uncertainty estimation and in educational context.

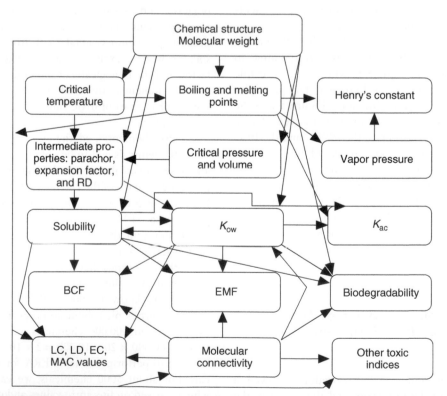

Figure 6 A network of estimation methods is shown. It illustrates the possibilities of using general estimation methods, but as discussed in the text, the general estimation methods have a higher uncertainty. BCF, biological concentration factor; EMF, ecological magnification factor; K_{ac}, the concentration in soil with 100% organic matter relative to the concentration in water in equilibrium with the soil. EC, effect concentration; LC, lethal concentration; LD, lethal dosis; MAC, maximum allowable concentration; RD, molar refraction.

The advantage of EEP is that it is fast to use, but the disadvantage, as mentioned above, is that the uncertainty is high in some cases. If an ERA has to be developed for a chemical, it is necessary to use all accessible estimation methods simultaneously and it is furthermore recommendable to determine a few important properties by laboratory examinations.

It is always recommendable to try to get as many different determinations, including literature values and values estimated by EEP or other estimation tools, as possible. It is particularly important for the two most crucial parameters: K_{ow} (the ratio of solubility in fat tissues to solubility in water) and the biodegradability. High K_{ow} values indicate that the compound is soluble in fat tissues, which implies that it will bioaccumulate and biomagnify. Low K_{ow} values indicate that the compound is readily soluble in water and therefore very mobile. As a result, it will be easily disseminated to adjacent ecosystems. It will inevitably be difficult to control the effect of a harmful compound with a high water solubility. A compound with a high biodegradability is, however, decomposed before it may cause ecotoxicological problems by bioaccumulation, biomagnification, and high mobility. If, on the other hand, the compound has a low biodegradability, it will stay for a long time in the environment (e.g., DDT or PCB) and if it has at the same time either a high or a low K_{ow}, it will easily cause ecotoxicological problems. However, to get a more complete picture, we need to know the entire spectrum of properties mentioned above.

Part A of this book presents the basic concepts of ecotoxicology. It contains the nomenclature, the most focal topics, the history and the three very fundamental concepts of dose–response, body residues, and assimilative capacity. Further details and ecotoxicological core concepts of environmental management (ERA) are given in the last two chapters of Part A.

Part B, which comprises 17 chapters, presents detailed information about the properties of the ecotoxicological chemicals. Part C has 22 chapters and these are devoted to the presentation of ecotoxicologically most important chemicals – from the point of their effects and widespread uses. Parts B and C give two different overviews of ecotoxicology: one is based on the harmful properties and the other is based on the harmful chemicals.

Appendix 1 presents the periodic table from an ecotoxicological point of view. In Appendix 2, the two overviews given in Parts B and C are combined by a matrix to identify the harmful properties of ecotoxicologically important chemicals and the relationship between the two.

- which properties do you find by which chemicals? and
- which harmful properties have the ecotoxicologically important chemicals?

These questions are very complex if the answer should reveal all possible toxic effects on all possible organisms. The matrix gives, however, a useful overview of the information that can be found in Parts B and C.

Further Reading

Jørgensen SE (2009) *Introduction to Ecological Modelling*, 205pp. Southampton: WIT.

Jørgensen SE and Bendoricchio G (2001) *Fundamentals of Ecological Modelling*, 525pp. Amsterdam: Elsevier.

Jørgensen *et al.* (1998) *Handbook of stimation Methods for Ecotoxicology and Environmental Chemistry*, 230pp. CRC Baton Rouge: Florida.

Jørgensen SE, Fath BD, Bastianoni S, *et al.* (2007) *A New Ecology: Systems Perspective*, 288pp. Amsterdam: Elsevier.

Mitsch W and Jørgensen SE (2003) *Ecological Engineering and Ecosystem Restoration*, 386pp. New York: John Wiley.

Relevant Website

http://146.107.217.178/lab/alogps – This address has links to estimate K_{ow} and other physical-chemial properties of chemicals

http://chemfinder.cambridgesoft.com/result.asp – This page allows the estimation of K_{ow} and other physical-chemial properties of chemicals

http://ibmlc2.chem.uga.edu/sparc – This link can be used to convert the CAS number to SMILES code, that is often needed by application of data bases for chemical and toxicolgical properties of chemicals.

http://www.logp.com – This site has links to estimate K_{ow} and other physical-chemial properties of chemicals

http://www.nlm.nih.gov – It is a very useful homepage with links to 'United States – National Libery of Medicine' developed under the title Toxicology and Environmental Health Information Program (TEHIP). There are links to HSDB (Hazardous Substance Data Bank), TOXLINE (toxicological properties), ChemIDplus (chemical structures), IRIS (Integrated Risk Information System), ITER (human health risk assessment), CCRIS (cancer database), GENE-TOX (genetic toxicology test results) DART/ETIC (reproduction-toxicologi).

http://search.epa.gov – This site has several links to human toxicology and ecotoxicology.

http://www.syrres.com/esc/est_kowdemo.htm – This page has links to estimate K_{ow} and other physical-chemial properties of chemicals

http://toxnet.nlm.nih.gov – This site has links to toxicological informatic about chemicals and drugs.

Ecotoxicology Nomenclature

M C Newman and Y Zhao, Virginia Institute of Marine Science, Gloucester Point, VA, USA

Introduction
Regression-Derived Effect Metrics
Effect Metrics Derived with Hypothesis Tests

Inferring Consequences of Exposure
Further Reading

Introduction

Ecotoxicology is a relatively new science concerned with contaminants in the biosphere and their effects on constituents of the biosphere, including humans. Although relevant effects range from the molecular to the biospheric levels of biological organization, most measures of effect generated by ecotoxicologists are designed to infer adverse effect at the level of the individual organism. This is a consequence of core methods adoption from classic toxicology, a discipline appropriately concerned with effects on individuals. This has created a bias in the ecotoxicology literature that can compromise inferences about effects at higher levels of biological organization such as the population, ecological community, or ecosystem levels. Despite this bias, the methods described herein can and are used pragmatically to infer effects including those occurring at higher levels.

Effect metrics are derived for different kinds of exposures, most notably acute exposures to high concentrations and chronic exposures to low concentrations. Acute exposures are defined in various ways but all definitions reflect a relatively brief and intense exposure scenario. By recent convention, chronic exposures are defined as those exceeding 10% of an individual's life span. However, the definition of chronic exposure varies in literature and depends on the specific methodologies used to produce an effect metric. As an example, some chronic toxicity tests for aquatic species use an exposure duration of 28 days regardless of the longevity of the test species.

Regression-Derived Effect Metrics

The two general approaches, regression-based and hypothesis testing-based methods, to quantify adverse effects were established early in the history of ecotoxicology. Most regression-based methods are intended to predict the intensity of effect associated with exposure to a toxicant concentration for a specified duration although regression models incorporating exposure concentration and duration simultaneously are becoming increasingly more common. Hypothesis testing-based methods were originally designed to generate some measure of effect in situations in which a regression model cannot be fit with sufficient goodness of fit. However, they have gradually become the effect metrics of choice for chronic exposures that commonly, but not always, produce data less amenable to regression modeling than do acute exposure studies.

Regression-based metrics of effect are generated with well-established test designs. The most widely applied design includes a series of exposure concentrations or doses. There are tests, notably effluent toxicity tests, with slightly different treatments. In the case of an effluent test, the effluent is mixed with different amounts of clean water to generate a series of effluent dilutions. The diluent water is either taken from the receiving water or standard synthetic water is used. The treatment intensity is expressed as (effluent volume)/(effluent volume + dilution water volume) × 100%. Regardless, subsets of test individuals are either exposed to constant levels of toxicant in their surrounding media or food, or given a specific dose of toxicant, perhaps by injection, topical application, or gavage. In the case of a lethal effect, the number of individuals affected at each concentration or dose treatment is tallied after a specified duration (**Figure 1**). The data pairs (concentration, dose, or dilution vs. intensity of effect) generated for the series of concentration, dose, or dilution treatments are then fit to one of several candidate models using conventional regression methods. Most of the commonly applied models are sigmoid functions that accommodate lowest and highest possible limits for effect intensity (**Figure 1**). For lethal effects, the lowest level might be zero or some baseline level of natural ('spontaneous') mortality, and the upper limit of effect is often 100% mortality for the exposed group of individuals.

The most commonly fit sigmoid function is the log normal (probit or normit) model although others such as the log logistic (logit), Weibull (Weibit), or Gompertz (Gompit) often provide excellent fit to these types of data. The conventional probit and probit with spontaneous mortality are the following:

$$p = \Phi(a + b(\ln \text{ dose})) \qquad [1]$$

$$p = S + (1 - S)(\Phi(a + b(\ln \text{ dose}))) \qquad [2]$$

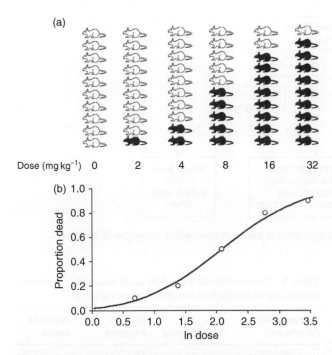

(a)

Dose (mg kg⁻¹)   0   2   4   8   16   32

(b)

Figure 1   (a) A typical experimental design used to generate regression-based metrics of effect and (b) the associated sigmoid dose–response curve . The organisms that died in response to the treatment are denoted in black and those still living are denoted in white.

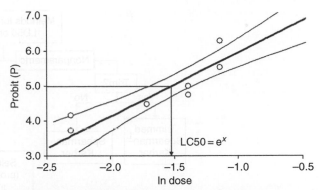

Figure 2   The probits of the predicted and observed proportions vs. the natural log of the effluent dilution, and the predicted 95% confidence intervals for the data from **Table 1**. The LC50 is estimated by exponentiating the x values corresponding to the probit for 50% mortality.

where $p=$ the probability of or proportion of exposed individuals dying, $S=$ the probability of or proportion of unexposed individuals dying, $\Phi=$ normal cumulative function, $a=$ an estimated regression intercept, and $b=$ estimated regression parameter accounting for the influence of ln dose on p. Spontaneous mortality may be included in the model because laboratory culturing conditions are such that some unavoidable mortality occurs or because the longevity of the organism is short relative to the length of the test, natural mortality is to be expected during the test. In either case, the assumption is made that the spontaneous mortality does not influence the relationship between dose/concentration and associated mortality. This may not be an acceptable assumption in some cases.

Such regression models were initially used by laboratory toxicologists to estimate threshold doses below which no effect was expected. However, because the error associated with such estimates was very large, doses predicted to produce certain p's eventually became the norm. Because the prediction error tends to be lowest toward the center of the predicted curve (**Figure 2**), prediction of the concentration producing 50% effect ($p=0.50$) became the conventional effects metric in mammalian toxicology and was adopted by early ecotoxicologists. Another advantage of prediction for this proportion is that the effects metrics derived by the most common functions (probit and logit) produce very similar results at 0.50. The median

lethal dose (LD50) and median lethal concentration (LC50) predicted after a specified duration of exposure are currently the primary metrics of acute lethal effects. For nonlethal effects fit by regression to concentration– or dose–effect models, the median effective dose (ED50) or concentration (EC50) are predicted instead. Despite this convention of predicting median effect levels, a trend has begun that draws the focus of effect assessments more and more often toward lower levels of mortality or effect, that is, LDx or LCx where $x < 50$. We anticipate that this trend will continue, resulting in some changes to the methods described below for predicting LDx and LCx values and their confidence limits. As the emphasis in ecotoxicology shifts downward on the dose/concentration–response curve, more attention will be required in selection of the best among the candidate sigmoid models. Effect metrics are similar for the models at the center of the curve but predictions from the most commonly applied models (e.g., probit, logit, or Weibull) diverge as one makes predictions toward the tails of the distributions.

Predictions of LD50 or LC50 using data from dose/concentration–effect experimental designs can be made using a variety of means. The parametric methods shown in **Figure 3** are most often executed by maximum likelihood estimation (MLE) because it is common to have 0.00 and 1.00 proportions responding (i.e., all individuals in a treatment survived or died) in the data set but the sigmoid functions to which these data are applied never reach 0.00 or 1.00. Which method is applied depends on the data qualities. A model that incorporates spontaneous or natural mortality might be adopted if such baseline mortality is obvious in the data set. In many software packages, the natural mortality can be either specified by the modeler or estimated using various methods by the software. The best sigmoid model fitting the data set can be selected using a goodness-of-fit statistic such as a

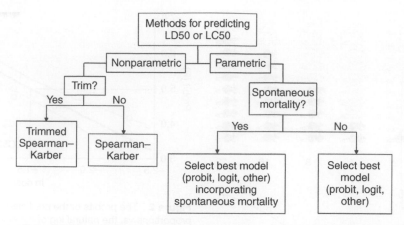

Figure 3 Methods for predicting LD50 or LC50 values using data from dose or concentration–effect experimental designs.

χ^2 statistic. The advantage of these models is the ease with which estimates and associated confidence intervals can be generated for different p values. A nonparametric approach can be applied instead if the data set does not fit any model acceptably or if the iterative MLE method fails to converge on an acceptable solution. The Spearman–Karber technique with or without trimming of data from the distribution tails is the most commonly applied technique in such a case to generate a LD50 or LC50 estimate and the associated 95% confidence limits.

Data from a mysid shrimp experiment can be used to illustrate the regression method for predicting an LC50 value. Juvenile mysid shrimp were randomly assigned to a series of diluted refinery effluent solutions with ten shrimps per tank. There were seven treatments including a control with duplicate tanks per treatment. Concentrations were expressed as the percentages of total exposure volume made up of the effluent. The mortalities were checked after 48 h of exposure (**Table 1**). These data were fit to candidate models of log normal ('probit') and log logistic ('logit') by an interative maximum likelihood estimation. The associated Pearson χ^2 statistics ($\chi^2 = 4.81$ for log normal, $\chi^2 = 5.01$ for log logistic) indicated that the log normal model provided a better fit than the log logistic. **Figure 2** shows the probits of the predicted and observed proportions dying at 48 h (expressed in probit units) versus the natural log of effluent dilution, and the predicted 95% confidence intervals. (The observed data for the 0% and 100% mortalities were not shown because they do not have corresponding probit values.) Thus the LC50 and the associated lower and upper 95% confidence intervals can be estimated by exponentiating the x values corresponding to the probit of 50% mortality (5), which are 21.9%, 18.4%, and 25.8%, respectively.

The original use of LD50 and LC50 estimates in classic toxicology was as a measure of toxicity. For example, a mammalian toxicologist might use a set of LD50 values to

Table 1   Dose–response data of an acute toxicity test exposing juvenile mysid shrimp to a simulated refinery effluent

Concentration (% effluent)	Replicate	Number exposed	Number dead
Control	1	10	0
	2	10	0
10%	1	10	2
	2	10	1
18%	1	10	3
	2	10	3
25%	1	10	5
	2	10	4
32%	1	10	9
	2	10	7
56%	1	10	10
	2	10	10
100%	1	10	10
	2	10	10

Modified from table 2 of Buikema AL, Niederlehner BR, and Cairns J (1982) Biological monitoring, Part IV-toxicity testing. *Water Research* 16: 239–262.

determine the relative toxicities of a series of poisons or to assess how different factors influence the toxicity of a single poison or drug. In such applications, the exposure durations would be set for convenience, for example, acute toxicity after 96 h of exposure because a 96 h test fits conveniently in a workweek, and still generates a meaningful metric of toxicity. So, a p of 0.5 and 96 h test duration might be used for statistical and logistical convenience, not because they reflect pivotal values relative to an acceptable or unacceptable effect to humans.

These regression-derived effect metrics were borrowed by ecotoxicologists who then attempted to apply them to making decisions about the concentration or amount of a chemical that should be a concern if present in an environmental media. Given the multiple levels of biological organization that an ecotoxicologist must consider in such a decision, it should be no surprise that these effect metrics do not provide all of the information needed to make an informed decision. Often the duration selected for LDx or LCx estimation is different from that of interest; so extrapolation is required to predict the p associated with an exposure duration other than that used in the test. Such extrapolation can generate unacceptable, or minimally, undefined uncertainties in predictions. The associated uncertainty can be reduced by noting the proportions responding in a series of durations during the test and estimating several LDx/LCx for these different durations or by applying survival time regression models instead. Another shortcoming of these effect metrics is that mortality occurring during the period following exposure is rarely measured. Some toxicant exposures produce considerable post-exposure mortality that is important to consider when making predictions of effects to populations exposed in the environment to the chemical of interest. A third shortcoming is not as much one of the regression-related metrics but rather of the decision-making process using these metrics. Often the responsible risk assessors or decision makers lack the expertise or information to determine what level of predicted effect (p) should be used as a cutoff for unacceptable adverse effect to exposed individuals, populations, or ecological communities. This can make the regression-related metrics less appealing to

assessors and managers than the hypothesis test-based methods described in the following section.

Effect Metrics Derived with Hypothesis Tests

Developed initially to cope with dose/concentration–response data for which an acceptable model could not be developed, hypothesis test-based methods now are applied heavily in tests of chronic or subtle effects. As will be shown, the intent is to estimate a threshold concentration or dose above which an observable effect might be expected. Most, but not all, relevant statistical methods are conventional hypothesis tests.

The general approach (**Figure 4**) is similar to that shown in **Figure 1** but the variance within and among treatments are assessed instead of developing a dose/concentration–effect model. A series of dose, dilution, or concentration treatments are established with replication within each treatment. After a specific duration, the level of effect manifesting within each treatment is scored and that for each treatment compared statistically to that in the reference treatment. As shown in **Figure 4**, each treatment for which the effect is statistically significantly different from that of the reference treatment is identified (denoted with an asterisk in the figure). The lowest treatment concentration with a response statistically different

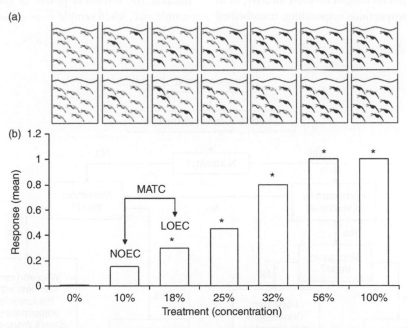

Figure 4 (a) The experimental design of hypothesis testing-based methods and (b) the determination of NOEC, LOEC, and MATC (data from **Table 1**). The organisms that died in response to the treatment are denoted in black and those still living are denoted in white. The treatments for which the effect is statistically different ($\alpha = 0.05$) from that of the reference treatment are denoted with an asterisk. In this example, the effect was death after the exposure although other responses can, and often are, used in these kinds of experiments.

from that of the reference (e.g., 0%) is called the 'lowest observed effect concentration' (LOEC). The highest treatment concentration with a response that is not significantly different from the reference response is called the 'no observed effect concentration' (NOEC). Although formally a dubious inference from hypothesis testing, the NOEC and LOEC are pragmatically treated in ecotoxicology as the lower and upper bounds for the 'maximum acceptable toxicant concentration' (MATC), that is, a threshold concentration presumed to be 'safe'. Extending this pragmatic approach, the geometric mean of the NOEC and LOEC is sometimes used as the best estimate of the MATC. Considerable debate continues about the acceptability of such interpretations of these hypothesis test-derived metrics.

A range of hypothesis tests are commonly applied to NOEC and LOEC estimation including parametric and nonparametric tests (**Figure 5**). These tests differ in their underlying assumptions and consequent ability to detect a significant difference if there was one, that is, their statistical power. The tests carrying the most assumptions are generally the most powerful. However, the differences in power can be trivial or critical depending on the specific tests being compared and the qualities of the data. As important examples in **Figure 5**, the parametric tests are generally more powerful than the nonparametric tests and tests assuming a monotonic trend with treatment concentration are more powerful than those that do not assume a monotonic trend. With the hypothesis testing approach, the data (concentration, dilution, or dose vs. effect level for each treatment replicate) might be used directly, or as commonly done for proportions responding, transformed

in order to meet assumptions of the subsequent hypothesis tests. Formally, the parametric methods can be applied if the data show no evidence of non-normality or heterogeneity of variances among treatments. A powerful parametric trend test (Williams's) can be used if an additional assumption of a monotonic trend (increase or decrease in response) with increasing concentration or dose is justifiable. In some cases such as in the presence of hormesis, a monotonic trend would not be expected. If the assumptions allowing use of the parametric tests are not met, the less powerful nonparametric methods can still be used. If a trend is assumed, then the Jonckheere–Terpstra test can be applied. If not, the less powerful Wilcoxon rank sum test with a Bonferroni adjustment of experiment-wise error rate or the Steel's many-to-one rank test can be used. These last two tests tend to be the least powerful of the hypothesis tests described to this point because they carry the fewest assumptions.

The formal assumptions of and hypotheses tested by these methods differ in important ways. The most important assumption to be met for all is that individuals be randomly assigned to treatments. The results of the hypothesis tests are questionable if this fundamental assumption is not met. The parametric tests further require that the data be normally distributed although most are robust to moderate violations of this assumption. The normality is often tested with a statistic such as the Shapiro–Wilk statistic (W). A small value of W (or a p value less than a predetermined α level such as 0.05) leads to the rejection of the null hypothesis of normality. Because the statistical power of the test increases with sample size, when sample sizes are small, a higher α level

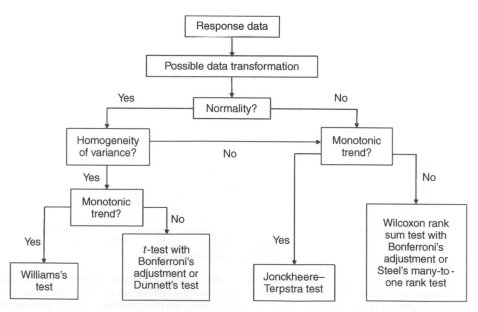

Figure 5 Parametric and nonparametric hypothesis tests that are commonly applied for NOEC and LOEC estimation.

may be applied in tests of normality. These methods also require that the treatments have the same variances, that is, homogeneity of variances, although again, the methods are robust to moderate deviations from the homogeneity of variances assumption. This assumption can be formally tested with Bartlett's, Levene's, or one of several similar tests. Caution should be taken with the commonly applied Bartlett's test because it can be inaccurate even if the data deviate slightly from being normally distributed.

The common parametric tests differ slightly relative to the exact hypothesis they test. The hypothesis assessed by the t-test with Bonferroni adjustment of experiment-wise error rates and Dunnett's test is simply that the mean responses of the treatments are not significantly different from the mean of the reference (control) treatment. Williams's test carries an additional assumption of a monotonic trend (consistently increasing or decreasing effect) with dose/concentration. It tests the null hypothesis that there is no monotonic trend.

The nonparametric methods do not require data normality or homogeneity of variances. With the Wilcoxon rank sum test with Bonferroni adjustment of experiment-wise error rates or Steel's many-to-one rank sum test, the null hypothesis is that observations in the treatments come from the same population. The Jonckheere-Terpstra test is the nonparametric equivalent of the Williams's test in that it has an alternate hypothesis of a monotonic trend. Formally, the null hypothesis for this test is no different in the distribution of responses among the treatments.

The mysid shrimp data can be used again to illustrate the hypothesis testing method (**Figure 4**), although normally more replicates would be recommended. After testing for normality and homogeneity of variance, the data without transformation are tested with Dunnett's one-tailed t-test, with the null hypothesis being that the mean response of each treatment is not significantly higher than the control mean (experiment-wise $\alpha = 0.05$). The results show that 10% effluent is the highest concentration whose response is not significantly higher than the control, and 18% effluent is the lowest concentration with the response significantly higher than the control; the NOEC and LOEC were determined to be 10% and 18%, respectively. Accordingly, the MATC could be estimated as the geometric mean of the NOEC and LOEC, or 13.4%. If the log normal (probit) model generated previously had been used to estimate the proportion dead at the NOEC level (10% effluent), the prediction would be 8.0% mortality at the NOEC. The results generally agree between the regression and hypothesis testing approaches although such is not always true.

One shortcoming of the approach of applying these various methods to produce NOEC and LOEC values has already been discussed in the preceding paragraphs – the NOEC and LOEC values can vary for the same data set as a function of the chosen hypothesis test. Also, the NOEC and LOEC values depend heavily on the experimental design and statistical aspects of the calculations that influence statistical power. The power of any test will depend on the number of observations per treatment, number of treatments, and variability in the background response. Literature surveys have demonstrated that the designs normally applied in effects testing have sufficient power to detect an approximately 5–10% effect difference in mammalian toxicology studies and 10–34% effect difference in ecotoxicology studies. But effects less than these levels can have unacceptable consequences. A final shortcoming is that these hypothesis testing methods were not initially designed to infer a biological threshold concentration or dose. A threshold estimated from a test of statistically significant difference is not necessarily a good estimate of a significant biological effect threshold.

Inferring Consequences of Exposure

Inferring consequences from these effect metrics is challenging but essential. Typical for human effects studies and increasingly common for ecotoxicological studies, the NOEC can be adjusted in a conservative manner to estimate a reference dose (RfD), reference concentration (RfC), or acceptable daily intake (ADI). A common set of adjustments is the following for chronic human exposure:

$$RfD = \frac{NOEC}{UF1 * UF2 * UF3 * UF4 * MF} \quad [3]$$

where UF1 = uncertainty adjustment accounting for variation in natural sensitivity within human populations, UF2 = uncertainty adjustment if extrapolation was performed from animal data to effects to humans, UF3 = uncertainty adjustment if the NOEC comes from a subchronic test data set, UF4 = uncertainty adjustment if the LOEC is used in the calculation instead of an NOEC, and MF = an additional adjustment based on professional judgment. The value for any of these factors can range from 1 to 10 depending on the associated level of uncertainty. This or a similar equation is used to estimate an RfD in the following manner. The literature is searched for all relevant NOEC/LOEC data for the toxicant of interest. The study with the lowest relevant LOEC is identified and the associated NOEC used in the calculation (UF4 = 1). If only the LOEC is available, then the LOEC is applied instead with an UF4 = 10. In the case of chronic human exposure, the RfD is used to estimate the dose level thought to be below that which will cause an adverse effect during chronic exposure. Several types of RfDs are relevant to environmental exposures including short term, subchronic, chronic, or developmental RfDs. The RfD or RfC values may also be developed for different routes of exposure.

With sufficient knowledge, dose- or concentration-effect models can also be applied to estimation of RfD or RfC values. The benchmark dose (BMD) approach uses regression model predictions for a specified effect level (benchmark response) instead of an NOEC or LOEC to estimate the RfD or RfC. Often the lower 95% confidence limit for the estimated BMD (BMDL) is used instead of the NOEC in the above equation to estimate a BMD-based RfD. The UF and MF values can be the same or lower than those used for the NOEC-based approach. Taking the mysid shrimp data as an example, the BMD_{10} could be used to estimate a certain RfC. The BMD_{10} is predicted to be 7.2% effluent. This is the predicted lower 95% confidence interval of the LC10 (10.9% effluent) generated from the log normal (probit) model. As another example, the Environmental Protection Agency (EPA) applied such a BMD approach to determine a human chronic oral methylmercury exposure RfD. Using information from several epidemiological studies, the BMD associated with the lowest 5% of methylmercury-exposed children (BMD_5) was chosen as the basis for calculation of the RfD. The primary advantage of this BMD-based approach is that it avoids many of the shortcomings described earlier for the hypothesis test-related effect metrics.

Regardless of how it is calculated, an RfD is used with information about exposure (e.g., inhalation rates, ingestion rates, bioavailability, and exposure duration) to calculate a maximum allowable concentration (MAC, maximum permitted concentration in a particular source such as food, air, drinking water, or soil) or level of concern (LOC, the concentration in the relevant medium above which an adverse effect could manifest).

Protection of human health is facilitated with a set of RfD or RfC values for various exposure scenarios such as acute, prolonged, lifetime, or developmental exposure. Relative to ecotoxicological testing, calculations associated with estimating 'safe' or acceptable exposures are not as straightforward, requiring consideration of consequences at different stages of life cycles of many species and several levels of biological organization. Partial and complete life cycle tests have emerged to address this requirement. A series of tests are conducted at each major life stage of a species, quantifying important effects notionally linked to an individual organism's fitness. The lowest effect metric for the various tests in such a complete life cycle test is used to generate regulatory limits or goals. The cost and difficulty of performing a complete life cycle test has given rise to a less inclusive set of tests (partial life cycle tests) that assess only the life stages thought to be most sensitive. Often these are the early

life stages, leading to a battery of tests called early life stage tests. The emphasis during the interpretation of partial or complete life cycle tests is on protection of the individual; however, the EPA stresses the importance of considering population protection for most nonendangered or nonthreatened species existing in ecological communities. That the conventional interpretation of life cycle test-generated effect metrics does not directly address population or community level consequences of exposure is seen as a significant shortcoming in this approach as currently practiced in ecotoxicology. Fortunately, resolving this incongruity between metrics generated with current ecotoxicity tests and prediction of population- and community-level consequences is currently a very active area of research.

See also: Acute and Chronic Toxicity.

Further Reading

Bliss CI (1937) The calculation of the time-mortality curve. *Annals of Applied Biology* 24: 815–852.

Buikema AL, Niederlehner BR, and Cairns J (1982) Biological monitoring, Part IV-toxicity testing. *Water Research* 16: 239–262.

Crane M and Newman MC (2000) What level of effect is a no observed effect? *Environmental Toxicology and Chemistry* 19: 516–519.

Faustman EM and Omenn GS (1996) Risk assessment. In: Klaassen CD (ed.) *Casarett and Doull's Toxicology,* 5th edn., pp. 75–88. New York NY: McGraw-Hill.

Finney DJ (1978) *Statistical Method in Biological Assay,* 3rd edn. London: Charles Griffin and Company.

Gad SC and Weil CS (1988) *Statistics and Experimental Design for Toxicologists.* Caldwell, NJ: Telford Press.

Hamilton MA, Russo RC, and Thurston RV (1977) Trimmed Spearman–Karber method for estimating median lethal concentrations in toxicity bioassays. *Environmental Science and Technology* 11: 714–719.

Newman MC (1995) *Quantitative Methods in Aquatic Ecotoxicology.* Boca Raton, FL: CRC/Lewis Publishers.

Salsburg DS (1986) *Statistics for Toxicologists.* New York: Dekker.

Sprague JB (1969) Measurement of pollutant toxicity to fish. I. Bioassay methods for acute toxicity. *Water Research* 3: 793–821.

Sprague JB (1971) Measurement of pollutant toxicity to fish. III. Sublethal effects and 'safe' concentrations. *Water Research* 5: 245–266.

Suter GW, II (1993) *Ecological Risk Assessment.* Boca Raton, FL: CRC/Lewis Publishers.

US EPA (2002) *Methods for Measuring the Acute Toxicity of Effluents and Receiving Waters to Freshwater and Marine Organisms,* 5th edn. EPA-821-R-02e012. US Environmental Protection Agency, Washington, DC.

US EPA (2002) *Short-Term Methods for Estimating the Chronic Toxicity of Effluents and Receiving Waters to Freshwater Organisms,* 4th edn. EPA-821-R-02-013. US Environmental Protection Agency, Washington, DC.

Zhao Y and Newman MC (2003) Shortcomings of the laboratory derived LC50 for predicting mortality in field populations: Exposure duration and latent mortality. *Environmental Toxicology and Chemistry* 23: 2147–2153.

Ecotoxicology: The Focal Topics

S M Bard, Dalhousie University, Halifax, NS, Canada

Further Reading

Ecotoxicology is a relatively young field that was first defined by René Truhaut in 1969 as "the branch of toxicology concerned with the study of toxic effects, caused by natural or synthetic pollutants, to the constituents of ecosystems, animal (including human), vegetable and microbial, in an integral context." Ecotoxicology is multidisciplinary and aims to primarily combine the study of ecology (species richness, abundance, and distribution) and toxicology (toxic effects caused by anthropogenic or natural substances). The scope of ecotoxicology encompasses the interaction, transformation, fate, and effects of xenobiotics ('foreign compounds') on the organism, population, community, and ecosystem, from the regional to global level. To elucidate such broad scientific questions, the study of ecotoxicology incorporates concepts contributed from diverse fields including analytical and environmental chemistry, biochemistry, molecular biology, microbiology, immunology, physiology, behavioral ecology, soil science, limnology and oceanography, atmospheric science, environmental and chemical engineering, economics, public environmental policy, and other disciplines. At the level of the organism, ecotoxicological studies include elucidating the cellular and molecular defense mechanisms that wild species deploy against toxicants. Such defense mechanisms include biotransformation of toxicants by drug-metabolizing enzymes (e.g., phase I cytochrome P450, phase II glutathione S-transferase, etc.) and transmembrane toxin efflux transporters (e.g., multidrug resistance proteins) that facilitate the elimination toxicants and their metabolites via bodily excretion. Other ecotoxicological endpoints include endocrine disruption, genotoxicity, immunomodulation, and other biological responses that are considered biomarkers of exposure to toxicants. Examining these measures in large numbers of individuals can permit an evaluation of the health effects of toxicants across a population. In highly toxic environments, some populations can develop chemical resistance either through adaptive response or genetic selection in which only the hardiest individuals survive. Elucidating the influence of toxicants on whole communities involves assessing biological diversity (species richness and abundance) along a pollution gradient. Near polluted sites, a few hardy species can tolerate the extreme chemical conditions and may even thrive; at sites of intermediate exposure, sensitive species begin to appear and species diversity increases; but

many species are pollution intolerant and can only survive at uncontaminated sites. Understanding the pollution tolerance of species within a community can permit development of biological indices to rate pollution exposure at sites based on the species present. In addition to the toxicological sensitivity of different species, the trophic role of individual species plays a strong role in structuring communities, in particular the presence of keystone predators (e.g., *Pisaster* seastar), species that provide physical habitat for other species (e.g., mussel beds or leafy seaweed cover), and organisms that facilitate other species' breeding success (e.g., seagrass nurseries). Ecotoxicological studies at the ecosystem level include examining the biomagnification of persistent organic pollutants through the food web. Lipophilic ('fat-loving') anthropogenic contaminants can be absorbed from the environment and bioaccumulate in oily deposits in low-trophic-level organisms. These compounds bioconcentrate through the food web such that top predators such as marine mammals, polar bears, and humans in turn accumulate the highest levels of these contaminants in their own bodies and can discharge them in fat-rich breast milk. The biomagnification of persistent organic pollutants is not restricted to one region, as these chemicals can be transported thousands of kilometers from the tropics to concentrate in the polar regions by both atmospheric processes and migratory species to create a chemical source and sink problem of global significance.

See also: Biogeochemical Approaches to Environmental Risk Assessment; Body Residues; Ecotoxicological Model of Populations, Ecosystems, and Landscapes.

Further Reading

Carson R (1962) *Silent Spring*. New York: Mariner Books.

Davis D (2002) *When Smoke Ran like Water. Tales of Environmental Deception and the Battle against Pollution*. New York: Basic Books.

Hoffman DJ, Rattner BA, Burton GA, Jr., and Cairns J (eds.) (2002) *Handbook of Ecotoxicology*, 2nd edn. Boca Raton, FL: CRC Press.

Klaassen CD (ed.) (2001) *Casarett and Doull's Toxicology: The Basic Science of Poisons*, 6th edn. New York: McGraw-Hill.

Klaassen CD and Watkins JB (eds.) (2003) *Casarett and Doull's Toxicology: The Essentials of Toxicology*, 6th edn. New York: McGraw-Hill.

Landis WG and Yu M (eds.) (2004) *Introduction to Environmental Toxicology, Impacts of Chemicals upon Ecological Systems*, 3rd edn. Boca Raton, FL: Lewis Publishers.

Newman MC and Unger MA (eds.) (2002) *Fundamentals of Ecotoxicology*, 2nd edn. Boca Raton, FL: CRC Press.

Plant N (2003) *Molecular Toxicology*. New York: Garland Science.

Rand GM (ed.) (1995) *Fundamentals of Aquatic Toxicology: Effects, Environmental Fate and Risk Assessment*. Boca Raton, FL: CRC Press.

Timbrell J (ed.) (2003) *Principles of Biochemical Toxicology*, 3rd edn. Oxford: Taylor and Francis.

Walker CH, Hopkins SP, Sibley RM, and Peske DB (eds.) (2005) *Principles of Ecotoxicology*, 3rd edn. Oxford: Taylor and Francis.

Yu M (ed.) (2005) *Environmental Toxicology: Biological and Health Effects of Pollutants*, 2nd edn. Boca Raton, FL: CRC Press.

Ecotoxicology: The History and Present Directions

M C Newman, College of William and Mary, Gloucester Point, VA, USA

Introduction
Scales of Study

Current Trends in Ecotoxicology
Further Reading

Introduction

Ecotoxicology is the science of contaminants in the biosphere and their effects on constituents of the biosphere, including humans. (The inclusion here of effects to humans in the purview of ecotoxicology is consistent with the original definition of Truhaut but atypical of recent definitions.) The impetus for this new science was the need to understand and make decisions about environmental contaminants. From the close of World War II and into the 1960s, several pollution events occurred with consequences universally acknowledged to be unacceptable. These watershed events included population crashes of raptor and piscivorous bird species due to DDT effects on reproduction, widespread water pollution, and epidemics of mercury (Minamata disease) and cadmium (Itai–Itai disease) poisoning. Expertise for dealing with such issues became essential to society and several practical sciences coalesced into the nascent science of ecotoxicology.

Ecotoxicology is a synthetic science that combines causal explanations (paradigms) and information from many sciences, particularly biogeochemistry, ecology, and mammalian, aquatic, and wildlife toxicology. The integration of paradigms and data from these disciplines is presently incomplete. Chief among remaining challenges is establishment of congruency among theories and data emerging from different levels of biological organization. Because ecotoxicology is an applied science, ecotoxicologists take on different roles that are also not fully integrated. Some ecotoxicologists are concerned chiefly with scientific goals, that is, organizing facts around explanatory principles. Others focus on the technical goals, that is, developing and applying tools to generate high-quality information about ecotoxicological phenomena. Still others focus closely on resolving specific, practical problems such as assessing ecological risk due to a chemical exposure or the effectiveness of a proposed remediation action. Associated activities overlap but are presently performed inconsistently in many instances. For example, the ecotoxicity tests applied today focus on effects to individual organisms, but predictions of consequences to populations and communities are a very high priority for ecotoxicologists. A major theme in ecotoxicology today is finding the best way of achieving scientific, technical, and practical goals while organizing a congruent body of knowledge around rigorously tested explanations.

Some general trends exist in ecotoxicology relative to different levels of the biological hierarchy (**Figure 1**). Causes of lower-level phenomena such as biomolecular effects tend to be easiest to identify and relate to immediately adjacent levels such as to cells or tissues (top panel of **Figure 1**). Techniques for study of lower-level effects often have the advantage of documenting quicker responses than those occurring at higher levels (middle panel of **Figure 1**). Unfortunately, the ecological relevance of change at the lowest levels is more ambiguous than that for higher-level changes. This creates a dilemma for ecotoxicologists attempting to develop better technologies. The ecotoxicologist tries to avoid measuring precisely the wrong effect or imprecisely the right effect. Another trend is that lower-level effects tend to be used proactively and those at higher levels are applied reactively by ecotoxicologists trying to solve specific problems (bottom panel of **Figure 1**). Lower-level effects tend to be more tractable than those at the higher levels.

Scientific activities of ecotoxicologists rely on conventional methods although conventions vary among scientists focused at different levels of organization. Controlled experiments tend to be practiced more during studies of

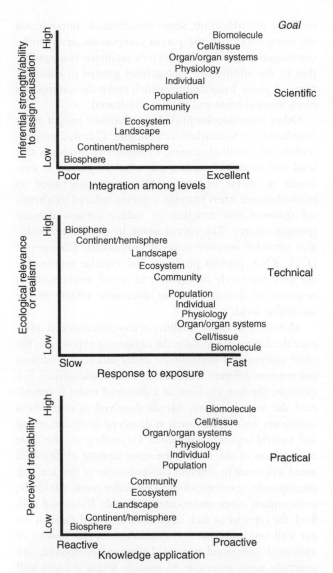

Figure 1 Qualities of ecotoxicological knowledge based on the three goals to which it is applied and the biological levels for which it is generated.

lower-level effects such as biochemical shifts; whereas, higher-level effect studies rely more on observation and natural experiments such as accidental toxic releases. Regardless of the manner in which insight is obtained, the scientific intent is to organize facts around rigorously tested paradigms. Some ecotoxicologists work to produce more precise or detailed information around existing paradigms while others work to rigorously test existing paradigms or to propose novel ones. Both of these activities are essential to the growth of ecotoxicology as a science.

Activities to develop new or enhance existing technologies produce tools with which to understand contaminant fate and effects in the biosphere. Ecotoxicological technologists develop analytical instruments, procedures for studying regional impact, and specific tools for

documenting exposure or effects. As examples, biomarkers are continually developed and improved so effects from the biochemical to individual levels can be documented. Biomarkers (cellular, tissue, body fluid, physiological, or biochemical changes in individuals) are also useful for documenting effect or exposure even in the absence of any discernible adverse effect. Also important at higher levels of organization are biomonitors, changes in organisms or groups of organisms used to infer adverse impact of contaminant exposure. Qualities valued in ecotoxicological technologies are biomarker or biomonitor effectiveness (including low cost and ease of application), precision, accuracy, appropriate sensitivity, consistency, and capacity to generate clear results.

The goal of practical ecotoxicology is the use of existing science and technology to document or solve specific problems such as remediating harm done by a chemical spill. Much of practical ecotoxicology is currently done within the ecological risk assessment (ERA) framework (**Figure 2**). ERAs can be retroactive, predictive, or comparative. A retroactive ERA estimates the risk from an existing situation such as a contaminated site, whereas a predictive ERA predicts the same for a future situation such as the proposed licensing of a new agrochemical. A comparative risk assessment might be done if the risks of two or more alternative actions are to be contrasted during environmental decision making. Regardless of the kind of ERA, the best available science is applied to formulate the problem, that is, define plausible consequences of exposure. The best available science and

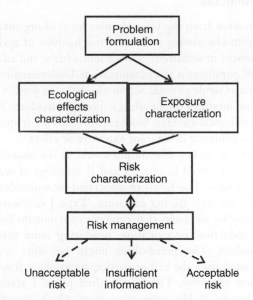

Figure 2 The ecological risk assessment paradigm including problem formulation, exposure characterization, ecological effects characterization, and risk characterization. This paradigm was derived from that developed by the National Academy of Science.

technologies are applied next for ecological effects and exposure characterizations. A computer model might predict movement of the contamination or the contaminant concentration might be measured in the relevant media. The best information is gathered to relate the contaminant concentrations to possible effects to valued ecological entities. All of this information is combined in the last stage of ERA to produce a risk statement. This information is shared with risk managers who decide on the most appropriate action. This decision might also mandate a remediation which would draw on ecotoxicological science and technologies.

Scales of Study

Ecotoxicological subject matter spans a wide range of biological levels (**Figure 1**). Levels drawing heavily on classic toxicology extend from the biomolecular to the individual. Some population issues addressed by ecotoxicologists also benefit from the work of human epidemiologists. Associated themes and information at these levels of organization could be described as autecotoxicology, just as similar subjects in ecology are classified as autecology. Issues associated with higher levels would then be issues of synecotoxicology: considerable biogeochemical and ecological knowledge are applied in synecotoxicological studies. A brief sampling of major autecotoxicological and synecotoxicological research themes is provided below.

Biomolecule

Information from the biomolecular level of organization is key to elucidating molecular mechanisms of toxicity, differences in sensitivity among individuals, and adaptation of populations to contamination. Understanding the molecular mode of toxic action also helps to predict how toxicant mixtures might affect exposed individuals. From a technological vantage, biochemical shifts are frequently used as evidence of toxicant exposure or effect.

Perhaps the best illustration would be the biomolecular shifts involved in phase I and II reactions of organic contaminants. The levels of associated biomolecules can quickly increase during exposure. Type I reactions are mediated by enzymes that catalyze contaminant hydrolysis, reduction, or oxidation, producing more reactive metabolites. The metabolites might be more readily eliminated from the cell or participate in phase II detoxification reactions. The best-studied phase I system is cytochrome P-450 monooxygenase which transforms contaminants such as polycyclic aromatic hydrocarbons (PAHs), chlorinated hydrocarbons, polychlorinated biphenyls (PCBs), hydrocarbons, dioxins, and dibenzofurans. Although phase I transformations are intended to facilitate detoxification, some transformed contaminants are more toxic than the parent compounds or might be carcinogenetic. Phase II enzymes facilitate conjugation, that is, the addition of endogenous groups to contaminants or phase I metabolites which make the compounds more water soluble and readily eliminated.

Other biomolecules provide mechanistic insight and a foundation for biomarker technologies. Elevated concentrations of metallothioneins, cysteine-rich proteins that bind and sequester metals, are often employed as evidence of metal exposure. Also commonly used as biomarkers are stress proteins, proteins induced by chemical stressors that function to reduce protein damage (proteotoxicity). The recent surge in genetic technologies provides another suite of biomarkers. Changes in DNA, RNA, protein products, and cellular metabolites are used separately or together to reveal mechanisms of response or damage, and to document effects at the molecular level.

Molecular and ionic qualities of contaminants also influence the nature of exposure. An organism's exposure to the same amount of a contaminant under different conditions can result in different realized doses and consequences. For example, the free ion form of a dissolved metal is considered the most bioactive. Metals dissolved in water form complexes with ligands such as dissolved inorganic anions and natural organic compounds. Depending on the ionic composition of the waters, the same amount of dissolved metal will result in different concentrations of free ion and, consequently, concentrations of bioactive metal. Similarly, some organic contaminants are weak acids. If ingested with food, the capacity of such contaminants to pass through the gut wall and cause harm is dependent on the amount of unionized compound present. Unionized compounds are generally more amenable to passage across the gut wall than ionized compounds. Under different pH conditions, different amounts of such a weak acid would be unionized as can be easily estimated with the Henderson–Hasselbalch relationship:

$$f_{\text{unionized}} = \frac{1}{1 + 10^{\text{p}K_a - \text{pH}}}$$

where $f_{\text{unionized}}$ is the fraction of the compound present that is unionized, and $\text{p}K_a$ the $-\log_{10}$ of the compound's ionization constant (K_a). Exploring the influence of molecular factors like the two just described is an active area of exposure research in ecotoxicology today.

Cells and Tissues

Toxicant-induced changes in cells and tissues are useful biomarkers. Some changes reflect a cell's failure to remain viable in the presence of toxicants and others reflect partially successful attempts to maintain homeostasis. For example, histological examination of the liver from an exposed organism might reveal many dead (necrotic) cells. In the same tissues, inflammation might be occurring in an attempt to isolate, remove, and replace damaged cells. Both necrosis and inflammation are common biomarkers. Other changes such as the cellular accumulation of damaged biomolecules or cells modified to cope with toxicant damage are also good histological biomarkers.

Cancer is a cellular response to carcinogen exposure that is carefully studied by ecotoxicologists. Several ecological studies have demonstrated the role of environmental toxicants on cancer etiology. For example, Puget Sound English sole (*Parophrys vetulus*) taken from sites with elevated sediment contamination showed high prevalence of liver cancers. Another case of elevated cancer prevalence (27% of dead adults) involved beluga whales (*Delpinapterus leucas*) inhabiting a contaminated reach of the St. Lawrence estuary.

Exposure studies at this level focus on the routes of contaminant movement into and out of cells, and differences in accumulation in various tissues. Generally, contaminant movement into and out of cells involves (1) simple diffusion across the membrane lipid bilayer or through an ion channel, (2) facilitated diffusion involving a carrier protein, (3) active transport, or (4) endocytosis. Some of these routes are designed for other purposes such as ATPase active transport of cations but facilitate movement of contaminants such as cadmium. Other mechanisms are more specific. For example, the multixenobiotic resistance (MXR) mechanism specifically removes moderately hydrophobic, planar contaminants from the cell.

Organ and Organ Systems

Toxicant effect on organs and organ systems is another major theme in classical toxicology that also has a role in ecotoxicology. Organs can be targets of toxicant effects as in the case of the liver cancer mentioned above or can be routes of toxicant entry into the body as in the case of the integument, breathing organs, and digestive tract.

Contaminant effects on organs and organ systems are diverse. Pyrethoid pesticides modify essential ion exchange across amphibian skin. Fish gills are changed by exposure to low pH or high metal concentrations in such a way that normal ion and gas exchange are altered. Some contaminants (teratogens) can cause abnormal organ development. For example, fish embryos develop cardiovascular abnormalities if exposed to high concentrations of PAH. Still other toxicants compromise immunological competency, increasing susceptibility to infection or infestation. These examples represent only a few of the possible organ or organ system effects of contaminants on nonhuman species.

An issue attracting considerable attention at the moment is the ability of some environmental contaminants to modify endocrine functions such as those essential for sexual development and viability, or optimal metabolic activity. For example, the presence of the anti-fouling paint constituent, tributyltin, caused pervasive imposex (imposition of male features such as a penis or vas deferens on females) in whelk populations along the English and Northeast Pacific coasts. Contaminants that act as estrogen include DDT and its replacement, methoxychlor, nonylphenol from surfactant and detergent synthesis, and synthetic hormones from birth control pills that enter waterways from sewage treatment plants. Still other endocrine modifiers such as ammonium perchlorate from military munitions disrupt thyroid function.

Exposure studies at this level of biological organization emphasize target organs. Some organs or organ systems are more prone to toxicant impacts due to their intimate contact with environmental media, location relative to blood circulation, or specific function. For example, the gills of aquatic organisms are often target organs for dissolved contaminants because of their intimate contact with the surrounding water. The liver or analogous organs in invertebrates are often sites of harmful effects because of their prominent detoxification function, that is, the liver cancer noted above in English sole was caused by contaminant activation during phase I reactions in the liver.

Whole Organism

Effects to individuals are used to make inferences about contaminant impacts on individual fitness, and indirectly, on populations and communities. The most commonly measured qualities are mortality, development, growth, reproduction, behavior, physiology, and bioenergetics.

Lethal effects are measured under different exposure scenarios. They might be measured during acute (4 days or shorter) or chronic (longer than 10% of an individual's lifespan) exposures. They might also be measured for contaminant exposure via different media such as water, air, food, and sediment. Most are studied in the laboratory in such a manner that physical, chemical, and biological factors influencing response to exposure are controlled. Therefore, mortality predicted for a particular exposure concentration might not completely define the mortality that would occur in the field where exposed individuals must successfully forage, compete with individuals of other species, avoid predators, and interact with individuals of the same species in order to remain alive.

Exposure studies that involve whole organisms emphasize bioaccumulation, the net accumulation of contaminant in an organism from water, air, or solid phases of its environment. Mathematical bioaccumulation models range from simple ones such as the one compartment model shown below to multicompartment, pharmacokinetic models:

$$C_t = C_{\text{Source}} \frac{k_u}{k_e} \left[1 - e^{-k_e t} \right]$$

where C_t is the concentration in the organism at time, t, C_{Source} is the concentration in the source, k_u is uptake clearance, and k_e is elimination rate constant. Most studies attempt to understand and quantify the influence of extrinsic (e.g., food type containing the contaminant) and intrinsic (e.g., animal sex or size) factors on bioaccumulation.

Population

Knowledge of effects to individuals is valuable, but insufficient, for predicting population-level impacts. Consequently, a growing number of ecotoxicologists study population-level effects directly. Such studies emphasize vital rates such as birth, death, stage change, or migration rates. Demographic models based on vital rates improve our ability to project consequences such as a drop in the population growth rate or increase in local population extinction risk.

Some population studies treat the population as one in which individuals are uniformly distributed in the area of interest but others consider the population (metapopulation) to be composed of subpopulations inhabiting habitat patches of different qualities, including different levels of contamination. The differences in vital rates, including exchange rates among patches, are used to project contaminant exposure consequences. With metapopulation models, effect manifestation at a distance from a contaminated patch can be explored: a population member can be exposed in one patch yet the effects might manifest in an uncontaminated patch after migration.

Community

Community ecotoxicology explores the consequences of contaminant exposure of and movement of contaminants within ecological communities. The majority of such studies are field studies either addressing scientific questions or applying knowledge to assess risk or define remediation action for a contaminated system. Like the biomarkers applied at lower levels of biological organization, bioindicators are applied by community ecotoxicologists. Bioindicators might be particularly sensitive species whose absence suggests an adverse impact. A community metric such as species richness, evenness, or diversity might also be used as an indicator of an adverse exposure consequence. Any study in which biological systems are applied to assess the structural and functional integrity of ecosystems is referred to as a biomonitoring study.

Exposure within communities is often explored in the context of contaminant trophic transfer. Depending on its properties, a contaminant can increase (biomagnify), decrease (trophic dilution), or not change in concentration with progression through a food web. Contaminants such as methylmercury or persistent organic pollutants (POPs) such as DDT biomagnify. Biomagnification can lead to adverse consequences to higher trophic level species such as the raptors and piscivorous birds mentioned above. Studies of food webs including omnivorous members require a measure of trophic status for species. A convenient measure of trophic status is provided by the nitrogen isotopic fractionation that occurs with each trophic exchange: the amount of heavy (^{15}N) nitrogen increases in tissues relative to light (^{14}N) nitrogen with each trophic exchange. The $\delta\,^{15}$N is the conventional metric for expressing relative N isotopic abundances:

$$\delta^{15}\text{N} = 1000 \left[\frac{\left[^{15}\text{N}_{\text{Species }i} \right] / \left[^{14}\text{N}_{\text{Species }i} \right]}{\left[^{15}\text{N}_{\text{Air}} \right] / \left[^{14}\text{N}_{\text{Air}} \right]} - 1 \right]$$

Graphs (e.g., **Figure 3**) or quantitative models of contaminant concentration in species within the subject community versus $\delta\,^{15}$N facilitates prediction of contaminant movement in communities.

Ecosystems

Ecosystem-level studies vary widely in their spatial and temporal scales. Often ecosystem modeling techniques are applied to an easily definable ecosystem such as a contaminated lake or watershed. Fate and movement of

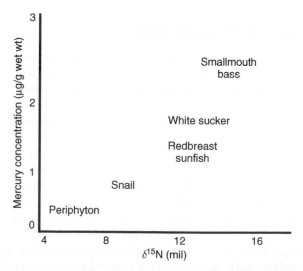

Figure 3 Mercury biomagnification in the South River (Waynesboro, Virginia, USA) illustrated with periphyton, a grazing snail (*Leptoxis carinata*), two intermediate fish species (redbreast sunfish, *Lepomis auritus* and white sucker, *Catostomus commersoni*), and a top predator species (smallmouth bass, *Micropterus dolomieu*). Author's unpublished data.

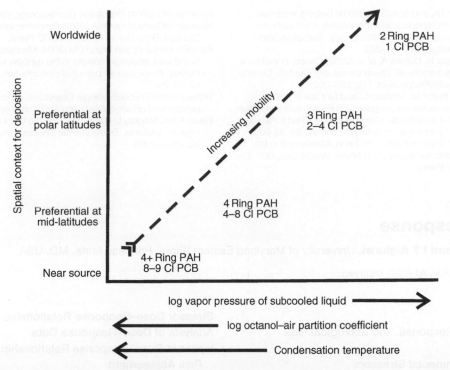

Figure 4 Global movement of a POP is determined by several qualities including its (subcooled liquid) vapor pressure and condensation temperature which, together, determine its tendency to move into and remain in the atmosphere. A more volatile compound will be transported more by atmosphere movement than one that is less volatile. Its condensation temperature influences the latitudinal limits to which it might move. The octanol–air partition coefficient is also important because it reflects the POP's tendency to remain associated with the solid and liquid phases of the Earth versus the atmosphere. Here, polycyclic aromatic hydrocarbons (PAHs) with different numbers of aromatic rings and polychlorinated biphenyls (PCBs) with different numbers of chloride atoms are used to illustrate trends for global deposition of POP. This biospheric process is called global distillation.

contaminants are then modeled by computer or measured in extensive sampling programs. Larger-scale studies are required for contaminants amenable to wide spatial dispersion via atmospheric transport such as mercury from coal power plants or contaminants used widely by society such as atrazine, an herbicide applied in the North American Corn Belt. Often geographical information system (GIS) and remote sensing technologies are essential in these types of studies. In still other instances, a global perspective is required to adequately grasp the ecotoxicological consequences of contaminants. Current global issues are ozone depletion in the stratosphere due to chlorofluorcarbon (CFC) release, global warming due to greenhouse gas emissions, and global movement of persistent organic pollutants (**Figure 4**). More and more frequently, large-scale issues are emerging as critical ones in ecotoxicology.

Current Trends in Ecotoxicology

A working knowledge of the movement and effects of environmental contaminants is recognized worldwide as essential to maintaining an acceptable quality to life. Ecotoxicology has emerged as the applied science that addresses the central issues of contaminants in the biosphere. Major challenges in this young science include the following: (1) the integration of causal explanations and knowledge arising at different levels of biological organization into a coherent whole; (2) integration of scientific, technical, and practical goals of ecotoxicologists; and (3) consideration of ecotoxicological issues at increasingly wider spatial and longer temporal scales.

See also: Bioaccumulation; Biogeochemical Approaches to Environmental Risk Assessment; Biomagnification; Ecological Risk Assessment.

Further Reading

Alvarez MMS and Ellis DV (1990) Widespread neogastropod imposex in the Northeast Pacific: Implications for TBT contamination surveys. *Marine Pollution Bulletin* 21: 244–247.

Bryan GW and Gibbs PE (1991) Impact of low concentrations of tributyltin (TBT) on marine organisms: A review. In: Newman MC and McIntosh AW (eds.) *Metal Ecotoxicology. Concepts and Applications*, pp. 323–361. Chelsea, MI: Lewis Publishers.

Cassano G, Bellantuono V, Ardizzone C, and Lippe C (2003) Pyrethroid stimulation of ion transport across frog skin. *Environmental Toxicology and Chemistry* 22: 1330–1334.

Clements WH and Newman MC (2002) *Community Ecotoxicology*. Chichester: Wiley.

Incardona JP, Collier TK, and Scholz NL (2004) Defects in cardiac function precede morophological abnormalities in fish embryos exposed to polycyclic aromatic hydrocarbons. *Toxicology and Applied Pharmacology* 196: 191–205.

Matineau D, Lemberger K, Dallaire A, *et al.* (2002) Cancer in wildlife, a case study: Beluga from the St. Lawrence estuary, Quebec, Canada. *Environmental Health Perspectives* 110: 285–292.

Myers M, Landahl J, Krahn M, Johnson L, and McCain B (1990) Overview of studies on liver carcinogenesis in English sole from Puget Sound: Evidence for a xenobiotic chemical etiology. Part I: Pathology and epizootiology. *Science of the Total Environment* 94: 33–50.

National Academy of Science (NAS) (1983) *Risk Assessment in the Federal Government: Managing the Process.* Washington, DC: National Academy Press.

Newman MC (2001) *Population Ecotoxicology.* Chichester: Wiley.

Newman MC and Unger MA (2003) *Fundamentals of Ecotoxicology*, 2nd edn. Boca Raton, FL: Lewis/CRC Press.

Pane EF, Haque A, and Wood CM (2004) Mechanistic analysis of acute, Ni-induced respiratory toxicity in the rainbow trout (*Oncorhychus mykiss*): An exclusively branchial phenomenon. *Aquatic Toxicology* 69: 11–24.

Truhaut R (1977) Ecotoxicology: Objectives, principles and perspectives. *Ecotoxicology and Environmental Safety* 1: 151–173.

Wania F and Mackay D (1996) Tracking the distribution of persistent organic pollutants. *Environmental Sciences and Technology* 30: 390A–396A.

Dose–Response

A B Ishaque and I T Aighewi, University of Maryland Eastern Shore, Princess Anne, MD, USA

Introduction

The concept of dose–response is central to the scientific study of poisons or toxicology. It is also well known in toxicology that the dose of a substance is what separates a poison from a remedy. All biological organisms including humans can potentially be exposed to a range of chemical, physical, radiological, and biological doses in the environment that can trigger a variety of responses. The process of quantitatively assessing the dose received and response by a biological entity produces a dose–response relationship which is usually represented graphically as a dose–response curve. Such a curve has traditionally been one of the steps involved in conducting human health risk assessment of exposure to various chemical, physical, and radiological contaminants in the environment, including drugs and medicine.

Risk assessment provides the scientific or rational basis for managing and regulating toxic substances in the environment. While traditional human health risk assessment focuses mostly on the effects of chemicals and xenobiotics on humans with respect to definite endpoint or health outcome such as cancer, endpoints are not so clear-cut in risk assessment dealing with ecosystem health or ecological risk assessment. In fact, multiple endpoints are more frequently encountered and must be defined for particular ecosystems under consideration. Furthermore, 'stressor dose' rather than chemical dose is

the main emphasis in ecological risk assessment and will be used throughout this article.

This article is meant to provide a basic background on stressor dose–response within the context of traditional dose–response in ecological risk assessment, drawing occasionally from human risk assessment as background. The article begins with a discussion of stressor dose–response and ecological risk assessment, followed by a detailed review of stressor dose, biological responses, and stressor dose–response relationship. The article concludes with some knowledge gap in the use of stressor–dose relationships.

Stressor Dose–Response, and Ecological Risk Assessment

Risk assessment is a general approach established for independent, neutral, science-based evaluation of the probable likelihood of harm (response) from exposure (dose of stressor) to deleterious elements in the environment. Risk assessment is borne out of the need to manage risks of any such negative occurrence in order to protect public health and our ecosystems. While ecological risk assessment is not entirely new, it definitely lags behind human health risk assessment. However, the growing public awareness and concern about the consequences of major environmental events of our times at both local and global levels, for example, acid precipitation, global warming, biodiversity loss, ozone depletion, etc., has created renewed interest and urgency for appropriate approach for predicting these human-induced stressors in both terrestrial and aquatic ecosystems. Consequently, risk-assessment methodologies, particularly for ecological systems, are constantly evolving and improving upon methods used in the past.

'Dose' is defined as the quantity of an agent to which an entity is exposed to in the environment. Since these agents – biological, chemical, or physical – exert stress on the entity receiving them, they are termed stressors. A stressor is thus aptly defined as a substance or condition that causes stress on an entity in an ecosystem, whereas a 'response' is the deleterious effect(s) manifested in the entity as a result of exposure to the stressor. Dose–response from an ecological context is the quantity of exposure to a stressor (chemical, physical, or biological agent) and the resulting changes in function or health (response) of a designated entity receiving the agent. A graphical representation of such a relationship is a stressor dose–response curve which is slightly different from the classical dose–response curves and models used for human health risk assessment.

Types of Environmental Stressors

Different types of environmental stressors exist in nature. They include biological, chemical, or physical stressors that can be broadly categorized as biotic (living) or abiotic (nonliving) stressors. Abiotic stressors consist of two major types: physical stressors and chemical stressors. Abiotic stressors originate from the ambient environment and may or may not have any biotic input. For example, fire outbreak can put stress on organisms in an environment; however, such a fire may be started by humans (biotic origin) or by lightning (abiotic origin). Unlike abiotic stressors, biotic stressors have only biological origin, that is, living organisms exerting stress on another biological organism. For these stressors, the removal or destruction of the organism exerting the stress usually results in the complete abatement of the effects with no carry-on effect unlike abiotic stressors that may have a carry-on effect.

Abiotic Stressors

Abiotic stressors may include light intensity, temperature range, pH level (acidity or alkalinity), water availability, dissolved gases, nutrient availability, radiation level, heavy metal contamination, etc. They are nonliving, physical, and chemical factors which could affect an entity from the molecular level to an entire ecosystem level. For example, too much light has the ability to interfere with an organism's survival and reproduction. This could lead to responses observed at individual organism level through population to community and to an entire ecosystem level. Abiotic stressors may vary in a given environment and may contribute in the determination of the types and numbers of organisms present in that environment. A major abiotic stressor that has attracted the interest of environmental scientists is the amount and nature of chemical contaminants in environmental media – soil, water, and air.

Physical stressors

Examples of physical stressors on terrestrial ecosystem include habitat loss or landtake such as the loss of a wetland to real estate development. This can affect shellfish and waterfowl that depend on such a habitat, or forest land conversion to real estate and the resultant loss of valuable plant and animal species. Habitat fragmentation which has led to the extinction of venerable species is also a form of physical stressor and is often initiated by humans from activities such as road construction that put much stress on organisms that require large areas for their survival. Physical stressors could also be in the form of a gradual disappearance of a stream, lake, or river due to siltation or dam construction. Global warming is a physical stressor that could result in significant shift in vegetation due to changes in climatic variables such as temperature and precipitation; such a shift in vegetation could also affect the distribution of other organisms in both terrestrial and aquatic ecosystems as well. Fire, radioactive isotopes, volcanic eruption, tornadoes, hurricanes, drought, tsunami, and mudslide can also be considered physical stressors in the environment.

Chemical stressors

Chemical stressors include various pesticides such as dichlorodiphenyltrichloroethane (DDT) which could lead to egg shell thinning in avian species such as the well-known case of the American bald eagle; it could be estrogenic chemicals such as 17β-estradiol from poultry litter that can lead to feminization of some male fish species. It could be a chemical such as polychlorinated biphenyls (PCBs) discharged into water systems that could induce mutation in some freshwater species. Toxins resulting from snake bite or heavy metal contamination by cadmium, arsenic, and mercury of soil, water, food, or air may constitute stressors for both plants and animals when they enter biological organisms via ingestion, inhalation, or dermal contact.

Biotic Stressors

Biotic stressors could be pathogenic, parasitic, or predatory in nature and originate from living organisms. Both pathogenic and parasitic stressors have effect on the health of the entity hosting the pathogen/parasite. This effect starts at a molecular level, with the host radiating into a whole ecosystem level – especially if the host is a keystone species. For most pathogens, there is a positive relationship between the dose (the quantity of pathogens) and the response (reaction from the host) such that the occurrence and severity of the stressor (effect of the pathogens) on the host are proportional to the number of pathogens to which a host organism is exposed to. The above relationship holds true for parasite–host relationship. The commonality between pathogen and parasite relationships with their host is that in both situations the response could be observed at molecular, cellular, organ, and whole-organism level. Predation in an ecosystem is somewhat similar to the above description except that the response starts at population (prey) level. The prey is the population that is receiving the effect of the stressor (predation) coming from the predators, which are also individuals within a population.

Biotic stressors could also be any of the introduced invasive or exotic species in an ecosystem such as African honeybee (*Mellifera scutellata*), Asian tiger mosquito (*Aedes albopictus*), or kudzu (*Pueraria montana*) on land. In freshwater systems, animal species such as bighead carp (*Hypophthalmichthys nobilis*), lionfish (*Pterois volitans* (Linnaeus)), or plants such as common water hyacinth (*Eichhornia crassipes*) can be considered biological stressors. An epidemic of army worm infestation on a corn field or locust infestation of epidemic proportion on vegetation could impose severe stress on other organisms. In fact, direct human disturbance such as harassment or overharvesting of wildlife or fishes could be considered biological stressors in nature.

Stressor Doses

According to Paracelsus, "The right dose differentiates a poison and a remedy." In toxicology, however, dose has always been defined rather loosely. Strictly, dose applies to the administered levels of a drug or chemical to a test organism under controlled environmental conditions. In ecological context, a dose will be referred to as a quantity of stressor received by an entity or the quantity of exposure to a stressor – hence stressor dose. Since a stressor is any physical, chemical, or biological agent or condition that can induce adverse response in any entity of interest, a stressor dose therefore may include the amount of an environmentally stable chemical such as PCB or a physical agent such as radiation or a biological agent such as

bacteria that produces a response from the entity receiving it. The entity could be an individual species, population of species, community, or an ecosystem. Dose – chemical, physical, or biological – can sometimes be differentiated on the basis of the amount encountered in the environment (exposure dose), or the actual amount of the exposed dose that enters the organism's body (absorbed dose), or the quantity administered to an organism under controlled situations (administered dose), and the combination of all types (total dose). Note however that it may be practically difficult to administer or measure progressive doses of a stressor on ecosystems, unlike the animal models used in human health risk assessment under controlled laboratory situations.

In addition to differentiating the doses encountered by an organism(s), the frequency of exposure to the stressor is also an important consideration in ecological risk assessment. This becomes even more relevant if the stressor is seasonal or permanent in an ecosystem. Organisms, for example, a fish species exposed to frequent sublethal doses of a xenobiotic may respond differently than those that receive one high but infrequent lethal dose of the same stressor.

Response

The adverse ecological effects suffered as a result of exposure to stress is termed a 'response' (toxic reaction in toxicology). Reponses from an entity receiving a dose of a specific stressor could lead to changes in function or health of that entity receiving that stressor. Responses could be a reaction from either/both biotic and abiotic stressors.

Predation is a biotic stressor which has its response observed at a population (prey item) level. There are some similarities between response to a biological stressor and response to an abiotic stressor. For example, a disease could be a response to exposure to pathogenic bacteria or exposure to chemical pollutant. In both situations, the response could be observed at individual, population, community, and ecosystem level through the responses manifested at the molecular level of an individual. Responses could be detected at molecular, cellular, organ, organism, community, and ecosystem levels. An example of an entity responding at a molecular level from exposure to an abiotic stressor such as lead is the inhibition of aminolevulinic acid dehydratase (ALAD) in human and other organisms. This kind of response at a molecular level could be used as a biomarker of lead exposure and could serve as an early warning system for Pb contamination of an environmental medium. ALAD inhibition in organism is specific to lead exposure but other biomarkers are nonspecific. The nonspecific biomarkers indicate that harm is being caused but does not indicate which stressor is causing the harm. An example is the induction of monooxygenases. Monooxygenase enzyme is induced by a host of environmental chemicals including PCBs, DDT, dichlorodiphenyldichloroethylene (DDE), etc. Molecular-level responses could lead to response at cellular level, for example, cytotoxicity. Cytotoxicity can lead to the dysfunction of organs producing an organ-level response. An organ-level response can lead to response at the organism level. A typical example of organism-level response is decrease in reproduction; such a decrease in reproduction leads to a decline of the organism's population. Decrease in populations could lead to response observed at community level and finally community-level response could lead to responses observed at the ecosystem level.

Unlike the traditional dose–response in humans with definite assessed endpoint, for example, cancer resulting from exposure to environmental carcinogens, responses in ecological risk assessment vary considerably and can be a source of contention even when experimental evidence exists that links the stressor under consideration to the effects. The reason for this is not unconnected to the social and political/legal dimension that often comes to bear on ecological risk analyses, not the least of which are the variety of species, ecological communities, and ecological functions from which to choose, and the statutory ambiguity regarding what needs to be protected. Some known examples of ecological responses at the ecosystem levels include (1) 'forest decline' in Europe in the 1980s, resulting from a combination of stressors that included acid rain, elevated ozone levels, fogs, and frost following drought years; (2) the dwindling growth of maple seedlings attributed to acid rain as stressor in northeastern United States, inhibition of nitrogen-fixing ability of nitrogen-fixing bacteria species in the soil, and the consequent reduction in yields of legumes and (3) bird kill resulting from chemical spill, or from long-term bioconcentration of toxic chemicals or fish kill, resulting from the Exxon Valdez oil spill of the 1990s in Alaska. Regardless of these responses, however, it is important to establish true linkage between the environmental stressor and the endpoint. Recently, the US Environmental Protection Agency (USEPA) published a document that describes a set of endpoints known as generic ecological assessment endpoints (GEAEs) that can be considered and adapted to specific ecological risk assessments for improving the scientific basis for ecological risk-management decisions.

Types of Responses

From a strictly ecological standpoint, organisms are usually not exposed to one stressor at a time but rather multiple stressors; hence, it is very difficult to decipher which stressor is producing which response. Generally,

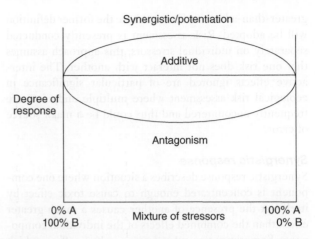

Figure 1 Main types of responses from exposure to mixture of stressors.

the resulting effect (response) from multiple stressors follows one of four patterns: potentiation, additive, antagonistic, and synergistic (see **Figure 1**). Pattern of responses among interacting environmental stressors could also be studied using the toxic units (TU) concept. TU is defined as the concentration of a chemical in a toxic mixture divided by its individual toxic concentration for the endpoint measured (e.g., its individual 48 h lethal concentration 50 (LC_{50}) value.). Mixtures with summed TU values close to 1.0 are considered to be additive in toxicity (**Figure 2**). Those with summed TU < 1.0 are less than additive in toxicity (antagonistic) and those with summed TU > 1.0 are greater than additive in toxicity (synergistic).

Potentiating response

A stressor which on its own will not trigger any response may in the presence of another stressor augment the response of the other to produce a response exceeding that of the sum of the two individual responses ($0 + 2 = 4$). Potentiating response is frequently used interchangeably with 'synergism', which describes a

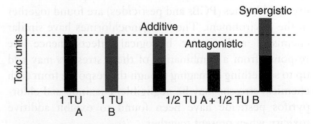

Figure 2 Combined effect of quantifiable mixture of stressors expressed as toxic units (TU). Both A and B are independent stressors acting as a composite mixture on an entity. The x-axis represents the quantity of the composite mixture and the y-axis represents the toxic units. Reproduced by permission of Taylor & Francis.

greater-than-additive effect, but here the former definition will be adopted. Risk assessment is presently conducted separately on individual stressors; this approach assumes that one risk does not interact with another. The interactive effects ignored are of particular significance in ecological risk assessment where multiple endpoints are frequently encountered and thus could be a major source of error.

Synergistic response

Synergistic response describes a situation where one component is concentrated enough to cause toxic effect by itself but the presence of another causes a much greater effect than the combined effects of the individual components. Sometimes two substances may have effects which reinforce each other so much that the combined effect is more than the additive. For example, tobacco smoke in combination with asbestos dust can cause lung cancer in humans more readily than exposure to either the cigarette smoke or asbestos dust alone. In fact, it has been documented that smokers are 40 times more likely than nonsmokers to get lung cancer following exposure to asbestos dust. In this case, the response is greater than additive toxicity $(2 + 2 = 6)$.

Another classical example of a synergistic effect is the formation of trihalomethane known to cause cancer in humans. Trihalomethane is a compound formed from chlorine used as disinfectants for microorganisms in water when it reacts with suspended particulates (solid). Other examples include the herbicide atrazine that has been found to produce synergistic toxicity in a binary mixture with methyl parathion.

Additive response

If the combined effect of two or more chemicals is simply the sum of the effects of the individual chemicals, the effect is said to be additive $(2 + 2 = 4)$. For example, if a dose of drug A that produces 25% of the maximum response is combined with a dose of drug B that produces 50% of the maximum response, then 75% of the maximum response is produced. This is possible when the stressors are sufficiently similar. For example, most stable organochlorines (PCBs and pesticides) are found together in the environment. These organochlorines have similar chemistry and similar biological effects; hence the response from a combination of these stressors may add up to something damaging though the response from each chemical stressor may be negligible. Diazinon and chlorpyrifos pesticide have been found to exhibit additive toxicity when present together.

Antagonistic response

There are also cases where potentially toxic substances may counter each other's effects. When the effect of an exposure to two or more chemicals is less than would be expected (if the known effects of the individual chemicals were added together), the effect produced is known as antagonistic, that is, less than additive toxicity $(2 + 2 = 3)$. For example, selenium reduces the toxic effects of mercury. Other antagonists that have been identified are methionine and vinyl chloride, arsenic and selenium, and zinc and cadmium. Less-than-additive toxicity has been reported for the combination of atrazine with methoxychlor pesticides.

Stressor Dose–Response

In real world scenarios, organisms are usually exposed to multiple environmental rather than individual stressors in the environment. Not only does this mixture of stressors differ qualitatively, they also differ in terms of quantity. In addition, a stressor to an entity may also constitute a stressor to another entity in the same ecosystem, directly or indirectly. To understand the nature of responses emanating from complex mixtures of stressors, for example, one therefore needs a clear understanding about the functioning of that particular ecosystem and the mechanism of interaction among the interacting environmental stressors. In addition, one needs to know how biotic systems interact with each other as well as with these complex mixtures of stressors in the ambient environment. The interaction between an entity and specific quantity of stressor(s) produces stressor dose–response relationship (simply dose–response in human toxicology), which when appropriately quantified or modeled will provide pertinent information establishing the extent of stressors and the corresponding magnitude of adverse effects in risk assessment.

Stressor dose–response models are graphical representation of quantifiable stressors and responses depicting real causal relationships between the two variables, that is, dose and response, and used as an indispensable component of human or ecological risk assessment for guiding policy and risk-management decisions. Stressor dose–response relationship is one of the major steps in ecological risk-assessment processes. It depicts the change in response upon exposure to differing levels of stressor. **Figure 3** describes processes involved in ecological risk assessment with particular reference to the stressor dose–response relationship as an integral part of the process; the others being exposure assessment, response assessment, and risk characterization. Response to a stressor depends on the quantity of the stressor and the type of biological organism receiving the stressor.

The assumptions on which the stressor dose–response relationships can be successfully based are (1) that the response is based on the knowledge that the response produced is actually due to known stressor/toxic agent(s), (2) that the response is in fact related to the dose of the stressor, and (3) that the stressor dose–response

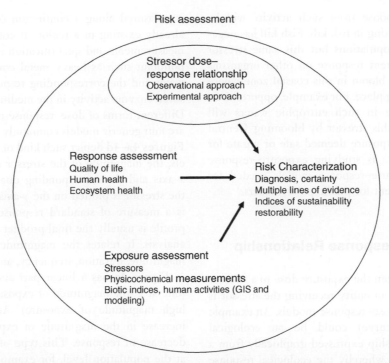

Risk assessment

Stressor dose–
response relationship
Observational approach
Experimental approach

Response assessment
Quality of life
Human health
Ecosystem health

Risk Characterization
Diagnosis, certainty
Multiple lines of evidence
Indices of sustainability
restorability

Exposure assessment
Stressors
Physicochemical measurements
Biotic indices, human activities (GIS and
modeling)

Figure 3 Risk-assessment process showing stressor–response relationship as a major component. Adapted from US Environmental Protection Agency (1992) Framework for ecological risk assessment. EPA-Risk Assessment Forum, EPA/630/R-92/001, Washington, DC.

relationship is based on the existence of a quantifiable method of measuring and a precise means of expressing the effect of the stressor. When a large quantity of a specific stressor is used for a short time, the corresponding response is normally a complete destruction of the entity. This type of exposure is known as acute. For example, natural disasters like earthquake, tornadoes, hurricanes, volcanic eruption, tsunami, etc., could be considered as acute events because they occur for a very short time and the effect is usually catastrophic. Another example of acute events is fish kill in highly eutrophic waters during very hot summers. On the other hand, if a small quantity of stressor is applied for a very long time, the exposure is referred to as chronic. In chronic exposure, there is no lethality or destruction but a major functional physiology of the entity could be affected. For example, egg shell thinning in birds exposed to low levels of DDE for a long period of time caused the bird's population to decline. On a global scale, the gradual increase in carbon dioxide concentration could be considered a chronic event and the plausible response is global warming. Global warming has been linked to sea-level rise resulting from polar ice melting; hence the response from the global ecosystem as the result of gradual increase in carbon dioxide concentration is the rise in sea level.

The LC_{50} is one way to measure the short-term potential effect of a stressor at a very high quantity. It refers to a concentration of a stressor in an ambient environment which kills 50% of a sample population. Although many

endpoints are quantitative and precise in human risk assessment, they are often indirect measures of toxicity. For example, changes in enzyme levels in blood can indicate tissue damage. Many direct measures of effects are not necessarily related to the mechanism by which a stressor produces its harm to an entity but have the advantage of permitting a causal relationship to be established between the agent and its action. For new chemicals, however, it is customary to use lethality as a starting point or index of toxicological evaluation.

Response from an abiotic component of an ecosystem could be a decrease in the suitability as a habitat for resident organisms. Since organisms obtain benefits such as ecosystem services including but not limited to food and waste assimilation from a properly functioning ecosystem, a decrease in suitability of such a habitat could jeopardize the existence of the organism. Another example of response from abiotic stressor is changes in land use such as a rural community becoming urbanized. This type of change in land use could adversely impact both the terrestrial and aquatic coastal ecology, particularly where intensive crop and animal husbandry are replaced by industry and real estate. Once there is a change in land use, the response associated with the new land use could be different depending on the nature of the new stressor.

For example, cultural eutrophication is a major problem associated with most coastal zones due to disposal of household waste into these zones. The stressor in this situation is increase in nutrients from human wastes living in the area. The response from such activity will be oxygen depletion resulting in fish kill. Fish kill has negative impact on fish populations but this same stressor could produce a different response on other organisms such as phytoplankton bloom in this coastal zone where eutrophication is taking place. For example, opportunistic plant species dwelling in such eutrophic waters will respond positively to this stressor by blooming. Central to defining levels of exposure deemed safe or unsafe for environmental stressors is studying exposure–response and developing exposure–response models needed for ecological risk assessment for public policymakers.

Stressor Dose–Response Relationship

The relationship between the exposure dose to a stressor and the response from an entity receiving the stressor is described by stressor dose–response models. An example of such a model (curve) could be an ecological dose–response relationship expressed graphically from a retrospective approach whereby the ecological response

is measured along a continuum of human disturbances already existing in a region. It could also be in form of measurements and quantification of levels of a chemical pollutant such as heavy metal contamination of soil or water and the corresponding response that can be measured enzyme activity in the medium under investigation. Different forms of dose–response models exist, but there are four generic models commonly used to describe them. **Figures 4a–4d** depict such kind of models. In these models, the magnitude of the stressor dose is plotted on the x-axis and the corresponding quantified response from the stressor is plotted on the y-axis as percentages and it is a measure of standard response. A stressor–response profile is usually the final product of a stressor–response analysis. It relates the magnitude of the effect to the magnitude, duration, frequency, and timing of exposure.

Figure 4a has a linear part and two threshold parts (one at a low magnitude of exposure and the other at a high magnitude of exposure). At the linear part, an increase in the magnitude of exposure brings about a decrease in response. This type of response is observed at the population level, for example, a decrease in prey

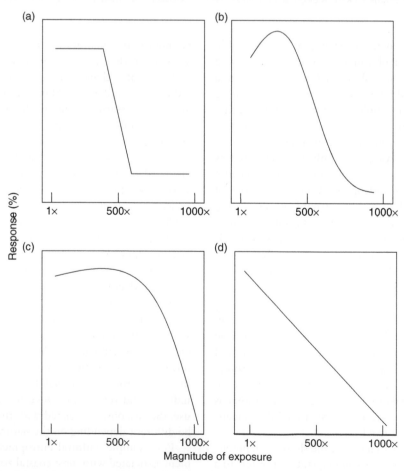

Figure 4 Dose–response curves showing (a) an abrupt change in response with dose, (b) subsidy at low doses that can serve as a practical threshold, (c) asymptotic with a practical threshold, and (d) no threshold. Reproduced by permission of Taylor & Francis.

items as the number of predators increases. An abrupt change in response with an increased magnitude of exposure is only observed where an increased in magnitude of exposure is slightly higher than the first threshold level. With the linear-part-type models, the stronger the exposure the steeper the linear part will be. Generally, at the threshold part of the curve, an increase in the magnitude of exposure has no effect on the response. At the second threshold level, an increase in exposure magnitude did not bring a change in response since the minimum response had already been achieved. **Figure 4b** has a practical threshold in the low levels of exposure which resulted in an increase in response. **Figure 4** showed asymptotic rate of change of response. Here, at low magnitude of exposure, the change in response rate was so small to be considered negligible but as the exposure magnitude increases there is also steep change in response. **Figure 4d** has just the linear part of **Figure 4a** with no threshold levels.

More often, causal evaluation is used to identify factors that are responsible for observed effects such as the criteria developed by Hill in 1965 for establishing causality (**Table 1**).

The integration of stressor–response and exposure profiles include: (1) comparing single effects and exposure value; (2) comparing distribution of effects and exposure; and (3) conducting simulation modeling. The choice of any of these depends on the original purpose of the assessment as well as data available and time. A typical stressor–response curve is shown in **Figure 5**. This curve shows the ecological response on the y-axis and the stressor on the x-axis. Note that unlike in traditional dose-response curve, the units for either of the axes vary depending on the assessment being conducted. Allowance is usually made for a predetermined threshold that is considered acceptable as shown in the graph.

According to the USEPA framework for ecological risk assessment, the stressor–response model may focus on different aspects of the stressor–response relationship depending on the objective of the assessment, the conceptual model, and the type of data used for analysis. Also crucial is the temporal and spatial distributions of the stressor in the experimental or observational setting. In the case of physical stressors, specific attributes of the environment after disturbance can be related to response.

Stressor (dose)–response relationships curves usually involve the derivation of indices of toxicity such as ecological dose (ED_{50}), or EC_{50} similar to the lethal dose (LD_{50}). Such descriptors convey useful information necessary for assessing the level of toxicity associated with the substance under investigation. These indices essentially describe the dose of the stressor or chemical necessary to produce 50% death or inhibition of the organism used for testing.

The LD_{50} is defined as the lethal dose at which 50% of the population is killed in a given period of time; an LC_{50} is the lethal concentration required to kill 50% of the population. In ecological risk assessment, EC_{50} may be used instead of LD_{50}; whereas LC_{50} is a measure of concentration (e.g., $mg\,l^{-1}$), it could also be a specific temperature or any other parameter in a different unit of measure. These bioassays involve subjecting several replicate groups of individuals to a range of concentrations (or doses) of a toxic compound and measuring the mortality after a defined time interval that can range from minutes to hours or days. The data are then plotted and the lethal dose of interest interpolated from the graph as shown in **Figure 5**.

The curve shows a range of response to varying stressor intensity; the points are commonly used levels of

Table 1 Hill's criteria for evaluating causal associations

1. Strength of effect: A high magnitude of effect is associated with exposure to the stressor
2. Consistency of occurrence: The association is repeatedly observed under different circumstance
3. Specificity of effect to a stressor: The effect is diagnostic of a stressor
4. Temporality: The stressor precedes the effect in time
5. Presence of a biological gradient: A positive correlation between the stressor and response
6. Plausible mechanism of action
7. Coherence: The hypothesis does not conflict with knowledge of natural history and biology
8. Experimental evidence
9. Analogy: Similar stressors cause similar response

Source: US Environmental Protection Agency (1992) Framework for ecological risk assessment. EPA-Risk Assessment Forum, EPA/630/R-92/001, Washington, DC.

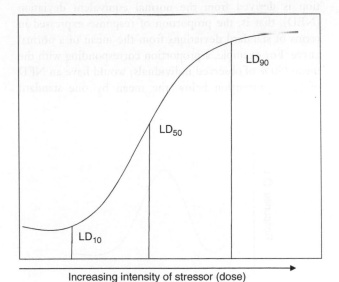

Figure 5 A typical sigmoidal curve for stressor–response relationship (dose/percentage of mortality).

effect estimating lethal dose (LD) fatal to a given percent of organisms in a population. More complex relationships typical of ecological risk assessments with multiple stressors and cumulative ecological effects may differ slightly and EC may be used in place of LD_{50}. The final outcome of the stressor dose–response relationship feeds the final stage of risk assessment involving the characterization of the toxin in human risk assessment or the preparation of a stressor profile in ecological risk assessment. Note that on a large scale such as ecological risk assessment conducted for major ecosystems, the stressor dose–response models or curves will only provide a scientific framework for establishing how bioindicators (when present) respond to increasing human disturbance. Other factors such as the establishment of thresholds has to invoke sociopolitical factors; for example, society must set the threshold by considering tradeoff between economic growth and the level of ecological risk society may be willing to accept.

Analysis of Dose–Response Data

The two common approaches to analyzing dose–response data quantitatively is by the probit and log logistic models. Log normal curve (**Figure 6**) formed the basis for probit model and logistic model is linked to processes such as enzyme kinetics, autocatalysis, and adsorption phenomena. Both models predict a sigmoidal curve (**Figure 5**). During dose–response analysis, data may be used directly or transformed. In most cases, the objective behind the transformations is to make linear the relationship between dose and response. The most common transformation performed is the logarithm scale. The probit transformation is derived from the normal equivalent deviation (NED), that is, the proportion of response expressed in terms of standard deviations from the mean of a normal curve. For example, a proportion corresponding with the mean (50% of observed individuals) would have an NED of 0. A proportion below the mean by one standard

deviation (16% of observed individuals) would have an NED of -1. To get rid of negative values, the number 5 is added to NED values, probit $(P) = \text{NED }(P) + 5$, where $P =$ proportion of observed individuals that responded to a treatment and NED = the normal equivalent deviation.

The log logistic model (logit) has the following form:

$$\text{Logit}(P) = \ln[P/1 - P]$$

A transformed logit is more commonly employed than that calculated by equation above because values of this transformed logit are nearly the same as probit values – except for proportions at the extreme ends of the curves (**Figure 7**),

$$\text{Transformed logit} = [\text{logit}(P)/2] + 5$$

where logit $(P) =$ logit value (estimated by logit $(P) = \ln[P/1 - P]$).

Slopes and intercepts of stressor–response lines provide valuable information about the relative sensitivity of biota to the same or different stressors. The slope quantifies the steepness of the line. It equals the change in response for each unit change in stressor intensity. It is expressed in the units of the response axis divided by the units of the stressor axis. If the slope is positive, response increases as stressor increases. If the slope is negative, response decreases as stressor increases. Imagine a second line with a much steeper slope intersecting the drawn line at the LD_{50} (**Figure 8**). Although the LD_{50} would be the same for both lines, a small change in stressor intensity has much more of an effect with one stressor (steeper slope) than the other (shallow slope). The response intercept is the response value of the line when stressor intensity equals zero. It defines the elevation of the line and indicates the potential threshold exposure.

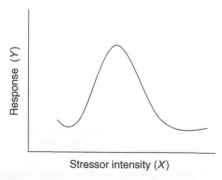

Figure 6 Typical log normal curve for stressor–response data.

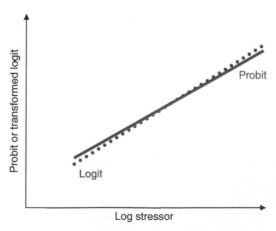

Figure 7 Lines resulting from the probit and logit transformations. Reproduced by permission of Taylor & Francis.

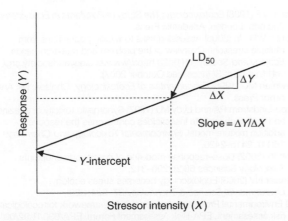

Figure 8 Stressor–response relationship with intercept and slope.

Issues in Dose–Response Relationship in Ecological Risk Assessment

A fundamental difference between dose–response relationship in human toxicology and ecology is that individuals studied may not necessarily be the individuals at risk. Second, they are usually vulnerable to confounding factors. For example, the response of soil microorganisms to heavy metal amendment using such measurable endpoints as soil dehydrogenase activity, antifreeze protein (AFP) content, microbial respiration, and microbial biomass measured in samples containing different concentrations of the heavy metals in sewage is an example of attempts to develop stressor dose–response relationship as an indirect approach for assessing the bioavailability of nickel to humans via ingestion of plants grown in such a soil. However, a stressor dose–response relationship established in this case addresses several toxicological issues. First, the microbial activities measured are meant to assess the effects of the heavy metals on the microbial population as well as plant uptake of nitrogen through the activity of nitrogen-fixing bacteria that may be present in the soil. The data obtained can be used to determine the level of ecological toxicity (ED$_{50}$) for guiding decision on heavy metal-laden sewage disposed in farmlands. The foregoing indicates that except clear-cut objectives that are established from the onset of an ecological risk assessment, the outcome may be of little or no practical use.

Ecological stressor/dose–response has often been applied to the soil ecosystem for evaluating the effects of heavy metals on the microbial activity through the use of enzyme (arylsulfatase) activity as a measurable endpoint. The effects are then fitted to a logistic dose–response model and graphical ED$_{50}$ determined like in human dose–response curves. Attempts have also been made to evaluate the ecological risks at the landscape level by developing a science-based ecological dose–response curve to help define appropriate and socially acceptable thresholds or limits of 'acceptable change' of an ecosystem by measuring the ecological response (i.e.,

abundance of species productivity) using a continuum of human disturbance already existing in a region. Statistical relationships were developed that tie the abundance of particular species to different levels of human disturbance.

Although the stressor dose–response concept provides the scientific framework for establishing the impact of increasing human and natural disturbances, it is believed that science cannot provide the acceptable threshold of human activity since threshold must be set by society by integrating tradeoff factors between economic growth and the level of ecological risk people are willing to accept. The stressor–response analysis, as opposed to traditional dose–response relationship, in human health risk assessment describes the relationship between the magnitude, frequency, or duration of the stressor and the magnitude of response in ecological risk assessment.

Ecological risk assessments evaluate ecological effects of chemical or physical stressors at the individual, population, community, ecosystem, and even landscape levels. Toxicological effects of exposure to xenobiotics must be well defined for dose–response to be meaningful. The stressor dose–response analysis may focus on different aspects of the stressor dose–response relationship depending on the assessment objectives, the conceptual model, and the type of data used for analysis. For example, it could be invoked in situations such as the accidental discharge of pesticides or chemicals on land by humans or in water systems, draining of wetlands, human-induced forest decline resulting from acid deposition, etc.

In summary, the following are unresolved issues related to stressor dose–response relationship needing further studies in ecological risk assessment:

- quantifying cumulative impacts and stress dose–response relationships for multiple stressors;
- methodology for predicting ecosystem recovery;
- improving the quantification of indirect effects;
- describing stressor–response relationships for physical perturbation;
- distinguishing ecosystem changes due to natural processes from those caused by man;
- models that reflect compensatory processes at population and evolutionary timescale;
- logical frameworks and guidance for conducting wildlife risk assessment to support a variety of environmental decision contexts;
- methods that allow extrapolation of effects across species and levels of biological organization; and
- data sets and systems needed for wildlife risk assessment, and mechanistic population models for particular species and classes of species that use these data.

See also: Biogeochemical Approaches to Environmental Risk Assessment.

Further Reading

Bailey HC, Miller JL, Miller MJ, et al. (1997) Joint acute toxicity of diazinon and chlorpyrifos to *Ceriodaphnia dubia*. *Environmental Toxicology and Chemistry* 16(11): 2304–2308.

Bailey HC, Miller JL, Miller MJ, Wiborg LC, and Konemann H (1981) Fish toxicity tests with mixtures of more than two chemicals: A proposal for a quantitative approach and experimental results. *Toxicology* 19: 229–238.

Burns LA and Baughman GL (1985) Fate modeling. In: Rand GM and Petrocelli SR (eds.) *Fundamentals of Aquatic Toxicology: Methods and Applications*, pp. 558–584. Washington, DC: Hemisphere Publishing Corporation.

Cohen BL (1990) Ecological versus case-control studies for testing a linear-no threshold dose–response relationship. *International Journal of Epidemiology* 19(3): 680–684.

Haanstra L and Doelman P (1989) An ecological dose–response model approach to short and long-term effects of heavy metals on arylsulphatase activity in soil. *Journal of Environmental Quality* 7: 115–119.

Hoffman DJ, Rattner BA, Burton GA, Jr., and Cairns J, Jr. (2003) *Handbook of Ecotoxicology*. Boca Raton, FL: Lewis Publisher.

Landis W and Yu M-H (2004) *Introduction to Environmental Toxicology*, 512pp. Boca Raton: FL: CRC Press.

Marking LL (1977) Method for assessing additive toxicity of chemical mixtures. In: Mayer FL and Hamelink JL (eds.) *Aquatic Toxicity and Hazard Evaluation, ASTM STP 634*, pp. 99–108. Philadelphia, PA: American Society for Testing and Materials.

Moriarty F (1988) *Ecotoxicology: The Study of Pollutants in Ecosystems*, 2nd edn. London: Academic Press.

Munns WR, Jr. (2006) Assessing risks to wildlife populations from multiple stressors: Overview of the problem and research needs. *Ecology and Society* 11(1): 23 http://www.ecologyandsociety.org/vol11/iss1/art23/ (accessed October 2007).

Newman MC (1998) *Fundamentals of Ecotoxicology*. Chelsea, MI: Ann Arbor Press.

Pape-Lindstrom PA and Lydy MJ (1997) Synergistic toxicity of atrazine and organophosphate insecticides contravenes the response addition mixture model. *Environmental Toxicology and Chemistry* 16(11): 2415–2420.

Slob W (2002) Dose–response modeling of continuous endpoints. *Toxicology Sciences* 66(2): 298–312.

Straaler NV (2003) Ecoloxicology becomes stress ecology. *Environmental Science & Technology* 37: 324–329.

US Environmental Protection Agency (1992) Framework for ecological risk assessment. EPA-Risk Assessment Forum, EPA/630/R-92/001, Washington, DC.

Walker CH, Hopkin HP, Sibly RM, and Peakall DB (1996) *Principles of Ecotoxicology*. Bristol, PA: Taylor and Francis.

Wolfe CJM and Crossland NO (1991) The environmental fate of organic chemicals. In: Cote RP and Wells PG (eds.) *Controlling Chemical Hazards: Fundamentals of the management of toxic chemicals*, 475pp. London: Unwin Hyman.

Body Residues

L S McCarty, LS McCarty Scientific Research & Consulting, Markham, ON, Canada

---

Body Residues in Environmental Toxicology   Further Reading

Body Residues in Environmental Toxicology

Body residue describes the amount, usually a concentration, of one or more chemical substances measured in and/or on (i.e., absorbed and/or adsorbed) the whole body, organs, or tissues of an organism that is associated with a particular adverse effect. Since the term body burden can imply an amount of substance per whole body, rather than a concentration, the term body or tissue residue is more commonly used to avoid such confusion.

Nevertheless, advances in understanding of the toxicological significance of body/tissue residues have been made using simple models in conjunction with limited residue measurements. Substantial insights have been achieved with residue-based interpretation of standard testing results, especially with small aquatic organisms. These include greatly improved understanding of the influences of bioavailability, bioaccumulation, and modes of toxic action. Regulatory applications of residue-based approaches have been shown to provide improved scientific justification and flexible application. Also, compilations of available environmental residue-effect information are available and continue to be updated.

The presence of a chemical in an organism does not necessarily represent an appreciable risk. The fundamental premise of toxicology, dose–response, or concentration–response, as more commonly formulated in environmental work, still applies. Adverse effects due to the presence of chemicals in the body are usually associated with elevated levels. Concentrations below real or risk-specific threshold levels are considered to represent little or no risk. For substances that are naturally occurring or produced in and/or regulated in the body of an organism, risks are usually associated with deficits, excesses, or levels above typical body-regulated concentrations, often in a specific subcompartment of the body.

The lack of knowledge about the toxicological significance of body/tissue residues is an historical artifact related to the development of operational definitions for dose. An 'exposure' or 'external' dose is related to the concentration of chemical in an exposure medium such as food, water, or air to which an organism is exposed via oral, respiratory, or dermal pathways. An 'internal' dose is related to the amount of chemical that has actually been absorbed into (and/or sometimes adsorbed on) the body during an exposure and represents a received or acquired internal dose. Thus, exposure dose metrics are reported in units of amount or concentration of substance in exposure media while internal-dose metrics (i.e., body/tissue residues) are reported in units of concentration of substance per volume or mass of whole body, organ, or tissue. Molar concentration units (e.g., $mM\,kg^{-1}$ or $\mu M\,l^{-1}$) are more appropriate for reporting as toxicological phenomena are typically related to the number of molecules present rather than their mass.

The toxicological significance of the concentration of a chemical in the body or some portion or compartment of it currently remains largely uncertain, both in terms of the effective concentration and its temporal character. Information on residue levels associated with adverse effects has not been readily available until recently. Also, residues are not directly comparable with the primary dose metric commonly available in existing toxicity data that are based on concentrations in exposure media.

In the past, chemical analysis for substances was difficult with poor sensitivity and high detection limits. There were also many analytical limitations and interferences associated with estimation of levels of chemicals in tissues. As toxicity-testing methods were being developed, exposure-based doses were first estimated by adding known amounts of pure substance to the exposure medium, for example, adding chemicals to the water in which aquatic organisms were introduced. As chemical analysis capability improved, routine measurements of exposure media became possible, enabling confirmation of exposure media concentrations. As most commonly used toxicity-testing protocols were developed decades ago, and standardized in the period from about 1950 to 1980, it is not surprising that exposure-based dose metrics are found in the bulk of currently available toxicity data.

It is generally agreed that the effective dose at the site(s) of toxic action in and/or on an organism is the 'true' measure of dose. This is rarely estimated as sites(s) of toxic action are largely unknown. Thus, both external- and internal-dose metrics are really surrogate measures for the 'true' dose. Ongoing developments in toxicology are directed toward resolving the relationship between external- and internal-dose metrics. Major issues include bioavailability from exposure media, absorption, distribution, metabolism, and excretion (ADME) by the organism, and the nature of the mode/mechanism of toxic action that is associated with the adverse effect of concern. Quantification of the influences of various chemical, physical, and biological factors that modify the expression of toxicity by altering the relationship between external and the received effective internal dose will alleviate difficulties associated with moving between internal- and external-dose metrics.

Organism-based dose metrics – body/tissue residues – is neither new nor revolutionary. Conceptually, toxicological testing is based on models that assume an effective dose is present in and/or on some portion of the exposed organism. The difference is that for external doses the assumption is implicit while for internal doses it is explicit. Making the internal-dose assumption explicit encourages and facilitates improvements in toxicological theory, test design, and interpretation, and applications of toxicological knowledge. This has particularly important implications for regulating chemicals in the environment.

Both hazard- and risk-based regulatory approaches rely on thorough interpretation and understanding of toxicity-testing data. Hazard-based approaches are valid only when it is certain that the testing results being evaluated all represent consistent and comparable measures of relative toxicity. Risk-based approaches build on hazard and add a key requirement; that is, that differences in the influence of each major toxicity-modifying factor be accounted for in the extrapolation from toxicity test exposure conditions to the scenario-specific exposure conditions being addressed by the regulatory policy. This is often termed the laboratory-to-field extrapolation problem.

Improved toxicological understanding based on better quantitative understanding of the relationship between external- and internal-dose metrics will facilitate a key toxicological objective of determining comparable measures of relative toxicity. It will improve estimation of both rank order and quantitative differences in toxicity within and between groups of substances, as well as better understanding of actual toxicological differences within and between species. When using toxicity-testing information in regulatory applications it can provide substantial quantitative information for improving extrapolation from laboratory to field exposure conditions.

In addition, improved understanding of external–internal-dose relationships will facilitate addressing the growing interest in human and environmental biomonitoring data. Chemical analysis technology can now quantify extremely low concentrations of substances in the tissues of various living organisms. However, as noted above, the toxicological significance of body/tissue monitoring results are largely uncertain as such biomonitoring residue data is neither directly nor readily comparable with standard toxicity test results and the development of residue–effect relationships is not yet well developed.

Given the large amount of exposure dose data that currently exists, and the unlikeliness of repeating large numbers of tests using internal-dose metrics, there is a strong incentive to quantify the relationships between external- and internal-dose metrics. As well as advancing basic toxicological theory and concepts, this will have immediate practical utility. In particular such knowledge will enable the large existing toxicity database and associated knowledge to be more effectively applied to both addressing the laboratory-to-field extrapolation problem and establishing the toxicological significance of body/tissue residue data obtained in biomonitoring programs.

See also: Dose–Response.

Further Reading

Escher BI and Hermens JLM (2004) Internal exposure: Linking bioavailability to effects. *Environmental Science and Technology* 38: 455A–462A.

Jarvinen AW and Ankley GT (1999) *Linkage of Effects to Tissue Residues: Development of a Comprehensive Database for Aquatic Organisms Exposed to Inorganic and Organic Chemicals.* Pensacola, FL: SETAC Press.

McCarty LS and Mackay D (1993) Enhancing ecotoxicological modeling and assessment: Body residues and modes of toxic action. *Environmental Science and Technology* 27: 1719–1728.

Meador J (2006) Rationale and procedures for using the tissue-residue approach for toxicity assessment and determination of tissue, water, and sediment quality guidelines for aquatic organisms. *Human and Ecological Risk Assessment* 12: 1018–1073.

Rand GM, Wells PG, and McCarty LS (1995) Introduction to aquatic toxicology. In: Rand GM (ed.) *Fundamentals of Aquatic Toxicology II: Effects, Environmental Fate, and Risk Assessment,* ch. 1, pp. 3–67. Bristol, PA: Taylor and Francis.

Relevant Website

http://el.erdc.usace.army.mil – Environmental Residue-Effects Database (ERED), US Army Corps of Engineers/US Environmental Protection Agency.

Assimilative Capacity

W G Landis, Western Washington University, Bellingham, WA, USA

Defining Assimilative Capacity

Application of Assimilative Capacity to the TMDL Process

Further Reading

Assimilative capacity is a term with a number of definitions. The term is used in context with the total maximum daily load (TMDL) calculation that is part of the compliance process for the Clean Water Act of the United States.

This article derives a definition of assimilative capacity that is in line with current understandings of aquatic ecology and policy making. The role of assimilative capacity in the process of calculating a TMDL is presented.

Defining Assimilative Capacity

Assimilative capacity has been defined in numerous ways. In summary, the assimilative capacity is defined as the amount of nutrients, sediments, or pathogens that an aquatic system (stream, lake, river, estuary) can absorb without exceeding a numeric criterion. The numeric criterion is set so that the waterbody meets the policy goals set for it by the statutory regulatory agencies.

In the United States, the United States Environmental Protection Agency (USEPA) is the overarching regulatory authority for the TMDL program and so its definition is the baseline. Assimilative capacity is defined in its role in determining a TMDL by the USEPA. The term 'assimilative capacity' represents the amount of contaminant load that can be discharged to a specific waterbody without exceeding water-quality standards or criteria. Assimilative capacity is used to define the ability of a waterbody to naturally absorb and use a discharged substance without impairing water quality or harming aquatic life.

Note that the USEPA definition has several discrete segments. The initial part of the document deals with discharges. The specific use of 'discharge' is because point sources or known discharges have been the focus of past regulation. However, assimilative capacity in the TMDL context often deals with nonpoint sources where the discharge site is not specific. Nonpoint sources may have diffuse input to the waterbody by small streams, groundwater sources, or atmospheric deposition. Assimilative capacity is also identified with specific water bodies, so this is meant to be a site-specific property and this is reflected in the application of the TMDL process. The last line "without exceeding water quality standards" can be translated as: without the receiving waterbody having qualities that do not meet management goals as set by public policy. This is the critical criterion for establishing an assimilative capacity.

This criterion is the keystone of the definition. It reflects cultural values about the use of a particular waterbody. These values include the use of the waterbody as a water supply, recreational use, the support of important ecological resources, and other uses. The criteria derived in order to meet these uses form the basis for deriving numeric values that set the limit in the assimilative capacity definition.

Table 1 presents several examples of the variables measured and the nutrient criteria set in order to protect a number of rivers. In Boulder Creek, CO, the criteria for unionized ammonia was set at $0.06\,\mathrm{mg\,l^{-1}}$, compared to $0.025\,\mathrm{mg\,l^{-1}}$ for Laguna de Santa Rosa in California. Phosphorus is also a common indicator for which criteria are set. Lake Chelan has a very low criteria for total phosphorus compared to the values for Truckee River or Clarke Fork River. For the nutrients the numbers are a maximum, but for dissolved oxygen the numbers are a minimum. Dissolved oxygen criteria also vary by waterbody; compare the value of $4.0\,\mathrm{mg\,l^{-1}}$ for Appoquinimink River to the $7.0\,\mathrm{mg\,l^{-1}}$ for Laguna de Santa Rosa. The Clarke Fork River also includes criteria for chlorophyll *a*, not a nutrient but as an indicator of the nutrient condition of the waterbody. The assumption is that the nutrients in the waterbody control the amount of algae and therefore the concentration of chlorophyll *a* in the water column. Since ensuring that total algal productivity in the lake is below a certain value, the managers are attempting to prevent the rapid eutrophication of the system.

Application of Assimilative Capacity to the TMDL Process

The derivation of the amount of nutrients, sediments, pathogens, or other stressors that can be added to the waterbody without exceeding the criteria depends upon a number of factors (**Figure 1**).

Without man-derived (anthropogenic) inputs, there will still be a variety of inputs to the receiving water from the watershed and atmospheric deposition. Nutrients (manure and decomposition products), pathogens, and naturally derived toxins are derived from biological processes. Runoff from the watershed can carry nutrients from soil or rock along with sediment. Natural outcroppings from the watershed can be sources of hydrocarbons, metals, nutrients, and other materials. Atmospheric deposition can deliver particulates and nutrients that may have been transported very long distances.

The receiving environment will also act upon the materials transported into the waterbody. Dilution by the addition of water from the watershed will occur. Current outside the study area can also transport the materials introduced to the river. Many materials can be biodegraded or biotransformed so that they no longer have the original activity. Particulates and the materials that attached to them can be trapped as sediment and not be available to the water column. Organics such as benzene and related aromatics can volatilize, escaping from the water column to the atmosphere. At higher pHs, metals and some other materials can combine with carbonates and precipitate from the water column. The combination of the rates of the inputs and outputs results in a baseline concentration of the nutrient or other material regulated by the TMDL process. The difference between this level and the established criteria provides an indication of the additional materials that could be added. Of course, this baseline is not a constant but depends upon the seasonal and natural changes that occur in the watershed. Times of low precipitation will result in a lower amount of runoff from the watershed. The lack of precipitation in a watershed will also result in low flow and reduce the amount of dilution and the transport of the contaminants from the waterbody. In cases where the data are available, it is possible to establish confidence intervals for the ranges that the input and output variables may take over time to provide a more

Table 1 Examples of indicators and criteria used in the TMDL process. Note that the criteria are set for each waterbody and vary in type depending upon the site and the intended use

Waterbody	Indicators and the criteria selected
Boulder Creek, CO	$0.06\,\mathrm{mg\,l^{-1}}$ unionized ammonia
Appoquinimink River, DE	$5.5\,\mathrm{mg\,l^{-1}}$ dissolved oxygen (daily average), $4.0\,\mathrm{mg\,l^{-1}}$ dissolved oxygen (instantaneous minimum)
Lake Chelan, WA	$4.5\,\mathrm{\mu g\,l^{-1}}$ total phosphorus
Truckee River, NV	$0.05\,\mathrm{mg\,l^{-1}}$ total phosphorus, $210\,\mathrm{mg\,l^{-1}}$ total dissolved solids
Clarke Fork River, MT	$100\,\mathrm{mg\,m^{-2}}$ chlorophyll *a* (summer mean), $300\,\mathrm{\mu g\,l^{-1}}$ total nitrogen, $20\text{--}39\,\mathrm{\mu g\,l^{-1}}$ total phosphorus specific to stretches on the river.
Laguna de Santa Rosa, CA	$0.025\,\mathrm{mg\,N\,l^{-1}}$ unionized ammonia, $7.0\,\mathrm{mg\,l^{-1}}$ dissolved oxygen (minimum)

Modified from USEPA (1999) *Protocol for Developing Sediment TMDLs*, 132pp. EPA 841-B-99-004. Washington, DC: United States Environmental Protection Agency, Office of Water (4503F).

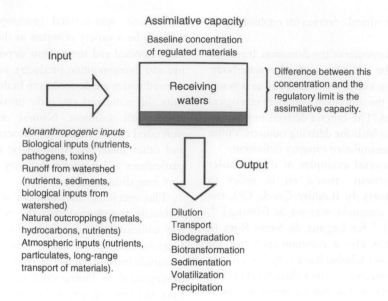

Figure 1   Diagram of the interactions that are part of determining the assimilative capacity of a receiving water.

realistic picture of the conditions of watershed and the waterbody.

Man-made nutrients, alterations to water flow, and other factors broaden the types of considerations (**Figure 2**). Point and nonpoint sources from human cultural activities add nutrients, contaminants, pathogens, and other materials to the receiving water. Points sources, such as outflows from manufacturing or municipal water treatment systems, can contribute elevated levels of

metals in a refined form, novel toxicants, an increase in temperatures, dyes, and materials not normally found in nature. The place of input is also localized and at a concentration. Only the output processes at the point of introduction and those downstream are available to process the materials. Nonpoint sources such as those from agricultural areas or residential zones introduce other unique materials. Pesticides, herbicides, and fertilizer runoff can come from both areas. Antibiotics have been

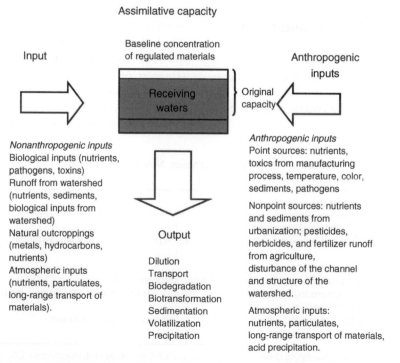

Figure 2   Anthropogenic inputs add to the loading of the receiving water and start to encroach on the regulatory limit.

detected in agricultural runoff while pathogens can be obtained from a variety of sources. Agriculture, residential areas, and manufacturing regions can also alter the structure of the waterbody by channelization and pave changing the hydrodynamics of the system. Atmospheric inputs can bring contaminants from outside the watershed depositing them as they fall out of the atmosphere as particulates or in precipitation. Rain and snow both can be contaminated. Organics can be found as part of snow pack even in remote mountainous regions. It is unlikely that any site exists that does not receive a detectable amount of an anthropogenic contaminant.

In response to these inputs, an increase in the rates of degradation and other factors controlling the output of the material from the waterbody may occur. As the inputs of nutrients and organics increase, biodegradation and biotransformation rates of toxicants may also increase. However, the rates may reach a maximum depending upon temperature, oxygen concentration, flow rates, or other factors. In situations where the receiving water is already above the criteria, the factors that control the removal of the contaminant are likely to already be at a maximum.

In order to estimate the loading limits that will not exceed the criteria set for the receiving water, an expression formally connecting the features in **Figure 2** in a causal relationship should be derived. **Figure 3** illustrates the tools that have been used in order to accomplish this linkage. The goal is to be able to connect the loading to the final concentration of the contaminant in the water body. In this fashion specific limits on the amount and rates of loading can be established to ensure that the water-quality goals for the receiving water can be met.

The tools that have been used fall into three categories. The first set of tools are the use of mechanistic mathematical models that have functions that describe the important features of the receiving water that control the concentration of the contaminant. Such a model includes input rates for the contaminant, degradation or sedimentation rates, volatilization rates, dilution factors, and other features that essentially turn **Figure 2** into an equation for a specific situation. These models have the potential to be accurate and can address a number of issues very quickly. The downside is that a complete process model can take a lot of time to construct and the data may not exist. It may also not be clear what factors control some types of contaminants. When sufficient knowledge of the process is not available to construct a process model, then it is necessary to use alternative approaches.

Empirical models use regression techniques to connect input loadings to final concentration in the receiving waters. In some instances, these models may have many different components and a multiple regression equation used to define the relationships. There are also models that describe the relationship between inputs to the system and a specific type of water quality index. Indexes are numbers that composite many kinds of data and may not clearly represent the criteria established for the system of interest. The accuracy of empirical methods depends largely on the amount of data available and its origin. Data for several different receiving waters may have to be combined in order to have enough data to derive a reasonable regression. There may also be regionally specific factors that may not be included, or different bioregional regions may require different regressions in order to provide accurate predictions. Assumptions of the models and the source of the data used to derive the regression should be stated as part of the reporting process.

Simulation models, often incorporating segments that are process derived and some with empirical backgrounds, are among the most commonly used tools in estimating loadings that do not exceed the assimilative capacity of the receiving water. **Table 2** lists some of the characteristics of those simulation models used for nutrients.

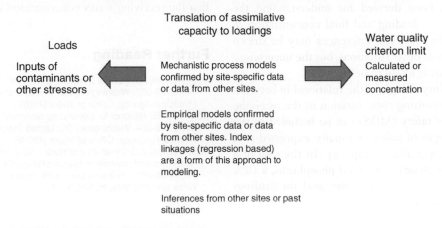

Figure 3   Tools for connecting loads and water quality criteria.

Table 2 Currently available models for calculating loadings and assimilative capacities for rivers, streams and lakes. The models and websites are current as of summer 2006

Source	Model	Comments	Webpage
US Army Corp of Engineers Waterways Experimental Station, Vicksburg MS	CE-QUAL-RIV1	One-dimensional, dynamic flow and water quality model for streams	http://el.erdc.usace.army.mil/ products.cfm?Topic=model&Type=watqual
	CE-QUAL-W2	Two-dimensional, vertical-longitudinal hydrodynamic water quality model for reservoirs	
	BATHTUB	Steady-state water and nutrient balance calculations in a spatially segmented hydraulic network which accounts for advective and diffusive transport and nutrient sedimentation	
	FLUX	Program allows estimation of tributary mass discharges (loadings) from sample concentration data and continuous flow records	
	PROFILE	Data reduction and analysis of water quality data. Includes several eutrophication response variable calculations	
	CE-QUAL-ICM	Two-dimensional (horizontal) and three-dimensional water quality model for coastal systems currently restricted to US Corp of Engineers use	
USEPA Ecosystems Research Division	BASINS/QUAL2K	This stream and river water quality model that assumes that the stream is well mixed	http://www.epa.gov/ATHENS/wwqtsc/html/ qual2k.html
	WASP7	Model includes the water column and the benthos of lakes	http://www.epa.gov/athens/wwqtsc/html/ wasp.html

It is important that the models that are used are as transparent as possible, that is, that the underlying assumptions, constants, and calculations be available for review. This requirement precludes against the use of proprietary models that are not open source and have not met peer review. The model should also be as simple as meets the requirements of setting the loading limits.

In some instances there may be no mathematical relationship that has been derived for understanding the relationship between loading and final concentration in the water body. In this case, inferences may be drawn from the past or from other situations, but the uncertainty in these predictions is likely to be high.

When uncertainty is high in the relationship between loadings and the resulting concentration in the receiving water, a margin of safety (MOS) can be included in the process. This margin of safety is usually expressed as a percentage of the assimilative capacity. In the case of a regulatory limit of $20 \, \mu g \, l^{-1}$ for total phosphorus, a 10% margin of safety would result in the goal for loadings would be not to exceed $18 \, \mu g \, l^{-1}$.

In summary, assimilative capacity is a means of tying loading from a number of sources to site-specific regulatory limits for receiving water. Assimilative capacity is tied directly to the TMDL process. The calculations that tie loading to in-stream concentrations are typically done by models that are designed for this purpose and are specific for the type of receiving water. Uncertainty in the predictive model is dealt with using an MOS to ensure that the receiving water concentration is not exceeded.

Further Reading

USEPA (1999) *Protocol for Developing Nutrient TMDLs*, 135pp. EPA 841-B-99-007. Washington, DC: United States Environmental Protection Agency, Office of Water (4503F).

USEPA (1999) *Protocol for Developing Sediment TMDLs*, 132pp. EPA 841-B-99-004. Washington, DC: United States Environmental Protection Agency, Office of Water (4503F).

USEPA (2004) *A GIS Inventory of Pacific Northwest Lakes and Reservoirs and Analysis of Historical Water Quality Data*, 36pp. EPA 910-R-04-009. US Environmental Protection Agency, Office of Water and Watersheds, Region 10.

Ecological Risk Assessment

S M Bartell, E2 Consulting Engineers, Inc., Maryville, TN, USA

Introduction	Exposure–Response Relationships
Definition of Ecological Risk	Risk Characterization
Problem Formulation	Uncertainty
Analysis – Exposure Characterization	Further Reading
Analysis – Effects Characterization	

Introduction

Managers and decision makers are challenged to solve complex environmental problems associated with the increasing pressures placed on vital natural resources by human activities. These challenges are made difficult by the sheer number and diversity of human disturbances and exacerbated by the complexity of imperfectly understood natural ecological systems. The process of ecological risk assessment (ERA) addresses ecological complexity and incorporates uncertainty in characterizing the impacts of natural and man-made disturbances on ecological resources.

ERA integrates ecology, environmental chemistry, environmental toxicology, geochemistry, hydrology, and other fundamental sciences in estimating the probabilities of undesired ecological impacts. In theory, ERA can be viewed as a subset of basic disturbance ecology. In practice, ERAs derive from specific needs to assess human-induced impacts on the environment. Many ERAs conducted in the United States are motivated by legislation, including the National Environmental Policy Act (NEPA), the Toxic Substances Control Act (TSCA), and the Comprehensive Environmental Response, Compensation, and Liability Act (CERCLA or Superfund). ERAs are also undertaken by private industry to determine future risks and liabilities associated with the development, use, and disposal (i.e., life cycle) of new or existing products (e.g., herbicides, pesticides, and industrial chemicals).

Several different approaches for performing an ERA have been developed internationally. No single methodology has been officially sanctioned. However, the approach developed by the United States Environmental Protection Agency (US EPA) guides many ERAs performed in the United States. The following discussion emphasizes this methodology.

Definition of Ecological Risk

Risk is defined as the probability that an undesired event will occur. Correspondingly, ecological risk refers to the probability of the occurrence of an undesired ecological event. Alternative definitions of risk include an evaluation of the consequences of the undesired event along with estimation of its occurrence. For the most part, risk pertains to the probability of occurrence and this definition will serve this presentation.

ERA originally focused on the undesired ecological effects of toxic chemicals. As ERA evolved, the set of stressors has expanded to include physical, geological, hydrological, and biological stressors. Examples of these kinds of stressors include physical habitat degradation, erosion of soils or sediments, drought/floods, and introductions of exotic species. One testament to the conceptual soundness of this approach to ERA has been its successful application to nonchemical stressors.

Problem Formulation

This initial and perhaps most important part of the assessment defines the nature and scope of the ERA, describes the sources of potential risk (stressors), identifies the undesired ecological impacts (endpoints), considers the nature of the ecological impacts in relation to the stressors, and produces a conceptual model of the overall assessment. Thus, problem formulation essentially encapsulates the entire ERA process. Performing this step requires collaboration among risk managers and risk assessors to define the assessment objectives and develop the corresponding conceptual model. This model should be viewed as dynamic and subject to change throughout the ERA in relation to modifications to the objectives and the development of new data and information.

The initial interactions between risk managers and risk assessors might also involve other organizations and concerned members of the public (stakeholders). Initial discussions can help ensure that all important aspects of the assessment are identified, included as part of the problem formulation, and represented in the conceptual model. Such interactions can also ensure that the kinds of results produced by the ERA can be used effectively in the process of risk management and decision making.

Following construction of the conceptual model, problem formulation continues by developing a plan to implement the conceptual model. The resulting analysis plan further characterizes the stressors, identifies specific ecological effects of concern, and identifies applicable data, as well as measures or models that can be used to quantitatively relate the stressors to the expected ecological effects.

Upon completion of risk estimation, risk managers, risk assessors, and stakeholders may reconvene to discuss the nature and interpretation (e.g., conclusions, assumptions, caveats) of the results in the context of the overall assessment objectives. Possible outcomes of these interactions include revisions to the conceptual model, collection of new data, and subsequent iterations of risk estimation. Once the requirements of risk managers and decision makers are fulfilled and documented, the process ends.

Following the problem formulation, the ERA continues with analyses that characterize exposure to the stressor(s) and the ecological effects of concern. The subsequent derivation of functional relationships between exposure and effects sets the stage for risk estimation.

Analysis – Exposure Characterization

Exposure is characterized by identifying the processes and mechanisms that bring organisms into contact with the stressor(s) of concern and quantifying the frequency, magnitude, and duration of such contact. The nature of the stressor(s) and the kinds of ecological effects of concern will strongly influence the exposure analysis. Each identified stressor will suggest a relevant spatial and temporal scale for analysis. The scales might be local and relatively short term, as for accidental spills of toxic, yet readily degraded or volatilized chemicals that result, for example, from hazardous waste management. Conversely, some stressors (e.g., fire, climate change) can exert ecological impacts over large expanses and for durations that greatly exceed the generation time of most organisms. Stressors are also evident at intermediate scales, for example, major oil spills and certain exotic species (e.g., gypsy moth, zebra mussel, Asian long-horned beetle).

The nature of specific stressor(s) can provide information concerning the processes or mechanisms of exposure that should be evaluated in an ERA. Chemical contaminants introduced into the environment are naturally transported by the movements of wind and water. Certain chemicals can accumulate in organisms and be transmitted throughout complex food webs. Some organic chemicals are comparatively insoluble in water and are rapidly adsorbed to soils and sediments, while other chemicals remain in solution and are subsequently transported by water. In contrast, movements of biological stressors such as invasive species might be augmented by private and commercial transportation. Corresponding characterization of exposure might emphasize delineation of transportation networks in place of physical–chemical processes.

The kinds of ecological effects included in the conceptual model can also provide insights into exposure analysis. Organisms occupy certain dimensions in space and time. Habitats have measurable spatial extent; ecological processes exhibit characteristic rates. Such observations can guide the analysis of exposure. For example, knowledge of the timing and duration of a sensitive life stage (e.g., eggs, larvae) can focus the corresponding measurement of stressors of concern and provide more meaningful quantification of exposure than longer-term averages or monitoring that might completely miss the critical time period for exposure. Similarly, seasonal changes in light, temperature, precipitation, and other physical factors can result in spatial–temporal variability in exposure. The important point is that variability in both the processes that influence the stressor and the characteristics of the ecological entities of concern should be addressed in performing a meaningful analysis of exposure.

Alternative approaches can be used in a sequential manner to assess exposure. Worse-case scenarios can be developed that assume maximum values of the stressor. For example, end-of-pipe concentrations of toxic chemicals can be used without accounting for physical dilution, chemical alterations, or biological degradation that would otherwise reduce the concentrations experienced by the organisms of concern. This approach is biased toward overestimating exposure and risk. If acceptable risks result from these extreme exposures, the assessment process might reasonably stop. As an alternative to worse-case scenarios, exposures might be measured. Actual measures of exposure are undoubtedly the most easily defended scientifically (presuming competent sampling and analysis) and the most realistic inputs to an ERA. Finally, exposures might be estimated using physical (e.g., microcosms, mesocosms) or mathematical models.

Exposure characterization generates an exposure profile. For chemicals, the profile includes the nature of the source; pathways of exposure; identification of environmental media of concern (e.g., soils, water, sediments, contaminated biota); estimates or measures of exposure concentrations (magnitude, timing, duration, recurrence); and uncertainties associated with these concentrations. Analogous exposure profiles are developed for nonchemical stressors addressed by an ERA.

Analysis – Effects Characterization

The large number and different kinds of ecological effects that are of potential concern distinguish, in part, ERA from more traditional human health risk assessment. The diversity of effects reflects the comprehensive nature of ecology and the environmental sciences. Ecologists have recognized several levels of organization as being useful in describing the natural world. These traditional levels include individual organisms, populations, communities, and ecosystems. Within recent decades, these levels have expanded to include macromolecules and landscapes. ERAs commonly identify more than one kind of ecological effect of concern in problem formulation. The ecological effects of concern identified during problem formulation should be ecologically important, sensitive to the stressor(s), and relevant to risk management.

Endpoints in ERA can include several effects at different levels of organization. An ERA might address alterations in basic physiological processes (e.g., photosynthesis, respiration) and corresponding lethal or sublethal (i.e., reduced growth) effects on individual organisms. In rare instances, impacts on individual organisms might be selected as endpoints. In these cases, the individuals are likely to represent small populations of endangered species.

ERAs routinely emphasize impacts on populations. The dynamics of populations has been a subject of basic ecological study for more than a century and it is no surprise that population-level endpoints have become a norm in ERA. In practice, the effects of interest emphasize decreases in the population sizes of one or more species of interest. It is not uncommon for species officially designated as threatened or endangered to be the focal points for ERA. Population-level endpoints include, for example, reductions in population size, impaired reproduction, alterations in genetics, and the likelihood of local extinction. Note that for socially or economically undesirable species (e.g., pests, invasives, toxic blue-green algae), the endpoint would be the probability of a population increase. Population models have been developed extensively and these models are being used increasingly to estimate ecological risks.

Ecologists recognize that individual populations do not persist in an ecological vacuum. The number of species, their absolute and relative abundances, and their correlation of occurrence in space and time define community structure. A variety of concepts and methods for describing community structure have been developed by ecologists. Subsequently, alterations in community structure have been introduced as endpoints in ERA. Such alterations have taken the form of reductions in species diversity or changes in community similarity in response to the stressor of interest. Indices of biotic integrity and measures of community similarity have been introduced into the assessment of ecological risk.

Ecosystem ecology addresses important feedback mechanisms between biotic and abiotic processes that determine ecosystem structure and function. The effects of stressors on fundamental ecosystem processes (e.g., primary production, total system respiration, decomposition, nutrient cycling) are becoming increasingly important as

endpoints in ERA. The ecosystem concept also emphasizes the scale dependence and asymmetry of ecological interactions. Even in highly complex systems, not all components and processes are of equal importance. Delineation of critical scales and feedbacks can help define the relevant spatial–temporal scales in designing ERAs.

More recent recognition of stressors that operate at larger scales (e.g., acid precipitation, climate change) has led to the consideration of landscape-level impacts in ERA. Landscape endpoints in risk assessment include alterations in the spatial distribution and extent of different habitat types within landscapes. Changes in the size, shape, and proximity of similar habitat areas (patches) can be measured and modeled.

Exposure–Response Relationships

This component of the overall ERA methodology develops the functional relationships between the stressors and the ecological responses of concern. The exposure–response functions are central to ERA. Fundamentally, ERA can be described as the development and application of uncertain exposure–response functions in assessing ecological impacts. For a given stressor, these functions estimate the severity of the expected ecological response in relation to the magnitude, frequency, and duration of the exposure. The derivation of exposure–response functions depends on the quantity and quality of available data.

Sources of data that might be used in the construction of exposure–response functions include: the results of toxicity tests (acute, chronic) performed under controlled laboratory conditions, direct measures of exposure and response in controlled field experiments, and the application of statistical relationships that estimate the biological effects of chemicals based on physical or chemical properties of specific toxicants. The order of preference among these sources of data identifies field observations as the most valuable, followed by laboratory toxicity tests, and finally by the use of empirical relationships. In the absence of directly relevant data, the development of exposure–response functions may require extrapolations among similar stressors or ecological effects for which data are available. For example, effects might have to be extrapolated from an available test species to an untested species of concern. Similarly, toxicity data might be available only for a chemical similar to the specific chemical stressor of concern. An extrapolation from the known chemical to the unknown would be required to perform the assessment.

Exposure–response functions generally increase monotonically and are nonlinear. Some evidence exists for certain stressors (e.g., ionizing radiation) that actually result in a positive response for very low magnitudes of exposure. This uncommon phenomenon of hormesis remains the exception to the usual sigmoid-shaped function. Depending on the nature of the stressor, the functions can exhibit a threshold magnitude of exposure required before any ecological response is observed. This value is termed a lowest observed effects concentration or LOEC. The LOEC can be used as an endpoint for comparison with exposure estimates in assessing risk. Another exposure value often used as an endpoint is the concentration (or dose) that produces 50% of the maximum response. For example, the concentration that results in 50% mortality during a prescribed period of exposure (e.g., 48, 96 h) defines the LC_{50} (lethal concentration that produces 50% mortality). An EC_{50} defines an exposure that results in a 50% decrease in an endpoint other than mortality, for example, growth.

Risk Characterization

Risk characterization combines the exposure profiles with the exposure–response relationships to estimate ecological risks in ERA. A variety of methods and tools are available for risk estimation. For assessing risks posed by toxic chemicals, one simple method simply divides the exposure concentrations by the toxicity reference values (TRVs). A LOEC is an example of a TRV, as is an LC_{50} or EC_{50}. Quotients equal to or greater than 1.0 imply risk; quotients less than 1.0 suggest minimal or no risk. Such quotients can prove useful in initial screening-level assessments to reduce the number of stressors that should be analyzed in greater detail. The screening assessments may be particularly effective if exposure estimates used in risk characterization are biased toward overestimating risk.

Depending on the availability of data, distributions of exposure and toxicity can be constructed and compared. Risk can be estimated by statistically comparing the degree of overlap between these distributions: the greater the overlap, the higher the risk. Using comparisons of distributions in screening-level assessments can extend the single-value quotient approach by incorporating more information, including uncertainty, in the risk estimation.

Experiments under field conditions or more controlled conditions in the laboratory (e.g., microcosms, mesocosms) can be used to characterize ecological risks. Experimental systems provide opportunities to physically impose the stressors of interest on the ecological resources of concern. Such experiments may be the only practical method for assessing risks posed by stressors not intended to be introduced into the environment. This approach may also prove essential in assessing risks posed by stressors that are virtually unknown or whose attributes are proprietary.

Mathematical and computer simulation models can be used to estimate ecological risks. Following decades of model construction in support of basic ecological research

and development, it stands to reason that some of these models might prove useful in estimating ecological risks posed by various stressors on individual organisms, populations, communities, and ecosystems. To be useful in characterizing risk, the selected ecological model must include some representation of one or more of the assessment endpoints as a dependent variable. The model must also represent the stressor as an independent variable. The remaining critical aspect in selecting or adapting models for assessing risk is the ability to derive exposure–response relationships for the stressor(s) and ecological impacts of interest.

Uncertainty

Risk implies uncertainty. ERA was designed expressly to include uncertainty as an integral component of the assessment process. Sources of uncertainty include natural variability in ecological and environmental phenomena, as well as bias and imprecision associated with the exposure–response functions. This latter source of uncertainty can be exacerbated if extrapolations were involved in the derivation of the functions (e.g., laboratory to field, across species). Incomplete and imperfect understanding of baseline ecological phenomena also adds uncertainty to ERA.

Uncertainties inherent to the risk assessment process can be quantitatively described using, for example, statistical distributions, fuzzy numbers, or intervals. Corresponding methods are available for propagating these kinds of uncertainties through the process of risk estimation, including Monte Carlo simulation, fuzzy arithmetic, and interval analysis. Computationally intensive methods (e.g., the bootstrap) that work directly from the data to characterize and propagate uncertainties can also be applied in ERA. Implementation of these methods for incorporating uncertainty can lead to risk estimates that are consistent with a probabilistic definition of risk.

Methods of numerical sensitivity and uncertainty analysis can be used to examine uncertainty and identify the key sources of bias and imprecision in quantitative estimates of risk. Once identified, limited resources (e.g., time, funding) can be efficiently allocated to obtain new information and data for those major sources of uncertainty and reduce uncertainty. These analyses can be repeated until uncertainties associated with the risk estimates are of an acceptable degree or until uncertainties cannot be further reduced. Importantly, application of these methods for analyzing uncertainty will identify whether unacceptably high estimates of risk derive from the inherent severity of the stressor or from high uncertainty.

See also: Bioaccumulation; Bioavailability; Biodegradability; Biodegradation.

Further Reading

Bartell SM, Gardner RH, and O'Neill RV (1992) *Ecological Risk Estimation*. Chelsea, MI: Lewis Publishers.

Calabrese EJ and Baldwin LA (1992) *Performing Ecological Risk Assessments*. Chelsea, MI: Lewis Publishers.

Kaplan S and Garrick BJ (1981) On the quantitative definition of risk. *Risk Analysis* 1: 1–11.

Kolluru R, Bartell S, Pitblado R, and Stricoff S (1996) *Risk Assessment and Management Handbook for Environmental, Health, and Safety Professionals*. Boston: McGraw-Hill.

Pastorok RA, Bartell SM, Ferson S, and Ginzburg LR (2002) *Ecological Modeling in Risk Assessment*. Boca Raton, FL: Lewis Publishers.

Paustenbach DJ (ed.) (2002) *Human and Ecological Risk Assessment – Theory and Practice*. New York: Wiley.

Suter GW, II (1992) *Ecological Risk Assessment*. Boca Raton, FL: Lewis Publishers.

UNEP (2000) *IECT Technical Publication Series 14: Technical Workbook on Environmental Management Tools for Decision Analysis*. Osaka, Japan: Division of Technology, Industry, and Economics, United Nations Environmental Programme.

Biogeochemical Approaches to Environmental Risk Assessment

V N Bashkin, VNIIGAZ/Gazprom, Moscow, Russia

O A Demidova, Moscow State University, Moscow, Russia

Introduction	Critical Load and Level Approach for Assessment of Ecosystem Risks
Environmental Risk Assessment	Uncertainty in IRA and ERA Calculations
Biogeochemical Approaches to Environmental Risk Assessment	Summary
	Further Reading

Introduction

Quantitative assessments of environmental risks are at present being conducted at a variety of international and local levels. However, the use of natural mechanisms in managing risk processes is not widely understood and therefore is not being applied to the degree possible. In addition, there exists among many stakeholders a high level of uncertainty about risk estimates. It is now known that the sustainability of modern technogeosystems, within the confines of existing economic systems, can be determined by natural biogeochemical cycles, transformed to different degrees by anthropogenic activity. An understanding of fundamental principles in the management of pollutant fluxes in the biogeochemical food web will allows us to use current concepts in quantitative risk assessment and to apply technological solutions for managing these risks within the given economic structure.

This article summarizes the important current research being done in this field and provides a basis for various problem solutions to both practitioners and students of environmental risk management.

Environmental Risk Assessment

Traditionally, risk assessment (RA) has been focused on threats to humans posed by industrial pollutants. In recent times there has been a shift to other types of hazards and affected objects. Environmental risk assessment (ERA) has already evolved into separate methodology under the general risk assessment framework.

When applied to a particular site and/or project, ERA procedures include several generic steps such as 'hazard identification', 'hazard assessment', 'risk estimation', and 'risk evaluation'. Despite rapid development of ERA guidance and wide support for the idea of tools integration, ERA is rather exclusion in environmental impact assessment (EIA) practice. In fact, the formal risk assessment follows the 'bottom-up' approach to assessing ecosystem-level effects. The assessor depends mainly on findings of laboratory toxicity testing that are extrapolated to higher levels of natural system hierarchy (from organisms to communities and even ecosystems) using various factors. Meanwhile, too many assumptions put a burden of high uncertainty on final quantitative risk estimates. Moreover, ecosystem risk assessments of this type are rather experiments than established practice. High costs and lack of required data are among key reasons for avoiding this approach by practitioners.

As a result, an EIA practitioner faces considerable difficulties while assessing impacts on ecosystems. On the one hand, there are legal requirements to assess fully ecological effects and best practice recommendations to undertake quantitative assessments where possible. On the other hand, many assessors lack tools and techniques to undertake estimations with a high degree of confidence and prove them to be scientifically defensive. Of

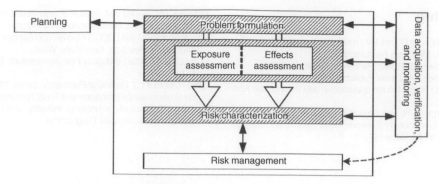

Figure 1 The framework for ecological risk assessment.

importance, there are formal RA techniques for tackling the uncertainty (first, data uncertainty) in a clear and explicit manner and its quantification, to increase impact predictability (the two most widely known are sensitivity analysis and Monte Carlo error analysis).

As to assessment of ecosystem impacts, the proposed integration model implies using formal ERA methodology. The general ERA framework suggested by the US Environmental Protection Agency is depicted in **Figure 1**. It is similar to schemes followed by other counties.

Ecological risk assessment in EIA is to evaluate the probability that adverse ecological effects will occur as a result of exposure to stressors (stressor is a chemical, physical, or biological agent that can cause adverse effects in nonhuman ecological components ranging from organisms, populations, and communities, to ecosystems) related to a proposed development and the magnitude of these adverse effects. A lion's share of site-specific ERAs was concerned with chemical stressors – industrial chemicals and pesticides.

In formal ERA framework, three phases of risk analysis are identified: problem formulation, analysis, and risk characterization followed by 'risk management'. The analysis phase includes an 'exposure assessment' and an 'ecological effects assessment' (see **Figure 1**).

Biogeochemical Approaches to Environmental Risk Assessment

It is well known that biogeochemical cycling is a universal feature of the biosphere, which provides its sustainability against anthropogenic loads, such as acid forming compounds of S and N species, heavy metals and persistent organic pollutants (POPs). Using biogeochemical principles, the concept of 'critical loads' (CLs) has been firstly developed in order to calculate the deposition levels at which effects of acidifying air pollutants start to occur. A UN/ECE (United Nations/Economic Committee of Europe) working group on sulfur and nitrogen oxides under long range transboundary air pollution (LRTAP) convention has

defined the critical load on an ecosystem as: "A quantitative estimate of an exposure to one or more pollutants below which significant harmful effects on specified sensitive elements of the environment do not occur according to present knowledge." These critical load values may be also characterized as "the maximum input of pollutants (sulfur, nitrogen, heavy metals, POPs, etc.), which will not introduce harmful alterations in biogeochemical structure and function of ecosystems in the long-term, i.e. 50–100 years."

The term 'critical load' refers only to the deposition of pollutants. Threshold gaseous concentration exposures are termed 'critical levels' and are defined as "concentrations in the atmosphere above which direct adverse effects on receptors such as plants, ecosystems or materials, may occur according to present knowledge."

Correspondingly, transboundary, regional, or local assessments of critical loads are of concern for optimizing abatement strategy for emission of polutants and their transport (**Figure 2**).

The critical load concept is intended to achieve the maximum economic benefit from the reduction of pollutant emissions since it takes into account the estimates of differing sensitivity of various ecosystems to acid deposition. Thus, this concept is considered to be an alternative to the more expensive best available technologies (BAT) concept. Critical load calculations and mapping allow the

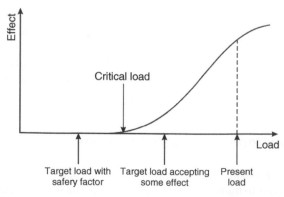

Figure 2 Illustration of critical load and target load concepts.

creation of ecological–economic optimization models with a corresponding assessment of minimum financial investments for achieving maximum environmental protection.

In accordance with the above-mentioned definition, a critical load is an indicator for sustainability of an ecosystem, in that it provides a value for the maximum permissible load of a pollutant at which risk of damage to the biogeochemical cycling and structure of ecosystem is reduced. By measuring or estimating certain links of biogeochemical cycles of sulfur, nitrogen, base cations, heavy metals, various organic species and some other relevant elements, sensitivity of both biogeochemical cycling and ecosystem structure as a whole to pollutant inputs can be calculated, and a 'critical load of pollutant', or the level of input, which affects the sustainability of biogeochemical cycling in the ecosystem, can be identified.

Critical Load and Level Approach for Assessment of Ecosystem Risks

As has been mentioned above, the critical load and level (CLL) concept was introduced initially for emission control at an international scale. From the beginning, it has been applied for regional and local assessments of ecological effects. The latest advances and trends in developing the CLL concept encouraged researchers to consider if critical loads and their exceedances could be applied in EIA for assessing effects on ecosystems. Critical loads and levels are measurable quantitative estimates showing the degree of tolerable exposure of receptors to one or more pollutants. According to present knowledge, when this exposure remains below CLL thresholds, significant harmful effects on specified receptors do not occur. They serve as reference points against which pollution levels can be compared and potential risks to environmental components can be estimated.

The most common shortcomings include:

- failure to analyze impacts beyond development site boundaries,
- failure to quantify ecological impacts (vague descriptive predictions are the norm),
- failure to identify or measure cumulative ecological effects,
- failure to mitigate important ecological impacts (proposed mitigation measures are inappropriate and implementation is not mandatory), and
- lack of monitoring or follow-up (actual outcomes are not known and no corrective action can be taken, e.g., in the event of mitigation failure).

The CLL concept is an important element for emission control policies in Europe. It has become the internationally agreed scientific underpinning for setting targets in controlling SO_2, NO_x, and NH_3 emissions; development of critical loads and levels and similar pollution abatement strategies for heavy metals and POPs is currently in the making.

Initially, the United Nations Economic Commission for Europe (UNECE) introduced the CLL approach into the control of transboundary air pollution under the convention on long-range transboundary air pollution (CLRTAP). In 1994, critical loads of acidity served as inputs to the second 'sulfur protocol'. More recently, European critical load maps were central to the development of the Gothenburg protocol on acidification, eutrophication, and ground level ozone adopted by the executive body of the UNECE CLRTAP in November 1999. Critical load calculating and mapping has been currently undertaken worldwide at national levels including countries, which are not bound with CLRTAP obligations, for example, Korea, India, China, and Thailand.

Over time, there has been growing interest in defining critical loads at a regional level to define sensitivity of particular areas to inputs of pollutants and to set specific threshold exposure values. Most of the research on critical loads and levels is concentrated in regions sensitive to sulfur and nitrogen pollution to generate input data for mapping critical loads and levels following common methodology developed under the convention framework.

More and more research publications on critical loads of pollutants for natural (specific lakes, their catchments, forests) and urbanized ecosystems within defined boundaries are appearing. The following strengths of the CLL approach in the context of ERA/EIA are summarized below.

Quantitative Nature of the CLL Approach

Numerical tolerable exposure levels for pollutants of concern are defined to establish quantitative thresholds for risk characterization; therefore the CLL approach provides a basis for quantitative ecosystem risk and damage assessment.

Scope of the CLL Approach

Critical loads and levels can be calculated for various specified "sensitive elements of the environment." However, terrestrial and aquatic ecosystems are most frequently referred to as receptors in this effect-based approach. In addition, specific parts of ecosystems (e.g., populations of most valuable species) or ecosystem characteristics and (biogeochemical food webs) can be defined as receptors as well. Such flexibility and established provisions for ecosystem assessment makes the CLL concept a promising solution for ecosystem risk assessment and a potential substitute for site-specific chemical RA following the bottom-up approach.

CLL Approach and Ecosystem, Risk Analysis

This approach provides insights on assessment and measurement endpoints for ecosystem-level ERA since it has a set of environmental criteria to detect the state of ecosystems; critical load itself can be treated as a criterion for ecosystem sustainability. Moreover, one can derive 'spatial' ecosystem risk estimates based on the percentage of ecosystems protected/potentially at risk under the current and predicted level of pollutant loads.

CLL Approach and EIA Baseline Studies

While calculating and mapping critical loads, an assessor reviews and systematize most of the data on current state of environment in the site vicinity; the clear and illustrative picture of receptors and their sensitivity to potential impacts is an output of this process.

CLL Approach and Impact Mitigation

Critical loads are particularly useful for elaborating more focused and impact-oriented environmental monitoring programs; mapping critical loads and their exceedances highlights ecosystems (or areas) being damaged by actual or potential pollutant loads giving hints on siting environmental monitoring locations. In turn, critical levels provide a basis for defining maximum permissible emissions to substantiate the development of mitigation measures.

CLL mapping is extremely useful in 'communicating' findings of environmental impact studies both for general public and decision makers.

Input Data Requirements

Critical loads and levels are estimated with the help of biogeochemical models that require a great deal of input data on parameters of biogeochemical turnover and pollutant cycling in ecosystems. Ideally, an assessor should use findings of field studies aimed at measuring all necessary parameters with appropriate extent of accuracy and at appropriate scale. For regions with underdeveloped networks of environmental monitoring (like vast areas of the Russian Federation or China), lack of required data would be a key obstacle for applying CLL within EIA. At the same time, simplified algorithms for CLL calculation have already been elaborated. One of these methods allows for defining critical loads through internal ecosystem characteristics and derived environmental criteria including soil properties, vegetation type, and climatic data. Therefore, an assessor is able to select a CL algorithm among those available bearing in mind input data availability (both empirical, modeled, and literature data) and selected highest degree of uncertainty.

Credibility of the CLL Approach Is Relatively High

Today the CLL approach is a widely known internationally agreed effect-oriented methodology applied worldwide; this aspect is meaningful in communicating research findings on effects and making decisions on risk management.

Progressive Update and Improvement

Even those who criticize the theoretical soundness of this approach acknowledge efforts to validate and improve the CLL concept for increasing degree of confidence of critical loads and levels. UNECE CLRTAP provided an organizational and scientific framework for CLL elaboration having established a program dealing with collecting input data for the CLL calculation (EMEP), and a number of programs under the working group of effects (WGEs) of CLRTAP focused on processing collected data while calculating CLL for specific receptors (forest ecosystems, aquatic ecosystems, and human health, materials) as well as respective international cooperative programs (ICPs). In addition, there are ICPs engaged in developing methodologies and improving practice of mapping and modeling and environmental monitoring. The recent trend in developing CLL methodology is introducing a dynamic approach into modeling.

Usability of CLL

There are plenty of practical guidelines on calculating critical loads and levels including the constantly updated Manual on Methodologies and Criteria for Modeling and Mapping Critical Loads & Levels and Air Pollution Effects, Risks and Trends (Modelling and Mapping Manual, 2004). Moreover, many research groups engaged in biogeochemical model development make them available as 'freeware'. Annual reports published by the National Focal Centers of the LRTAP convention provide insights on methodologies and partially input data for the CLL calculations.

The key shortcoming of the CLL approach from an EIA practitioner's perspective is data uncertainty – a 'sore subject' for any predictive exercise. This is especially true for a simplified algorithm for critical load calculation (see above). Both assessors and reviewers will ask the following questions:

1. Do critical loads really protect ecosystem health?
2. Do applied models provide scientifically defensive results?
3. Are current models capable of acceptive relevant data?

The uncertainty analysis that is a part of formal ERA methodology is designed to ensure adequate estimation of ecological effects based on a state-of-the art scientific basis. Moreover, if applied on a local scale for site-specific assessments, with the use of empirical input data as biogeochemical parameters, the CLL approach is likely to provide results with a higher degree of confidence than the formal ERA model.

In the authors' opinion, even if imperfect, the CLL approach is preferable to apply for ecosystem risk assessment than a qualitative EcoRA based mainly on expert judgment. In response to the need for more consistent

treatment of ecological effects resulting from development projects, the current paper proposes a structured framework for introducing the CLL concept as an approach to ecosystem risk assessment into EIA. The model of the 'integrated' process depicted in **Figure 3** represents the widely accepted idea of 'embedding' risk assessment into EIA. It is organized according to the sequence of generic EIA stages: screening, scoping, impact prediction and

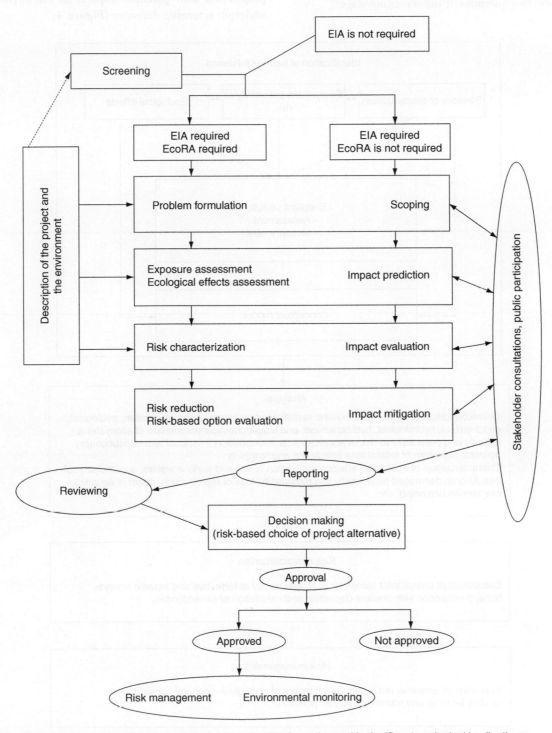

Figure 3   The model for assessment of ecosystem risks in the EIA for projects with significant ecological implications.

evaluation, mitigation, reporting, decision-making, and post-project monitoring and evaluation (EIA follow-up) with public participation and consideration of alternatives potentially incorporated at all stages of the process. The CLL methodology is considered as a quantitative approach to assessing ecological effects. Proposed CLL inputs into the EIA process are discussed below.

In the proposed model project, appraisal starts with addressing two questions at the screening stage:

● Is EIA necessary? and
● Is EcoRA within EIA necessary?

It is the responsibility of the appointed environmental consultants to undertake preliminary investigations and decide if a proposed development may result in significant ecological effects. Data on 'risk agents' including ecosystem biogeochemical stressors associated with the project and their potential impacts on the environment underpin screening decisions (**Figure 4**).

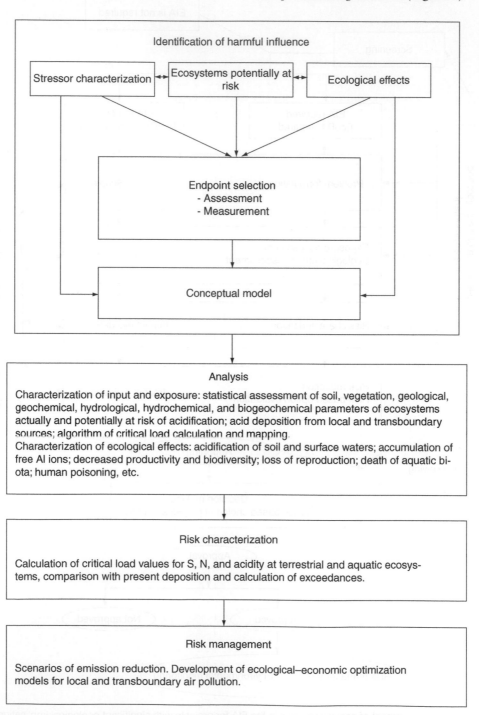

Figure 4 Comparative application of application of CLL and ERA analysis of acidification loading at ecosystems.

Scoping should include defining project alternatives, compiling the list of project impacts, which should be subject to comprehensive impact assessment and planning the further steps of the assessment process. In the formal ERA framework this step is related to problem formulation. A separate task of this stage is to select methods and procedures for dealing with particular impacts. For ecosystem effects, available information on stressors, effects, and receptors is analyzed to define risk assessment biogeochemical endpoints (assessment and measurement endpoints) and possible conceptual models. In addition, policy and regulatory requirements, available budget, and an acceptable level of uncertainty are considered in developing a plan for ERA. Here the assessment team may consider applicability of the CLL concept to project ecological effects and develop a plan of specific studies for calculating and mapping critical loads. The outcome of the scoping is to be an EIA 'terms of reference' (ToR) referring to all above-mentioned issues.

The next step is impact prediction that requires detailed quantitative information about the sources of risk agents, exposure models, the receptors and possible changes in the state of these receptors caused by the defined agents. If the CLL concept was selected for assessment ecosystem effects, it should firstly be utilized for impact baseline studies or assessing the 'do-nothing' scenario. In this context CLL calculation includes the following steps:

- characterizing receptors that are potentially affected by the proposed development,
- defining environmental quality criteria,
- collecting input data for CLL calculations,
- calculating critical loads (CLs),
- comparing CLs with actual loads to calculate the exceedances as a stock at risk.

When the environmental baseline is established, one can proceed with predicting the magnitude of potential impacts onto receptors at risk for 'exposure assessment' in ERA terms. This includes:

- quantifying emissions of pollutants of concern,
- modeling their transport in the environmental media,
- estimating the predicted exposure levels, and
- estimating predicted loads.

Under the CLL approach, 'ecosystem effect assessment' means comparing critical loads with predicted loads of pollutants. Of importance, this may be limited to an ecosystem as a whole without further evaluating adverse effects on specific ecosystem components. CL mapping with the help of GIS is especially useful for this purpose.

Impact prediction should cover all project alternatives selected at scoping (either spatial or technological) and project phases (construction, operation, closure and post-closure are the main subdivisions). Moreover, exposure assessment should cover both normal operation and accidental conditions.

Significance of the predicted impacts should be assessed in the process of impact evaluation or interpretation. At this stage the health risk estimates (quantitative and qualitative) are analyzed in terms of their acceptability against relevant regulatory and/or technical criteria: environmental quality standards or exposure limits.

Critical load exceedances may serve as the basis for interpreting ecological impacts as ecological risks (or rather changes in the level of current risk to 'ecosystem health'). This would refer to the process of ecological risk characterization.

There are a number of approaches to measuring risks depending on assessment and measurement endpoints selected. At ecosystem level, one can propose a percentage of the affected area with CLs exceeded as an acceptable quantitative parameter for ecosystem risk magnitude. In pristine areas, actual state of the environment may be taken as a reference point for risk characterization.

As to risk significance, the degree of alteration in the current environment should be amended with qualitative and semi-qualitative criteria. Ecological impact significance should be considered in terms of:

- ecosystem resilience to particular impacts,
- principal reversibility of potential ecosystem damage,
- threats to valuable ecosystem components, for instance, biogeochemical food webs, etc.

The estimation of accuracy of quantitative predictions and the degree of uncertainty of the assessment findings should be attempted as well.

The results of impact prediction and evaluation are used for designing impact mitigation measures that aim to prevent or reduce the adverse effects associated with the projects and restore or compensate the predicted damage to the environment. Impact mitigation should firstly involve risk reduction measures: (1) control of the source of risk agents; (2) control of the exposure; (3) administrative/managerial improvements; (4) risk communication allowing for more comprehensive risk perception. The selection of appropriate mitigation measured would benefit from using risk-benefit analysis (with formal quantification of residual risks for every option if applicable).

Following the logic of the CLL approach, impact mitigation in EIA is to derive critical limits of exposure (concentrations of pollutants in exposure media) and based on these values calculating maximum permissible emissions that ecosystems in the site vicinity would sustain during the lifetime of the proposed facility. Therefore, any technology that allows for not exceeding CLs for potentially affected ecosystems should be acceptable from the environmental viewpoint, not exclusively the BAT as often recommended.

Uncertainty in IRA and ERA Calculations

One can identify two major categories of uncertainty in EIA: data (scientific) uncertainty inherited in input data (e.g., incomplete or irrelevant baseline information, project characteristics, the misidentification of sources of impacts, as well as secondary, and cumulative impacts) and in impact prediction based on these data (lack of scientific evidence on the nature of affected objects and impacts, the misidentification of source-pathway-receptor relationships, model errors, misuse of proxy data from the analogous contexts); and decision (societal) uncertainty resulting from, for example, inadequate scoping of impacts, imperfection of impact evaluation (e.g., insufficient provisions for public participation), 'human factor' in formal decision making (e.g., subjectivity, bias, any kind of pressure on a decision maker), lack of strategic plans and policies, and possible implications of nearby developments.

Some consequences of increased pollution of air, water, and soil occur abruptly or over a short period of time. Such is the case, for instance, with the outbreak of pollution-induced diseases, or the collapse of an ecosystem as one of its links ceases to perform. Avoiding or preparing for such catastrophes is particularly difficult when occurrence conditions involve uncertainty.

In spite of almost global attraction of the critical load concept, the quantitative assessment of critical load values is connected till now with some uncertainties. The phrase 'significant harmful effects' in the definition of critical load is of course susceptible to interpretation, depending on the kind of effects considered and the amount of harm accepted. Regarding the effects considered in terrestrial ecosystems, a distinction can be made in effects on (**Figure 5**):

- soil microorganisms and soil fauna responsible for biogeochemical cycling in soil (e.g., decreased biodiversity);

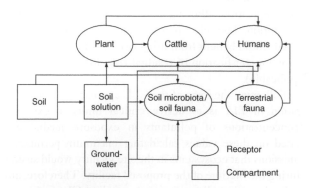

Figure 5 A simplified biogeochemical food web in the terrestrial ecosystems.

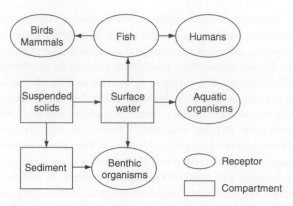

Figure 6 A simplified biogeochemical food web in the aquatic ecosystems.

- vascular plants including crops in agricultural soils and trees in forest soils (e.g., bioproductivity losses);
- terrestrial fauna such as animals and birds (e.g., reproduction decrease);
- human beings as a final consumer in biogeochemical food webs (e.g., increasing migration of heavy metals due to soil acidification with exceeding acceptable human daily intake, etc.).

In aquatic ecosystems, it is necessary to consider the whole biogeochemical structure of these communities and a distinction can be made accounting for the diversity of food webs (**Figure 6**):

- aquatic and benthic organisms (decreased productivity and biodiversity);
- aquatic plants (e.g., decreased biodiversity, eutrophication);
- human beings who consume fish or drinking water (surface water) contaminated with mobile forms of heavy metals due to acidification processes (e.g., poisoning and death).

Summary

Therefore, the CLL concept based on the biogeochemical approaches is a valuable methodology for ecological impact and risk assessment and is easily adjusted to the formal EIA procedure. The proposed framework could be applied to EIAs of development projects with high ecological implications that can potentially affect the environment both on local and regional scales. The model may be applicable to developments that involve releases of acidifying and eutrofying compounds, heavy metals and POPs into the environment in areas with high ecosystem vulnerability and/or pristine areas.

Ecological effects are often treated inadequately in the assessment of environmental impacts of proposed developments, while lack of quantitative ecological impact predictions is mentioned among key drawbacks of the current EIA practice. The idea of integrating ERA into EIA for improving the quality of the relevant studies has been supported by many EIA practitioners. At the same time, formal ecological risk assessment has significant limitations for assessing ecosystems risks related to proposed developments.

To improve addressing ecological implications of human activities, the authors have attempted to incorporate the CLL approach, an established methodology for assessing effects of industrial pollution on ecosystems and their sensitive components, into the EIA process. Benefits of and obstacles to applying that approach to assessing ecosystem effects within EIA were analyzed. Finally, a structured framework for CLL application for ecosystem risk assessment in EIA aimed at integrating three assessment tools was presented and key CLL inputs into impact assessment stages were discussed.

The proposed model of integrated assessment process is suggested for testing in EIAs for development projects with high ecological implications: those associated with releases of pollutants covered by current CLL calculating and mapping methodology and located in areas particularly sensitive to the selected indicator chemicals.

See also: Ecological Risk Assessment; Risk Management Safety Factor.

Further Reading

Arquiaga MC, Canter L, and Nelson DI (1992) Risk assessment principles in environmental impact studies. *Environmental Professional* 14(3): 201–219.

Bashkin VN (1997) The critical load concept for emission abatement strategies in Europe: A review. *Environmental Conservation* 24: 5–13.

Bashkin VN (in cooperation with Howarth RW) (2002) *Modern Biogeochemistry*, 572pp. Dordrecht–London–Boston: Kluwer Academic Publishers.

Bashkin VN (2005) *Environmental Risk Management*, 450pp. Moscow: Scientific: World Publishing House.

Bashkin VN and Park S (eds.) (1998) *Acid Deposition and Ecosystem Sensitivity in East Asia*, 427pp. New York: Nova Science Publishers.

Carpenter RA (1996) Risk assessment. In: Vanclay F and Bronstein DA (eds.) *Environmental and Social Impact Assessment*, pp. 193–219. Chichester: Wiley.

Demidova O and Chep A (2005) Risk assessment for improved treatment of health considerations in EIA. *Environmental Impact Assessment Review* 25(4): 411–429 (available at www.sciencedirect.com).

De Vries W and Bakker DJ (1998) *Manual for Calculating Critical Loads of Heavy Metals for Soils and Surface Waters*. DLO Winand Staring Centre, Wageningen, The Netherlands, Report 165, 91pp.

De Vries W and Bakker DJ (1998) Manual for calculating critical loads of heavy metal for terrestrial ecosystems. *Guidelines for Critical Limits, Calculation Methods and Input Data*. SC report 166, DLO Winand Staring Centre. 144pp.

Gregor H-D and Bashkin VN (eds.) (2004) *Proceedings of 6th Subregional Meeting of ICPs and Training Workshop on the Calculation and Mapping of Critical Loads for Air Pollutants Relevant for UN/ECE Convention on LRTAP in East and South East European Countries.* Moscow-Pushchino.

Posch M, Hettelingh J-P, and Slootweg J (eds.) (2003) Manual for dynamic modelling of soil response to atmospheric deposition. *Coordination Center for Effects, RIVM Report 259101012*, Bilthoven, Netherlands, 71pp. (www.rivm.nl/cce).

Posch M, Hettelingh J-P, Slootweg J, and Downing RJ (2003) Modelling and mapping of critical thresholds in europe. *Status Report 2003. Coordination Center for EffectsNational Institute for Public Health and the Environment Bilthoven, Netherlands. RIVM Report No. 259101013/2003*, 139pp.

Smrchek JC and Zeeman MG (1998) Assessing risks to ecological systems from chemicals. In: Calow P (ed.) *Handbook of Environmental Risk Assessment and Management*, pp. 417–452. Oxford: Blackwell Science.

UNECE CLRTAP (UNECE Convention on Long Range Transboundary Air Pollution) (2004) *Manual on Methodologies and Criteria for Modeling and Mapping Critical Loads & Levels and Air Pollution Effects, Risks and Trends* (available at www.icpmodelling.org).

US EPA (United States of America Environmental Protection Agency) (1998) *Guidelines for Ecological Risk Assessment*. Washington, DC: U.S. Environmental Protection Agency.

Relevant Website

http://www.icpmodelling.org – Modelling and Mapping Manual, 2004.

PART B

Ecological Effects of Toxic Substances

Acute and Chronic Toxicity

W T Waller, University of North Texas, Denton, TX, USA

H J Allen, US EPA, NRMRL, Cincinnati, OH, USA

Introduction

This article presents an overview of the relationships between acute and chronic toxicity and ecotoxicology. The examples used come from the field of freshwater aquatic ecotoxicology although the principles are applicable to other systems.

Ecotoxicology is a combination of the terms ecology and toxicology. Ecology is the study of the relationships between plants and animals and their abiotic environment while toxicology is the study of poisons. The term ecotoxicology then may be defined as the study of the effects of poisons on ecological systems or components thereof. Components of an ecosystem include individuals, populations, communities, and the abiotic environment in which they are found. The ultimate goal of all toxicity testing is to provide data that can be used to establish biologically safe concentrations for toxicants. When sufficient data have been gathered experts in toxicology may develop a criterion for a toxicant. If criteria are incorporated into standards by regulatory agencies, they become enforceable legal limits. To that end the list of toxicity tests (listed in order from least to more complex) shown in **Figure 1** have been developed so that their successful completion provides the data that can be used to develop criteria/standards.

There are advantages and disadvantages associated with each of the test methodologies shown in **Figure 1**. As indicated in the figure, as one moves from single species toxicity tests (which includes laboratory acute and chronic tests) to whole lake or natural system testing the complexity of the tests increases and therefore so do their expense. Generally, because of the characteristics of the tests, a tiered approach is used where testing begins with the simple tests and based on the data needs may progress across the tests. However, it is not necessary, nor is it likely that each test in the series shown in **Figure 1** would be performed to evaluate the potential impact of a chemical. If funds available for testing are limited the results from the simpler tests are often used to prioritize which chemicals receive more complex testing. In many cases criteria are based only on results from acute and chronic toxicity tests performed on a variety of species.

Principles

Toxicologists are guided by principles and three of those are:

1. you only find what you are looking for ...;
2. the dose makes the poison ...;
3. only living material can measure toxicity ....

Principle 1, you only find what you are looking for and you only find it if it is in concentrations high enough to be detected by the method being used to analyze for it, may best be explained by example. Chemical-specific analyses have been and still are used to monitor water quality under some regulatory programs. In some cities in some parts of the United States the sign shown in **Figure 2** can be seen as one enters a city. The city displaying the Superior Public Water Supply System sign was given the right to display the sign because when the water they are providing to the public was tested it was found to meet the standards for the parameters it was required to test. While the parameters the water provider was required to test for are considered important for public water supplies, they certainly do not include all the possible contaminants of the water supply. For example, historically the gasoline additive MTBE was not tested for in water supplies but has subsequently been found to be widely distributed. The same can be said for the herbicide atrazine among other chemicals. You only find what you are looking for!

The axiom given in principle 2 is attributed to the physician Paracelsus (**Figure 3**) who stated, "All substances are poisons, there is none which is not a poison.

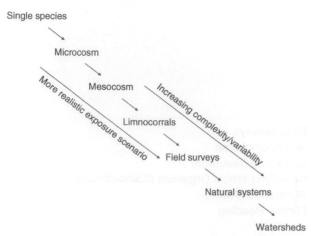

Figure 1 Aquatic toxicity tests used to establish biologically safe concentrations of potential toxicants.

Figure 2 Sign found at the city limits of Texas towns whose public water supplies meet state standards for acceptability.

Figure 3 Woodcut of Auroleus Phillipus Theostratus Bombastus von Hohenheim 'Paracelsus' (1493–1541). Credited with introducing opium and mercury in medicinal use. Also responsible for the often heard, the dose makes the poison.

Figure 4 United States Lincoln Penny minted in 1982. Composed of 97.5% zinc and 2.5% copper.

The right dose differentiates a poison and a remedy." The key part of this axiom is dose, the quantity of potential toxicants administered or consumed. The reason results in aquatic toxicology are expressed, as a concentration instead of dose, the quantity administered, is that the dose an organism receives in aquatic studies is often not known due to multiple routes of exposure. What is known is the concentration of a chemical in the water in which the organism(s) is exposed. Therefore, exposure in aquatic toxicology can be defined as the magnitude, duration, and frequency with which an organism(s) interacts with a biologically available toxicant. The concept of biological availability is important. Just because something is measured in the environment at 'high' concentrations does not *a priori* mean that it is toxic.

For example, the US Lincoln penny shown in **Figure 4** was minted in 1982. The composition of the 1982 penny is 97.5% zinc and 2.5% copper. Given the approximate weight of a penny it contains 59 500 μg of copper and 2 420 000 μg of zinc. The US Environmental Protection Agency's National Criteria for Copper states: "The procedures described in the Guidelines for Deriving Numerical

National Water Quality Criteria for the Protection of Aquatic Organisms and Uses indicate that, except where a locally important species is very sensitive, freshwater aquatic organisms and their uses should not be affected unacceptably if the 1-hour average concentration in (μg/L) does not exceed the numerical value given by the formula $e^{(0.9422[\ln(\text{hardness})]-1.464)}$ more than once every 3 years on average. For example, at hardness values of 50, 100, and 200 mg/L as $CaCO_3$ the safe 1-hour average concentrations are 9.2, 18, and 34 μg/L" (the formula is solved by entering the hardness of water that is to be protected and solving the equation: if a hardness value of 50 is entered into the equation the answer is 9.2 µg l^{-1} which would be considered a safe concentration for acute exposure in the 50 µg l^{-1} CaCO$_3$ water). The criterion for aquatic life for acute exposure to zinc states, for total recoverable zinc the criterion to protect freshwater aquatic life, as derived using the guidelines is (µg l^{-1}) should not exceed the numerical value given by e$^{(0.83[\ln(\text{hardness0}]+1.95)}$ at any time. For example, at hardness values of 50, 100, 200 mg l^{-1} as CaCO$_3$ the concentration of total recoverable zinc should not exceed 180, 320, 570 µg/L at any time. Clearly the amount of copper (59 500 µg) and zinc (2 420 000 µg) in a penny minted on or after 1982 far exceeds the safe concentration of these two essential elements if they were present in a liter of water that was otherwise acceptable to aquatic organisms. However, if you place aquatic organisms normally used to test for acute toxicity in a liter of otherwise acceptable water that also contains a penny, what happens? Nothing happens because the copper and zinc in the penny are not in a biologically available form. There is one additional concept that is important to understand with regards to this example. The criteria presented here represent the state of water quality criteria in the US in 1986 when the criteria for metals were based on Total Recoverable Metal. Current criteria are based on the Dissolved Fraction in the water column. The principle still holds i.e., just because you measure something in the environment it does not mean that it is necessarily bioavailable and toxic!

Single Species Toxicity Assays

The third principle is, no instrument has been devised that can measure toxicity. Toxicity is the degree to which a compound or mixture is capable of causing damage or death. Chemical concentrations can be measured with an instrument but only living material can be used to measure toxicity.

All individuals within a population of organisms do not respond to a compound at the same concentration or in the same period of time. Some individuals respond at lower concentrations while others respond at higher concentrations or not at all. The normal cumulative distribution of this response is often sigmoid in shape as depicted in **Figure 5**.

This plot can be interpreted to indicate that a small proportion of individuals exhibit toxic effects at lower and higher concentrations with the majority of individuals responding at the middle concentrations. For this reason, single compound toxicity tests are often performed at multiple concentrations. An accepted measure of population toxicity can be calculated from these results.

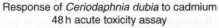

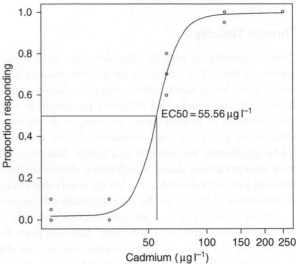

Figure 5 Dose-response frequency distribution and modeled EC50 concentration for *Ceriodaphnia dubia* 48 h cadmium acute toxicity assay.

Acute Toxicity Tests

Acute toxicity can be defined as toxicity that comes speedily to a crisis, that is, the toxicity is manifested over a short time period. In mammalian toxicity tests, where the contaminant is administered directly to the organism, the results of acute tests are expressed as an LD50 or the lethal dose to 50% of the test population that occurred over a given period of time (the length of the test). In aquatic toxicity tests the results are expressed as an LC50 or an EC50. The LC50 is the concentration that caused 50% mortality to the test population in a given period of time. The time is generally 48 h for invertebrate test organisms and 96 h for fish. The EC50 is the concentration that effectively killed 50% of the test organisms in 48 or 96 h. The difference in the LC50 and EC50 is that organisms in the EC50 concentration might still be alive at the end of the exposure period but are effectively dead, that is, if transferred to noncontaminated control water the organisms would not recover.

The result for an acute copper test with the cladoceran *Ceriodaphnia dubia* would be expressed as the 48 h LC50 or EC50. For copper and other divalent cationic metals the additional information presented would include the hardness of the water in which the test was carried out. This is important for metals because toxicity decreases as hardness increases. Hardness in natural waters is a measure of the amount of calcium and magnesium in the water expressed in terms of CaCO$_3$. As data for copper toxicity were developed for different species, sufficient data were gathered that allowed for the development of a criterion for copper. The expression of 'safe' concentrations for acute exposures for copper and zinc presented earlier both contain allowances for changes in hardness in the water.

Chronic Toxicity

Chronic toxicity is toxicity that develops over longer periods of time. The endpoint for chronic toxicity tests can be death but generally other endpoints such as reproductive effects (number of offspring produced or eggs laid); changes in growth rates; or changes in organism behavior or physiology are measured. These responses can be graphically represented in a similar fashion as the acute data presented above. Results from chronic toxicity tests can also be expressed as the lowest observable effects concentration (LOEC), or the no observable effects concentration (NOEC). These are calculated statistically as the lowest concentration significantly different from the control and the highest concentration not statistically significantly different from the control group, respectively. For the majority of chemicals the concentrations causing chronic effects are less than the concentrations causing acute effects. However, there are cases where too little as well as too much can cause problems for organisms.

Single species chronic tests are almost always performed in a closed system under laboratory conditions with temperature and photoperiod control. The organisms are fed a consistent ration and are maintained in well-defined water. To perform a short-term chronic test with *C. dubia* a control group is compared with treatment groups as described above. Ten neonates are placed one each in ten beakers. Over the course of the 7 day test period (which includes three reproductive events), the number of young (neonates) produced by the individual control replicates will be similar but not exactly the same. This variability is the inherent variability for the species under these experimental conditions. The experimental beakers are held under the same conditions as the control except each set of 10 experimental beakers contains a concentration of toxicant.

Numerous species have been used as test species in aquatic acute and chronic toxicity tests, including fish, invertebrates, macrophytes, algae, and bacteria. One group of invertebrates that have been extensively used is commonly referred to as water fleas. These organisms belong to the order Cladocera (Latreille 1829) and family Daphniidae (Straus 1820). The family includes, among others, *Ceriodaphnia dubia* (Dana 1853), and *Daphnia magna* (Müller 1785). *C. dubia* will be used as the example organism in the discussion that follows. **Figure 6** shows a picture of an adult *C. dubia*. The adult is approximately 1 mm in length. The specimen in **Figure 6** is carrying developing young or neonates in its brood pouch. This species reproduces by cyclic parthenogenesis, that is, females give rise to female offspring. Sexual reproduction

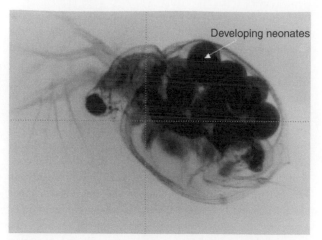

Figure 6 Mature parthenogenetic *Ceriodaphnia dubia* with developing embryos in the brood pouch.

only occurs when conditions become 'unfavorable' to the females in the population and males are produced. The trigger for sexual reproduction is not fully understood but food availability is probably an important factor. When sexual reproduction occurs females produce a robust resting structure called an ephippium (**Figure 7**) that contains a fertilized egg that begins development when the environment the ephippium is in becomes 'favorable'. When the ephippial embryo develops, a parthenogenetic female is produced that then produces female offspring. *C. dubia* is a popular test organism because in the laboratory at 25 °C and the proper diet, it will develop from a less than 12-hour-old neonate to an adult and have three reproductive events in a 7 day period of time. Tests with this organism that are allowed to complete the 7 day time period are referred to as short-term chronic tests. Comparable chronic tests with the cladoceran *D. magna* may take as long as 21 days and some chronic fish tests may take months to complete. Acute tests with *C. dubia* and *D. magna* take 48 h while acute tests with fish take 96 h. The advantage of chronic testing with *C. dubia* is obvious.

Water Quality Criteria

An assumption made when criteria/standards are developed is that ecological systems are protected when the criteria/standards are being met. This is not necessarily a realistic assumption. Almost, if not all standards for toxicants are based on single chemicals, that is, there is a standard for copper, a standard for zinc, a standard for atrazine, etc. However, this begs the question, how often is an ecological system or a component of that system exposed to a single toxicant? It is likely that this may not be a realistic assumption of exposure. Our knowledge of the effects of combined toxicants and our ability to regulate them is very limited.

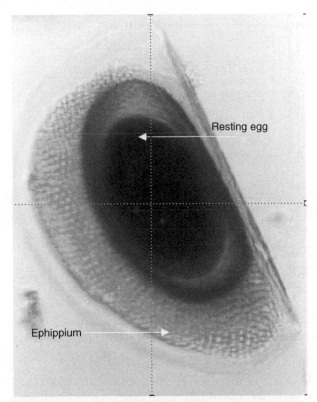

Figure 7 Ephippium produced by *Ceriodaphnia dubia*. The ephippium contains a resting egg.

The Clean Water Act and its subsequent amendments established the National Pollutant Discharge Elimination System (NPDES) to be administered by the states under direction of the US Environmental Protection Agency to protect aquatic systems receiving discharges from municipal or industrial entities. The NPDES uses a water-quality-based rather than a chemical-specific approach for assessment of effluents. All discharges to waters of the United States are required to have a permit. These permits generally have limits on specific chemicals or parameters known to be associated with that type of industry. In addition, most dischargers must perform toxicity tests on the discharge. The principles are the same for both marine and freshwater systems but the organisms used in the tests differ.

C. dubia is often used for testing effluents in NPDES permitting tests for freshwater discharges. *C. dubia* is considered a highly sensitive organism and these tests use its most sensitive life stage, early instars, to provide a conservative estimate of toxicity deemed to be protective of organisms in receiving waters. The design of the test is such that the performance (survival and number of neonates produced) of ten control organisms not exposed to effluent is compared to the performance of ten replicates of organisms exposed to various effluent dilutions. If the survival and neonate production of the organisms exposed to the effluent are not significantly different from the control survival and neonate production, then it is assumed

that the effluent is not having a toxic impact on organisms in the receiving system. If the organisms exposed to effluent dilutions that would be expected to occur in the receiving system at the critical dilution show lower survival and/or neonate production, then it is assumed that the organisms in the receiving system may also be impacted. In an example in which the organisms at the critical dilution do show a significant difference from the control organisms what do we know about the effluent? We know that the organisms at the critical dilution are performing significantly different (more death and/or fewer neonates produced) but we have no idea what is causing the difference. Without knowing what is causing the difference fixing the problem becomes very difficult. Fortunately, methods have been developed to help us figure out what chemical or chemicals are causing the toxicity. These procedures are called toxicity identification evaluations (TIEs).

Toxicity Identification Evaluation

The process of a TIE involves treating a toxic sample using a variety of techniques in an effort to reduce or remove toxicity as indicated by follow-up assays. If any manipulation(s) remove or reduce the toxicity, information regarding the contaminant causing toxicity can be deduced from the expected chemical activity of the manipulation. **Figure 8** shows the kinds of manipulations that might be applied to a toxic effluent in an attempt to identify the toxicant(s). First, it is important to note that TIEs are not performed unless there is a toxic effluent. Second, the tests are performed on waters that have had their pH adjusted above and below the initial pH. pH adjustment can affect the speciation, solubility, polarity, stability, and volatility of a compound and hence its bioavailability and toxicity. The tests that are used to segregate toxicants into groups of similar chemicals include the EDTA chelation test that is designed to bind cationic metals and make them less biologically available. The 'oxidant reduction test' is designed to reduce toxicity due to chlorine. However, ozone and chlorine dioxide are also removed, as are some chemicals formed during chlorination such as mono- and dichloramines, bromine, iodine, manganous ions, and some electrophile organic chemicals. Aeration tests are designed to determine how much toxicity is associated with volatile, sublatable, or oxidizable compounds. The C_{18} 'solid phase extraction test' removes nonpolar organics and metal chelates that are relatively nonpolar. Filtration tests are designed to remove toxicants associated with filterable material. Toxicants associated with suspended materials may be less biologically available, although ingestion of these particles provides another route of exposure. The graduated pH test is designed to test for ammonia toxicity. The unionized form of ammonia is the more toxic form and the proportion of the total ammonia in water is a function of

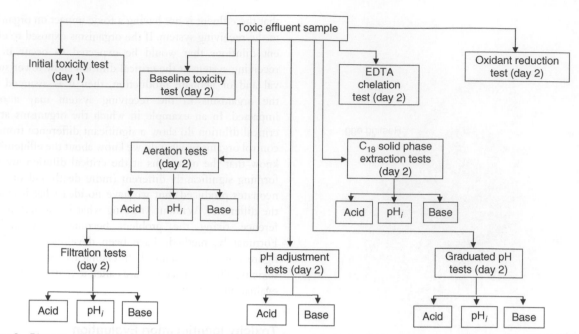

Figure 8 Diagram of the steps often used in a toxicity identification evaluation to determine the cause(s) of toxicity in effluents.

temperature and pH. There is more unionized ammonia at higher pH's.

After each of these procedures is performed the manipulated sample is retested to determine if it is toxic. If the toxicity is removed, then there is an indication that the chemical(s) causing toxicity belong to the class of chemicals the manipulation was designed to remove. For example, if the EDTA test removed toxicity in the sample, there is strong indication that a metal or metals were involved in the toxicity. The question then becomes which metal or metals. Unraveling which metal(s) is causing the toxicity is beyond the scope of this article but having narrowed the search for the toxic chemical(s) down to metals has eliminated many potential classes of toxicants increasing the likelihood of finding the causative toxicant. In some effluents, multiple contaminants may be causing or contributing to the toxicity.

TIE methods were originally designed for use with effluents but the methodological concept has more recently been applied to porewater (water occupying the spaces between particles in sediment), ambient water (water taken directly from a lotic (flowing) or lentic (still)) system, and marine waters.

Model Ecosystems

Model ecosystems are an effort to recreate some of the complexity found in natural systems, and by their very nature they are more complex than single species toxicity tests. Results of experiments using model ecosystems are intended to indicate potential population and community level effects. Because they are more complex model ecosystems are also usually more expensive to build, often requiring significant area and supporting infrastructure resulting in greater associated costs and difficulty in performance over single species tests. Results from replicates of model ecosystems are often more variable than replicates of single species tests. For example, referring to the chronic assay protocol described above, in order to find a negative effect due to a toxicant, the average number of neonates produced in any concentration of toxicant must be statistically significantly less than the average number in the controls. Whether or not a difference in average neonate production is found between the control and the experimental beakers is in part dependent on the variability in neonate production or variance. Variance is by definition a function of the number of replicates in an experiment. If an investigator wants to decrease the variance associated with the average number of neonates produced and thereby increase the ability to detect differences, the number of replicates can be increased. In the single species assays, the number of replicates in the treatment is ten. In most cases, model ecosystem replication will be limited to at most three, and in some cases there are no replicates. Increasing the number of replicates in a laboratory toxicity test is much less expensive than increasing the number of replicate model ecosystems. Cost on the other hand cannot be the overriding consideration in determining how many and what kinds of tests shown in **Figure 1** should be performed in evaluating the potential effects of a chemical.

When data collected from model ecosystems are analyzed, it is still common to break down the massive amount of data collected into distinct groups of organisms. The structure of aquatic insect communities or algal communities might be compared between the controls and the various treatment levels as might the numbers of each of the groups as well as overall species richness (the number of different species present) and evenness (the distribution of individuals among the species present). While this can provide valuable information, these analyses alone do not take full advantage of the community data collected. The development and application of multivariate statistical methods analyzes all the available data from the study and provide managers, regulators, and ecotoxicologists with powerful tools to visualize and present impacts at the community and ecosystem level.

Microcosms

The factors that increase the 'realism' of microcosm tests over single species tests are that there are multiple species in a microcosm. Because there are multiple species, structural as well as functional endpoints can be evaluated. However, the structure of a microcosm is not such that the system can support all trophic levels found in larger 'cosms'. On the other hand, microcosms can be replicated more easily than larger mesocosm and natural systems. By one definition outdoor microcosms are experimental tanks/ponds that contain less than 15 m^3 water volume. **Figure 9** shows a series of microcosms and mesocosms at the University of North Texas used to study the effects of pesticides on community structure and function.

Mesocoms

Mesocosms, as the name suggests, are larger than microcosms. By one definition model outdoor experimental tanks/ponds greater than 15 m^3 water volume are considered mesocosms. **Figure 9** shows a series of pond mesocosms at the University of North Texas Field Station. Mesocosms generally have a better developed community structure than microcosms but usually are still not large enough to maintain some top level predators and do not have all the complexities of natural systems such as larger ponds, lakes, or streams. The endpoints measured in mesocosms include changes in structure and function of the developed communities and through the use of multivariate techniques all the data collected in a study. In some cases organisms in cages have been added to mesocosms to measure the direct effects of the tested chemical on a particular caged species.

Artificial Streams

Artificial streams depending on their size are the lotic (flowing water) equivalent of lentic (nonflowing) microcosms and mesocosms. Artificial streams less than 15 m in length are considered microcosms while those 15 m or

Figure 9 Aerial photograph of the 52, 10000 l fiberglass microcosms and 46, 0.1 ha mesocosms located at the University of North Texas Water Research Field Station. In addition, there are 24 smaller 1000 l microcosms not shown in the photograph. Photograph courtesy of Dr. James H. Kennedy, Director of the Field Station. Microcosm facility at the University of North Texas Aquatic Research Facility.

greater are considered mesocosms. Artificial streams are generally constructed in a manner that allows them to be fed through a headbox by natural stream or river water.

Figure 10 shows a series of artificial streams located at the University of North Texas. As shown in the figure the streams are fed from a common headbox that gets its water from a stream and/or, in this case, a wastewater treatment plant. As is typical for these kinds of systems a substrate of gravel and rocks are added to the artificial stream bottom. Organisms that enter from the headbox or from the surrounding environment colonize the streams. The US Environmental Protection Agency (EPA) operates an experimental stream facility in Cincinnati, OH that is designed with riffle and pool stages. After the period of colonization the streams are randomly assigned as controls and experimental replicates dosed with different concentrations of toxicants or in this case different dilutions of effluent. The endpoints measured are the same as those measured in micro- and macrocosms. In

Figure 10   Experimental streams facility at the University of North Texas. Each stream is fed from a common headbox. Each stream has a run of shallow water and riffles followed by a terminal still water tank. Photograph courtesy of Dr. Tom La Point, Director of the Institute of Applied Sciences, University of North Texas.

some cases organisms in cages have been added and direct effects on those organism can also be determined.

Limnocorrals

Limnocorrals are enclosures placed in natural lakes. Multiple corrals in a lake have been used to study the fate and effects of pesticides on the populations enclosed in the limnocorrals. Unlike mesocosms that require a period of time for the development of resident communities of organisms, limnocorrals take advantage of the already developed communities in the lake. Randomly chosen limnocorrals are assigned to a control group while others are assigned to various treatment groups. Changes in the structure and/or function of the assemblages of organisms in the limnocorrals are measured, as are the overall effects through the use of multivariate statistics. In some cases organisms in cages have been added to limnocorrals and direct effects on those organisms can also be determined.

Experimental Lakes

While there are not a large number of experimental lakes research facilities in the world, the Canadian and Ontario governments have established one such area in Northwestern Ontario. The experimental lakes area (ELA) includes 58 small lakes and their drainage basins, plus some additional stream segments. Studies on nutrient enrichment in two basins of a lake divided by a plastic curtain showed that in this lake limiting levels of phosphorus could control eutrophication. The picture shown in **Figure 11** is of a lake that was divided into two sections by a plastic divider. The

Figure 11   One of the lakes from the Canadian Experimental Lakes Area partitioned with a plastic curtain. The upper part of the lake was dosed with carbon, nitrogen, and phosphorus, the lower part was dosed with carbon and nitrogen. Reproduced by permission of Fisheries and Oceans Canada.

lower section of the lake received environmentally relevant concentrations of carbon and nitrogen, while the upper section received carbon, nitrogen, and phosphorus. The bright green scum of algae on the surface of the upper lake was the result of the phosphorus additions. This study along with other monitoring studies by Ontario's Ministry of the Environment led Canada to become the first country to ban phosphate detergents. Studies have also been carried out in the ELA on acid precipitation, heavy metals, and the effects of flooding of vegetation as a result of reservoir construction. More recently, studies have been undertaken to measure the effects of ethynylestradiol on fathead minnow and pearl dace populations (see the section on 'endocrine disruption').

Field Surveys

Field surveys have been used to evaluate whether or not an ecosystem has been impacted. If one is interested in whether or not an ecosystem has been impacted, then why not just monitor the ecosystem in question? This can be done and should be done but just as toxicity in an effluent tells you that the effluent is toxic, it does not tell you what is causing the toxicity, the same is true for changes found in ecosystems, that is, what caused it, was it due to toxicity, and is the causative agent(s) still present and still causing an impact? Methods for applying the TIE-like manipulations discussed earlier to parts of ecosystems (porewater and ambient water) have been developed but whether or not toxic conditions found in an ecosystem can be linked directly to ongoing toxicity or compared to toxicity that has occurred previously is a difficult task. A classic example of the use of field surveys is to measure the status of an ecosystem above and below an outfall. There are several problems with this approach: first, and one that is sometimes hard to avoid, is the statistical requirement of independence, that is, the downstream sites are not independent of the upstream sites. Second, as discussed earlier measuring an ecosystem response is often much more difficult than measuring the toxicity of a sample. How much change in the ecosystem parameter measured is considered too much: 1%, 5%, or 50%? The answer to this question, that is widely debated, dictates the design of the study and its cost. For example, the number of samples required to detect a change of 1% is much greater than the number of samples required to detect a change of 50%. Third, what parameters are going to be used to judge degradation? The methods used to measure structure in ecosystems have been fairly well developed and include measuring things such as species diversity, evenness, similarity, richness, and biotic integrity. The methods used to measure function are less well developed and include things such as photosynthetic rate, community respiration, organic degradation rates, and energy transfer.

More recently, and in part as an attempt to avoid the lack of independence of samples above and below a discharge, the use of reference systems has been employed. A reference system is defined as the least impacted system in a region that all systems in that region of equal physical and background chemical conditions should resemble. The system above and below an outfall should have a structure and function that look like the reference system. Of course, an important question that must be addressed is, are the least impacted systems in a region the appropriate target of acceptability?

Bioaccumulation

The impact of some chemicals on ecosystems may not be measured directly by the types of toxicity tests discussed above. Some chemicals are known to bioaccumulate either directly from the water (bioconcentrate) or through steps in the food chain (biomagnifiy). The concentrations of chemicals with these characteristics may not be acutely toxic and the classic endpoints of chronic toxicity tests may not show the impacts. The classic example of this type of toxicity is the eggshell thinning from biomagnified chlorinated hydrocarbons that resulted in lowered hatching success for certain bird species, most notably the Bald Eagle. The concentrations of chlorinated hydrocarbons in the environment did not suggest direct toxicity and this pathway for toxicity was not understood for a long time.

Endocrine Disruptors

The endocrine system is comprised of a series of ductless glands that produce hormones that are released into the blood stream. The major hormone-producing glands of the endocrine system include the pituitary, thyroid and parathyroids, hypothalamus, adrenals, pancreas, ovary, and testes. These systems are necessary for normal bodily functions and are instrumental in regulating growth and development, mood, metabolism, sexual function, and reproductive processes. The brain, heart, lungs, kidneys, liver, thymus, skin, and placenta also produce hormones. In order for the hormones to carry out their functions, there must be receptors in the body that respond to the hormones. When a hormone comes in contact with its receptor, it fits together like a key in a lock and the hormone sends a signal to the cell to carry out some function. Hormonal signals may also cascade through complex pathways to an ultimate site of action, with interference at any point having the potential of blocking or changing the intended result.

There are chemicals in the environment that mimic or modify the behavior of hormones. Some of these chemicals act like a hormone and fit the receptor for that hormone stimulating the body to perform some function. Other chemicals can block the receptor so that the hormone cannot reach its target, while others interact with the hormone itself or the gland producing the hormone making it either ineffective in performing its normal function or interfering with the timing of critical events during development. Endocrine disruptors are a diverse group of chemicals including some pesticides, flame retardants, chemicals used in plastics production, cosmetic ingredients, pharmaceuticals, natural products such as plant-derived estrogens and many more. The Prague Declaration on Endocrine Disruption during May 2005 concluded that the existing safety assessment framework for chemicals is ill-equipped to deal with endocrine disruptors. Testing does not account for the effects of simultaneous exposure to many chemicals and may lead to serious underestimations of risk.

An example of the kind of response that researchers have found in studies of endocrine disruption includes those of male fish exposed to wastewater from sewage treatment plants. In this example male fish were found to produce elevated concentrations of the egg yolk precursor vitellogenin. This is of interest because while male fish can produce vitellogenin, they normally do not because as males they never receive the signal from their endocrine system to produce eggs. Since the male fish normally never receive the signal to produce vitellogenin, the signal must be coming from some exogenous source. It has been known for sometime that pharmaceuticals and personal care products (PPCPs) that are excreted from our bodies through urine and feces contain traces of the drugs and personal care products we are exposed to. One of those products routinely found in wastewater is ethynylestradiol, a synthetic estrogen found in birth control pills and other estrogen therapies. Laboratory studies of male fish exposed to environmentally relevant levels of ethynylestradiol have shown elevated levels of vitellogenin. While this does not point directly to ethynylestradiol in the effluent as the causative endocrine disruptor, it does show that it is one possible chemical causing the observed elevation of vitellogenin. Therefore, the production of vitellogenin in male fish is a biomarker for exposure to some chemical that is stimulating the production of vitellogenin. The real question then becomes, does the elevated concentration of vitellogenin in the male fish have any negative implications for the populations of fish. Studies of a dosed lake in the experimental lakes area (see the section 'Experimental lakes') that contained well-defined populations of lake trout, white sucker, fathead minnow, and pearl dace showed that males and female fathead minnow and pearl dace showed elevated whole body concentrations of vitellogenin within 7 weeks of the addition of environmentally relevant concentrations of ethynylestradiol to the lake. Egg development was delayed in the fathead minnow and the pearl dace, testes development was severely impaired and testes-ova (testes containing ovarian tissue) were observed in males of these species. Reproductive failure was observed in both of these minnow species during the second year of ethynylestradiol addition, answering, in this example, that population effects were indeed observed.

Real-Time Whole Organism Biomonitoring

Real-time whole organism biomonitoring involves the use of organisms as sentinels in the environment. One underlying disadvantage of the methods previously discussed is their dependence on temporally discrete or a composite of discrete sampling of waters to be assayed. Contamination of source waters is often episodic; therefore, monitoring must be continuous and 'time relevant' to provide valuable information to stakeholders. The techniques have been variously described as on-line, continuous, real time, and time relevant. Essentially, some observable physiological or behavioral parameter of a group of organisms or single cells is measured and analyzed using computer technology. The historical example of this is the canary in the coal mine.

The science of real-time biomonitoring has become more sophisticated and useful, as technology has progressed. The concept is based on principle 3, only living material can measure toxicity. Many of these systems rely on changes in the behavior of the test organisms to signal

that the environment has changed. For example, one such system relies on the gape behavior of bivalves to monitor the environment. Under nonstressful conditions, bivalves tend to behave in an uncoordinated way over brief periods of time. However, all bivalves share the same defensive behavior of isolating vulnerable tissues by closing shells. It is this coordinated response to changing environmental conditions that is the trigger resulting in further action (**Figure 12**).

In addition to bivalves, systems based on fish, algae, and water fleas have been developed. Current available online toxicity monitors use biota ranging from single cells to whole organisms. Spectroscopic methods are used to measure fluorescence in monocultures of the luminescent bacteria *Vibrio fischeri* and algae *Chlorella vulgaris*, and indigenous algal communities. Swimming behavior in the Cladocera *D. magna* and various species of fish is used as an endpoint. Myoelectric action potentials are measured in the fish *Lepomis macrochirus*. Results of single contaminant laboratory exposures indicate that exposures of short duration (1–2 h) elicit responses similar in concentration to those of longer-term (48–96 h) acute assays, and sometimes approach chronic values.

A system response can be used to signal a water sampler to start taking samples of the water associated with the change in behavior. The water samples can then be analyzed using the TIE methods discussed earlier and if the response is due to the increased level of a toxicant, it may be possible to trace the likely sources of the toxicant and take some corrective action to reduce or remove the toxicant. This approach works best if a watershed is instrumented so the potential sources can be narrowed to a certain subwatershed. These techniques received an elevated visibility following a massive chemical spill on the Rhine River in

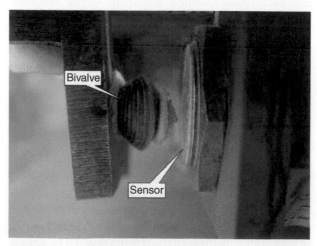

Figure 12 Bivalve mounted on a rack positioned in front of an industrial proximity sensor. A steel washer is attached to the free-moving valve of the mussel. When the mussel moves the small electromagnetic field emitted by the proximity sensor is disturbed and based on that disturbance researchers can tell the position of the valve, that is, is it open or closed.

1986 that resulted in the virtual death of the river from the Swiss/German border to its mouth in The Netherlands. European authorities invested heavily in the development of the Rhine early warning system that is now a model for current source water and distribution system protection efforts in the US and elsewhere.

Watershed (TMDL)

Ecotoxicologists view the watershed as the ultimate aquatic ecosystem level. The watershed is defined as the geographical area that drains to a common point. For example, the Mississippi River watershed is the area of the United States and southern Canada that lies between the Rocky and Appalachian mountain chains and channels surface water to the Mississippi delta in Louisiana. Watersheds range in scale from major river systems, that is, the Mississippi, to small single order streams that may drain only a few hectares. Watersheds, in fact, entail not only the aquatic system but also the land that drains into the aquatic system. This framework recognizes the relationships between land-use, geological, ecological, and societal factors that can influence water quality. The land is a major source of pollutants that are carried into the system. Managing land uses to reduce surface flows is one strategy to maintain water quality. This is a daunting task given the often large land areas and varied uses, including agriculture, industry, and residential. It is sometimes difficult to separate water-quality from water-quantity management. Historically, lands, including wetlands, have been managed to drain rapidly. This lack of residence time results in water laden with particulates and dissolved contaminants. Natural systems have a great assimilative capacity if given a minimal contact time. Tiled agricultural lands, impervious surfaces in developed areas, and direct drainage of storm water increase the overall volume of water flowing through systems and increase the range of minimal and maximal flows. From an ecotoxicological perspective, natural systems are stressed by this variability in flow and accompanying water-quality dynamics.

Implementation of the watershed approach to water quality requires methods of data collection relevant to this different way of thinking about the land and water. Since all parts of a watershed are connected both spatially and temporally by the water that runs through it, data need to be collected in a way that reflects the dynamic nature of the system. Data must be collected from different points throughout the system to provide a comprehensive picture of water quality, and at a rate that allows for relationships to be drawn between areas separated geographically, but linked by the flow of water. Data regarding water and habitat quality must be collected at a scale relevant to an understanding of a large dynamic system. As water flows through a watershed, its quality must be tracked and the information presented to interested parties in a manner that would allow for the maintenance of water quality and quantity to meet both habitat and drinking water needs. A time-relevant, continuous, water-quality monitoring system is a necessary tool for successful implementation of the watershed paradigm.

Total maximum daily loads (TMDL) is a strategy promulgated by the US EPA in an effort to address water-quality issues at larger system scales. This approach establishes a pollution budget for a segment of receiving water based on identification of impairment(s). If a segment of a receiving system is found to be impaired, a TMDL must be established and the waste load is then allocated to all contributors through the NPDES process. The intent is to identify the total amount of pollution a system can assimilate and retain functionality and then split that total amount among all polluters.

Further Reading

Butterworth FM, Gunatilaka A, and Gonsebatt ME (eds.) (2000) *Biomonitors and Biomarkers as Indicators of Environmental Change* 2. New York: Kluwer Academic/Plenum.
Colborn T, Dumanoski D, and Myers JP (1996) *Our Stolen Future*, 306pp. New York: Penguin Books.
Dell'Omo G (ed.) (2002) *Behavioural Ecotoxicology*. Chichester: Wiley.
EWOFFT Organizing Committee (1994) In: Hill IR, Heimbach F, Leeuwangh P, and Matthiessen P (eds.), *Freshwater Field Tests for Hazard Assessment of Chemicals*, 561pp. Boca Raton, FL: Lewis Publisher/CRC Press.
Hoffman DJ, Rattner BA, Burton GA, Jr., and Cairns J, Jr. (eds.) (2003) *Handbook of Ecotoxicology*. Boca Raton, FL: CRC Press.
Landis WG and Ming-Ho Y (1999) *Introduction to Aquatic Toxicology: Impacts of Chemicals upon Ecological Systems*. Boca Raton, FL: CRC Press.
Norberg-King TJ, Ausley LW, Burton DT, *et al.* (eds.) (2005) *Toxicity Reduction and Toxicity Identification Evaluations for Effluents, Ambient Waters, and Other Aqueous Media*, 496pp. Pensacola, FL, USA: Society of Environmental Toxicology and Chemistry (SETAC).
Ostrander GK (ed.) (1996) *Techniques in Aquatic Toxicology*. Boca Raton, FL: CRC Press.
Sparks T (ed.) (2000) *Statistics in Ecotoxicology*. Chichester: Wiley.
USEPA (1985) *Methods for Measuring the Acute Toxicity of Effluents to Freshwater and Marine Organisms*, EPA/600/4-85/013. Cincinnati, OH: USEPA Environmental Monitoring and Support Laboratory.
USEPA (1985) *Short-Term Methods for Estimating the Chronic Toxicity of Effluents to Freshwater Organisms*, EPA/600/4-85/014. Cincinnati, OH: USEPA Environmental Monitoring and Support Laboratory.
USEPA (1991) *Methods for Aquatic Toxicity Identification Evaluations: Phase I Toxicity Characterization Procedures*, 2nd edn., EPA/600/6-91/003. Washington, DC: USEPA Office of Research and Development.

Relevant Websites

http://www.epa.gov/ecotox – This site contains a searchable toxicology database.
http://www.epa.gov/water – This site contains good information ranging from watershed protection to water quality with links to many programs in the US EPA.

http://www.umanitoba.ca – This link takes you to the University of Manitoba. (refer to Experimental Lakes Area in Canada where whole lake and watershed research is carried out by researchers from around the world).

http://www.edenresearch.info – This site takes you to the document produced as the Prague Declaration on Endocrine Disruption.

Bioaccumulation

K Borgå, Norwegian Institute for Water Research, Oslo, Norway

Definition	Importance and Use
Assessing Bioaccumulation	Further Reading
Processes Affecting Bioaccumulation	

Definition

Bioaccumulation describes the accumulation and enrichment of contaminants in organisms, relative to that in the environment. Bioaccumulation is the net result of all uptake and loss processes, such as respiratory and dietary uptake, and loss by egestion, passive diffusion, metabolism, transfer to offspring, and growth (**Figure 1**).

Bioaccumulation thereby comprises the more specific processes of bioconcentration and biomagnification. Bioconcentration is the process of direct partitioning of chemicals between the water and the organism, leading to elevated concentrations in the latter. Biomagnification is the result of contaminant uptake from the diet leading to higher concentrations in the feeder than in the diet. Subsequently, biomagnification leads to increased chemical concentration with higher trophic position in the food web.

This article focuses on organic contaminants; however, there are also other chemicals that are accumulating such as various metals and radionuclides. Examples of bioaccumulating substances are halogenated hydrocarbons such as polychlorinated biphenyls, brominated flame-retardants, perfluorinated compounds, and polyaromatic hydrocarbons.

Assessing Bioaccumulation

Bioaccumulation of a chemical is often reported by bioaccumulation factors (BAFs, $1\,\text{kg}^{-1}$ lipid; eqn [1]), to describe the increase of contaminants such as persistent organic pollutants (POPs) from water to biota due to uptake from all exposure routes:

$$BAF = \frac{[POP_{BIOTA}]_{\text{LIPID CORRECTED}}}{[POP_{WATER}]_{\text{DISSOLVED}}} \quad [1]$$

where $[POP_{BIOTA}]$ is the concentration of the contaminant in the organism, corrected for the animal or plant's lipid content, and $[POP_{WATER}]$ is the dissolved concentration of the contaminant in water. Whereas BAFs include uptake from all exposure routes, the bioconcentration factor (BCF) describes only the exposure from the abiotic environment, and uptake due to equilibrium partitioning of contaminants between the surrounding environment and the organic phase in the biota. Various chemical management programs categorize contaminants with BCF or BAF higher than 5000 (wet weight basis) as

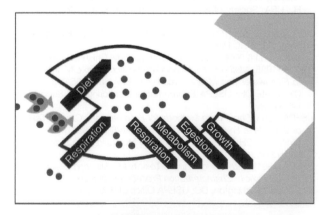

Figure 1 Bioaccumulation of contaminants (dots) to an organism (fish) as a net result of uptake and loss processes (arrows). Uptake is direct from the water by respiration and indirect through the diet. Examples of loss processes are respiration, metabolism, egestion, and growth dilution. Bioaccumulating chemicals increase more than 5000 times from the water to the animal. As the total biomass decreases per trophic level in the food chain (while the contaminants remain), the contaminant concentrations increase moving up the food chain.

bioaccumulative. BAFs and BCFs are corrected for lipid content, as this is the organic phase into which the organic contaminants dissolve. Therefore, lipid-normalized BAF and BCF values (eqn [1]) are more useful when comparing across animals, as the variation due to variable lipid content is eliminated.

BAFs can be estimated by empirical or mechanistic models when empirical data are not available. Several studies show a relationship between a chemical's relative solubility in lipids compared to that in water, as measured by the octanol–water partitioning coefficient (K_{ow}), where octanol and lipids are assumed to have similar properties. BCF on a wet weight basis can be predicted empirically from the chemical's K_{ow}. Although there is a theoretic 1:1 relationship between BAF or BCF and K_{ow} on a logarithmic scale, empirical studies show a high degree of variability in BAFs, often 1–2 orders of magnitude.

Bioaccumulation of organic contaminants in organisms can also be predicted by kinetic, mechanistic models that are based on hypotheses about the exposure and elimination processes involved in bioaccumulation:

$$\frac{\Delta[\text{POP}_{\text{ANIMAL}}]}{\Delta t} = k_1[\text{POP}_{\text{WATER}}] \\ + k_{\text{DIET}}[\text{POP}_{\text{DIET}}] - (k_2 + k_M + k_E + k_G) \\ \times [\text{POP}_{\text{ANIMAL}}] \quad [2]$$

where $\Delta[\text{POP}_{\text{BIOTA}}]/\Delta t$ is the change in contaminant concentrations in the animal over time, k_1 is the respiration and passive diffusion uptake rate, $[\text{POP}_{\text{WATER}}]$ is the dissolved concentration of the contaminant in water, k_{DIET} is the uptake rate from diet, k_2 is the elimination rate due to respiration and passive diffusion, and k_M, k_E, and k_G are the elimination rates due to metabolism, egestion, and growth, respectively. An advantage of mechanistic models is that they quantify different processes of varying importance for bioaccumulation, such as respiration and feeding rates, growth dilution, and biotransformation. This is important not only to assess bioaccumulation, but also to consider which major ecological and chemical factors influence the accumulation of various substances. In addition, the source of contaminants can be identified for animals that are living both in the free water masses and that are occasionally exposed to contaminants through the sediment-related benthic food web.

Even if a contaminant is being taken up by an organism, it does not automatically result in bioaccumulation. An organism can modify the absorbed mixture of contaminant; some chemicals are retained, whereas others that are more water soluble or degradable are eliminated from the body, resulting in no net accumulation. In the food web, animals thereby show very different bioaccumulation of various chemicals, both in levels and in the relative composition.

Processes Affecting Bioaccumulation

Bioaccumulation is not a passive arbitrary process, but a process that is influenced by the molecular properties of the chemical, the amount of particulate matter in the water, and the properties of the algae or animal such as size and lipid content.

Compound Specific – Chemical Factors

For a chemical to be bioaccumulated by the organism, it must be biovailable; available for uptake. For passive uptake from water, this means that the chemical must be truly dissolved, and not associated to particles. However, if not available for passive uptake by respiratory surfaces, or by adhering to the surface, it may be available for dietary uptake, by ingestion of particles. The partitioning of a chemical in the dissolved or particulate phase of water is determined by the chemical's physicochemical properties, especially its hydrophobicity, reflected in their K_{ow}. As a general rule, chemicals that bioaccumulate have log K_{ow} higher or equal to 5.0; however, some persistent contaminants with log K_{ow} lower than 5 but higher than 3.5 also show bioaccumulation.

One group of bioaccumulating contaminants is halogenated organic compounds, in which K_{ow} increases with increasing halogenation degree. Along with increasing halogenation, the molecule grows larger and may no longer be as available (due to steric hinderance) for uptake over biological membranes as are smaller molecules. If absorbed, however, the higher degree of halogenation which increases the hydrophobicity, makes the chemical more difficult to eliminate from the animal. The elimination is more difficult, not only from passive diffusion, but also from enzymatic breakdown, as the higher degree of halogenation leaves fewer positions available to the enzyme to attack.

Animal Specific – Biological Factors

In addition to the chemical-specific properties that affect bioaccumulations, the animals themselves differ in the degree a chemical is accumulated. Some of the most important factors that differ among animals that result in differences in bioaccumulation (and BAF) are lipid content, feeding ecology, habitat use, reproduction, age, biotransformation ability, and energy demand.

Lipid content is important as this is the organic phase into which the organic compounds dissolve. However, some compounds such as metals and fluorinated chemicals are associated to the animal's proteins in the muscle. The animal's feeding ecology is important as this determines the exposure to the contaminants from the diet, just like habitat use is important in an environment where the chemicals are not evenly distributed, but may depend on water masses, depth, sediment type, etc. Reproduction is important as an elimination pathway, especially for female mammals, due to maternal transfer of lipid-soluble contaminants from the mother to the offspring. Often this results in a buildup of contaminants in male mammals with age, as they do not have this pathway of elimination, whereas females reach a steady-state level sooner. Biotransformation ability is the ability of an animal to transform the accumulated chemical into another, preferentially a more-water-soluble compound that can be eliminated. This biotransformation ability depends on the enzymatic activity of the animal, and this is highly species specific. The difference in biotransformation results in a pattern of contaminants that differ widely from the lower end of the food web to animals that occupy higher trophic levels. In general, coldblooded species have a poor ability to biotransform the mixture of chemicals they have absorbed from the diet or from the water, and they therefore often reflect the pattern of chemicals seen in the water or the diet. Warmblooded species on the other hand have a greater enzymatic ability, and can modify to different degrees the accumulated contaminant mixture, resulting in a contaminant pattern of persistent compound and persistent metabolites formed in the biotransformation process. Warmblooded animals also have a higher energy demand than coldblooded species, due to their requirement of a high and stable body temperature. This higher energy demand results in a higher bioaccumulation than in coldblooded species; for a fish and a bird with the same diet and body size, the bird will accumulate higher contaminant levels than the fish.

Importance and Use

The understanding of bioaccumulation of chemicals is important as the accumulated contaminants may lead to toxic effects in the organism and alteration of the animal's normal physiology and ecology. A chemical's BAF (given empirically or estimated from K_{ow}) is used by legislation and decision makers to evaluate regulation and use of new chemicals. The BAF is very useful to compare the difference in bioaccumulation potential among chemicals. Bioaccumulation aids in explaining how contaminants are distributed in the ecosystem and the importance of ecological processes in this respect. Understanding bioaccumulation is closely linked to both ecology and environmental chemistry.

See also: Assimilative Capacity; Bioavailability; Biodegradability; Biomagnification; Body Residues; Dioxin; Ecotoxicological Model of Populations, Ecosystems, and Landscapes; Ecotoxicology: The Focal Topics; Exposure and Exposure Assessment; Food-Web Bioaccumulation Models; Halogenated Hydrocarbons; Persistent Organic Chemicals; Polychlorinated Biphenyls.

Further Reading

Borgå K, Fisk AT, Hoekstra PF, and Muir DCG (2004) Biological and chemical factors of importance in the bioaccumulation and trophic transfer of persistent organochlorine contaminants in Arctic marine food webs. *Environmental Toxicology and Chemistry* 23: 2367–2385.

Connell DW (1988) Bioaccumulation behaviour of persistent organic chemicals with aquatic organisms. *Reviews in Environmental Contamination and Toxicology* 101: 117–154.

Gobas FAPC and Morrison HA (2000) Bioconcentration and biomagnification in the aquatic environment. In: Boethling RS and Mackay D (eds.) *Handbook of Property Estimation Methods for Chemicals: Environmental and Health Sciences*, pp. 189–231. Boca Raton, FL: Lewis Publishers.

Mackay D and Fraser A (2000) Bioaccumulation of persistent organic chemicals: Mechanisms and models. *Environmental Pollution* 110: 375–391.

Bioavailability

K A Anderson and W E Hillwalker, Oregon State University, Corvallis, OR, USA

Introduction

Ecological toxicology, ecotoxicology, is the study of relationships between organisms and environmental contaminants in order to predict biological consequences. The adverse response of organisms to environmental contaminants requires an understanding of exposure processes and contaminant dose. The bioavailability concept originates from the fact that the detrimental effects in exposed organisms and ecosystems are not caused by the total amount of chemical released to the environment, but rather only a certain fraction, the bioavailable fraction. Bioavailability is not an intrinsic property of the contaminant; rather bioavailability reflects the response of a biological system to many contaminant-integrated processes. The bioavailable fraction is characterized only under a defined set of conditions and depends on the physical–chemical characteristics of the contaminant. For example, organic chemicals and metals are different classes of contaminants with unique bioavailability properties, influenced differently by many environmental parameters. Understanding bioavailability allows one to reduce the uncertainty in predicting toxicity of environmental contaminants. The bioavailability of contaminants is therefore important to understand, both to ensure protection of ecosystems and to effectively implement remediation strategies.

Defining Bioavailability

Biological availability, bioavailability, is a dynamic concept that considers physical, chemical, and biological processes of contaminant exposure and dose. In the ecotoxicology literature, bioavailability has been used nonspecifically, in part because no single definition is recognized. Bioavailability incorporates concepts in environmental chemistry and ecotoxicology integrating contaminant concentration, fate, and an organism's behavior in the environment.

As shown in **Figure 1**, bioavailability incorporates many complex processes and embedded in these processes are many parameters. Because bioavailability includes numerous processes, there have been many descriptions and interpretations, each focusing on different aspects. Definitions have ranged from the restrictive, to more integrative, to conceptually splitting each process. The latter approach determining the fate of the bioavailable contaminant may divide the bioavailable definition into two primary concepts: bioaccessibility and toxicological bioavailability. Bioaccessibility incorporates physical–chemical processes in the environment, that is, 'environmental availability', and physiological uptake processes, that is, 'environmental bioavailability'. Toxicological bioavailability considers an organism's

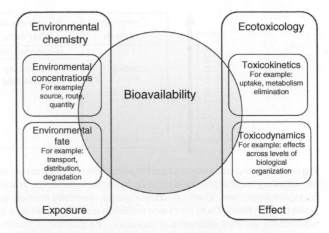

Figure 1 Bioavailability is a concept that incorporates many integrated processes. Bioavailability bridges environmental chemistry and ecotoxicology, integrating contaminant concentration, fate, and behavior, as well as the effect of contaminants within the environment.

internal distribution and metabolic effect processes, also known as 'pharmacological bioavailability' (**Figure 2**).

Although this differentiation at a functional level is helpful for understanding the complexities of bioavailability, an integration of these concepts is also useful. A synthesis of these processes is presented as an integrating definition.

- Bioavailability describes many complex processes, including the mass transfer and uptake of contaminants into organisms, which are determined by substance properties, compartment properties, the biology of organisms, and climatic influences.
- The bioavailable contaminant fraction represents the relevant exposure dose for organisms.

This definition illustrates that bioavailability is a set of processes that incorporate site-specific parameters including space and time. Although a useful ecotoxicological definition has been provided, other definitions and discipline-specific definitions will continue to abound. Unfortunately, this semantic stumbling block may confound issues and impede the implementation of bioavailability in risk analysis. Therefore, defining the bioavailable processes may afford a better perspective by focusing on a mechanistic-based understanding of bioavailability.

Bioavailability Processes

A scientific basis for predicting contaminant impacts on ecosystems requires a mechanistic understanding of bioavailability processes. A mechanism is a physical or chemical process involved in bioavailability. For example,

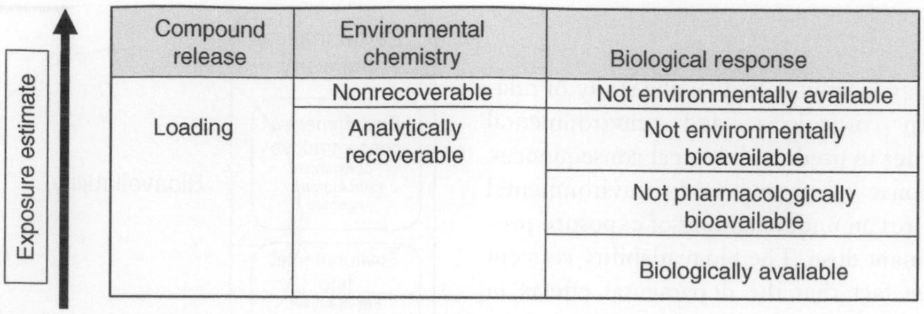

	Compound release	Environmental chemistry	Biological response
		Nonrecoverable	Not environmentally available
	Loading	Analytically recoverable	Not environmentally bioavailable
			Not pharmacologically bioavailable
			Biologically available

(Left vertical axis label: Exposure estimate)

Figure 2 Analytical methods used in environmental chemistry may be conceptually divided – the analytically recoverable fraction representing traditionally used rigorous extractions methods. Traditional analytical contaminant methods rarely synchronize with biological responses. Ideally an environmental chemistry method would synchronize with the biologically available fraction, discussed in the section titled 'Tools for characterization and measurements of bioavailability'. Different conceptual approaches can lead to significantly different estimates of exposure. The addition of fractions that are not bioavailable increases exposure estimates.

in environmental chemistry, a reaction mechanism is the step-by-step sequence of reactions by which an overall chemical change occurs. The physical, chemical, and biological interactions that define bioavailability as the exposure dose include the bioaccessible mechanisms of contaminant association/dissociation processes within and between environmental matrices, transport processes of both free and bound contaminants to the biological membranes, and uptake/passage through biological membranes. Another process defining bioavailability is the distribution, metabolism, and accumulation at the target organ where toxicity may occur, although this last process does not determine whether a contaminant is bioavailable.

In **Figure 3**, the bioavailability process shown in 1 refers to the physical and chemical process of binding and unbinding of contaminants with other compartment components. Environmental compartments include air, water, soil, and sediments. Unbound contaminants are

Bioavailability processes: 1, 2, 3, and 4

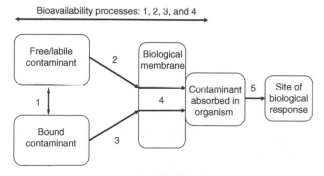

Figure 3 Process 1 illustrates contaminant interactions within environmental compartments, processes 2 and 3 illustrate transport of contaminants to the organisms, and process 4 illustrates passage of contaminants across the biological membrane. Process 5 illustrates circulation within the organism where accumulation in a target organ and toxic effects may occur. Bioaccessability includes processes illustrated in steps 1–3, while bioavailability comprises the first four processes through the biological membrane.

often described as free or labile. Contaminants may bind and unbind on different timescales; some contaminants may bind and unbind quickly while others take years. The kinetics or time frames of these processes are another component of the bioavailable process. Binding mechanisms may include sorption into/onto particles or precipitates. For example, the organic contaminant chlordane, an organochlorine insecticide, may bind and unbind with naturally occurring dissolved organic carbon in aquatic systems. The solubility of chlordane can increase by several hundredfolds in waters containing even modest amounts of dissolved organic carbon. The enhanced solubility results from partitioning of this hydrophobic insecticide into the dissolved organic carbon fraction. However, the increased water solubility does not necessarily indicate an increase in bioaccessibility for transport through an organism membrane. Therefore, dissolved organic carbon may increase transport and mobility of chlordane in the water column, but reduce bioavailability.

In another example of bioavailable process 1 (**Figure 3**), copper, an inorganic contaminant, may bind and unbind with organic carbon or dissolved inorganic components, such as carbonate, bicarbonate, or chloride. Metals may exist in different valence states, and forms of complexes, depending on the metal and environmental site conditions. Each metal valence state or complex will react differently depending on site conditions. The free metal ion, unbound, is considered the most likely metal species for transport through biological membranes, for example, bioavailable. When metals like copper are bound, they are generally inhibited from passing into the biological membrane and are considered less bioavailable. Binding processes may also occur within other environmental compartments such as soils and sediments. Binding may include bonding with particles or precipitation into nonsoluble fractions. Contaminants bound to particles may become unbound by many processes including reduction/oxidation reactions, as well as the processes discussed above. Contaminants bound to solids include many different types of interactions, and the strength of these interactions can vary, ultimately affecting their bioavailability or lack thereof.

Bioavailable processes 2 and 3, shown in **Figure 3**, involve the transport of bound and free contaminants to the biological membrane of an organism. Unbound, or free/labile, contaminants in the gas or aqueous phase are subject to transport processes, such as diffusion, dispersion, and advection. Particles may be transported by moving air or water advection processes including resuspension, bioturbation, and diffusion. Contaminants bound to particles may also be immobilized via precipitation and physical entrapment.

Bioavailable process 4 involves the mechanisms associated with the movement of contaminants through the biological membrane. There are many organisms in the environment and their physiologies differ; however, one common feature among all organisms is the presence of a cellular membrane that separates the cytoplasm, cell interior, from the external environment (**Figure 4**). Most contaminants must pass through the biological membrane before toxic effects on the cell or organism can occur. Processes of contaminant passage through the membrane include passive diffusion, facilitated diffusion, or active transport. Metals must generally pass through ion channels of specific type and diameter, while organic contaminants generally have sufficient lipophilic character to pass through the cell membrane relative to the physical pore size.

Bioavailable process 5 involves the mechanisms occurring after a chemical has crossed the biological membrane. These may include metabolism, storage, and elimination.

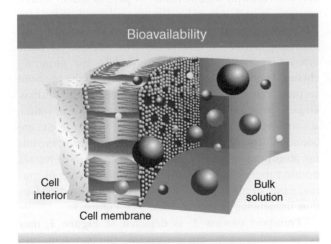

Figure 4 Cartoon of a biological membrane. Illustrated are several features of biological membrane bioavailability, including ion channels, lipophilic character, and pore size. All of these features affect whether a contaminant will progress into the cell interior where toxic action may occur. Contaminants are idealized as the spheres in the bulk solution; some spheres are bioavailable, shown as crossing into and through the cell membrane, whereas others, illustrated by the larger spheres, are not able to cross the biological boundary.

The contaminant may be metabolized to a form that is less toxic, resulting in no observed effect. Conversely, exposure may result in accumulation of the contaminant to levels that are lethal. Between these two extremes, other harmful effects may occur, such as endocrine disruption, reproductive impairment, or other fitness failures. In addition to the direct effects of exposure on the organism, bioaccumulation or biomagnification through food webs may pose serious environmental consequences, discussed in the section titled 'Bioavailability in ecological risk assessment'.

Although it is useful to consider bioavailability processes in isolation (**Figure 3**, 1–5), it is also important to recognize that the processes occur in concert and are often interdependent. The following sections provide additional detail on the effect of environmental chemistry on contaminant bioaccessibility, processes that determine what makes a contaminant bioavailable, factors that influence bioavailability, and tools developed to characterize and measure bioavailability.

Physical–Chemical Properties of Bioavailable Contaminants

Contaminants where bioavailability is considered an important component for assessing impacts to ecosystems are typically defined as persistent, bioaccumulative, and toxic (PBT). The list of PBT contaminants continues to grow with developing technologies. New emerging contaminants, such as personal care products, pharmaceuticals, brominated flame retardants, and fluorinated surfactants are released as uncontrolled, non-point sources into the environment. Characterization of their persistence, bioavailability, and toxicity are generally not always well understood. To assess bioavailability, a compound's physical–chemical characteristics can be plugged into models, such as quantitative structure–activity relationships, QSARs, and the biotic ligand model. The breadth in physical–chemical characteristics and their influence on bioavailability are briefly described next, using examples for common organic and metal contaminants.

Millions of synthetic and naturally occurring organic compounds have been dispersed in the environment by the production of fuels, foods, personal products, and other manufactured goods. Polycyclic aromatic hydrocarbons (PAHs) are naturally occurring and anthropogenic compounds generated by the combustion of organic material such as coal, wood, garbage, and gasoline. PAHs are a class of chemical compounds that consist of many congener types having two or more fused aromatic rings. Examples of PAHs are naphthalene, pyrene, and benzo[a]pyrene. Polychlorinated biphenyls (PCBs) represent a class of 209 individual chemical congeners originating from multiple sources, such as components in industrial hydraulic oils, electrical capacitors, carbonless paper, and from forest fires. PCBs are a group of chemical compounds that consist of two fused benzene rings and containing between two and ten chlorine atoms. The PAH compounds have water solubilities that span 4 orders of magnitude, while the range of K_{ow} for PCBs spans 3 orders of magnitude resulting in a large distribution of lipophilicity characteristics. Some, though not all, PAH and PCB congeners are persistent and toxic. As illustrated by PAHs and PCBs, even within a single chemical class, the breadth of physical–chemical properties is large, affecting distribution in the environment and thus bioavailability.

Inorganic contaminants, also known as elements or metals, represent nearly half of the 25 most detected hazardous substances in water. The most frequently detected metals include arsenic, beryllium, cadmium, copper, chromium, lead, mercury, selenium, silver, thallium, and zinc. Metals in the environment may occur from natural deposit redistribution due to mining activities; other sources of metals include storm water runoff, batteries, plating, wood preservation, paints, and pesticides. The chemical form, or speciation, of inorganic contaminants plays a major role in determining their bioavailability. The speciation of metals includes their valence state and complexation with natural ligands. An important property of metals is the ability to bind with organic matter, cations, and anions in waters, soils, and sediments. Examples of natural ligands include fulvic acid, carbonates, chlorides, and sulfides. The ability of metal ions to chelate, that is, bind strongly, with humic-type organic matter or colloids reduces metal bioavailability.

The distribution of a compound between environmental compartments may be estimated with the use of partition coefficients. A partition coefficient is defined as the ratio of the compound concentration in two specific phases under constant environmental conditions. For example, one of the most widely used physical–chemical characteristics of an organic compound is the n-octanol/water partition coefficient, defined as

$$K_{ow} = [\text{concentration}]_{\text{octanol}} / [\text{concentration}]_{\text{water}}$$

This partition coefficient parameter is important as it may mimic in part the biota lipid/water distribution process. Octanol is somewhat similar to biota lipids and the K_{ow} can provide some indication about the likely distribution of organic compounds between environmental compartments and their general lipophilicity. However, a high K_{ow} may indicate sorption to both biological membranes, indicating increased bioavailability, and also sorption to natural organic matter, resulting in reduced bioavailability. Therefore, it is important to recognize that knowledge of K_{ow} does not always predict bioavailability. Examples of other physical–chemical contaminant properties that

can influence organic and metal environmental distribution and thereby affect bioavailability include pK_a, pH, oxidation–reduction potential, vapor pressure, and partition coefficients, such as with organic carbon, K_{oc}.

Effect of Environmental Chemistry on Contaminant Bioavailability

Regardless of the level of biological complexity, ecotoxicological mechanisms primarily begin with and depend upon the bioavailability of the environmental contaminant. As discussed earlier, the distribution of contaminants is strongly influenced by environmental parameters, thus determining bioavailability. In aquatic systems, the bioavailable fraction will be influenced by all the site parameters, such as pH, hardness, alkalinity, anions, temperature, oxygen content, and organic carbon content. Even the type of organic carbon, for example, fulvic acid, humic acid, humin, kerogen, coal, soot, and black carbon, affects the contaminant distribution and bioavailable fraction differently. In soils and sediments, contaminant bioavailability will also be affected by site-specific conditions such as organic matter, clay content, pH, and cation-exchange capacity. This section describes in more detail the effects of the environmental chemistry on each of the bioavailable processes shown in **Figure 3**.

Bioavailable process 1, depicted in **Figure 3**, is an important factor affecting environmental distribution and bioavailability. In the environment, the contaminant may interact with organic matter or particulates in waters, or organic and mineral solids in soils and sediments. These interactions are characterized as the contaminant becoming bound or free/labile. The strength of these interactions is not only characterized by the physical–chemical partitioning properties discussed in the previous section, but also by the environmental conditions. Organic processes are typically dominated by adsorption, absorption, and partitioning. Inorganic contaminants are bound to particulates via many types of processes, including absorption and adsorption and precipitation. Overall, a wide variety of mechanisms exist, which results in contaminants becoming bound/unbound with solid phases, thus influencing processes 2 and 3 in **Figure 3**.

Transport process 3, as depicted in **Figure 3**, may occur within any environmental compartment: air, water, soil, and sediment. Often, transport in water or gas phase are considered most important for bioavailable exposure scenarios, but contaminants may also be transported via soil and sediment-borne particles. These particles may transport to the biological receptor, biological membrane, via suspension in air, or water. Transport process 2 considers the released, free contaminant, typically transported within the fluid phases by fluid advection processes. Even in nonflowing fluids, contaminants are still transported by the smaller-scale molecular diffusion processes. Free contaminants may also volatilize and move into the atmosphere where they are transported long distances. For example, the persistent PCB contaminants have been discovered on a global-scale distribution due to long-range atmospheric transport. It is worth noting that during transport, chemical transformation may occur and affect contaminant bioavailability. Transport transformation may include photochemical reactions, oxidation–reduction reactions, hydrolysis reactions, and acid–base reactions.

Passage of contaminants into the biological membrane is depicted as process 4 in **Figure 3**, and illustrated in **Figure 4**. Although in ecotoxicology there are many different types of receptors, such as plants, animals, and microorganisms, they may all be conceptualized by the cell membrane-cell interior. Biological systems depend in part on the presence of the biological membrane to separate the organism from the environment. Yet, the biological membrane must allow some compounds to move through it while preventing others. The selectivity of the membrane is important. In **Figure 4**, the biological membrane is composed of phospholipids arranged in a bilayer, as shown by the small ball with two dangling tails. The hydrophobic portion of the molecule is the tail directed toward the center of the membrane. The hydrophilic portion points toward the outward sides of the cell membrane. The surface of the cell therefore interfaces with water, that is, the bulk solution, while the centers are lipids. Proteins are embedded in the membrane, which create pores depicted as channels in the figure, where small chemicals can move into or out of the cell. The main processes by which a chemical can move across the cell membrane are passive diffusion, facilitated diffusion, and active transport.

Once contaminants have entered the organism, the fate of the contaminant is complex and may have deleterious effects. Processes that occur include accumulation, distribution, metabolism, and excretion. As descibed above, the bioavailability of contaminants is governed by a wide range of physical, chemical, and biological processes. These processes occur in concert and may be interdependent. Bioavailability of contaminants is a function of site-, chemical-, and organism-specific conditions and processes as well as climatic/time influenced.

Effects of Time on Bioavailability in Ecotoxicology

The effect of time on bioavailable processes cannot be neglected. Time influences bioavailability processes in several ways. Aging of contaminants affects their bioavailability, often decreasing frequency or amount of

contaminant available for processes 2, 3, and 4 in **Figure 3**. Contaminant aging in solid phase environments has been shown to be an important aspect affecting bioavailability. Both organic and metal contaminants typically become less bioavailable with the aging process, as they diffuse or sorb into/onto mineral lattices and organic matrices in soils and sediments. The longer the contaminant is in contact with a sorbent, such as organic matter, the greater is the extent to which these processes occur. However, at present there is little ability to predict the changes in bioavailability with any specific contaminant at any specific site over time.

In general, the longer a contaminant is in the environment the more it is subject to transformations affecting bioavailability. Both metal and organic contaminants may be microbially degraded to different products which may be more or less bioavailable; the microbial degradation process may also be time dependent. For example, microbial degradation often transforms inorganic or elemental mercury to methylmercury, a much more bioavailable form of mercury resulting in increased food web transfer and bioaccumulation.

Exposure time is also important when considering organism species in their natural habitats versus organism species in biological tests. Time can be a varying factor in exposure frequency for different organisms. Potential bioavailability and actual bioavailability have been proposed to address this issue of time with regard to an organism's actual exposure time versus potential exposure concentration over a longer period of time. However, there are considerable ambiguities with the approach, and generalization at the organism species level still affords important predictive bioavailable information. In addition to exposure time, other effects of time on organisms include seasonal habitat changes and life span.

Bioavailability in Ecological Risk Assessment

Within the ecological risk-assessment framework, bioavailability processes are taken into account in exposure intake equations. Ecological risk assessment is complex as numerous organisms and physical–chemical processes must be considered to predict the impact of contaminants on the ecosystem. Within contaminant exposure and intake equations, typically two pathways are considered; direct contact with the environment and dietary intake. The influence of basic partitioning processes for metals or organic contaminants between the different environmental phases has been used to estimate exposure. These processes were reviewed in the section titled 'Effects of environmental chemistry on contaminant bioavailability'.

Dietary intake pathways incorporate all dietary exposure. Accumulation of contaminants from water into organisms is called bioconcentration. Bioconcentration is defined as the partitioning of a contaminant from the aqueous phase into an organism; typically, this occurs when uptake is greater than elimination. A consequence of contaminant storage by an organism is bioaccumulation. Bioaccumulation is the total amount of contaminant in the organism, the route of uptake including all forms of exposure such as dietary, water, and dermal. Depending on the storage mechanism, bioaccumulated contaminants may be transferred to higher-trophic-level organisms through predator–prey interactions. Biomagnification can occur when the concentration of the original contaminant available from the environment is less than the concentration found in the animal. Biomagnification occurs when bioaccumulation causes an increase in tissue concentration from one trophic level to the next. In contrast, biodilution, regulation of contaminant uptake through the food web, has been observed for some contaminants, especially metals (such as cadmium). Organism distribution and metabolism processes are of great importance in determining the overall effect of contaminants to an organism, and bioaccumulation and biomagnification processes relate to contaminant distribution on ecosystem-level scales. Approaches to determine the bioavailability of contaminants in lower-order animals, such as invertebrates, are rare. Generally, they are assumed to be 100% bioavailable. To efficiently generate quantitative exposure estimates and to accurately characterize risks posed by contaminants, bioavailability needs to be considered in the risk-assessment process.

Tools for Characterization and Measurements of Bioavailability

Because bioavailability processes are embedded in ecosystem health risk frameworks, the development of tools that quantitate bioavailability is important. Environmental assessment tools capable of spatially and temporally resolving contaminant variation are important for assessment of exposure frequencies and levels, episodic spills, and natural and anthropogenic remediation. Conventional methods for ecological exposure assessment involve measuring contaminant concentrations in the ambient environment and extrapolating to toxicological endpoints as well as measuring the concentration of parent compounds or their metabolites in biological samples. However, measures of total ambient contaminant concentrations represent only a rough estimate of exposure and do not reflect the bioavailable fraction. Biomonitoring provides at best only a transient estimate of exposure. Conventional 'snapshot' techniques are cost intensive, lack time-integrated information, and are not effective long-term solutions.

Regulatory agencies typically rely on analytical methods that entail vigorous extraction of matrices with organic solvents for organic contaminants such as PCBs and PAHs, and the use of strong acids for metals. The relevance of such methods to toxicity is often not considered; thus, decisions are based on data that are often irrelevant for prediction of potential exposures and risk. Current analytical methods that measure analytically recoverable concentrations include 'biologically unavailable' fractions, possibly overestimating the magnitude of environmental risk from these pollutants. The total contaminant concentration, 'analytically recoverable', is the amount quantified after vigorous extraction, and includes particle-bound contaminants that are generally not available for uptake. The evidence is compelling that the quantities recovered by vigorous extraction/digestion fail to predict bioavailability of the compounds. Regulatory agencies have recognized the importance of determining bioavailable versus total contaminant concentration. Some regulatory agencies have allowed certain regions to develop site-specific criteria based on bioavailable levels of priority pollutants; however, they report the limited availability of real world bioavailable data.

More recently, the scientific and risk-management communities have concluded that, when available, research tools to determine contaminant bioavailability should be used or considered. In particular, bioavailable approaches and tools that involve mechanistic approaches are most useful. Bioavailable tools may be broadly divided into physical, chemical, and biological tools. Hundreds of bioavailable tools and models have been developed. A few examples within each group are described briefly.

Physical–chemical characterization of solid phases has been used to generally characterize bioavailability. This entails measuring such properties as organic carbon, particle size, and cation exchange capacity to name a few. More specific types of matrix characterization tools include the use of nuclear magnetic resonance (NMR) to characterize soil sorption quality and capacity and have been related to bioavailability. XRD, or X-ray diffraction, has been used to characterize the crystalline structure of solids, and when coupled with scanning electron microscopy (SEM) to identify morphology of soils these characterizations may then be related to bioavailability. Other types of physical–chemical characterization tools include infrared (IR) absorbance, petrography, and elemental analysis. Operationally defined extractions of environmental matrices including traditional conventional extractions, discussed in **Figure 2**, rarely relate to bioavailability. Other sequential extractions have found limited usefulness.

The use of solid-phase and membrane-based *in situ* approaches has been a rapidly developing field for new bioavailable analytical approaches. Passive sampling devices (PSDs) are finding widespread use to assess organism exposure to bioavailable contaminant fractions in soils, sediments, water, and air. The tools essentially sequester unbound contaminants to a solid phase on a membrane. Advantages of the PSD technique are the ability to distinguish between free and bound contaminants. The free, unbound fraction of the contaminants is often related to mobility, bioavailability, and toxicity. Also, passive integrative samplers act as infinite sinks for accumulated residues as no significant losses of sequestered residues occur even when ambient chemical concentrations fall during part of an exposure. There are many different types of PSDs, successfully shown to sequester metals, others for nonpolar organic compounds, and still others for semipolar contaminants.

PSDs are thought to mimic key mechanisms of bioconcentration including diffusion through biomembranes and partitioning between organism lipid and the surrounding medium. One group of organic PSDs, for example, the semipermeable membrane device, consists of a polyethylene tube. The polyethylene tube is normally thought of as nonporous. However, random thermal motions of the polymer chains form transient cavities with maximum diameters of approximately 10 Å. The diameters are very similar in size to cell membranes pores estimated at about 9.5 Å. Because these cavities are extremely small and dynamic, hydrophobic solutes are essentially solubilized by the polymer, as illustrated in **Figure 5**. The cross-sectional diameters of nearly all environmental contaminants are only slightly smaller than the polymeric cavities. Therefore, only dissolved,

bioavailable, labile organic contaminants diffuse through the membrane and are concentrated over time.

PSDs are deployed in the environment, removed after a period of time, and the organic, nonpolar and semipolar, bioavailable contaminants are easily extracted. The extracts may be quantitated by standard chromatographic methods or used in bioassays.

Another PSD technique developed for sequestration of bioavailable metals, the diffusive gradient thin films (DGTs), has the unique advantage of quantitating unbound metals *in situ*. The DGT PSD employs a layer of Chelex resin impregnated in a hydrogel to bind the metals. A diffusive layer of hydrogel and filter overlies the resin layer. Free and bound metal have to diffuse through the filter and diffusive layer to reach the resin layer where only the ion will be irreversibly bound to the chelex. The DGT device is deployed for a known time and then the mass of metal on the resin layer is measured after elution with acid. An attribute of this tool is the ability to distinguish between bound and free metals, an important mechanistic character often related to bioavailability. **Figure 6** illustrates the deployment and extraction procedure for this tool.

Another attribute in general with the PSD technique is its ability to be deployed over a specified time period. The time-integrated nature of PSD addresses some of the temporal variability in an ecosystem that conventional 'snapshoot' sampling cannot address. *In situ* methods also avoid environmental equilibrium problems, generated when removing samples from a dynamic system; the emphasis on *in situ* types of techniques is likely to increase. Other types of PSDs include tenax, XAD, and C-18. PSDs have found some success in predicting bioavailability. However, correlation studies comparing contaminant uptake into various types of PSDs and organisms are still limited. Validation of PSD with biotic endpoints is still necessary.

Normalization techniques have been employed to predict bioavailable fractions. An example of a technique often used to estimate partitioning of metals in sediments is based on normalizing their concentration in sediment to acid-volatile sulfides. Acid-volatile sulfides are an operationally defined concentration of sulfides present in sediments. The acid-volatile sulfide normalized metal concentrations are hypothesized to more closely relate to bioavailability in sediment pore water than the total measured metal concentration in sediment. Another example of a normalization technique often used to estimate partitioning of organic contaminants in sediments is based on the biota–soil/sediment accumulation factor (BSAF). The BSAF is an empirically defined ratio calculated from the chemical concentration measured in tissue relative to the chemical concentration measured in soil or sediment. However, like many empirically defined techniques, the BSAF values are dependent on the physical–chemical properties of both the contaminant and the soil or sediment, as well as the lipid

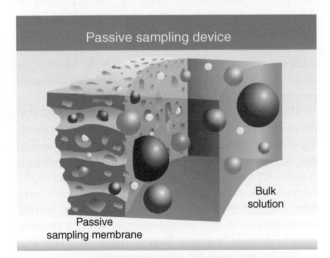

Figure 5 Example of a bioavailable *in situ* analytical tool. Depicted is a cartoon of a polyethylene membrane PSD. Illustrated are several features of PSDs, including lipophilic character and pore size. All of these features affect whether a contaminant will progress into the polyethylene membrane. Contaminants are idealized as the spheres in the bulk solution; some spheres are bioavailable shown as crossing into the polyethylene, whereas others, illustrated by the larger spheres, are not able to cross into the PSD.

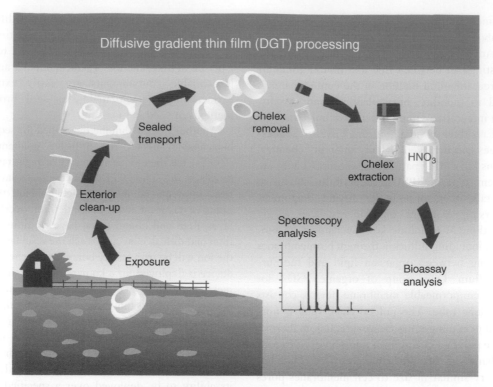

Figure 6 An example of an *in situ* bioavailable analytical tool for determining bioavailable metals in soils and sediments. DGTs are deployed, the devices are removed after a designated period of time, and the metals are easily extracted by acid dissolution. The extracts may be quantitated by standard spectroscopy techniques or used in bioassays.

nature of the organism. Therefore, the BSAF is site and species specific. Both these normalization approaches have a great deal of uncertainty and at best only provide an indication of bioavailability of contaminants.

Another bioavailable approach is the use of equilibrium partitioning theory (EPT), also called the pore water hypothesis. The theory assumes a thermodynamic equilibrium distribution of contaminants between soil particles, soil water, and organisms. Contaminant concentrations in soil pore water are calculated using the soil–water partitioning coefficients, K_d. The values are related to biological effects observed with bioassays and compared with the measured contaminant concentration. Bioconcentration factors (BCFs) are used to determine the uptake of organic contaminants from pore water and compared with the values directly measured in the organisms. The important parameter needed for this type of analysis is the organic carbon partition coefficient, K_{oc}. Both K_{oc} and BCF are dependent on K_{ow}, which as stated earlier does not always accurately represent bioavailability properly. A further refinement of the EPT approach is the 'biotic ligand model' (BLM). The BLM is a model that incorporates metal physical–chemical characteristics into a distribution scenario and relies on site-specific water chemistry information, such as pH and concentrations of ions and inorganic and organic ligands. This model has been used with some success for metal speciation distribution in waters and is currently under development for soil environments.

Biological approaches to measuring bioavailability include bioassays, assimilation and elimination efficiencies, and biomarkers. Bioassays may be at the cellular or whole-organism level. Examples of whole-organism bioassays may include plants, invertebrates, and fish; depending on the contaminant and study goals, various toxic or fitness endpoints may be used. Biomarkers are another approach used to evaluate bioavailability. The biomarker may be a metabolite of the parent toxicant, for example, found in the blood or urine of the organism that represents a biological response to a contaminant exposure.

Many bioavailability tools have been developed that differ in their definition and application. An understanding of the dynamic processes that make up bioavailability, and definitions for contaminant-, site-specific conditions, is necessary before selecting a tool that best describes the relevant risk endpoint. Until bioavailability tools have been validated relative to both biological and site-specific considerations, it may be necessary to select a range of tools to provide 'multiple lines of evidence' about bioavailability processes for a site assessment.

Summary

Bioavailability integrates environmental chemistry and ecotoxicology concepts. The bioavailable contaminant fraction represents the relevant exposure dose for organisms. Many complex processes describe bioavailable exposure dose, including the mass transfer and uptake of contaminants into organisms, which are determined by substance properties, compartment properties, the biology of organisms, and climatic influences. Expansion of bioavailability considerations into ecological risk-assessment decision tools will eventually improve decision making. Development of an ecological risk-assessment framework that incorporates bioavailability considerations can reduce uncertainty and provide better predictive power. However, the lack of validated analytical tools and models limits our current understanding of bioavailability. As bioavailable analytical tools and models continue to be developed and mechanistic-based understanding of bioavailable processes increases, confidence in the use of bioavailability will continue to increase.

See also: Acute and Chronic Toxicity; Dose–Response; Ecological Risk Assessment; Ecotoxicological Model of Populations, Ecosystems, and Landscapes.

Further Reading

Alexander M (2000) Aging, bioavailability, and overestimation of risk from environmental pollutants. *Environmental Science and Technology* 34(20): 4259–4265.

Allan IJ, Vrana B, Greenwood R, et al. (2006) 'Toolbox' for biological and chemical monitoring requirements for the European Union's Water Framework Directive. *Talanta* 69(2): 302–322.

Ares J (2003) Time and space issues in ecotoxicology: Population models, landscape pattern analysis and long-range environmental chemistry. *Environmental Toxicology and Chemistry* 22(5): 945–957.

Boudou A and Ribeyre F (1997) Aquatic ecotoxicology: From the ecosystem to the cellular and molecular levels. *Environmental Health Perspectives* 105(supplement 1): 21–35.

Connell DW (1990) *Bioaccumulation of Xenobiotic Compounds.* Boca Raton, FL: CRC Press.

Escher BI and Hermens JLM (2004) Internal exposure: Linking bioavailability to effects. *Environmental Science and Technology* 38(23): 455A–462A.

Landrum PF, Hayton WL, Lee HI, et al. (1994) Synopsis of discussion session on the kinetics behind environmental bioavailability. In: Hamelink JL, Landrum PF, Bergman HL, and Benson W (eds.) *Bioavailability Physical, Chemical and Biological Interactions,* pp. 203–219. Boca Raton, FL: CRC Press.

Landrum PF, Reinhold MD, Nihart SR, and Eadie BJ (1985) Predicting the bioavailability of organic xenobiotics to *Pontoporeia hoyi* in the presence of humic and fulvic materials and natural dissolved organic matter. *Environmental Toxicology and Chemistry* 4: 459–467.

Luoma SN (1996) The developing framework of marine ecotoxicology: Pollutants as a variable in marine ecosystems? *Journal of Experimental Marine Biology and Ecology* 200: 29–55.

Luthy RG, Allen-King RM, Brown SL, et al. (2003) *Bioavailability of Contaminants in Soils and Sediments; Processes, Tools, and Applications,* 420pp. Washington, DC: The National Academies Press.

Macrae JD and Hall KJ (1998) Comparison of methods used to determine the availability of polycyclic aromatic hydrocarbons in marine sediment. *Environmental Science and Technology* 32(23): 3809–3815.

Namiesnik J, Zabiegalal B, Kot-Wasik A, Partyka M, and Wasik A (2005) Passive sampling and/or extraction techniques in environmental analysis: A review. *Analytical and Bioanalytical Chemistry* 381: 279–301.

Niyogi S and Wood C (2004) Biotic ligand model, a flexible tool for developing site-specific water quality guidelines for metals. *Environmental Science and Technology* 38(23): 6177–6192.

Paquin PR, Gorsuch JW, Apte S, et al. (2002) The biotic ligand model: A historical overview. *Comparative Biochemistry and Physiology* 133C: 3–35.

Peijnenburg WJGM and Jager T (2003) Monitoring approaches to assess bioaccessibility and bioavailability of metals: Matrix issues. *Ecotoxicology and Environmental Safety* 56: 63–77.

Sethajintanin D and Anderson KA (2006) Temporal bioavailability of organochlorine pesticides and PCBs. *Environmental Science and Technology* 40(12): 3689–3695.

Tessier A and Turner D (1995) *Metal Speciation and Bioavailability in Aquatic Systems.* New York: Wiley.

Biodegradability

B R Zaidi, University of Puerto Rico, Mayaguez, PR, USA

S H Imam, Western Regional Research Center (ARS-USDA), Albany, CA, USA

Introduction
Climatic Conditions
Ecotoxicology

Abiotic Factors
Biotic Factors
Further Reading

Introduction

Industrialized and large developing countries, especially China, India, and Brazil, produce staggering quantities of agricultural and industrial chemicals. Some of these chemicals are either toxic to start with or become toxic after their use in the industrial operations. Many of these chemicals are intentionally or unintentionally discharged into the environment, thus contaminating water, soil, and sediments. Incidents range from industrial chemical waste contamination at Love Canal, New York, to halogenated hydrocarbons and pesticide in groundwater to oil spills in Prince William Sound, Alaska. In addition, shipping of huge quantities of organic chemicals to different parts of the world has the potential of causing a worldwide environmental problem. While some degradation of organic chemicals may be due to abiotic mechanisms, for example, photochemical reactions in aquatic environments, most of the degradation of organic chemicals is by indigenous microbial populations.

In natural environment, a chemical may be present at a level, which on its own would cause no harm, but upon interaction with other chemicals may become much more toxic. For example, in production of petrochemical smog, ultraviolet light from the Sun, in the presence of oxygen, hydrocarbons, and nitrogen oxide interact to form peroxyacyl nitrates that are much more toxic than hydrocarbons or nitrogen oxide alone; this is known as synergism. On the other hand, there are also cases when potentially toxic substances may interact to counter each other's effect known as antagonism.

Climatic Conditions

Climatic conditions can greatly influence the movement and the ultimate fate of many potentially toxic substances in the environment. For example, ultraviolet light from the Sun could accelerate the breakdown of many organic chemicals and at the same time also be injurious to microorganisms. Increased temperature due to bright sunlight especially in tropical climate may result in increase vaporization of chemicals into atmosphere, creating respiratory hazard for local communities. Increased temperature also decreases excretion through kidneys in mammals, which may promote the accumulation of toxic substances in their bodies. In natural waters, increased temperature decreases the oxygen content causing fish deaths and making the surviving fish more susceptible to other environmental stresses. Increased water in soil increases soil biological activity but may also lead to anaerobic conditions, particularly in unperturbed soil, causing biodegradation activity to slow down considerably.

Air movement increases loss of volatile chemicals from exposed surfaces and can move contaminants in the air far away from site of production. For example, emissions including sulfur dioxide and nitrogen oxide from British coal-fired power stations are carried across the North Sea to Scandinavia where it contributes to the acidification of lakes and resultant fish kill. The acid rain can also dissolve metals from the rocks, soil, and sediments and these metals may reach toxic levels in affected soil, lakes, streams, and rivers. Another good example is the air movement seen during the summer months in the Caribbean island of Puerto Rico carrying sand particles originated thousands of miles away in Sahara causing respiratory ailments especially in children and aged population. Increased evaporation due to hot air leads to salination of irrigated lands ultimately rendering them useless for crop cultivation. This situation is quite peculiar to some Southeast Asian countries such as India, Pakistan, and Bangladesh, where millions of acres of productive land have been lost due to high salination of soil.

Ecotoxicology

Ecotoxicology deals with the ecological effects of toxic substances in the environment. Internationally acceptable methods for testing of organisms referred to as 'biological monitoring' are being developed. Biological monitoring is the use of living organisms to determine presence, amounts, changes in toxic compounds, and effects of both abiotic and biotic factors in the environment. Biological monitoring provides information that is useful for setting standards which can provide a warning for any changes that may occur in an ecosystem. This information is deemed critical in making policy decisions and developing sound strategies for controlling pollution.

Ecotoxicology in Aquatic Ecosystems

If nothing is known about the environment, repeated surveys are required before commencing a full-fledged biological monitoring. The first survey is typically considered baseline survey, which is followed by a more detailed survey. In view of later survey, final decisions are made with respect to methods, species, and site for monitoring. Monitoring study typically includes a control site, a suitable area that has remained relatively unimpacted. For example, if one wants to study the effects of effluents from a particular industry, an area farther from discharge of effluents will be used as control. In this case, some effects will be common to both the study area and the control, for example, climatic changes, whereas others will be due to discharge of effluents observed only in the study area.

Indicator Organisms

Different classes of indicator organisms are used for 'biological monitoring', where each class may offer a very different response to pollution. Certain organisms have capacity to accumulate pollutants, whereas others are sensitive to the presence of pollutants and react either negatively or positively. Following are some different classes of indicator organisms used in environmental monitoring:

1. Sensitive organisms introduced into the environment to test for early warning of pollution, for example, canaries in coalmines.
2. Naturally occurring species that are very sensitive to pollution. Some species may show changes in growth, reproduction, or behavior and, if very sensitive, they might just disappear. For example, lichens are very sensitive to pollution.
3. Organisms whose presence indicates the presence of pollution. These organisms become abundant in polluted areas due to lack of competition, for example, *Enteromorpha*, following oil spills.
4. Organisms that take up and accumulate pollutants, for example, mollusks accumulating heavy metals or accumulation of hydrocarbons in mussels.
5. Some organisms can be used in the laboratory to detect the presence or concentration of toxic pollutants.

Ecotoxicology in Soil

Ecotoxicology is more developed for aquatic ecosystems than for soil. Methods of testing aquatic invertebrates, fish, and algae were developed more than 30 years ago and large toxicity database is available for aquatic ecosystem. Recently, database for soil has also grown. A major complicating factor in soil ecotoxiclogy is that most of the polluting substances are bound to the soil particle; thus, bioavailable fraction of the free or unbound toxic substance is much smaller. In soil one has to deal with three compartments (soil, pore water, and organisms) compared to just two (water and organisms) in aquatic ecosystems. Additionally in soil ecotoxicology studies, factors such as soil chemistry (sorption to soil, partitioning, and speciation) have to be taken into consideration. Bioavailability and soil chemistry are the principal factors determining changes in toxicity and biodegradation rates.

In ecological risk assessment, acceptable risk of substances in the environment has to be defined first. This is done simply by exposing organisms to series of concentrations and effects are measured at each concentration. The concentration corresponding to the maximum acceptable effect is then estimated from the results. This may be expressed as, for example, EC10 (or 10% effect concentration). However, in soil assessment, due to many site-specific modifying factors the maximum risk standards may not be applicable. Thus, site-specific estimate of risk is needed in soil treatment decisions.

Fate of Organic Chemicals in the Environment

It is practically impossible to predict the fate of a potentially toxic chemical when it enters the natural environment. However, continued postdisposal chemical and biological monitoring of such toxic chemical may provide information on its fate in that environment. Under some conditions, if a chemical accumulates in an environment, then chemical monitoring may detect and provide an early warning. Biological monitoring is biological assessment of exposed organisms in order to detect adverse effects, which may indicate their exposure to the levels of toxicity due to such substances in their environment. Species diversity may be reduced due to increase in pollution. Rare species particularly those that are sensitive to chemical contamination could be wiped out completely, which is detrimental to the overall ecological balance in the ecosystem. The classical example of ecotoxicological biological monitoring is the observation of declining populations of birds that led to the discovery of the food chain biomagnification of dichlorodiphenyltrichloroethane (DDT) and its effects on eggshell thickness and reproductive failure. Death of honeybees reported by beekeepers may be an early warning sign of excessive use of pesticide. Different species of lichens have different sensitivities to sulfur dioxide and their environmental distribution reflects the pollution load of the gas in the environment.

Volatilization and atmospheric transport are the major processes responsible for distributing synthetic organic chemicals throughout the biosphere. In surface and groundwater, the chemical is transported in soluble form, adsorbed to particles or the chemical may move through the food chain. The major sinks are the atmosphere, soils, sediments, oceans, and highest members of the given food chain.

Biodegradability

During the past 70 years, a wide variety of synthetic organic compounds have been produced. While some of these compounds were similar to naturally occurring compounds and were slowly degraded by microorganisms, others had molecular structures microorganisms were never exposed to before and were not recognized by then. These synthetic organic chemicals called xenobiotics (foreign to biological systems) are also resistant to degradation and accumulate in the environment.

Chemical structure of a compound can give certain clues as to its biodegradability, but similar compound can still be biodegraded at different rates and to variable extents. For example, many of organophosphorus pesticides have very similar structures, but show very different biodegradation rates. Aerobic biodegradation is faster than anaerobic biodegradation, but some chlorinated compounds are only degraded anaerobically. Some generalization about chemical structure of a synthetic compound and its persistence in the environment can be made and are described here.

Unusual substitutions. Unusual substitutions can alter the synthetic organic compound in a way that it either becomes partially or wholly resistant to degradation. For example, addition of a single Cl, NO_2, SO_3H, Br, CN, or CF_3 to a readily degradable substrate may increase their resistance to biodegradation. Similarly, addition of two identical or different substitutions may make organic chemicals even more resistant to degradation. The position of substitution greatly influences biodegradation of a compound. For example, if Cl is substituted in phenol at *meta*-position in soil, degradation is slow; but if substitution is not at *meta*-position then degradation is faster.

Unusual bonds or bond sequences, for example, tertiary and quaternary carbon atoms. In this regard much has been learned from the detergent industry. When alkyl benzyl sulfonates (ABS) detergents were first manufactured and used by consumers worldwide, it was not realized until much later that these compounds persisted in the environment. Their persistence in the lakes and rivers led to foaming of water and causing damage to the environment. Later on public concern forced detergent industry to investigate the cause of such persistence. Researchers quickly found that extensive methyl branching interfered with biodegradation process. Thus, switching to linear ABS detergents that were more easily biodegraded alleviated this problem (**Table 1**). Methyl branching is also associated with persistence of aliphatic hydrocarbons. The nonbranched alkanes are easily biodegraded in the environment than alkanes having multiple methyl branching.

Excessive molecular size. Biodegradation of long chain *n*-alkanes declines with increasing molecular weight. Synthetic polymers, for example, polyethylene, polyvinyl chloride, and polystyrene, have high molecular weight and are virtually nonbiodegradable. Many naturally occurring microorganisms have the ability to aerobically degrade polycyclic aromatic hydrocarbons (PAHs), but the process of biodegradation is inversely proportional to ring size of PAH molecule (**Figure 1**). Lower-molecular-weight PAHs are degraded much more rapidly in soil than higher-molecular-weight PAHs when oxygen is present.

Chlorinated aromatic hydrocarbons are degraded aerobically by a variety of mechanisms. Chlorine can be removed by ring cleavage in one- or two-chlorine-substituted compounds. In highly chlorinated compounds, chlorine can be removed by hydrolytic and oxidative reactions. In anaerobic environments, removal of chlorine is by reductive dechlorination reactions. The initial step in anaerobic biodegradation of these compounds is often reductive removal of chlorine atom from aromatic ring.

Pesticides. Most pesticides have simple hydrocarbon backbone bearing a variety of substituents such as halogens, amino, nitro, hydroxyl, and others. Aliphatic carbon chains are initially degraded by the *B*-oxidation and then by tricarboxylic acid cycle. Substituents on

Table 1 Chemical structures of nonlinear and linear alkyl benyl sulfonates (ABS) detergents

Nonlinear ABS detergent resistant to biodegradation

$$H_3-CH(CH_3)-[CH_2CH(CH_3)]_2-CH_2-CH(CH_3)-C_6H_4-SO_3Na$$

Biodegradable linear ABS detergent

$$H_3C-CH_2-[CH_2]_8-CH_2-CH_2-C_6H_4-SO_3Na$$

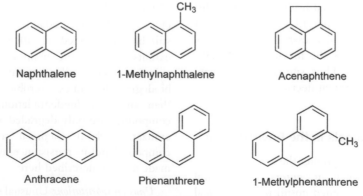

Figure 1 Chemical structure of some biodegradable polycyclic aromatic hydrocarbons.

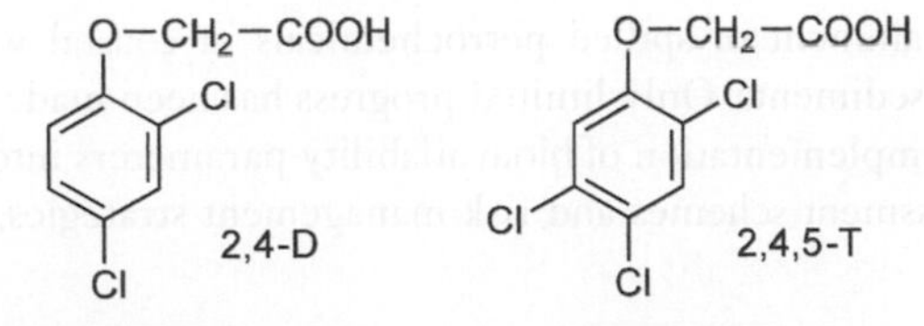

Figure 2 Structures of 2,4-dichloro-phenoxyacetic acid (2,4-D) and 2,4,5-tricholorophnyoxyacetic acid (2,4,5-T).

aromatic ring structures are first removed and then the ring is metabolized by dihydroxylation and ring cleavage mechanisms. Pesticides with substitutions such as halogens, nitro-, and sulfonates that are not common in nature are resistant to biodegradation. Often just a single additional chlorine substitution can make a pesticide quite recalcitrant. For example, 2,4-dichloro-phenoxyacetic acid (2,4-D) is biodegraded within a few days, but 2,4,5-tricholorophnyoxyacetic acid (2,4,5-T), on the other hand, is highly resistant to biodegradation and will persist in the environment. The difference is one additional Cl substitution at *meta*-position in 2,4,5-T (**Figure 2**).

Recalcitrant Compounds

Many organic compounds persist for long periods in the environment. A persistent or recalcitrant toxic chemical may travel to thousands of miles from the application site. The distances it may travel depend on particle size, solubility, and, in some cases, on atmospheric forces. In traveling long distances, the toxic chemical may become diluted. However, even at very low concentrations toxic chemicals may cause harm because of increased concentration of the compound due to its accumulation over time in the food chain. This process is called biomagnification. Therefore, it is not surprising that the concentration of DDT in aquatic environment is roughly around 0.3 ppb in water, but in plankton and fish due to the biomagnification process DDT levels have been found to exceed 30 and 300 ppb, respectively. Unfortunately, under such circumstances DDT levels are expected to continue to rise in higher trophic levels in the food chain. The toxic compounds that increase in concentration in food chain are both persistent and lipophilic. Because the compound is lipophilic, small dissolved amounts are partitioned from water into the lipids of the microorganisms. Grazing of these microorganisms by protozoa leads to further concentration of these toxic chemicals in protozoa in amounts much higher compared to microorganisms and so on. In higher trophic levels (predator fish, carnivores) the concentration of the pollutant may exceed by a factor of 10^4–10^6. Another example of persistent compound is

polychlorinated biphenyls (PCBs) with one to ten chlorine atoms per molecule. They were once used, for example, as plasticizers in polyvinyl polymers and as insulators in transformers. By law their production and use is now banned worldwide. However, even now these compounds are found in the environment; PCBs have also been found in US population with no exposure history. Though highly toxic and recalcitrant, DDT has been the most effective pesticide in controlling mosquitoes that carried malaria-causing larvae, and with re-emergence of malaria epidemic in some African countries, UN is now considering allowing the production and use of DDT once again.

Cometabolism

Some microorganisms can degrade an organic compound without using the substrate as carbon or energy source. This is called cometabolism. The microorganisms obtain no nutritional benefit from the substrate they cometabolize. In this case, the microorganism may be growing on a second substrate. However, presence of that substrate is necessary for cometabolism to occur. A large number of chemicals, for example, PCBs, chlorophenols, and pesticides, may undergo cometabolism in culture. Some of the species of bacteria that can cometabolize organic compounds include *Pseudomonas*, *Acinobacter*, *Bacillus*, and *Arthrobacter*. *Penicillium* and *Rhizoctonia* are some of the fungi that also cometabolize organic compounds.

Threshold Concentration

In aquatic environments, chemicals that are accumulated through biomagnification may eventually become toxic to higher organisms as well. The lowest substrate concentration that is required to sustain growth of a species is generally referred to as 'threshold' concentration. In biodegradation, it is the lowest toxic substrate concentration below which a microorganism cannot degrade the toxic substrate any further. Definitive proof of existence of threshold substrate concentration was obtained from biodegradation studies where one bacterium isolated from environmental samples was capable of degrading a toxic substrate at certain concentration, but failed to degrade the same substrate in quantities below their threshold concentration. However, the other bacterium isolated from same environment degraded the same chemical at considerably lower concentrations, indicating that different bacteria have different threshold values.

Acclimation Period

In biodegradation studies there is an initial period when little or no biodegradation takes place; this period is called acclimation period. During this time the concentration of the toxic substrate remains unchanged. This acclimation period may be short for readily degradable compounds and long for others, but the ultimate rate and extent of degradation depends on the chemical in question and the environment itself. The length of acclimation period is critically important for risk assessment purposes. Longer the acclimation phase, longer the period of exposure in humans, animals, and plants. From an environment perspective, if a chemical is introduced into a river or groundwater, the impact will be considerable as the chemical may potentially move long distances unchanged and risk exposure to population on a much wider scale. In anaerobic environments, the acclimation phase is especially long for some compounds.

End of acclimation period is indicated by start of detectable biodegradation. This period can be shortened if higher inoculums of bacteria are used. This may indicate that acclimation period is the period required for bacteria to grow in sufficient numbers to start degradation of the chemical. Interestingly, once degraded, if the same toxic chemical is added a second time, little or no acclimation period is observed. The acclimation of the microbial community to one substrate may also result in acclimation to some related compounds and is retained for some time.

Abiotic Factors

Generally, microorganisms exhibit different minimum, maximum, and optimum upon exposure to variety of ecological factors such as temperature, pH, presence of toxins, and moisture level in soil, etc., all of which impact their growth. In marine environment, along with temperature, pressure and salinity are also important factors to consider for microbial growth. Since there may be several species in the environment capable of degrading a toxic chemical, the tolerance level is also expected to be much broader.

Bioavailability

Some compounds with passage of time may become less available for biodegradation. The compound may also become unavailable when it undergoes changes and is bound to animate and inanimate surfaces and other residues. This is particularly true in the aquatic environment. Pictures of dead fish and seagulls soaked in spilled oil are a vivid reminder of such hazards. In some cases, alternate strategies such as the separation and/or extraction of toxic compounds from soil water and sediments have been developed for minimally available toxic compounds. However, such procedures have limitations and a narrow window of operation. Such procedures are also cumbersome and quite expensive, and have been applied with a limited success, particularly for cleaning and/or

containment of spilled petrochemicals in coastal waters and sediments. Only limited progress has been made with the implementation of bioavailability parameters into risk assessment schemes and risk management strategies.

Accessibility

Bioavailability is commonly looked at as a portion of total concentration of a contaminant that is available, whereas some residues remain bound. This accessible portion can be applied to estimate as to how much of the contaminant is available for biodegradation. The accessibility can be limited physically by obstruction as well as chemically by strong inter- and/or intrabinding interactions, resulting in a slow desorption rate for such chemicals.

Bioavailability in Soil

When marked decline in rate of biodegradation with time is observed, it indicates reduced availability of the substrate. A compound that is slowly degraded has more time to interact with physical and chemical parameters in soils, sediments resulting in alteration of its behavior. The changed material may become highly persistent and is called bound residues. Especially in soils, many insecticides, herbicides, fungicides, and other classes of chemicals undergo such changes, which result in the formation of bound residues. Many of the bound residues are in fact due to complexing of chemicals with humic materials in soils.

Shape and surface architecture of the material to which a toxic compound is adsorbed plays an important role in determining bioavailability. For example, a chemical may not be available for biodegradation if it is deposited or encapsulated in a micropore that is inaccessible to microorganisms. Soils and sediments are composed of particles of various sizes; between these particles are large and small pores. Therefore, the movement of the chemical from micropore to a site containing a bacterium that is able to degrade that chemical is by diffusion. However, if the environment has small particles, the movement of the chemical from the micropore to the bacterium may take much longer. The longer the path, the slower will be the biodegradation. There is some evidence that suggests that organic compounds are protected from microbial attack this way. The low bioavailability of polyaromatic aromatic hydrocarbons (PAHs) in soil has been attributed to their presence in inaccessible sites in soil matrix. The longer some chemicals remain in soil, the more resistant they become to desorption and to degradation.

Temperature

Microbial activity usually increases with increase in temperature. In temperate climate, almost no degradation is observed in environments at low temperatures that are typical of the winter season. Little or no degradation is also observed at the beginning and at the end of winter season. In a tropical environment, however, typically the temperature does not fluctuate much and the degradation of toxic chemicals is not much affected. However, in tropical coastal marine environment the microorganisms have shown little tolerance for temperature fluctuations. In frozen soils, the toxic chemical persists for long periods due to low metabolic activity of microorganisms.

Effect of pH

Biodegradation is generally faster at moderate pH values. Both, extreme low or high pH values, tend to adversely affect microbial growth, thus slowing microbial breakdown of toxic chemicals. However, because of the presence of a diverse group of microorganisms in a given environment, it is expected that biodegradation will take place over a wide range of pH values.

Inorganic Nutrients

In order to grow, the microorganisms, in addition to an organic carbon that serve as a source of carbon and energy, require inorganic nitrogen, phosphorus, and an electron acceptor. The electron acceptor is O_2 for aerobes, but it may be nitrate, sulfate, CO_2, ferric iron, or organic compounds for specific bacteria. Many microorganisms also require low concentrations of growth factors, for example, amino acids, vitamins, or other molecules. Absence of any of these essential nutrients from an environment prevents growth of microorganisms adversely affecting the biodegradation of toxic chemicals.

Soils, sediments, and natural waters may contain sufficient organic carbon, but most of the carbon exists in complex form and only a small portion is available to support microbial growth. Under these circumstances, it is likely that availability of carbon may become a limiting factor for sustained microbial growth, even if the other nutrients are present in abundant quantities. However, if an easily degradable toxic chemical is introduced in such an environment at high concentrations, then after some time supply of other nutrients may also be exhausted and they become a limiting factor in microbial growth. For example, after oil spill, the environment surrounding oil becomes high in carbon concentration, resulting in high microbial growth. This may lead to depletion of inorganic nitrogen and phosphorus.

Addition of nitrogen and phosphorus to groundwater contaminated with gasoline stimulates growth and degradation of gasoline by bacteria. Similar approaches have shown positive results in seawater after crude oil spill where enhanced degradation of oil was observed after addition of inorganic nitrogen and phosphorus fertilizers.

In some environments, adding just nitrogen or phosphorus but not both enhances degradation. However, there are cases when microbial growth is simultaneously limited by both nutrients.

Alternate Substrates

Natural environments contain many organic compounds that are used by microbes. These substrates can be used either by microorganisms simultaneously or, in some cases, preferentially as carbon source. If the concentrations of two substrates are different, that is, one high and one very low, then the substrate with high concentration is used first.

In some cases one substrate may both enhance and slow down the degradation of other substrates. Both effects have been observed. Sometimes the suppression by one compound of the metabolism of the second is sequential, that is, first substrate disappears only after the second is mostly destroyed. An example of this is sequential destruction of linear alkanes by microorganisms. The reasons for sequential degradation of substrate in natural environments are not known.

Biotic Factors

Synergism

In some cases, one species alone cannot degrade a toxic chemical and may require the cooperation of more than one or more additional species. This is called synergism. Several mechanisms for synergism are proposed. (1) One or more species provide growth factors such as B vitamins and amino acids. (2) One species grows on the toxic chemical and causes incomplete degradation yielding one or several intermediate products, and the second species then completely degrades the products formed due to incomplete degradation by the first bacterium. In the absence of second bacterium, these products tend to accumulate in the environment. (3) The first species cometamobolizes the toxic chemical to yield a product it cannot degrade and the second species then metabolizes this product. In this instance, the first species does not use the chemical as a carbon source. (4) The first species converts the substrate into a toxic product and only proceeds rapidly in the presence of second species. Presence of second species destroys the toxic compound. If second species is not present, the transformation either slows down or stops.

Predation

Natural environments contain predators and parasites that feed on bacteria. Protozoa are predators and are found in natural environments. Protozoa feed on many bacterial genera. When bacteria multiply rapidly, new cells formed may replace the cells lost to protozoa grazing. However, if the bacterium is growing slowly and the chemical concentration is low, the cells eliminated by protozoa grazing will not be replaced. Hence, the species is suppressed or eliminated even as the total bacterial community is maintained. If the bacterium that is capable of degrading a toxic chemical is adversely affected by grazing, then rate of degradation of that chemical will also be affected.

Testing of Biodegradability

Because of concern that chemicals may accumulate in the environment and may be detrimental to the health of the environment, United States Environmental Protection Agency (USEPA) has put the burden of proof on the manufacturers to show that their potentially dangerous chemicals are biodegradable within a reasonable amount of time without causing any harm to the environment. Toxic Substances Control Act of 1976 also authorized the US government to regulate the manufacture and distribution of potentially dangerous chemicals. A permission to produce toxic chemical by the industry will depend on biodegradability studies conducted by the industry at its own cost.

Biodegradability is generally tested by using pure culture of bacterium that is capable of degrading a toxic chemical in presence of test substance as the sole carbon source. This method is useful only to show that the test substance is biodegradable under controlled conditions, but one has to be careful drawing unrealistic conclusions as the conditions in the environment may differ largely from those in the laboratory. Some compounds may not be degraded by pure culture but are still amenable to degradation in the environment because of the presence of a complex microbial community and other easily utilizable carbon sources. On the other hand, a chemical that is degraded in pure culture may not be degraded in the environment. Some of the reasons why a toxic chemical may be degraded in pure culture and not in the environment are discussed under explanation of failure of bioremediation.

Methods of Biodegradability Assessment

Enrichment cultures are used to isolate organisms with degradative capabilities. The substrate to be tested is supplied as the sole carbon source in a medium to which mixed culture of microorganisms is added. Mixed culture source could be municipal sludge, soil, and river, or ocean water. Source of the microorganism is selected from the environment most likely contaminated with the substrate of interest. Only organisms with ability to degrade the

substrate grow and become dominant population. These organisms are then isolated and purified. The purified microorganisms are then added back to a medium containing the substrate of interest as sole carbon source. Degradation of the substrate by these microorganisms is demonstrated generally using radiolabeled substrates. Identification of isolate also permits the result of such studies to be analyzed with other studies. These bacteria are deposited in the culture collection for use worldwide.

Since conditions in pure culture are different from those in the natural environment, the results obtained by pure culture studies may not be representative. It is common practice now to introduce the test chemicals at higher concentrations in samples collected from the environment, such as soil or water, and incubated under conditions similar to the environment.

It is also important to identify the metabolites or by-products resulting from the substrate biodegradation. For this purpose gas chromatography (GC) is used. GC is a sophisticated separation technique that requires low sample concentration. This technique is highly sensitive, accurate, and reproducible. The sample may be in solid, liquid, or gaseous phase as long as it is volatilized at the operating temperature of the instrument. The sample to be analyzed is injected along with the gas, which carries the sample along a column packed with inert particles coated with a liquid. The solutes in the sample are distributed between the liquid and gas phases according to the solubility of solute in the liquid. Solutes of low solubility move through the column at faster rates. As the band exits the column, they are recorded as peaks with retention time related to their partition coefficients. The peaks can be compared with the retention times of standard substrates. The solutes separated by GC can be analyzed via mass spectrometer (MS) to determine molecular structures of the compounds.

Use of radiolabeled compounds is another very sensitive technique to monitor biodegradation of substrates. In this method, biodegradation can be stopped and remaining quantity of the compound can be measured by counting the radioactive emission of the solution in liquid scintillation counter. Use of radiolabeled elements (e.g., ^{14}C) is very useful to demonstrate whether biodegradation is complete or partial. In complete degradation (mineralization), CO_2 is evolved. Mineralization of substrate results in accumulation of radioactive CO_2, which can be quantified.

Bioremediation: Enhancement of Degradation

Enhancement of degradation or bioremediation may be achieved either by (1) adding microorganisms capable of degrading the test compound (inoculation) into the environment or (2) adding inorganic nitrogen and phosphorus to stimulate degradation by indigenous bacteria present in the environment.

Soils, sediments, marine, and freshwater all contain readily metabolizable organic matter. However, when a potentially degradable pollutant at sufficiently high concentration is introduced into the environment, the environment may become deficient in nutrients other than C. Addition of inorganic nitrogen and phosphorus then stimulates the microorganisms present in that environment to degrade the pollutant at a faster rate. In situations like these bioremediation may be achieved by just adding fertilizers containing N and P.

Inoculation

Most of the reports of bioremediation by inoculations have been on pesticides and hydrocarbon constituents of oil. There are several reports indicating that inoculation of microorganisms successfully enhanced degradation of organic chemicals in natural waters. For example, a hydrocarbon-degrading bacterium obtained from an estuary significantly enhanced the degradation of oil spilled into a saline pond. Addition of oil-degrading bacteria to seawater also degraded a substantial part of the crude oil that was added to water.

Failures of Bioremediation

There are several reports when efforts in bioremediation have failed. Some of the reasons are discussed here.

Explanation of Failures

Bioremediation research is a complex undertaking, which requires in-depth understanding of all interacting factors. Most of the information on the biodegradation of toxic chemicals has come from studies using pure cultures grown at high substrate concentrations in laboratory media under controlled conditions. However, in nature microorganisms are exposed to different conditions. They may be exposed to insufficient supply of inorganic nutrients, different temperatures, and pH values that may result in their loss of viability. They may be benefited or harmed by the presence of alternate substrates and by the activities of other microorganisms that are present in the environment. Thus, results obtained by pure culture studies have their limitations and cannot always be extrapolated to nature. It is therefore extremely important to have not only *in vitro* information on biodegradation but also understanding of environmental stresses microorganisms are exposed to in nature. Hence, to increase the likelihood of a microorganism's bringing about a reaction in nature that it can perform in axenic culture in laboratory media, the identities of abiotic and biotic stresses and

the means to overcome them must be established. The information can then be used to construct a genetically engineered microbial strain that is not only capable of degrading toxic chemicals at a faster rate but is also able to withstand environmental stresses when introduced back into the environment.

There are several reasons for the failure of microorganisms to enhance biodegradation when inoculated back into the environment. Even microorganisms that are successful in one environment may not be successful in another environment. These reasons for failure often reflect ecological constraints on the introduced organism. Some of the ecological constraints responsible for the failure of inoculation to enhance biodegradation have been mentioned under biodegradability. Briefly, they are: (1) alternate substrate, (2) presence of toxins in environment, (3) bioavailability of the substrate, (4) temperature, (5) predation, and most importantly (6) competition from ecologically established microbial flora in the environment.

Potential of Genetically Engineered Microorganisms in Bioremediation

One of the most successful projects of large-scale bioremediation was undertaken at the beaches of Prince William Sound, Alaska. Enhanced degradation of oil by application of inorganic nutrients (slow-release fertilizers) and inoculation of microorganisms capable of degrading oil convinced government officials that bioremediation was a viable alternative for at least some hazardous waste problems. Fertilizer applications stimulated the indigenous microorganisms to degrade the oil contaminating the beaches at faster rates.

Molecular biology provides highly useful techniques to modify microorganisms so that they have the desired properties. The process may involve separation of DNA from the cell, its treatment with specific restriction endonuclease to cleave the DNA, the rejoining of the DNA fragments with DNA ligase to give a new sequence of nucleotide bases, and the reintroduction of this hybrid molecule into a suitable bacterial cell in which it is replicated and expressed. These genetic modifications provide considerable promise to affect bioremediation of even the most recalcitrant compounds. Genetically engineered microorganisms introduced into natural environments face the same stresses as those affecting the existing organisms.

Some of the hazardous chemicals present in a contaminated site are degraded slowly, limiting bioremediation of sites contaminated with these chemicals. For example, microorganisms degrade some PCBs and PAHs very slowly, thus limiting the use of these bacteria in bioremediation of the sites contaminated with these chemicals. Genetically engineered microorganisms that can degrade these compounds at a faster rate can be of much value.

Isolation of bacteria that are resistant to high concentrations of toxic chemicals and the use of them as hosts in genetic engineering experiments would provide an opportunity for on-site bioremediation even when the sites are highly contaminated.

See also: Halogenated Hydrocarbons; Polycyclic Aromatic Hydrocarbons.

Further Reading

Alexander M (1994) *Biodegradation and Bioremediation*. New York: Academic Press.

MacGillivary AR and Cerniglia CE (1994) Microbial ecology of polycyclic aromatic hydrocarbons. In: Chaudhry GR (ed.) *Biological Degradation and Bioremediation of Toxic Chemicals*, pp. 125–147. Portland, OR: Dioscorides Press.

Prince RC (1992) Bioremediation of oil spills, with particular reference to the spill from the Exxon Valdez. In: Fry JC, Gadd GM, Herbert RA, Jones CW, and Watson-Craik IA (eds.) *Microbial Control of Pollution. Society for General Microbiology Symposium 48*, pp. 19–34. Cambridge: Cambridge University Press.

Reichenberg F and Mayer P (2005) Two complimentary sides of bioavailability: Accessibility and chemical activity of organic contaminants in sediments and soils. *Environmental Toxicology and Chemistry* 25(5): 1239–1245.

Van Straalen NM (2002) Assessment of soil contamination – A functional perspective. *Biodegradation* 13: 41–52.

Zaidi BR, Hinkey LM, Rodriguez NR, Govind NS, and Imam SH (2003) Biodegradation of toxic chemicals in Guayanilla Bay, Puerto Rico. *Marine Pollution Bulletin* 46: 418–423.

Zaidi BR and Imam SH (1999) Factors affecting microbial degradation of polycyclic aromatic hydrocarbon phenanthrene in the Caribbean coastal water. *Marine Pollution Bulletin* 38(8): 737–742.

Zaidi BR, Mehta NK, Imam SH, and Green RV (1996) Inoculation of indigenous and non-indigenous bacteria to enhance biodegradation of *p*-nitrophenol in industrial wastewater: Effect of glucose as second substrate. *Biotechnology Letters* 18(5): 565–570.

Biodegradation

S E Jørgensen, Copenhagen University, Copenhagen, Denmark

Introduction	Toxic Substances
Biodegradation Rate	Units
Wastewater Treatment	Further Reading

Introduction

Biodegradation is a very important property for toxic chemicals, because if the biodegradation rate is high, the concentration and thereby the toxic effect will be reduced rapidly, while very persistent chemicals will maintain their toxic effect for a very long time.

The range of biodegradation rates is very wide – from readily biodegraded compounds as for instance monomer carbohydrates, low molecular alcohols, and acids to very refractory compounds that have a biological half-life of several years as for instance DDT and dioxins.

In principle, biodegradation is carried out by many organisms, but in most cases we consider microbiological biodegradation for the most important from an environmental point of view. The biodegradation rates in water and in soil by microorganisms are of particularly interest. It is, however, not a characteristic value that can be used as a constant for a compound, because the biodegradation is strongly dependent on the conditions for the microorganisms in the water and in the soil. The biodegradation is furthermore dependent on the presence or absence of oxygen; it means aerobic or anaerobic conditions. Environmental degradation rates can, however, be found in the literature and in environmental handbooks, but they are always indicated as ranges. The biodegradation rate of the same compound in water or soil may vary orders of magnitudes from one type of aquatic or terrestrial ecosystem to the next. The half-life of methyl methacrylate in soil, to mention a typical example, is in the literature indicated as 168–672 h.

Biodegradation Rate

Biodegradation rates can be expressed in several ways. With good approximation, microbiological biodegradation can be expressed as a Monod equation:

$$\mathrm{d}c/\mathrm{d}t = -\mathrm{d}B/Y\mathrm{d}t = g_{\max}BC/Y(km + c)$$

where c is the concentration of the considered compound, Y is the yield of biomass B per unit of c, B is the biomass concentration, g_{max} is the maximum specific growth rate and km is the half-saturation constant. $I\ c \ll km$ the expression is reduced to a first order reaction scheme: $dc/dt = -K'Bc$. $K' = g_{max}/km\ Y$. In water, B is highly dependent on the concentration of suspended matter and in soil B is dependent on the porosity of the soil. B may therefore often be considered a constant. It implies that $dc/dt = k\ c$. k has the unit $1/h$, $1/24\,h$, $1/w$, or $1/y$. If the biological half-life is applied, we get the following: $\ln 2 = 0.7 = kt_{1/2}$, where $t_{1/2}$ is the half-life.

Wastewater Treatment

Biodegradation in wastewater treatment plans is of environmental importance, because we need to know how much the concentration of organic matter is reduced by the treatment. In this context, it is often used to indicate the percentage of the theoretical oxygen demand (ThOD), which is the amount of oxygen used by a complete oxidation. BOD is an abbreviation for biological oxygen demand. It is often used with indication of the considered time, for instance, BOD_5, which is the biological oxygen demand for period of 5 days. BOD_5 is often used to indicate the concentration of organic matter, because the environmental interest in water is the oxygen depletion for a reasonable period. Five days was chosen, when BOD was introduced in UK more than hundred years ago, because all British rivers reach the sea within 5 days. The stoichiometric equation organic matter + oxygen = carbon dioxide + water shows that for typical municipal wastewater, 1 g of organic matter will require approximately 1.4 g of oxygen. The usual biodegradation rate, k, for the organic matter in municipal wastewater, which is relatively readily biodegraded, is 0.35 $1/24\,h$. It means that if we indicate the loading of organic matter expressed by oxygen demand by L, we have

$$BOD_5 = L_0 - L_5 = L_o(1 - e^{-0.35 \times 5})$$

If BOD_5 is for instance $150\,mg\,l^{-1}$, L_o is $181\,mg\,l^{-1}$. Notice that $L_0 = BOD_{infinite} = ThOD$.

Toxic Substances

For most toxic substances, an acclimatization period of 1–2 months is needed for the microorganisms. We distinguish, therefore, between primary and ultimate biodegradation rates. Acclimatized microorganisms are utilized to remove toxic substances from soil by *in situ* application of ecotechnology.

The following rules can be used as a first rough estimation of the biodegradability of organic compounds:

1. Higher molecular weight implies reduced biodegradability.
2. Aliphatic compounds are more biodegradable than aromatic compounds.
3. Substitutions by halogens and nitro will decrease the biodegradability.
4. Introduction of double and triple bonds will generally imply increase in the biodegradability.
5. Oxygen and nitrogen bridges will decrease the biodegradability.
6. Branches are generally less biodegradable than the corresponding primary compounds.

Units

The biodegradation rate can be expressed by application of a wide range of units:

1. By the first-order reaction rate, k, see the equation above, for instance $1/24\,h$.
2. As half-life, for instance, hours or days or years.
3. Milligram per gram sludge per 24 h to indicate the biodegradation rate in a biological treatment plan, milligram per gram 24 h. Gram bacteria per gram sludge can be used to convert this unit into 4.
4. Milligram per gram bacteria per 24 h to indicate the biodegradation rate, milligram per gram 24 h.
5. $BOD_x/BOD_{infinite}$. This ratio indicates the biodegradation in oxygen demand after x days compared with the complete biological degradation.
6. BOD_x/COD. This ratio indicates the biodegradation in oxygen demand after x days compared with the complete chemical degradation expressed in oxygen demand, COD.

See also: Biogeochemical Approaches to Environmental Risk Assessment.

Further Reading

Howard PH, Boethling RS, Jarvis WF, Meyland WM, and Michalenko EM (1991) *Environmental Degradation Rates*, 725pp. Ann Arbor, MI: Lewis Publishers.

Jørgensen SE (2001) *Principles of Pollution Abatement. Pollution Abatement for the 21st Century*, 520pp. Amsterdam: Elsevier.

Jørgensen SE, Halling Sørensen B, and Mahler H (1997) *Handbook of Estimation Methods in Ecotoxicology and Environmental Chemistry*, 230pp. Ann Arbor, MI: Lewis Publishers.

Biomagnification

K G Drouillard, University of Windsor, Windsor, ON, Canada

Introduction
Definitions and Terminology Related to
   Biomagnification
Empirical Field Data Supporting Biomagnification

Mechanism of Biomagnification
Summary
Further Reading

Introduction

This article provides an overview of biomagnification as it applies to bioaccumulation of hydrophobic organic contaminants in ecosystems. While the term biomagnification has been applied to other pollutant classes including certain metals, the origins of the term, major case studies, and advancements in the mechanistic interpretation of biomagnification have largely been focused on hydrophobic organohalogen compounds. The first section of this article defines the term biomagnification, related terminology, and traces the origins of the term. The second section provides a brief overview of major empirical evidence documenting biomagnification in various food webs, and the last sections provide a summary of alternative mechanisms that have been used to explain the biomagnification phenomena.

Definitions and Terminology Related to Biomagnification

The term biomagnification has classically been defined as the condition where the contaminant concentration in an organism exceeds the contaminant concentration in its diet when the major chemical exposure route to the organism is from food. By extension the term food web biomagnification has been defined as the increase in contaminant concentration with increasing trophic status of organisms sampled from the same food web.

Biomagnification and food web biomagnification were originally coined from observations of chlorinated pesticide bioaccumulation in aquatic food webs. However, the term biomagnification (see Persistent Organic Pollutants) has been applied to other contaminants including mercury, heavy metals, and certain compounds of biogenic origin. The first demonstration of biomagnification was described for dichlorodiphenyldichloroethane (DDD), closely related to the pesticide dichlorodiphenyltrichloroethane (DDT), in Clear Lake California. Rachel Carson subsequently used the term 'biological magnifiers' in her book, *Silent Spring*, to describe how earthworms

concentrate DDT residues from soil in their bodies and transfer these residues to robins who consume them which in turn achieve even greater concentrations of the pesticides than worms. The term 'biological magnification' was later used by Woodwell to describe the 'systematic increase in DDT residues with trophic level' in his description of DDT trophodynamics in a salt marsh near Long Island, New York. Biological magnification subsequently became truncated to the commonly used term 'biomagnification' in later years.

The mechanism of biomagnification as applied to organic chemicals, particularly compounds demonstrating physical properties of low water solubility and high hydrophobicity, was intensely studied and vigorously debated in the 1980s and 1990s. During this period, biomagnification was conceptually distinguished from the process of bioconcentration which refers to chemical bioaccumulation (see Bioaccumulation) due to exposure of contaminant across respiratory exchange surfaces (i.e., gills and lungs). Research conducted in the 1960s and 1970s demonstrated how physical–chemical properties controlled environmental partitioning and diffusive flux of hydrophobic substances. Hydrophobic organic contaminants tend to distribute preferentially to organic phases which includes organic carbon of soils and sediments and lipid phases of organisms. Equilibrium partitioning theory (see Food-Web Bioaccumulation Models) was subsequently used to equate bioconcentration in animals to the equilibrium lipid/water distribution coefficient. Although equilibrium partitioning theory described laboratory bioconcentration data well and predicted laboratory bioconcentration factors (BCFs; defined as the lipid-normalized chemical concentration in the animal divided by the chemical concentration in water), it failed to fully account for elevated contaminant concentrations accumulated by upper-trophic-level animals in the field.

The failure to validate equilibrium partitioning as a theory explaining biomagnification prompted redefinition of the term, as applied to hydrophobic organic susbstances, to describe the thermodynamic context of biomagnification. Under this new definition, biomagnification refers to the condition where the chemical potential achieved in an animal's tissues exceeds the chemical potential in its food and its surrounding environment. Similarly, food web biomagnification was redefined as the increase in chemical potential of organisms with increasing organism trophic status. In practice, chemical potentials are not directly measured, but rather are compared relatively across different samples by normalizing the chemical concentration in a sample by the sample partitioning capacity for the chemical/sample matrix of interest. Since hydrophobic organic contaminants distribute primarily to neutral lipids within organisms, expression of lipid-normalized chemical concentrations have been used as surrogate measures of chemical potentials when comparing biomagnification between biological samples. Alternatively, chemical fugacity (see Food-Web Bioaccumulation Models) is used as a proxy for chemical potentials when comparing equilibration of contaminants between interacting abiotic and biotic samples. These data analysis methods apply to hydrophobic organic chemicals but do not necessarily apply to mercury or other contaminant classes which undergo biomagnification by the classic definition, but do not exhibit preferential internal distribution to lipids within animals.

The biomagnification factor (BMF) for organic contaminants is defined as the ratio of the lipid-normalized chemical concentration in the animal to the lipid-normalized chemical concentration in its diet. A BMF value greater than 1 indicates that the animal has achieved a greater chemical potential than its diet. Since organisms may include multiple food items in their diet, the BMF can also be expressed according to the weighted average lipid-normalized chemical concentrations in its various food items. Similarly, when the BCF value is shown to exceed the n-octanol/water partition coefficient (K_{OW}; a standard laboratory-measured partition coefficient used as a surrogate measure of the equilibrium lipid/water partition coefficient) this indicates that the chemical potential achieved in the animal exceeds that of water. Similar expressions can be derived for air-breathing animals by comparing the lipid-normalized concentration in the animal/air concentration ratio with the octanol/air partition coefficient.

Determination of food web biomagnification requires establishment of the trophic level of different organisms included in the sampling program. Traditionally this has been carried out using diet analysis and establishing discrete trophic steps (see **Figure 1**). More recently, emphasis has been placed on use of stable isotopes of carbon and nitrogen to define continuous trophic positions for different organisms in a sampled food web. The food web magnification factor (FWMF) has been defined as the slope generated from a regression of the logarithm of lipid-normalized chemical concentrations in biota expressed against trophic level on the independent axis.

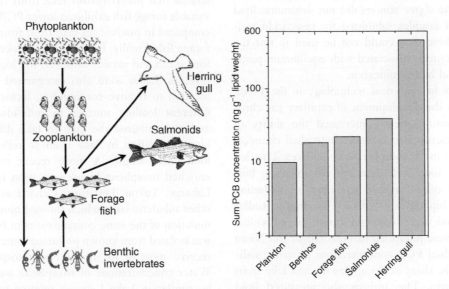

Figure 1 Food web biomagnification of polychlorinated biphenyls (PCBs) in Lake Ontario. Data for aquatic organisms collected in Lake Ontario during 1984 from Oliver BG and Niimi AJ (1988) Trophodynamic analysis of polychlorinated biphenyl congeners and other chlorinated hydrocarbons in the Lake Ontario ecosystem. *Environmental Science and Technology* 22: 388–397. Data on Herring gulls collected in Lake Ontario during 1985 from Braune BM and Norstrom RJ (1989) Dynamics of organochlorine compounds in herring gulls: III. Tissue distribution and bioaccumulation in Lake Ontario Gulls. *Environmental Toxicology and Chemistry* 8: 957–968.

Empirical Field Data Supporting Biomagnification

Hunt and Bischoff provided the first data demonstrating progressive bioaccumulation and increases in concentrations of the chlorinated insecticide DDD through an aquatic food web. DDD was applied to Clear Lake, California during three administration events in 1949, 1951, and 1957. Administrations were designed to achieve a nominal concentration of DDD in water of $50\,mg\,l^{-1}$, although reportedly water residues never achieved such levels. Mortalities of fish-eating birds were observed within months after the second and third applications, with the population of western grebes decreasing from 1000 pairs prior to DDD administration to less than 30 pairs in 1960. Food web sampling and residue analysis indicated phytoplankton achieved concentrations of approximately $5\,mg\,g^{-1}$, pelagic fish contained between 50 and $300\,mg\,g^{-1}$ and a brown bullhead contained $2500\,mg\,g^{-1}$ of DDD. DDD concentrations in western grebes and California gulls were reported at more than $2000\,mg\,g^{-1}$. Soon after, other studies began documenting DDT bioaccumulation in different food webs. Woodwell et al. determined DDT concentrations in water, soil, plankton, invertebrates, mussels, fish, and fish-eating birds in a salt marsh south of Long Island, New York. DDT increased from $0.04\,mg\,g^{-1}$ in plankton to $75\,mg\,g^{-1}$ in ring-billed gulls. Plankton concentrations were 800-fold higher than residues measured in water. Invertebrates and fish exhibited intermediate concentrations of DDT compared to plankton and birds, consistent with their trophic status. Unfortunately, the above studies did not determine lipid concentrations of samples submitted for insecticide residues. As such, these data could not be used to test the thermodynamic criteria associated with equilibrium partitioning theory and biomagnification.

Advancements in analytical technology in the 1980s, particularly with the development of capillary gas chromatography columns, greatly increased the ability of environmental scientists to examine individual chemical concentrations in more complex field matrices. This led to a plethora of food web data sets documenting biomagnification of other organic contaminants including polychlorinated biphenyls (PCBs). Two major studies documenting food web biomagnification of individual PCB congeners were published in 1988. Oliver and Niimi measured individual PCB concentrations in water, sediments, amphipods, slimy sculpin, alewife, and lake trout from Lake Ontario. The authors also measured lipid contents in the biological samples allowing them to directly test predictions of the equilibrium partitioning theory. Their data demonstrated increases in lipid-normalized PCB concentrations with increasing trophic

status (see **Figure 1**). Salmonids were also shown to have fivefold higher lipid-normalized concentrations than predicted from equilibrium partitioning theory based on residues in water. Herring gulls from the same lake collected 1 year later in another study demonstrated lipid-normalized PCB concentrations that were tenfold higher than measured for salmonids by Oliver and Niimi. Conolly and Pederson also demonstrated that the fugacity ratio of rainbow trout/water exceeded a value of 1 for PCBs having a log K_{OW} value of 4 and greater in Lake Ontario. The authors demonstrated that the trout/water fugacity ratio for PCBs increased with increasing chemical K_{OW} up to values from 10 to 100 for PCBs having log K_{OW} values of 6 and higher. The same authors also demonstrated progressive increases in the animal/water fugacity ratio for PCBs with animal trophic status in the Lake Michigan food web. PCB animal/water fugacity ratios ranged from 3 to 5 for white fish and chub occupying a trophic level of 2 and up to a value of 14 for fish occupying a trophic level of 4. Similar case studies of food web biomagnification using lipid-normalized data sets have subsequently been demonstrated in several other aquatic systems including all five Great Lakes, Lake Baikal, and in agricultural and arctic terrestrial ecosystems.

Other data sets have shown the relationship between hydrophobic organic contaminant bioaccumulation and food chain length or number of trophic steps within the system. Using data generated from the Ontario sport fish contaminant surveillance program, Rasmussen et al. examined PCB bioaccumulation in lake trout from a large number of lakes in Ontario, Canada. The authors demonstrated that planktivorous lake trout from lakes lacking suitable forage fish exhibited lower PCB bioaccumulation compared to piscivorous lake trout from lakes containing forage fish. Finally, lake trout from lakes containing both forage fish and mysids achieved the highest contaminant residues. Lakes were also categorized and analyzed by location to remove confounding factors associated with different loading sources to individual systems. The authors attributed the lake to lake differences in PCB bioaccumulation by lake trout to reflect differences in food chain length. A more recent study documented enriched toxaphene bioaccumulation in fish from Lake Labarge, Yukon Territory, Canada as contrasted with other subarctic lakes which showed much lower bioaccumulation of the same contaminants in fish. Lake Labarge was isolated from known pollutant sources and thought to receive most of its inputs via atmospheric deposition. Water concentrations of toxaphene were also found to be similar in Lake Labarge relative to the other lakes which showed lower toxaphene bioaccumulation in fish. The major difference noted for Lake Labarge lake trout, burbot, and lake whitefish was that fish from this lake were feeding at higher trophic levels as revealed both by diet analysis and trophic enrichment of stable nitrogen isotopes. This study, similar to that of Rasmussen's work provided the empirical linkage between ecosystem structure, number of trophic links, and magnitude of biomagnification realized in top predator fish.

Mechanism of Biomagnification

Dietary Exposure

A number of mechanisms have been proposed to describe the biomagnification process as it applies to persistent, hydrophobic organic compounds. The first model published to describe biomagnification of the insecticide DDT described the lipid co-assimilation mechanism. In this model, both lipids and contaminants are efficiently assimilated from food; however, a smaller fraction of lipids are retained as a result of metabolism of these nutrients to satisfy energetic requirements. Recalcitrant contaminants are retained in tissues and over time, in conjunction with the number of feeding events, magnify in concentration over that of ingested food. Under this mechanism, the maximum biomagnification potential in nondeterminant growing animals is inversely related to growth-conversion efficiency (i.e., rate of tissue growth relative to food consumption) when contaminant elimination from the animal approaches a value of 0. For determinant growers, biomagnification will continue to increase with age as a function of number of feeding events. In practice, most environmental contaminants do exhibit elimination, which will attenuate biomagnification in proportion to the magnitude of the elimination rate coefficient. In this case, the steady-state biomagnification factor will be positively related to the feeding rate of the animal and chemical assimilation efficiency from the diet and inversely proportional to the elimination rate coefficient and growth rate.

The lipid co-assimilation mechanism was also used to explain food web biomagnification. In this case, biomagnification as achieved in top predator organisms is assumed to correspond to the inverse of ecological efficiency. Thus, the low energy transfer efficiency across trophic levels (<10%) coupled with efficient chemical transfer efficiencies and high chemical retention among different organisms results in progressively increased residues through successive trophic steps. This model indicates that the number of trophic levels and trophic transfer efficiencies across each step specify the maximum contamination achieved for top predators whereas growth and elimination by individual organisms attenuates food web biomagnification.

Although lipid co-assimilation as a mechanism of bio-magnification is consistent with concepts of bioenergetics and trophodynamics, environmental chemists were quick to point out that such a mechanism is not consistent with chemical bioaccumulation through passive partitioning mechanisms as is thought to occur for hydrophobic organic chemicals. Lipids, fat-soluble vitamins, and hydrophobic organic contaminants cross biological membranes by passive diffusion. In the case of the proposed lipid co-assimilation model of biomagnification, net chemical diffusive flux would have to proceed in a direction of low concentration (ingested food and digesta of the gastrointestinal (GI) tract) to high concentration (animal tissues). The GI magnification model was subsequently developed as a competing model with lipid co-assimilation. This model is able to account for biomagnification while preserving chemical diffusion as the major mechanism of chemical flux between the organism and its gut contents.

Gastrointestinal magnification was first described in 1988. The model considers the GI tract and its contents as a separate compartment from animal tissues. The premise of the GI magnification model is that nutrient and lipid absorption occur independent of contaminants in the GI tract. The absorption of nutrients from digesta decreases both the volume and partitioning capacity of gut contents as digestion proceeds along the length of the intestines. The failure of mass balance within the GI tract compartment and loss of partitioning capacity raises the chemical potential of digesta in the GI tract above that of ingested food, providing the necessary gradient on which net diffusion can proceed from GI tract to animal even if the animal has a higher chemical potential than the food it has ingested. The animal thus equilibrates with the elevated chemical potential of its GI contents rather than its food.

Under the GI magnification model, the change in chemical partitioning capacity of feces relative to food and the volume reduction of feces produced relative to food consumed (i.e., diet absorption efficiency) provides the upper limit for the maximum biomagnification potential that can be achieved in an animal. These limits may vary according to the diet absorption efficiency for a given diet type and/or differences in digestive physiology between different species feeding on similar diets. The GI magnification model has been subject to experimental and field validation through studies that demonstrated increases in chemical potential of animal GI contents above that of ingested food. Careful laboratory measurements indicate maximum biomagnification potentials on the order of 3–12 in fish which appear to be consistent with field biomagnification factors experienced by fish. The GI magnification model has subsequently been adopted in numerous food web bioaccumulation models applied to hydrophobic organic contaminants. Food web bioaccumulation models allow consideration of relative exposures and biomagnification potentials in animals exposed to multiple diet items as defined by the diet matrix established for each species being modeled.

Recent studies have suggested amendments to the GI magnification model to account for additional physiological factors which may increase the biomagnification potential of an animal beyond diet assimilation and partitioning changes of feces relative to food. One of the simplifying assumptions applied in the original GI model solution was that the mass transport parameter describing chemical flux from gut contents to animal is equal to the mass transport parameter describing flux from animal to its gut contents. This asserts that uptake and elimination flux is diffusion-controlled and coupled throughout the length of the GI tract. However, evidence on hexachlorobenzene bioaccumulation in rats suggested that uptake occurs primarily in the upper GI tract whereas fecal elimination occurs predominately in the colon. Experimental studies on ring doves showed that PCB dietary assimilation efficiencies in ring doves during the uptake phase were higher than feces/animal exchange efficiency measured during the depuration phase. Similar observations were documented in humans.

The fat-flush model and micelle-mediated diffusion model have been suggested as potential submodels for use to augment the GI magnification model. Both the above models are used to explain the phenomena of decoupling of the site and timing of uptake and elimination processes in the GI tract. According to the fat-flush model, fatty acids assimilated by enterocytes of the intestinal mucosa of the small intestine become resynthesized into triglycerides and incorporated into growing chylomicron vesicles. This growth in lipid content of enterocytes increases the partitioning capacity of these cells and temporarily dilutes their chemical potential relative to blood and other body compartments. The lower chemical potential of enterocytes then favors chemical assimilation by diffusion. This process maximizes chemical absorption and minimizes losses of chemical from small intestine cells back to the lumen of the GI tract. When chylomicrons reach a critical size, they are released by active transport from the enterocyte into the circulatory system which again causes a temporary dilution of the blood compartment relative to other body tissues. Following the absorption of dietary lipids (and assimilated contaminant) from blood by other tissues, the chemical potential of blood once again re-equilibrates with other body compartments and becomes maximized. During the fasting state, when chemical potential in blood is highest, fecal elimination becomes more pronounced. At this time, the gut contents are found primarily in the large intestine where fecal elimination has been shown to take place. The fat-flush model therefore describes decoupling in both the site and timing of contaminant assimilation compared to fecal elimination.

The micelle-mediated diffusion model focuses on the physiological role of mixed micelles as vectors for lipid, hydrophobic vitamin and contaminant uptake in the GI tract. Mixed micelles are produced in the intestine as a result of the interaction of bile salts and fatty acids. These amphiphilic vesicles diffuse through the unstirred water layer (UWL) between the gut lumen and intestinal mucosa. Mixed micelles are capable of dissolving long-chain fatty acids, fat-soluble vitamins, as well as other hydrophobic compounds including contaminants in their interiors and transporting these compounds across the UWL. Recent physiological evidence indicates that mixed micelles are unidirectional in their movements between the lumen to enterocyte. A pH microgradient stimulates the breakup of mixed micelles at the interface of the intestinal mucosa of the small intestine. Thus, mixed micelles appear to be involved in the efficient assimilation of hydrophobic contaminants in the upper part of the digestive tract but do not facilitate elimination of chemical from enterocytes back to the lumen of the gut compartment.

Both the mixed-micelle and fat-flush models reflect extensions of the GI magnification model that bring about physiological realism to the digestive process. Current calibration of these models in birds and humans suggests that maximum biomagnification factors may be higher by a factor of 2–4 (i.e., total biomagnification factors ranging from 15 to 20 or higher) than predicted by the original GI magnification model. Calibration of the fat-flush model or mixed-micelle models in fish is yet to be completed. Further research to calibrate maximum bioaccumulation potentials in a wider variety of animal species as well as calibrated animal to gut and gut to animal transfer efficiency terms are required to substantiate these new model predictions and adopt them into food web bioaccumulation models as has been performed with the GI-magnification model.

Nondietary Mechanisms Explaining Biomagnification

The earliest critics of biomagnification identified equilibrium partitioning as the main mechanism of bioaccumulation and suggested that the phenomena of food web biomagnification could be explained primarily by differences in whole-body lipid content, and hence chemical partition capacities, of upper-trophic-level animals relative to lower-trophic-level organisms. As described above, food web bioaccumulation data sets generated in the late 1980s and 1990s provided lipid-normalized chemical concentration data and these studies were consistent with the thermodynamic definition of biomagnification. However, other alternative mechanisms of biomagnification, which do not involve special properties associated with dietary exposures, have been proposed.

Differences in spatially integrated exposures of sampled animals arising due to habitat size and/or differences in migration movements of organisms could potentially result in similar observations as biomagnification particularly in environments where the contaminant distribution in sediments and water is heterogeneous or subject to point sources. Smaller animals are likely to exhibit small spatial movements and be more reflective of contamination conditions at the local site of capture. Larger animals such as piscivorous fish may exhibit larger spatial movements and consequently integrate chemical exposures over broader spatial scales. Birds may carry residues over very long distances across their migration routes. Indeed the phenomena of biological vectors of pollution related to major spawning migrations of fish and seabirds flying to breeding sites have been recently described. Food web sampling programs for contaminants rarely consider the spatial scale of sample collections as it relates to the potential movements of organisms included within their collections. Spatial movements can confound interpretation of biomagnification factors when all animals are collected from the same location. If animals are collected at a highly contaminated site, food web biomagnification may appear attenuated as a result of high locally accumulated residues in benthic invertebrates and zooplanktons. Similarly, biomagnification trends may appear exaggerated when animals are sampled at relatively clean locations but are situated near enough hot spots that some of the larger animals are affected by the more distant contaminated areas.

Similar to the spatial scale described above, temporally explicit exposures may also confound biomagnification observations. Chemical elimination rate coefficients for negligibly biotransformed contaminants are inversely related to body size. Under conditions of pulses in environmental loadings, smaller, lower-trophic-level organisms are more likely to reflect equilibrium with water whereas larger organisms may exhibit lags in their ability to equilibrate with water during or after a pulse. For example, following reductions in water contamination after a seasonal pulse in inputs, as may be experienced during spring melt, phytoplankton and zooplankton may be capable of depurating their residues at a sufficient rate to maintain equilibration with the drop in water concentrations. However, larger fish will take longer to depurate their residues to water and will exhibit both higher concentrations and higher chemical potentials than their zooplankton/plankton counterparts. If the frequency of environmental pulse inputs is faster than the steady-state time of larger fish then this disequilibrium condition may be maintained. For some contaminants which do not achieve steady state in organisms, different animal ages also need to be considered when comparing residues among populations or between populations of species having different age structures.

Another confounding factor arises due to rapid changes in animal lipid contents, either through growth or weight loss. When an animal loses weight and lipids at a faster rate than it can lose contaminant, it concentrates its tissue residues and raises its chemical potential above its previous state even though net chemical flux proceeds in the direction of elimination. Rapid lipid depletion, and subsequent tissue concentration of contaminants, is likely to be common in animals that undergo seasonal cycles of weight gain and lipid loss or in animals that exhibit bioenergetic bottlenecks at critical times in their life history. Such observations were reported in depuration experiments involving birds and fish. In the case of birds, contaminant residues were found to become concentrated in blood following weight losses experienced by the animals during spring warming. The opposite was noted for warm-water fish, the yellow perch, where winter weight losses due to prolonged fasting caused an increase in chemical potential in animal tissues despite the fact that the study was measuring chemical elimination. Other examples documenting rapid weight and lipid loss during specific life-history points and subsequent tissue magnification of PCBs include metamorphosis in amphibians and pipping (hatching) of chicks.

The opposite condition of solvent depletion occurs during rapid growth. Extremely rapid turnover times demonstrated by algae during peaks in primary production result in growth rates that are faster than chemical-uptake coefficients. This causes phytoplankton to exhibit chemical potentials that are lower, and less than equilibrium, compared to water. Rapid growth dilution as experienced in juvenile animals will also reduce biomagnification factors due to high growth-conversion efficiencies experienced during these life stages.

Summary

Biomagnification is a well-documented phenomenon where persistent hydrophobic organic contaminant concentrations in an animal become elevated over and above its food. This increase in contaminant concentration propagates through successive trophic steps in a food web. Later definitions of biomagnification and food web biomagnification have imposed thermodynamic criteria specifying that biomagnification reflects a nonequilibrium process in which the chemical potential in an animal is elevated above that of its diet and environment. The GI magnification model and the new amendments to this framework explain how exposures through the diet can raise the chemical potential of the animal above that of its food and environment. The above mechanisms apply specifically to hydrophobic organic contaminants. However, other contaminants such as mercury are assimilated and distributed through tissues by different processes. Thus, while mercury conforms to the classic definition of biomagnification and food web biomagnification it is difficult to evaluate the behavior of this compound in the context of the thermodynamic definition of biomagnification. Further understanding of uptake, assimilation, and elimination mechanisms and the free energy relationships associated with these processes are needed to provide a unifying theory of biomagnification across other contaminant types.

See also: Bioavailability; Exposure and Exposure Assessment.

Further Reading

Blais JM, Macdonald RW, Mackay D, *et al.* (2007) Biologically mediated transport of contaminants to aquatic systems. *Environmental Science and Technology* 41: 1075–1084.

Braune BM and Norstrom RJ (1989) Dynamics of organochlorine compounds in herring gulls: III. Tissue distribution and bioaccumulation in Lake Ontario Gulls. *Environmental Toxicology and Chemistry* 8: 957–968.

Carson R (1962) *Silent Spring*. Greenwhich, CN: Fawcett Publications.

Connell DE (1989) Biomagnification by aquatic organisms – A proposal. *Chemosphere* 19: 1573–1584.

Connolly JP and Pedersen CJ (1988) A thermodynamic based evaluation of organic chemical accumulation in aquatic organisms. *Environmental Science and Technology* 22: 99–103.

Gobas FAPC, Muir DCG, and Mackay D (1988) Dynamics of dietary bioaccumulation and faecal elimination of hydrophobic organic chemicals in fish. *Chemosphere* 17: 943–962.

Gobas FAPC, Wilcockson JB, Russell RW, and Haffner GD (1999) Mechanism of biomagnification in fish under laboratory and field conditions. *Environmental Science and Technology* 33: 133–141.

Hamelink JL, Waybrant RC, and Ball C (1971) A proposal: Exchange equilibria control the degree chlorinated hydrocarbons are biologically magnified in lentic environments. *Transactions of the American Fish Society* 100: 207–214.

Harrison HL, Loucks OL, Mitchell JW, *et al.* (1970) Systems studies of DDT transport. *Science* 170: 503–508.

Hunt EG and Bischoff AI (1960) Inimical effects on wildlife of periodic DDD applications to Clear Lake. *California Fish and Game* 46: 91–106.

Kidd KA, Schindler DW, Muir DCG, Lockhart WL, and Hesslein RH (1995) High-concentrations of toxaphene in fishes from a sub-arctic lake. *Science* 269: 240–242.

Mackay D (1981) Calculating fugacity. *Environmental Science and Technology* 16: 274–278.

Oliver BG and Niimi AJ (1988) Trophodynamic analysis of polychlorinated biphenyl congeners and other chlorinated hydrocarbons in the Lake Ontario ecosystem. *Environmental Science and Technology* 22: 388–397.

Rasmussen JB, Rowan DJ, Lean DRS, and Carey JH (1990) Food-chain structure in Ontario lakes determines PCB levels in lake trout (*Salvelinus namaycush*) and other pelagic fish. *Canadian Journal of Fisheries and Aquatic Sciences* 47: 2030–2038.

Schlummer MG, Moser GA, and McLachlan MS (1998) Digestive tract absorption of PCDD/Fs, PCBs and HCB in humans: Mass balances and mechanistic considerations. *Toxicology and Applied Pharmacology* 152: 128–137.

Swackhamer DL and Skoglund RS (1993) Bioaccumulation of PCBs by algae: Kinetics versus equilibrium. *Environmental Toxicology and Chemistry* 12: 831–838.

Woodwell GM, Wurster CF, Jr., and Isaacson PA (1967) DDT residues in an east coast estuary: A case of biological concentration of a persistent insecticide. *Science* 156: 821–824.

Ecological Catastrophe

W Naito, National Institute of Advanced Industrial Science and Technology, Tsukuba, Japan

Introduction	Conclusions
Ecological Impact of Major Chemical Pollutions	Further Reading
Lessons Learned	

Introduction

'Ecological catastrophe' is defined as any disastrous event upon natural populations, communities, or ecosystems. Ecological catastrophes are caused by various stressors such as exposure to toxic substances, desertification, fire, and flood. Among various stressors this article presents the link between catastrophic events related to the release of toxic substances and ecological impacts. Environmental pollution and chemical accidents involving the massive amount of release of toxic chemicals have the potential to cause devastating ecological impacts (ecological

catastrophe). This article focuses especially on the historical environmental pollutions or chemical accidents involving the release of massive amount of toxic substances that have the potential to cause ecological catastrophes and presents an overview of the ecological consequences actually reported in such events.

In the light of history, several events in areas such as Minamata (Japan), Bhopal (India), Seveso (Italy), Chernobyl (Ukraine), Prince William Sound (Alaska), Sandoz (Switzerland), Aznalcóllar (Spain), and lately Jinlin (China) have highlighted such potential. The ecological impact observed in these areas was unique and differed greatly in the chemicals involved, the area covered, and the time periods affected. For example, in Minamata, there was chronic pollution of a rich fishery bay by the release of mercury from a chemical plant; in Bhopal and Seveso there was extensive contamination of areas with highly toxic chemicals due to an accident in chemical plants; in Chernobyl there was the most significant unintentional release of radiation into the environment due to the explosions at the nuclear power plant; in Prince William Sound, there were the significant marine oil spills due to the oil tanker Exxon Valdez accident; in Sandoz, there was pollution of the Lower Rhine River as far downstream as the Netherlands with a pesticides mixture occurring as a result of a fire at a chemical manufacturing plant; at Aznalcóllar, a part of Doñana National Park, 60 km downstream, was polluted with acidic water and a metal-rich sludge; and in Jilin, the Songhua River was polluted by the release of 100 of benzene and other compounds by an explosion at a petrochemical plant. Regarding the major chemical pollutions, in many cases, the effects on human health have been a primary concern and the related literature has focused primarily on human health issues. This article examines the various aspects of ecological impacts caused by the release of toxic substances in certain major environmental catastrophic events involving Minamata, Bhopal, Seveso, Chernobyl, Prince William Sound, Sandoz, and Aznalcóllar.

Ecological Impact of Major Chemical Pollutions

Release of Mercury at Minamata, Japan

One of the earliest and a significant environmental pollution occurred in the 1950s at Minamata Bay in southwestern Kyushu, Japan. In 1956, the first Minamata disease patient was reported initially as suffering from nervous symptoms of an unknown cause in Minamata City, which is located along the Yatsushiro sea coast in Kumamoto Prefecture, Japan. Minamata disease resulted from the release of mercury from an acetaldehyde plant into Minamata Bay. In the plant, acetaldehyde had been synthesized by the hydration of acetylene, and mercury oxide dissolved in sulfuric acid was used as a catalyst. A part of the inorganic mercury used as a catalyst had changed into methylmercury by a side reaction in the plant. Wastewater containing both inorganic mercury and methylmercury had been discharged into the bay, and methylmercury had accumulated in the fish. Minamata disease was caused by the consumption of fish or shellfish contaminated by methylmercury from the bay. The signs and symptoms of Minamata disease, a disorder of the central nervous system, in humans include sensory disturbances of the extremities, loss of coordination, and bilateral

concentric contraction of the visual field. As of March 1997, more than 2900 Minamata disease patients were certified, of whom about 2300 victims were located in the Yatsushiro Sea area. The ecological impacts in the Minamata area caused by massive mercury pollution are described below.

Minamata Bay received 70–150 t or more of inorganic mercury. Methylation takes place in sediment and in fish. Inorganic mercury in the environment can migrate into fish and accumulate as methylmercury. Almost all of the mercury found in the edible part of fish is methylmercury. In addition, it is estimated that approximately 600 kg of methylmercury had been discharged into the bay from an acetaldehyde process at the plant between 1938 and 1965. The transition of methylmercury emissions from the acetaldehyde plant in Minamata Bay as well as the production of acetaldehyde are shown in **Figure 1**. The release of methylmercury into the bay has markedly continued to increase after 1950, along with an increase in acetaldehyde production, but then began to decrease after 1960. These findings suggest that the sharp increase in methylmercury, probably along with an increase in inorganic mercury emissions into the bay, resulted in ecological damage to the bay as well as methylmercury poisoning to humans during the 1950s. The trend of methylmercury release correlated well with the mercury concentrations in bivalves and some fish in the bay (described later), indicating that the reduction in mercury discharges into the bay has led to a reduction in contamination levels in some species.

Once mercury was discharged into the bay it became concentrated to high levels in fish and filter-feeding shellfish by several routes such as bioconcentration and food chain biomagnification. Even before Minamata disease was first recognized by the public in 1956, the local residents had noticed a number of surprising incidents happening in animals and fish. The 'cat dancing disease,' as it was known by the local residents, was particularly shocking. Around 1953, numerous cats, as well as some

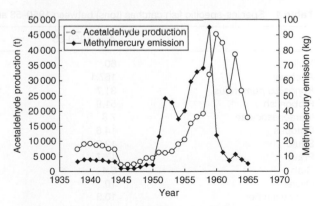

Figure 1 Transition of acetaldehyde production and the release of methylmercury to Minamata Bay from the factory. Data from Nishimura H and Okamoto T (2001) *Science of Minamata Disease*. Tokyo: Nippon Hyoron-sha (in Japanese).

pigs and dogs, went mad and died. Moreover, fish were found floating on the sea surface and some birds such as the crow and grebe suddenly fell into the sea and drowned.

Specific and qualitative testimonies regarding some of the mercury poisoning observed by local fishermen were summarized in literature. Around 1950, flounder, octopus, and sea bass were floating on the sea surface in the bay, and decrepit fish were caught by hand, indicating severely diminished locomotor activity and escape behavior. Seaweeds began to turn white, and their biomass decreased. Between 1951 and 1952, a wide variety of fish such as sea bream, rockfish, and gilthead began to be found floating on the sea surface; no small prawns were caught, and seaweed production decreased to one-third of previous levels. Furthermore, clam and mate shell decreased in the bay. Between 1953 and 1954, benthic fish and organisms were also found floating on the sea surface and shellfish near the drain of the factory in the bay were killed. Moreover, between 1955 and 1957, the numbers of fish floating on the sea surface in the bay were maximized and began to spread out of the bay. Shellfish inhabiting the bay side of an island completely died.

That a serious disaster occurred in the fish is reflected in the statistics regarding fish catches from Minamata Bay. Fish catches from some fisheries cooperative associations in Minamata are presented in **Table 1**. As seen in the table, the number of fish catches between 1950 and 1956 drastically decreased. Total fish catches in 1956 had decreased to almost one-fifth of that in 1950–53. From 1950–53 to 1956, the catches of mullet, ribbonfish, octopus, squid, and crab had decreased to almost one-third, whereas those of anchovy, prawn, and oyster had decreased to one-fifth. The number of species living in or near sediment decreased the most, implying that the conditions of sediment in the bay had markedly deteriorated and led to significant ecological damage such as a decrease in biomass. The decrease in catches for

migratory species such as anchovy, however, does not necessarily indicate a decrease in population caused by methylmercury. The population size of migratory fish varies according to the year. One reason for the drastic decrease in catches of anchovy, however, could be a reduction in the amount of zooplankton, which resulted from exposure to mercury in the bay.

Ecological damage such as the change in seaweed color and floating fish could be caused not only by methylmercury but also, to some extent, by hydrogen sulfide or an anoxic environment resulting from organics discharged from an acetyl cellulose factory located in the same area. An investigation conducted in 1957 showed that the organic matter content in sediment in the Minamata Bay area was very high and correlated well to sulfide. More than $1 \, mg \, g^{-1}$ dry weight of sulfide concentration was recorded in sediment at 1500 m within the drainage of the factory, and $0.2–0.5 \, mg \, g^{-1}$ dry weight was observed at the center of the bay in 1957.

Examining the transition of mercury residues in aquatic organisms from Minamata Bay could provide valuable information for understanding the significance of the pollution and its recovery trend. The concentrations of mercury in bivalve, *Hormomya mutabilis*, between 1959 and 1966 are shown in **Figure 2**. The highest mercury concentrations were observed in bivalve collected between 1958 and 1959, and the concentrations decreased to approximately 10 ppm after 1961. This decreasing trend of mercury levels in bivalve suggests that a drastic change in the water quality of the bay had occurred. In fact, the discharge of methylmercury decreased markedly during this period (see **Figure 1**). The sharp reduction in mercury levels observed in 1959 corresponded to the temporal stop of wastewater release from the plant to the bay between May and July of that year.

An investigation of the levels of localized mercury contamination in the adductor muscle of the mussel

Table 1   Species-specific fish catches (tons) between 1950–53 and 1956 in Minamata

	Avg. 1950–53	1954	1955	1956
Mullet	60	54.4	38	22.1
Anchovy	167.1	101.5	47	26
Konosirus pumctatus	31.7	6.8	6.1	1.2
Ribbonfish	51.9	29.7	24.5	20.2
Muraenesocidae	7.8	6.5	4.7	2.3
Octopus	14.6	9.1	7.6	4.4
Squid	12.3	9.4	5.6	3.9
Prawn	17.7	9.1	5.8	3.5
Crab	5.4	6.4	3.8	2.2
Oyster	10	7.4	5.4	1.6
Sea cucumber	10.3	8.6	6.1	2
Others	70.5	30.4	17.7	6
Total	459.3	279.3	172.3	95.4

Data from Nishimura H and Okamoto T (2001) *Science of Minamata Disease*. Tokyo: Nippon Hyoron-sha (in Japanese).

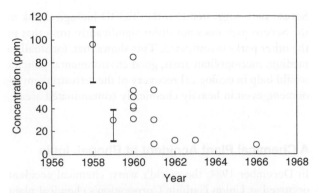

Figure 2 The mercury content in bibalve, *H. mutabilis*, from Minamata Bay. Data from Nishimura H and Okamoto T (2001) *Science of Minamata Disease*. Tokyo: Nippon Hyoron-sha (in Japanese).

Mytilus galloprovincialis collected from four sites around Minamata City from 1993 to 1995 showed that the mercury concentrations were higher (0.026–0.121 ppm, $n = 135$) at sites near the main fallout of wastewater from the chemical plant in Minamata Bay than at sites 1–5 km from the former sites (0.006–0.028 ppm, $n = 52$).

The mercury concentrations in tissue of floating fish from the bay caught in 1959 ranged from 1.09 to 52.3 ppm, and the concentrations in the guts were significantly higher than those in muscle. The mercury concentrations of fish caught from the bay were 7–23 ppm in 1961, 0.4–5 ppm in 1963, 0.4–12 ppm in 1965, 0.2–0.4 ppm in 1966, and 0.1–1.3 ppm in 1970. The mercury concentrations in fish have decreased markedly since 1966, and the concentrations of benthic fish are relatively higher than those of surface fish. There are close correlations between the mercury contents of zooplankton and suspended particulate matter, and of sediments and fish muscle, suggesting a pathway from sediment to fish by way of suspended matter and zooplankton. The mercury concentrations declined in fishes inhabiting affected surface waters after industrial discharges of mercury were reduced, although concentrations in fish have remained unacceptably high in some areas. The mercury concentrations in fishes (87 species) from the bay ranged from 0.01 to 1.74 ppm, and 16 species of fish contained greater than 0.4 ppm of mercury in 1989. Monitoring of the mercury concentrations of fish and other marine organisms continues even today in the area around Minamata Bay.

Laboratory studies indicate that fish with residues in muscle of 5–8 ppm in walleyes and 10–20 ppm in salmonids are sublethally affected. Symptoms of long-term dietary exposure to methylmercury include incoordination, inability to feed, diminished responsiveness, and starvation. The mercury concentrations of many fish collected in the bay until the early 1960s were above or similar to the levels at which some fish exhibit symptoms

of methylmercury intoxication, indicating that fishes in Minamata Bay were poisoned by methylmercury.

The total mercury content in the feathers of 95 stuffed, wild birds collected between 1955 and 1980 from the Minamata Bay area had been measured. The results indicate a strong relationship between the food of birds from Minamata and the mercury content in feathers; the content is highest in fish-eating seabirds and lowest in herbivorous waterfowl. A similar trend has been seen in birds collected from China and Korea, although the concentrations are significantly lower. Relatively high mercury levels were observed in the feather samples from Minamata till the late 1970s, although the discharge of wastewater from the factory stopped in 1968.

In order to deal with the massive mercury pollution in Minamata Bay, several measures have been taken. First of all, regarding the acetaldehyde plant, industrial wastewater containing methylmercury has not been released outside of the plant since 1966 due to the completion of a total circulation system, and the production of acetaldehyde was stopped in 1968.

Even after the release of methylmercury stopped, the bottom sediment in the bay contained considerable levels of mercury. Kumamoto Prefecture removed $1\,500\,000\,m^3$ of bottom sediment that contained greater than the regulatory standard (25 ppm of total mercury) by means of dredging and landfill between 1974 and 1990. This project created 58 ha of landfill area at a total cost of approximately 48 billion yen, with the company being responsible for a burden of 30.5 billion yen.

The dredging changed the pattern of mercury movement in the surface sediments of Yatsushiro Sea, and the mercury concentrations in the sediment in the bay were reduced markedly after the dredging. In the postexamination to assess the effects of removal, no more than 20 ppm of mercury was observed from the collected sediment samples in Minamata Bay. One prediction suggests a complete restoration of the surrounding marine environment, Yatsushiro Sea, by the year 2011 if the decrease trend observed between 1985 and 1990 continues.

A Chemical Plant Accident at Seveso, Italy

In July 1976, an accident occurred at the Industrie Chimiche Meda Societa Azionaria (ICMESA) chemical plant near Seveso, Italy, and 2,3,7,8-tetrachlorodibenzo-*p*-dioxin (TCDD) was released into the environment. The Seveso accident was caused by a runaway exothermic process in a trichlorophenol (TCP) batch reactor. The high temperatures led to a synthesis of TCDD, which was subsequently released into the atmosphere through a vent designed to relieve excessive pressure buildup in the reactor. The explosion released a toxic vapor cloud containing 1500 kg of various chemicals and upto a possible 30 kg of TCDD. The atmospheric deposition of the

TCDD contaminated a total area of 18 km^2, with levels in the most contaminated area around the factory reaching 20 mg m^{-2}. Impacts on human health were of primary concern, with much of the literature regarding this incident being related to human health issues. Ecological damage caused by the Seveso accident is considered below.

After the Seveso accident, thousands of small domestic animals died within a few weeks. Animal mortality and pathology had been observed after the accident and TCDD tissue levels found in animals from the Seveso area had been reported. At the time of the accident, more than 80 000 domestic animals, mainly rabbits and poultry, were present in areas that were subsequently contaminated. Most of the domestic animals were fed on fodder harvested close to or around the farm. The deaths started some days after the incident and increased markedly within the first 2 weeks, then decreasing in subsequent months. Massive mortality was observed among rabbits with 32% dying in the most contaminated zone, and 8.8% and 6.8% mortality was observed in the intermediate- and low-contaminated areas, respectively. A higher mortality was noted on the farms where animals were fed fodder obtained from the contaminated area. TCDD soil levels in the most-, intermediate-, and less-contaminated areas were up to 5477, 44, and 5 µg m^{-2}, respectively. Shortly after the accident, autopsies were performed on domestic animals from farms of the Seveso area, and TCDD tissue levels were determined for them. Of the 309 rabbits analyzed from the TCDD-contaminated area, 203 were TCDD-positive. Ninety-seven percent of the analyzed rabbits were positive with a range from 3.7 to 633 ng TCDD g^{-1} liver and a mean of 84.9 ng TCDD g^{-1} liver in the most-contaminated area. Autopsies showed various pathological signs such as hepatic lesions and hemorrhage.

The most-contaminated area was artificially reconstructed and transformed into a wood composed mainly of oaks with some scattered green fields and some bushy areas, the Bosco delle Querce urban park. The fauna of the Bosco delle Querce in comparison with 11 other urban and suburban parks in the Lombardy region have been analyzed. According to the analyses, a 4-year monitoring survey revealed that the present ecological and biological index of the park shows full ecological recovery as an urban park. Biocoenosis is well composed, and colonization of annelids, insects, amphibians, reptiles, birds, and mammals has occurred. The results of an action study of xenoestrogen-like molecules (i.e., gametogenesis and the gross morphology of genital organs in rabbits and house mice) show no signs of TCDD effects. The TCDD levels in liver of the animals from the Seveso park and controls were found to be 4.3 ± 0.4 and 7.2 ± 2.9 pg g^{-1} fat for rabbits, and 29.5 ± 13.8 and 41.3 ± 9.5 pg g^{-1} fat for mice, respectively. The contamination levels found in the

Seveso park suggested that the TCDD biological risk in the Seveso park does not differ significantly from that in the other parks investigated. This shows that, for small to medium metropolitan areas, good environmental policies would help in ecological recovery of the terrestrial environment, even in heavily chemically contaminated areas.

A Chemical Plant Accident at Bhopal, India

In December 1984, the world's worst chemical accident occurred at Union Carbide Corporation's chemical plant at Bhopal in the state of Madhya Pradesh, India. More than 40 t of methyl isocyanate (MIC) gas leaked from the tank of the plant and was released into the atmosphere. The immediate cause was a buildup of pressure in the tank due to an exothermic reaction caused by water in the tank. This pressure caused the safety valve to rupture and the gas to escape. This accident killed at least 3000 men, women, and children after breathing the lethal gases and caused significant morbidity and premature death for many thousands more. Extensive literature reports are available relating to the Bhopal accident, but only environmental and ecological issues are considered here for the purpose of this article.

Unlike TCDD in Seveso, MIC is highly reactive and rapidly hydrolyzed, so the initial effects of MIC were much more devastating than those of TCDD. The rapid degradation of MIC, unlike that of TCDD, however, meant that the long-term environmental impact of the accident was less serious than that at Seveso. The biodegradation of MIC and its anticholinesterase activities in different tissues of various species of fish have been studied and the concentrations of the degradation product of MIC, methylamine, in the water of the lakes from Bhopal over a period of 6 months after 3 weeks of MIC leakage into the environment have been estimated. There are two lakes in Bhopal. One of them, the Upper Lake, is the water source for the city and is far away from the accident site, whereas the Lower Lake is very close to the accident site and is completely eutrophic. The concentrations of methylamine in the Lower Lake decreased from the first sampling (23–30 December 1984) to the second sampling (25–26 February 1985), and an increase in concentrations with depth was observed. Methylamine was not detected during the third sampling (1–2 May 1985). Methylamine was detected in the fish samples of both lakes, and the concentrations in *Puntius ticto* from the Lower Lake were fairly high (9.2 µg kg^{-1} wet weight) at the first sampling, and methylamine was not detected during the third sampling. No methylamine was detected in other fish species from the Lower Lake. In the first sampling, the acetylcholinesterase activities of different tissues of fishes from the Lower Lake were significantly lower than those from the Upper Lake, implying the effects of MIC on their environment, but no clear trend in inhibition was determined.

Various aspects of the Bhopal accident, including its effects on animals and plants, have been reported. The animal death toll caused by MIC in this accident was massive, numbering over 1000 domestic animals, including 240 cows, 280 buffaloes, 18 bullocks, 84 calves, 288 goats, 60 pigs, 12 horses, 99 dogs, 2 cats, and 3 chicken. More than 7000 animals were reported to have some symptoms of varying degrees. Autopsies of the animals showed swollen livers and lymph glands, bloated digestive tracts, enlarged blood vessels or edema, and congestion of the heart and kidney. As for the damage to plants, broadleaf trees such as *Azadirachta indica* and *Ficus religiosa* showed total defoliation within 1 km of the accident site. Plants with thicker leaves and shrubs were less affected. Reaction with the leaf surface was the main cause of damage to the trees, whereas stems or other hardy parts of the tree were not affected.

An Explosion at the Chernobyl Nuclear Power Plant, Ukraine

In April 1986, the Chernobyl nuclear power plant located near Pripyat in Ukraine exploded. This accident was the most severe in the history of nuclear power industry, resulting in a huge release of radionuclides over large areas of Belarus, Ukraine, and the Russian Federation and changing the lives of more than 4 million people living in those areas. The lighter radioactive particles and mixture of gases released by the plant were carried to all countries in the Northern Hemisphere, deposited on the ground and on surface waters. For the purpose of this article, the environmental consequences at the time of the disaster and those still continuing today are described here.

The Chernobyl disaster has both short-term and long-term impacts in terms of radionuclide effects on environment. During weeks and months after the accident, levels of radioactivity in drinking water caused concern in the most affected areas of Ukraine. The initial levels were caused primarily by direct deposition of radionuclides on the surface of rivers and lakes. After this initial period, radioactivity in aquatic systems was generally below drinking water guideline limits. Bioaccumulation of radioactivity in fish showed the concentrations that were significantly above the permissible levels for consumption in the most affected areas. In some 'closed' lakes in Ukraine, Belarus, and the Russian Federation, these problems continue even today.

The radiation-induced acute effects on plants and animals were observed in the highly contaminated areas. No radiation-induced acute effects in plants and animals have been reported outside the 'exclusion zone'. After the disaster, 4 km^2 of pine forest in the immediate vicinity of

the nuclear plant turned ginger brown and died. In the worst-contaminated areas, some animals also died or stopped reproducing. Some domestic animals left on an island in the Pripyat River 6 km from the power plant died when their thyroid glands were destroyed by high levels of radiation doses.

During the first few years after the accident, the impact of irradiation in both somatic and germ cells have been observed in plants and animals of the exclusion zone. The relationship between the observed cytogenetic anomalies in somatic cells and ecological significance is now known.

Following the natural reduction of exposure levels due to decay of short-lived isotopes and migration, wildlife populations have been recovering from the acute radiation-induced effects. Population viability of plants and animals has recovered substantially due to the combined effects of reproduction and immigration from less-affected areas.

In the years since the disaster, the recovery of biota in the exclusion zone abandoned by humans has facilitated these areas become a haven for wildlife. As a result, populations of many species of wild animals and birds, which were never seen in the area prior to the disaster, are now plentiful and it has become a unique sanctuary for biodiversity, due to the absence of human activities.

A Fire at the Sandoz Agrochemical Warehouse, Switzerland

In November 1986, a fire occurred at the Sandoz agrochemical warehouse at Schweizerhalle (near Basel) in Switzerland. At the time of the fire, the warehouse contained *c.* 983 t of agrochemical products such as organophosphates, chlorinated organic compounds, and organic mercury compounds as well as 354 t of formulation auxiliaries and other chemicals. A majority of the stored chemicals were destroyed in the fire, but large quantities were released into the atmosphere, into the Rhine River through runoff of the fire fighting water (10 000–15 000 m^3), and into the soil and groundwater at the site.

The amount of chemicals released into the Rhine was not known but was estimated to be between 1% and 3% of the stored chemicals. The estimated discharge of the dominant compounds was 3000–8900 kg for disulfoton, 1200–3900 kg for thiometon, 160–1900 kg for propethamphos, 50–290 kg for parathion, 2.5–300 kg for fenitrothion, and 250–1900 kg for oxadixyl. The concentrations of chemicals in the Rhine River were measured on the day of the accident at Village Neuf, 173 km downstream of the Sandoz chemical warehouse. The concentrations of disulfoton, thiometon, propethamphos, parathion, and oxadixyl were present at 600, 500, 100, 200, and 80 µg l^{-1}, respectively. A comparison of water concentrations of the EC50s and LC50s suggests that certain chemicals could have contributed to the damage to biota, and cumulative or possibly even synergistic toxic effects resulted from the mixture of pesticides.

After the accident, the aquatic life in the Rhine was greatly damaged up to several hundred kilometers downstream of the spill site. The toxic plume destroyed the entire eel population (approximately 200 t) up to 400 km downstream of the incident along with inflicting severe damage to other fish species such as grayling, trout, and their food organisms. Populations of aquatic invertebrates were initially devastated, and the populations of a range of invertebrates were greatly reduced. Various chironomid populations were among the first to recolonize, followed by the recovery of caddis fly. It seems that for chironomid and caddis fly populations, no more than one generation was required for recovery and a second generation for complete recovery.

That the damage caused by the spill to the Rhine was not more extensive may be explained by the chronic contamination that had already eliminated many sensitive species. With regard to the ecotoxicological aspects of the spill, up to 400 km from the spill site, grayling and trout still died at concentrations 1000 times less than the LC50s of individual organophosphate compounds. It was probably the mixture of chemicals that was responsible for the fish kills, rather than the effect of any given compound. Some researchers concluded, in their analysis of the chemical spill into the Rhine River, that acute toxic effects occur at concentrations much smaller than LC50 values suggest and synergistic effects of the substances involved must be considered, as well as the fact that they added to the existing chronic pollution of the river water. They also postulated that environmental catastrophes, as long as they are not too big and do not occur too often, are less harmful on the whole than is the chronic intoxication of ecosystems.

As for lessons to be learned from the accident, new regulations have been established for large chemical plants requiring holding ponds to be built to retain fire water runoff.

The Exxon Valdez Oil Spill at, Prince William Sound, Alaska

In March 1989, the oil tanker Exxon Valdez ran aground on Blight Reef in northern Prince William Sound in Alaska and released approximately 42 million liters of crude oil into the sea and caused an ecological catastrophe. The oil spills of the massive amount contaminated at least 1990 km of the pristine shoreline and significantly raised wildlife mortality.

After the oil spill, short-term effects of wildlife were comprehensively studied. Marine mammals and seabirds were significantly affected due to their routine contact with the sea surface. The spilled oil on the sea surface puncture the fur and feather isolation coats present in these animals, and may lead to hypothermia, as well as

smothering, drowning, or ingestion of hazardous hydrocarbons. Thousands of animals were killed in the first days after the oil spill including 250 000 seabirds, between 1000 and 2800 sea otters, and 300 harbor seals. Destruction of billions of salmon and herring eggs were also reported. Due to thorough cleanup efforts, visual evidence of the oil spill hardly remained in areas just a year later.

In the long term, the consequences of the oil spill have been reported and reductions in population have been seen in various animals. The chronic exposure enhanced death rate in the following years in fish, sea otters, and ducks. After the spill, fish embryos and larvae were chronically exposed to partially weathered oil in dispersed forms that accelerated dissolution of multiringed polycyclic aromatic hydrocarbons (PAHs) which are toxic to pink salmon eggs at as low as 1 ppb. Many seabirds showed evidence of persistent exposure to residual oil after the spill, and some populations did not return to their pre-oil-spill population size. Some marine birds that forage on littoral benthic invertebrates showed induction of the CYP1A detoxification enzyme in the late 1990s. Otters born after the oil spill experienced the exposure to contaminated sediment and ingestion of contaminated benthic invertebrates. At heavily oiled areas, sea otters have remained at half the estimated pre-spill numbers with no recovery initiated by 2000, whereas population just doubled in the period 1995–98 in un-oiled areas. Higher levels of the CYP1A have been observed in individuals from heavily oiled areas in 1996–98.

Cascades of indirect effects were also present after the oil spill, where indirect interactions lengthened the recovery process on rocky shorelines for a decade or more. For example, dramatic initial loss of cover habitat, the rockweed, triggered a cascade of indirect effects, losses of important grazers and promoted blooms of unwanted ephemeral green algae and opportunistic barnacles.

A large oil spill that happened in the Arctic shows the contributions of delayed, chronic, and indirect effects of oil contamination in the marine environment. This suggests the importance of the inclusion of long-term effects of a petroleum activity when ecological impacts of such events need to be assessed.

Break of a Holding Lagoon of a Zinc Mine at Aznalcóllar

In April 1998, a holding lagoon of the Los Frailes zinc mine containing acid waste from the processing of pyrite ore collapsed and released 5 million m^3 of highly polluting sludge, rich in zinc (0.8% dry weight) and lead (1.2% dry weight) and arsenic (0.6% dry weight), among other metals and acidic water, with a pH of 2, into the Rio Guadiamar, SW Spain. The Rio Guadiamar ecosystem encompasses the World Heritage Site of Doñana National and Natural

Parks, one of the most important bird-breeding and over-wintering area in Western Europe. Following the accident, sludge and acid water moved into the Guadiamar River and reached 45 km further south to the edges of the Doñana National Park. The area affected by the sludge encompassed 4400 ha of crops, pasture land, and woodland.

The acid water lowered the pH of the wetland from 8.5 to 4.5 and caused massive metal contamination of the wetland, killing most of the fish and invertebrates in its path. In open water, Zn levels were recorded up to 270 mg l^{-1}, Pb up to 2.5 mg l^{-1}, and As up to 0.011 mg l^{-1} at the border of the Natural National Park, which has important white stork and black kite colonies. Zinc levels in solutions for the contaminated reaches were well above the LC50s for aquatic life. The levels of cadmium and lead were within the range of concentrations for the LC50s for a wide range of species for all trophic levels. These imply that the cumulative effects of the high levels of metals could have a large-scale impact on local ecosystems. The sediment contamination levels of cadmium, lead, and zinc found around the parks greatly exceeded the levels recorded previously, and was in cases higher than levels observed around mining areas.

This accidental spill resulted in catastrophic damage. Considerable fish and invertebrate populations disappeared, and habitat for waterfowl was destroyed. For aquatic life, burial, blows, gill blocking, and drastic changes in water properties were the main causes of animal death, with 37 t of dead fish being collected in the month following the accident. In the mud-polluted watercourses, all living crabs and shellfish disappeared. After the spill, the metal levels in the blood, liver, and eggs of birds in Doñana appeared to be elevated in relation to uncontaminated areas. The chemical analyses of bone and liver samples of waterfowl for As, Pb, Cu, Zn, and Se showed that metal concentrations were elevated in certain individuals, but they did not reach levels widely considered to be toxic. Due to the lack of detailed historical monitoring information in relation to metal levels, developing an accurate understanding of the impact of the spill on the ecosystem would be difficult. Long-term monitoring of the habitats and species combined with contaminant levels in the area will provide valuable information for understanding ecotoxicological impacts on ecosystems.

Lessons Learned

The major chemical pollutions or accidents presented in this article do not occur often, but the damage can be enormous when they do. Assessing the environmental impact of these types of events is difficult and complex due to the wide range of potentially toxic substances that may be released and the high degree of variability found in natural environment. As illustrated in the case of Minamata, Sandoz, and Aznalcóllar, not just a single substance but also chemical mixtures or environmental conditions such as anoxic sediment can determine the scale of the ecological damage. In many cases, neither ecological nor toxicological data are available. Interactions among chemical, environmental, and biological factors will also influence the scale of ecological damage. As illustrated by the Minamata, Exxon Valdez, and Aznalcóllar cases, the effects of pollution or accidents on ecosystems would be delayed and prolonged. The consequences of such disastrous events suggest the importance of the inclusion for long-term and indirect effects in addition to short-term effects when assessing or predicting ecological impacts of such catastrophic events.

The aftermath of such catastrophic events have represented a valuable opportunity to learn from misfortune. The scientific investigation of the aftermath in those highly contaminated areas can help us understand the environment, how it responds to catastrophic events, and how it recovers from the catastrophic events. As illustrated by Minamata, integrating a variety of patchy data regarding chemical release, environmental monitoring, and fishery statistics would provide insight into interpreting damage to the environment caused by chemical pollution. Long-term monitoring of the contaminated environment including residue levels in biota is an effective approach for initially assessing damage to an ecosystem, as was clearly demonstrated by the Minamata case. Monitoring of the habitats and species combined with contaminant levels in the area will provide valuable information to understand ecotoxicological impacts on ecosystems. Applying strategies learnt from the case histories could help with the design of both chemical and ecological monitoring program to improve the understanding of the relationship between ecological damage and chemical pollution and planning effective remediation approaches.

Conclusions

In this article we examined the ecological impacts caused by massive releases of toxic substances, including chemical accidents such as Minamata, Bhopal, Seveso, Chernobyl, Sandoz, Prince William Sound, and Aznalcóllar. Magnitude, scale, and the significance of ecological impacts differ from one case to the next. The ecological impact of disastrous chemical pollutions or accidents can be huge once they happen. In some cases, population size of the affected species has not returned to their pre-affected population size. Reducing catastrophic events resulting from human activities is essential to sustainable use of environment. In order to reduce chances for similar future catastrophes, the following points should be emphasized. Avoidance of accidents or minimization of the emission of toxic substances can be the best way to prevent ecosystem from catastrophic damage. Preparedness against various emergencies considering realistic worst-case scenarios and planning adequate response system should be incorporated. Investigations of the long-term environmental effects should be carefully undertaken to plan restoration action in a wiser fashion.

See also: Ecotoxicological Model of Populations, Ecosystems, and Landscapes; Food-Web Bioaccumulation Models.

Further Reading

Capel PD, Giger W, Reichert P, and Wanner O (1988) Accidental input of pesticides into the Rhine River. *Environmental Science and Technology* 22: 992–997.

Doi R, Ohno H, and Harada M (1984) Mercury in feathers of wild birds from the mercury-polluted area along the shore of the Shiranui Sea, Japan. *The Science of the Total Environment* 40: 155–167.

Environmental Agency of Japan (1997) Our intensive efforts to overcome the tragic history of Minamata disease. Environmental Health Department, Environment Agency, Tokyo.

Fanelli R, Bertoni MP, Castelli MG, *et al.* (1980) 2,3,7,8-Tetrachlorodibenzo-*p*-dioxin toxic effects and tissue levels in animals from the contaminated area of Seveso, Italy. *Archives of Environmental Contamination and Toxicology* 9: 569–577.

Fujiki M and Tajima S (1992) The pollution of Minamata Bay by mercury. *Water Science and Technology* 25: 133–140.

Garagna S, Zuccotti M, Vecchi ML, *et al.* (2001) Human-dominated ecosystems and restoration ecology: Seveso today. *Chemosphere* 43: 577–585.

Gupta RS, Sarkar A, and Kureishy TW (1991) Biodegradation and anticholinesterase activity of methyl isocyanate in the aquatic environment of Bhopal. *Water Research* 25: 179–183.

Güttinger H and Stumm W (1992) Ecotoxicology, an analysis of the Rhine Pollution caused by the Sandoz chemical accident, 1986. *Interdisciplinary Science Reviews* 17: 127–136.

IAEA (2006) Environmental consequences of the Chernobyl accident and their remediation: Twenty years of experience. Report of the Chernobyl Forum Expert Group 'Environment'. International Atomic Energy Agency, Vienna, 2006.

Meharg AA (1994) Industrial accidents involving release of chemicals into the environment: Ecotoxicology. *Environmental Technology* 15: 1041–1050.

Meharg AA, Osborn D, Pain DJ, Sánchez A, and Naveso MA (1999a) Contamination of Doñana food-chains after the Aznalcóllar mine disaster. *Environmental Pollution* 105: 387–390.

Nishimura H and Okamoto T (2001) *Science of Minamata Disease.* Tokyo: Nippon Hyoron-sha (in Japanese).

Pain DJ, Sánchez A, and Meharg AA (1998) The Doñana ecological disaster: Contamination of a world heritage estuarine marsh ecosystem with acidified pyrite mine waste. *The Science of the Total Environment* 222: 45–54.

Peterson CH, Rice SD, Short JW, *et al.* (2003) Long-term ecosystem response to the Exxon Valdez oil spill. *Science* 302: 2082–2086.

Savchenko VK (1995) *Man and Biosphere Series, Vol. 16: Ecology of the Chernobyl Catastrophe: Scientific Outlines of an International Programme of Collaborative Research.* Paris: UNESCO.

Singh MP and Ghosh S (1987) Bhopal gas tragedy: Model simulation of the dispersion scenario. *Journal of Hazardous Materials* 17: 1–22.

Taggart MA, Figuerola J, Green AJ, *et al.* (2006) After the Aznalcóllar mine spill: Arsenic, zinc, selenium, lead and copper levels in the livers and bones of five waterfowl species. *Environmental Research* 100: 349–361.

Wiener JG and Spry DJ (1996) Toxicological significance of mercury in freshwater fish. In: Beyer WN, Heinz GH, and Redmon-Norwood AW (eds.) *Environmental Contaminants in Wildlife: Interpreting Tissue Concentrations*, pp. 297–339. Boca Raton, FL: CRC Press.

Ecotoxicological Model of Populations, Ecosystems, and Landscapes

R A Pastorok, Integral Consulting, Inc., Mercer Island, WA, USA

D Preziosi, Integral Consulting, Inc., Berlin, MD, USA

D Rudnick, Integral Consulting, Inc., Mercer Island, WA, USA

Introduction
Rationale for Use of Ecotoxicological Modeling in Risk Assessment
Selecting Ecotoxicological Models
Example Applications of Ecotoxicological Models

Software for Ecotoxicological Modeling
Future Developments in Ecotoxicological Modeling
Conclusions
Further Reading

Introduction

Ecological risk assessment has developed over approximately the last 25 years as a scientific practice for assessing risks of toxic chemicals in the environment. Important

issues addressed by ecological risk assessment approaches include chemical contamination at industrial facilities or hazardous waste sites; the potential for chemical contamination of fish and wildlife from production and release of new chemicals; possible effects of toxic chemicals on endangered species; and biomagnification of chemicals in food chains. Ecological risk assessment has become an important tool for government agencies throughout the world in evaluating and regulating toxic chemicals.

Ecological risk assessment often deals with the potential effects of toxic chemicals in the environment by extrapolating toxicity data from laboratory experiments on test species to organisms, populations, and higher-level ecological systems in nature. Traditionally, ecotoxicology has used laboratory toxicity testing with single species to develop response thresholds indicative of no-effects or effects doses (or concentrations). Current methods for ecological risk assessment still focus on the endpoints of survival, growth, and reproduction of individual organisms because those endpoints are easily evaluated in laboratory tests. However, this focus has been questioned by ecotoxicologists who recognize the need to address higher levels of biological organization, such as populations, food webs, and ecosystems.

Need for Ecological Relevance in Risk Assessments

Most toxicity data are developed for biological endpoints at the level of the individual organism, such as mortality, fecundity, or physiological responses. Suborganismal endpoints such as alterations in enzymatic expression are becoming more common with increased research into biomarkers that can measure changes in these pathways. Typical risk assessments ignore effects above the level of the organism, or only qualitatively discuss risks to populations and higher levels of organization. Ecotoxicological models are important tools for addressing these higher levels of organization in an ecological risk assessment.

Ignoring population-level or higher-level effects and focusing only on organism-level endpoints may over- or underestimate risks, leading to possible errors in environmental management decisions. Thus, ecological risk assessments should address ecologically relevant endpoints, such as population growth, population age/size structure, recruitment, biodiversity, ecosystem productivity, and indices of landscape pattern. Ecological relevance is one aspect of ecological significance, which is a critical element of risk characterization in ecological risk assessment. Ecological significance is defined as the importance to population, community, or ecosystem responses (especially those that impact ecological structure and function). Several factors contribute to ecological significance, including the nature and magnitude of effects, the spatial and temporal extent of effects, and the recovery potential under partial or complete removal of a stressor.

For our purposes, an ecotoxicological model is a mathematical expression that can be used to describe or predict the effects of toxic chemicals on endpoints such as population abundance (or density), community species richness,

productivity, or distributions of organisms. Ecotoxicological models are therefore useful in evaluations of the ecological significance of perturbations of organism-level endpoints, such as survivorship or fecundity.

Higher-level models may be used to evaluate chemicals and other factors (e.g., habitat, nutrient enrichment) that affect population abundance and distribution, biological community structure, and ecosystem processes. For example, several researchers have concluded that the growth rate parameter for a population integrates potentially complex interactions among life-history traits and thereby provides a more relevant measure of toxicant impacts than organism-level endpoints. Thus, these researchers favor using population models for ecological risk assessment. Others favor using ecosystem or higher-level models to interpret chemical exposure–response data for organism-level endpoints in addressing more complex issues above the population level.

Available Ecotoxicological Models for Ecological Risk Assessment

Two broad classes of models are used in ecological risk assessment: exposure models and effects models. Exposure models address the co-occurrence of organisms and chemicals, uptake rates, and accumulation in tissues. Effects models deal with exposure–response relationships or extrapolation between effects endpoints. Ecotoxicological models, as defined here, are effects models with output parameters that directly apply as assessment endpoints above the level of individual organism effects (e.g., population abundance, biodiversity, ecosystem productivity). A mathematical model is used to define precisely (1) the relationship between measures of exposure and effects and the assessment endpoint of interest, and (2) assumptions and uncertainties in the extrapolation between endpoints. Many ecological models incorporate mechanistic functions to describe natural processes, such as nutrient and energy flows, organism growth, life-stage transitions, dispersal, competition, predation, and interactions between organisms and the environment. Specific kinds of ecotoxicological models are described later in this article.

Although many ecological models can be used to assess toxic chemical effects by modifying input parameters to mimic the effects of chemicals on organism survival, growth, and reproduction, vital rates of populations, or other ecological attributes, true ecotoxicological models incorporate exposure–response relationships for the toxic chemicals of interest in a risk assessment. Such models may simulate the structure and dynamics of populations, food webs, whole ecosystems, or landscapes. More complex ecological models, such as the aquatic ecosystem model AQUATOX, have built-in functions to account for toxic effects, as well as chemical fate and transport.

Ecotoxicological Models and Weight-of-Evidence Approaches

Ecotoxicological modeling may be one of several lines of evidence used in the risk characterization phase of an ecological risk assessment. Here, ecotoxicological modeling essentially translates information on measures such as fecundity, survivorship, growth, and distribution of individual organisms to a population context or higher level of organization. Empirical evidence from toxicity testing, or in some cases field surveys of populations or communities, may be available for the ecological risk assessment. These data need to be weighed in the ecological risk assessment along with the results of ecotoxicological modeling. In this case, a weight-of-evidence approach can help integrate information from empirical data and ecotoxicological modeling.

The weight-of-evidence approach considers the ecological relevance of each line of evidence, its relationship to the assessment endpoint, the strength of a demonstration of an exposure–response relationship, and the strengths and weaknesses of the model (or data) (e.g., spatial and temporal scope, amount and quality of data). Thus, the results of ecotoxicological modeling should be viewed in combination with all other information being used to assess risks.

Objectives of Article

The objectives of this article are to review the rationale for use of ecotoxicological models in chemical risk assessments, to describe factors useful in selecting ecotoxicological models to address a specific question, to show example applications of ecotoxicological models, and to review current software and trends in ecotoxicological modeling.

Rationale for Use of Ecotoxicological Modeling in Risk Assessment

Like most mathematical models, ecotoxicological models are used to describe quantitative relationships among variables to predict a given outcome. Ecotoxicological models build upon empirical observations of the effects of chemicals on individual organisms or groups of organisms (e.g., mesocosms) under laboratory settings and occasionally under field conditions. In conjunction with other lines of evidence, such as toxicity testing and field surveys, ecotoxicological models are an invaluable asset to progress beyond the traditional ecological risk paradigm and can provide increased ecological relevance in risk estimates.

Beyond the Traditional Ecological Risk Paradigm

Risks are quantified under the traditional ecological risk paradigm using the hazard quotient approach. A hazard quotient is defined as the ratio of an organism's estimated chemical exposure relative to a lowest-observed-effect level or a no-observed-effect level derived from laboratory toxicity testing or field data. The hazard quotient approach is rooted in human health risk assessment as a tool to measure the potential hazard (e.g., level of concern) posed by chemical exposure. As acknowledged by the US Environmental Protection Agency (EPA) in its Risk Assessment Guidance for Superfund (RAGS, Part A), a hazard quotient is in fact not a true measure of risk, but rather simply an indication of whether a threshold has been exceeded. This differs from EPA's approach toward carcinogenic risks to humans, which uses the linear low-dose cancer risk equation to estimate the probability of an individual's cancer risk over a lifetime. One proposed strength of the hazard quotient approach is its utility in ranking and prioritizing chemicals for more detailed risk characterization, but even this use is fraught with problems if implemented in a deterministic way. Nevertheless, this approach is included among other assessment methods for ranking of high-production-volume chemicals in the European Union (see regulation 1488/94 and directive 93/67) and the United States (as per the Toxic Substances Control Act).

For ecological assessments, hazard quotients are typically based on effect measures targeted at individual response levels (e.g., growth, survival) determined under laboratory conditions. This approach is limited in its ability to estimate risks to individual organisms, let alone estimate risks to groups of organisms. As described in numerous references within the open literature from the last decade, criticisms against reliance on the hazard quotients are vast. The criticisms are based on three chief arguments. First, hazard quotients do not denote risk *per se*, rather only potential hazard. Second, traditional hazard quotients provide limited insight, if any, to potential risks posed to populations and higher levels of organization. This is chiefly the result of a preponderance of laboratory toxicity tests based upon individual-level responses. Third, hazard quotients do not permit an explicit examination of potential impacts beyond standard laboratory exposure durations (generally from hours to days). Hazard quotients do not provide more meaningful expressions of impacts across multiple generations of organisms under field settings. While many researchers have tried to bridge the gap between hazard quotients to more meaningful expressions of ecological value described by assessment endpoints, few empirical data exist to conclusively support the continued sole reliance on a hazard quotient approach.

Uses of Ecotoxicological Models

In ecological risk assessments, ecotoxicological models are used to consider higher-level effects in diverse risk-based environmental management decisions. They can provide information for decisions related to environmental quality criteria and standards, assessments of potential impacts associated with past or proposed uses of chemicals, and evaluations of remedial actions or restoration options for cleanup of contaminated sites.

Ecotoxicological models are used for chemical risk assessments in the following ways:

- *Deriving environmental quality criteria.* For environmental regulatory programs, ecotoxicological models provide a tool to calculate risk-based quality criteria for surface water, sediments, soils, and air, based on protection of population- or ecosystem-level endpoints.
- *Assessing potential impacts associated with past or proposed uses of chemicals.* Chemical releases and uses involve risks to individual organisms, populations, and biological communities. Ecotoxicological models provide a method to assess potential risks of chemicals in the environment on endpoints at higher levels of ecological organization.
- *Evaluating remedial actions or restoration options for cleanup of contaminated sites.* For US EPA Superfund cleanup projects and related efforts, ecotoxicological models can provide information for decisions by answering questions such as which contaminated media or locations pose an unacceptable risk, which cleanup actions will reduce risk the most, what is the ecological risk associated with the cleanup actions themselves, and what is the residual chemical risk after remediation.
- *Evaluating ecological significance of estimated risks.* As part of the risk characterization phase of an assessment, ecotoxicological modeling can be used in a weight-of-evidence approach to evaluate the ecological significance of effects predicted or observed on organisms.

When to use an ecotoxicological model
Ecotoxicological modeling would not always benefit a risk assessment. For example, assessment of effects above the level of individual-organism endpoints is unnecessary when a worst-case analysis using individual-level endpoints shows that the risk is negligible. In worst-case analyses, endpoints should be carefully chosen to include potential sublethal effects on organisms, and the risk estimates should be interpreted conservatively. This approach minimizes the chance of missing cases where individual-level effects have impacts on the population or higher levels.

Higher-level endpoints are also not needed if an initial assessment shows, for example, that available evidence from quantitative field studies documents severe effects of chemical contaminants on the abundance of target species. Thus, ecotoxicological models may not be needed for a baseline risk assessment. In cleanup programs, however, such models may still prove useful for evaluating the mechanisms of population-level effects (i.e., which vital rate is affected or if indirect effects mediated through trophic interactions are likely), which would aid in designing remedial actions.

Selecting Ecotoxicological Models

Evaluating the usefulness of ecotoxicological models recognizes that choosing the right model for ecological risk assessment entails a tradeoff between flexibility (e.g., the power to address endpoints of concern) and practicality (e.g., data requirements). The characteristics of a model and its predictive power depend on the questions that are addressed. These questions can be considered in terms of the required endpoints of the model.

Classification of Ecotoxicological Models

Available ecotoxicological models can be classified in many ways. The most useful starting point considers levels of biological organization. Biological systems are organized as a hierarchy of components categorized by their level of organization, from smaller than an individual (e.g., including molecular and cellular structures), to groups of individual organisms comprising a population, to interactions of those populations to form a community, to interactions of communities with their biotic and abiotic environmental at the ecosystem, and finally to landscape levels of organization. **Figure 1** shows

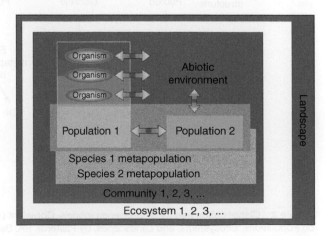

Figure 1 Hierarchy of ecological systems.

the hierarchy of ecological systems at and above the level of individual organisms.

Ecotoxicological models can also be classified by whether they address single species or multiple species, as in **Figures 2** and **3**. Single-species models are further classified according to whether they consider spatial structure of ecological systems explicitly, whether they consider individual organisms or only groups of organisms, whether they incorporate the age/stage structure of populations explicitly, and whether they include density dependence (**Figure 2**). Multispecies models are further classified according to whether they address abiotic variables as well as biotic entities, whether they consider spatial structure of ecological systems explicitly, and the kind of habitat they include (**Figure 3**).

How to Select a Model for a Risk Assessment

Depending on their complexity and ease of use, some ecological models are more appropriate for screening-level ecological risk assessments, whereas others are best reserved for detailed assessments. For example, simple models of population dynamics are most appropriate for screening-level ecological risk assessments. Ecosystem and landscape models are generally reserved for detailed ecological risk assessments because of the greater effort and expense involved in applying these models.

Ecotoxicological model selection is based on the objectives of the ecological risk assessment and the data available for modeling. The following factors should be considered when choosing an ecotoxicological model to address a specific risk issue:

● *Kind of question.* Is the ecological risk assessment assessing impacts of chemical contamination already present at a site or potential risks from uses of new chemicals? Is the ecological risk assessment addressing chemical releases and contamination, or cleanup/restoration issues?

● *Level of biological organization.* Does the assessment endpoint deal with populations, communities, ecosystems, or landscapes?

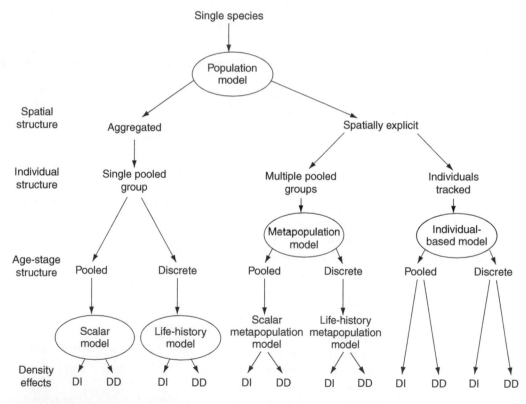

Figure 2 Classification tree for single-series models. Copyright (© 2002) *Ecological Modeling in Risk Assessment: Chemical Effects on Populations, Ecosystems, and Landscapes,* by Pastorok RA, Bartell SM, Ferson S, and Ginzburg LR (eds.). Reproduced by permission of Routledge/Taylor & Francis Group, LLC.

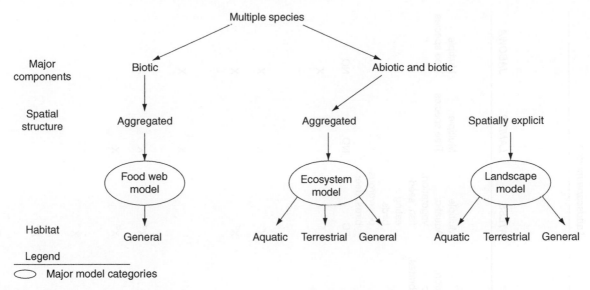

Figure 3 Classification tree for multispecies models. Coptright (© 2002) *Ecological Modeling in Risk Assessment: Chemical Effects on Populations, Ecosystems, and Landscapes,* by Pastorok RA, Bartell SM, Ferson S, and Ginzburg LR (eds.). Reproduced by permission of Routledge/Taylor & Francis Group, LLC.

- *Specific endpoints and ecological relevance.* Are ecological structures (e.g., habitat pattern, species abundances, or distributions) or processes (e.g., ecosystem productivity, population recruitment) of interest? How related are the model results to the endpoints that are used in ecological risk assessments? If the model results are not directly relevant endpoints, can the endpoints be easily calculated from model results?

- *Screening versus detailed.* Is the ecological risk assessment a screening-level assessment or a more detailed assessment?

- *Level of realism.* Does the model incorporate key processes known to be important in the system it simulates, as well as key factors that affect each of these resources? Are the assumptions realistic with respect to the ecology of the system? Is it necessary to model specific mechanistic processes? Should exposure-response relationships be explicitly incorporated into the model?

- *Spatially explicit or aggregated.* Is the spatial distribution of ecological entities of interest, so that a spatially explicit model is needed?

- *Individual organism or groups.* Is the assessment population a threatened or endangered species? Should the model simulate the distribution, behavior, and characteristics of individual organisms, or should it simulate groups of organisms (i.e., local populations or communities)?

Additional criteria for evaluating the suitability of a model for a specific problem include the following practical considerations:

- *Flexibility.* Can the model accept alternative formulations? For example, can the model be applied to species and systems other than those for which it was originally

developed? Can different life histories, distributions, or habitat types be easily modeled?

- *Treatment of uncertainty.* Does the model incorporate uncertainty? How easy is it to parse effects of natural variability and measurement uncertainties?

- *Degree of development and consistency.* How easy is it to understand what the model does, so that the underlying mechanisms can be checked for errors and bugs? Has the model been tested or validated?

- *Ease of parameter estimation.* How easy is it to estimate the required model parameters given the type of data available? Are accepted sampling or statistical methods for estimating model parameters included?

- *Regulatory acceptance.* How likely are regulatory agencies to accept results of an analysis with the model? Has the model been used by any regulatory agencies?

- *Credibility.* Does the model have scientific and technical credibility? Is it widely known? Has it been used in support of published research?

- *Resource efficiency.* How much time and effort would be needed to apply the model in a particular case?

Characteristics of Available Models for Ecological Risk Assessment

The characteristics of selected models of populations, ecosystems, and landscapes are summarized in **Tables 1–3.** Example applications of these three major types of models are presented in the next section.

To be useful in an ecological risk assessment, the output of ecotoxicological models must be linked to assessment endpoints and risk estimates. Models must

Table 1 Summary of biological variables in selected ecological models

Variable[a]	Population models — Scalar abundance (crucian carp)	Population models — Life-history matrix (fathead minnow)	Population models — Metapopulation – RAMAS-GIS (California gnatcatcher)	Ecosystem models — AQUATOX	Ecosystem models — CASM	Ecosystem models — IFEM	Landscape models — ATLSS	Landscape models — LANDIS	Landscape models — JABOWA
Community characteristics									
Number of species	One	One	One	Multiple	Multiple	Multiple	Multiple	Multiple	Multiple
Typical species	Fish			Phytoplankton, zooplankton, benthic infauna, fish	Phytoplankton, zooplankton, benthic infauna, fish	Zooplankton, benthic invertebrates, fish	Aquatic vegetation, fish, seed-eating birds, piscivorous birds, deer	Tree species	Tree species
Type of trophic interactions[b]	NO	NO	NO	FC	FC	PC	FC	NO	NO
Relative abundances in immigrant population									X
Individual population characteristics									
Abundance or biomass in native population	X	X	X	X	X	X	X	X	X
Age- or stage-specific abundance or biomass		X	X	X[c]			X[d]	X	X
Carrying capacity	X		X						
Density dependence	X	X	X						
Maximum age	X	X	X			X			
Population growth rate	X	X	X						
Overall mortality rate		X	X	X	X	X			X
Age- or stage-specific mortality rate		X	X	X[c]			X[d]	X[e]	X
Age of first reproduction		X	X				X[d]	X	
Overall frequency of reproduction	X	X	X						
Overall fecundity	X	X	X	X	X	X	X		
Overall fecundity (offspring per reproduction)			X						
Age- or stage-specific fecundity			X				X[d]		
Home range size			X						
Periodicity (seasonality) of presence in home range							X		

Variable				
Offspring dispersal distance	X		X[f]	
Immigration or emigration probability or rate	X			X
Habitat suitability coefficients (by habitat type)	X	X	X	X
Other characteristics			Shade tolerance, fire tolerance	Shade tolerance
Spatially explicit population data				
Locations inhabited	X	X	X	X
Abundances	X	X		X
Distribution of competitors				X
Distribution of predators		X		

[a] A blank cell indicates that the variable is not applicable to the model.
[b] None (NO), food chain or food web (FC), producer-consumer (PC).
[c] Two cohorts for fish species.
[d] Age-structured models are used only for aquatic species.
[e] For zero-age individuals only, representing sprouting ability.
[f] Both effective and maximum seed dispersal distance.

Adapted from Pastorok RA, Akçakaya HR, Regan HM, Ferson S, and Bartell SM (2003) Role of ecological modeling in risk assessment. *Human and Ecological Risk Assessment* 9(4): 939–972.

Table 2 Summary of environmental variables in selected ecological models

	Population models			Ecosystem models			Landscape models		
Variable[a]	Scalar abundance (crucian carp)	Life-history matrix (fathead minnow)	Metapopulation – RAMAS-GIS (California gnatcatcher)	AQUATOX	CASM	IFEM	ATLSS	LANDIS	JABOWA
Environmental media, variables, and species categories modeled									
Soil									X
Surface water				X	X	X	X		
Sediment				X		X	X		
Suspended particulates						X			
Soil biota									
Sediment in fauna				X	X	X			
Terrestrial vegetation									
Aquatic vegetation				X		X	X		
Plankton				X	X	X	X		
Fish				X	X	X	X		
Small mammals									
Birds							X		
Spatially explicit site data									
Location of immigrant population source			X						
Distribution of habitat types or land types			X				X		X
Distribution of food, water, nesting sites, etc.			X				X	X	

[a]A blank cell indicates that the variable is not applicable to the model.
Adapted from Pastorok RA, Akçakaya HR, Regan HM, Ferson S, and Bartell SM (2003) Role of ecological modeling in risk assessment. *Human and Ecological Risk Assessment* 9(4): 939–972.

Table 3 Summary of chemical toxicity and other disturbance variables in selected ecological models

Variable[a]	Population models — Scalar abundance (crucian carp)	Population models — Life-history matrix (fathead minnow)	Metapopulation – RAMAS-GIS (California gnatcatcher)	Ecosystem models — AQUATOX	Ecosystem models — CASM	Ecosystem models — IFEM	Landscape models — ATLSS	Landscape models — LANDIS	Landscape models — JABOWA
Chemical toxicity or other disturbances data									
Types of disturbances or stressors	Chemical, habitat reduction	Chemical (Mirex)	Cold, wet winters; fire	Chemical	Chemical	Chemical (PAH)	Low water levels	Fire, wind damage	Mortality (e.g., logging)
Contaminant mass balance modeled						Yes		No	
Soil contamination modeled									
Surface water contamination modeled				X	X	X			
Sediment contamination modeled						X			
Terrestrial vegetation contamination modeled									
Aquatic vegetation contamination modeled						X			
Bioaccumulation	X			X	X	X			
Growth rates/effects	X			X	X	X			
Mortality rates/effects		X				X			
Reproductive effects		X							
Other effects	Risk of extinction	Risk of population decline	Dispersal of juveniles, subpopulation characteristics, spatial correlations	Population size (biomass)	Population size (biomass)	Population size (biomass)		X	Community composition, species range and density
Exposure indicators									
Spatial location of disturbance								X	
Spatial extent of disturbance								X	
Time scale of disturbance[b]	C	C	C	C	C		P	I	

[a] A blank cell indicates that the variable is not applicable to the model.

[b] Continuous (C), periodic (P), irregular or any pattern (I).

Adapted from Pastorok RA, Akçakaya HR, Regan HM, Ferson S, and Bartell SM (2003) Role of ecological modeling in risk assessment. *Human and Ecological Risk Assessment* 9(4): 939–972.

also incorporate toxicological information either explicitly as exposure–response functions, or implicitly as key biological parameters capable of perturbation to mimic the effects of toxic chemicals (e.g., on processes such as survival, growth, and reproduction of organisms in a population model). In detailed risk assessments, it may be useful to link ecotoxicological models with fate/transport models or bioaccumulation models.

Linking model output with assessment endpoints and risk estimates

As discussed earlier, model output should match the assessment endpoints for an ecological risk assessment directly or be quantitatively related to the assessment endpoints. This objective is achieved by selecting the appropriate model. Ecotoxicological models may also provide risk estimates as output. A Monte Carlo analysis runs a model simulation multiple times with the input variables for each run selected from their prospective probability distributions. The output for a population model would show a series of population trajectories (e.g., graphs of abundance) over time or as a probability distribution. For example, the RAMAS Ecotoxicology software can be used to simulate multiple projections of population dynamics in a Monte Carlo analysis and the resulting output can be used to derive estimates of risk of extinction (or quasi-extinction), risk of a given percentage decline in a certain time period, and other estimates of population-level risk. An example of population trajectories derived from Monte Carlo analysis and the method for estimating quasi-extinction risk is shown in **Figure 4**. Essentially any projection of an output variable for an ecotoxicological model can be converted to a risk estimate using Monte Carlo analysis.

Incorporating toxicological information into ecological models

An ecotoxicological model includes consideration of toxicological effects on biological entities (i.e., organisms, populations, or communities). The most direct way to incorporate toxicological effects in an ecological model is to add an exposure–response function or a series of such functions for the various endpoints as part of the mathematical equations for the model or as a subroutine in a simulation. **Figure 5** shows an example exposure–response relationship derived from laboratory testing of PCB-contaminated sediments with mummichog (*Fundulus heteroclitus*) and interpretation of effects on survival and reproduction using a population life-history model to estimate population growth rate. Alternatively, toxicity data may be incorporated into a physiologically based or dynamic energy-budget model, which is linked with a demographic model.

Exposure data are summarized as chemical distributions in space and time for either environmental media relevant to toxic effects data (e.g., soil, sediment, water) or for tissue residues of chemicals in target species being modeled. Toxicity data may be based on dose– (or concentration–) response relationships, a toxicity threshold (e.g., no-observed-adverse-effect level (NOAEL)), or any field or laboratory data linking variation in toxic chemical concentrations with biological response endpoints. Data on toxicity and population or community parameters must be matched to the spatial–temporal scale of the ecotoxicological model.

Linking with fate/transport models

Models that predict the fate and transport of chemicals in the environment can provide the basic chemical distribution and concentration data needed for estimating exposure where measurements of exposure are not available (e.g., for a new chemical assessment) or impractical (e.g., for an

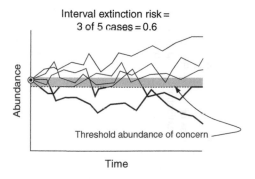

o = Initial condition

Figure 4 Example derivation of risk estimate from Monte Carlo analysis of a population model. Bold lines indicate abundances below threshold of concern. A small number of simulation runs are shown for clarity. In practice, >1000 runs could be used for a Monte Carlo analysis to derive a risk estimate based on frequency of runs in which abundance decreases below threshold. Reproduced by permission of Routledge/Taylor & Francis Group, LLc.

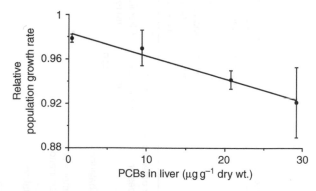

Figure 5 Example exposure-response relationship for PCBs and mummichog (*Fundulus heteroclitus*) derived from population modeling. Redrawn from Munns WR, Black DE, Gleason TR, *et al*. (1997) Evaluation of the effects of dioxin and PCBs on *Fundulus heteroclitus* populations using a modeling approach. *Environmental Toxicology and Chemistry* 16: 1074–1081, with permission from Society of Environmental Toxicology and Chemistry.

endangered species assessment). Linking the fate and transport model with the ecotoxicological model provides a comprehensive risk assessment approach. Bioaccumulation models are a special class of fate/transport models that estimate chemical uptake and sometimes distribution in target tissues of organisms from chemical data in the environment, organism distribution, chemical uptake and elimination rates by organisms, and other factors. Bioaccumulation models may incorporate food web exposure models. Many ecotoxicological models that predict ecosystem and landscape endpoints also include submodels that describe environmental transport and fate of chemicals in the environment, as well as exposure of organisms to those chemicals. Most population models do not incorporate fate/transport models, although linking such models may be a trend for future development of methods for ecological risk assessment.

Example Applications of Ecotoxicological Models

Applying an ecotoxicological model to an ecological risk assessment supports the integration of ecological information and chemical toxicity data to evaluate risk in terms of population-level endpoints. Model output for state variables such as organism abundance, age structure, and distribution can be analyzed in a Monte Carlo framework to estimate risks such as extinction risk, time to extinction, or interval decline risk. Ecosystem and landscape models can be applied to ecological risk assessments in a similar way, except that state-variable outputs from the model and corresponding risk estimates are specific to the level of biological organization under investigation. Additional ecological information specific to the system of interest can also be incorporated into the ecosystem or landscape model.

Below, we present several examples of applications of population, ecosystem, and landscape models to address ecological risk issues.

Population Model Case Study – Interpretation of Chronic Toxicity Test Data

One example of applying a population model to evaluate chronic toxicity is from the US Army Corps of Engineers' work with sediment toxicity tests using the estuarine amphipod *Leptocheirus plumulosus*. Percent survival and fecundity (number of offspring per female) were measured at weekly intervals in 30-week tests of chemically contaminated sediment from Black Rock Harbor (Connecticut, USA). Toxicity was measured in each of the three dilutions of Black Rock Harbor sediment under two food rations to evaluate nutritional effects on toxicity. Three different Black Rock Harbor sediment concentrations were used (0%, 3%, and 6% contaminated sediment

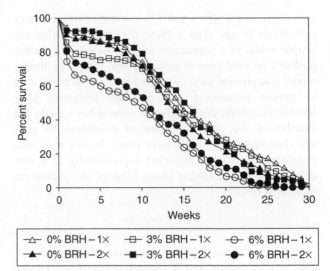

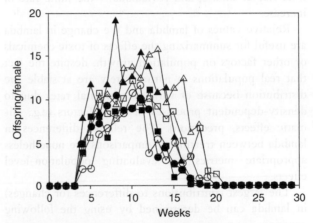

Figure 6 Survivorship and fecundity curves for *Leptocheirus plumulosus* during exposure to Black Rock Harbor (BRH) sediment under six combinations of different sediment dilutions (%) and food rations (X).

in a clean sediment matrix) and two levels of food ration were applied (normal food ration, 1×; and double the normal food ration, 2×). **Figure 6** illustrates the survivorship and fecundity data obtained from these sediment toxicity tests.

To evaluate the population-level effects of sediment toxicity to *Leptocheirus*, data for the toxicity test endpoints were entered into the following matrix model:

$$
\begin{bmatrix} n_1(t+1) \\ n_2(t+1) \\ n_3(t+1) \\ n_4(t+1) \\ .. \\ n_m(t+1) \end{bmatrix} = \begin{bmatrix} F_1 & F_2 & F_3 & F_4 & .. & F_m \\ S_1 & 0 & 0 & 0 & .. & 0 \\ 0 & S_2 & 0 & 0 & .. & 0 \\ 0 & 0 & S_3 & 0 & .. & 0 \\ .. & .. & .. & .. & .. & 0 \\ 0 & 0 & 0 & 0 & S_{m-1} & S_m \end{bmatrix} \begin{bmatrix} n_1(t) \\ n_2(t) \\ n_3(t) \\ n_4(t) \\ .. \\ n_m(t) \end{bmatrix}
$$

where $n_i(t)$ refers to the number of individuals in age class i at time t, F_i represents the fecundity, or birth rate, of individuals in age class i, and S_i refers to the survival rate of individuals in age class i. Note that this is a relatively simple model of a population without density-dependent feedback on vital rates of mortality and birth. The matrix model is applied to successive time steps of the population to project population size and age structure. After several time steps, the population approaches a stable age distribution (i.e., the proportion of individuals in each age class remains constant over time). In this stage, the population grows (or declines) exponentially at a rate determined by the largest eigen value of the population projection matrix:

$$N_t = N_0 \lambda^t$$

where $N_t =$ the total population size at time t, $N_0 =$ total size at time 0, and λ (lambda) = the finite rate of increase.

Relative values of lambda and the change in lambda are useful for summarizing the effects of toxic chemicals or other factors on population growth, despite the fact that real populations in nature rarely are at stable age distribution because of variations in vital rates due to density-dependent processes or other factors (e.g., climatic effects, predation). The relative differences in lambda between treatment comparisons are nonetheless appropriate metrics for evaluating population-level effects.

The largest contributions to differences (or changes) in lambda can be determined by using the following equation:

$$\lambda^e - \lambda^c \approx \sum_{ij} \left(a_{ij}^e - a_{ij}^c \right) \frac{\partial \lambda}{\partial a_{ij}} \Bigg|_{(A^e + A^c)/2}$$

where $a_{ij} =$ the element in the ith row and the jth column of the population projection matrix for either the experiment treatment (e) or control (c). Each term in the summation represents a contribution by a certain age-specific survivorship or fecundity term to the change in lambda. This approach was used to infer that the largest contributions to the difference in lambda between experimental treatments for *Leptocheirus* were generally due to changes in survival rates of age classes 1–3 and fecundities of age classes 4–8. As expected, increasing sediment concentration in the Black Rock Harbor sediment tests resulted in decreases in lambda and increasing food resulted in increasing lambda (**Table 4**).

In the model presented above, vital rates and population growth are density independent. In the more general case, a density-dependent model would be used. Population models with density dependence, either as a matrix formulation or a continuous process model (e.g., using differential equations), have been developed for burrowing mayfly (*Hexagenia* spp.) in Lake Erie (USA), Japanese crucian carp

Table 4 Changes in population growth rates (lambda) at stable age distribution for *Leptocheirus* exposed to Black Rock Harbor sediment at different concentrations (%) and different food ratios ($\times$)

Control	Treatment	Change in lambda
(0%, 1×)	(3%, 1×)	−0.071
(0%, 2×)	(3%, 2×)	−0.092
(0%, 1×)	(6%, 1×)	−0.221
(0%, 2×)	(6%, 2×)	−0.227
(3%, 1×)	(6%, 1×)	−0.150
(3%, 2×)	(6%, 2×)	−0.135
(0%, 1×)	(0%, 2×)	−0.081
(3%, 1×)	(3%, 2×)	−0.060
(6%, 1×)	(6%, 2×)	−0.075

(*Carassiun auratus*) in Lake Biwa (Japan), fathead minnow (*Pimephales promelas*) in the laboratory, song sparrow (*Melospiza melodia*) in the Mandarte Islands (Canada), harp seal (*Phoca groenlandica*) in the northwest Atlantic, and many other species. Metapopulation models are a special class of models that incorporate density-dependent processes by modeling a set of populations of the same species in the same general geographic area and including migration among populations.

Not all of the model applications discussed above were used for ecotoxicological assessments, but the models could be easily adapted for this purpose. In general, to apply a population model to a chemical risk assessment, the following data are needed:

- estimates of demographic parameters in absence of stress;
- toxicity data for selected endpoints, for example, (1) survivorship, (2) fecundity, (3) growth;
- estimate of parameters for density-dependent modeling, such as intrinsic rate of natural increase (r_{max}) and carrying capacity (K) or functions relating vital rates to density;
- initial age or stage structure for models of structured populations; and
- locations of suitable habitat patches, initial organism distribution (or presence–absence data), dispersal rates, and possibly time series of the distribution of habitat types for spatially explicit models, such as metapopulation models.

Examples of variables incorporated into population models are provided in **Tables 1–3**. Graphic forms of some commonly used population models are shown in **Figure 7**, including the Malthusian exponential growth model, the Malthusian model with uncertainty analysis (e.g., Monte Carlo), the density-dependent logistic growth model, and various spawner–recruit models that are popular in fisheries management.

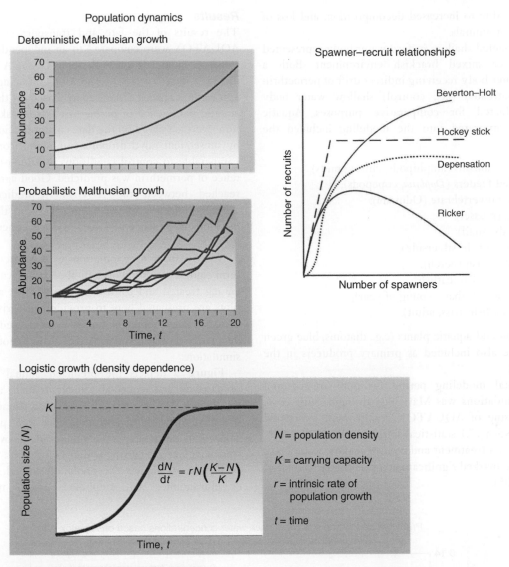

Figure 7 Forms of commonly used population models.

Ecosystem Model Case Study – AQUATOX Assessment of Pesticide Risks

AQUATOX is a simulation model for aquatic ecosystems developed under the auspices of the US EPA. The model predicts the fate of various stressors, such as nutrients, organic chemicals, and invasive species, and their effects on aquatic ecosystems, including those consisting of fish, invertebrates, and aquatic plants. Provided below is a case study example to describe the use of AQUATOX in the assessment of potential aquatic ecosystem-level effects associated with the use of permethrin. Permethrin is a synthetic pyrethroid that is most often aerially applied to areas surrounding water to control adult mosquitoes. Permethrin is used by vector control agencies in several eastern US states to control mosquito-borne pathogens, such as West Nile virus (WNV) and eastern equine encephalomyelitis virus (EEE).

Model setup

The focus of the assessment was on potential long-term impacts on abundance of populations and potential resultant impacts on community structure and function within a hypothetical shallow water body (e.g., a flooded marsh) following the aerial application of permethrin in Mastic Shirley, Long Island, New York.

The fate portion of the model, which especially applies to organic toxicants, includes the following: partitioning among organisms, suspended and sedimented detritus, suspended and sedimented inorganic sediments, and water; volatilization; hydrolysis; photolysis; ionization; and microbial degradation. The effects portion of the model includes the following: toxicity to the various organisms modeled and indirect effects such as release of grazing and predation pressure; increase in detritus and recycling of nutrients from killed organisms; dissolved oxygen sag due to increased decomposition; and loss of food base for animals.

The modeled shallow surface water body represented a freshwater mixed brackish environment. Both a shallow water body receiving indirect drift of permethrin and an untreated (i.e., control) shallow water body were evaluated for comparative purposes. Aquatic species incorporated into the modeling included the following:

- benthic organisms (amphipods, chironomids);
- suspended feeders (*Daphnia*, copepods);
- predatory invertebrate (Odonata);
- mollusks (mussel);
- gastropods (snail);
- small forage fish (silverside);
- large forage fish (perch);
- large bottom fish (catfish);
- small game fish (bass, young of year);
- large game fish (bass, adult).

Periphyhton and aquatic plants (e.g., diatoms, blue green algae) were also included as primary producers in the simulation.

The total modeling period for both treated and control simulations was May 2005 through April 2006. Postprocessing of AQUATOX results were performed using Statistica v7.1 statistical software. Observed differences between treatment and control shallow water body results were marked significant at a probability of error of 5% ($p \leq 0.05$).

Results

The results of the fate and transport component of AQUATOX were compared to an independent fate and transport model. **Figure 8** depicts AQUATOX's predicted surface water concentration following two aerial applications spaced 7 days apart. Based on the predicted aquatic persistence of permethrin, two peak concentrations were observed on the dates of applications followed by subsequent rapid dropoffs to nominal concentrations comparable to control conditions. No long-term persistence of permethrin was predicted. Good agreement was reached between the AQUATOX-predicted 14 day average concentration of $0.018\,\mu g\,l^{-1}$ and the independent model-predicted 14 day average concentration of $0.016\,\mu g\,l^{-1}$.

With respect to population-level impacts, no long-term significant differences in abundance were observed among treated and control organisms. Box and whisker plots are presented here to provide comparisons of predicted average annual abundances for treated (organisms denoted with 'P') and control (organisms denoted with 'C') simulations.

Figure 9 depicts the predicted annual abundances of *Daphnia* and copepods under treated and control simulations. No long-term differences in abundances were observed. Some short-term reductions were predicted for *Daphnia* in the treated simulation, with recovery to pretreatment levels occurring within 1–2 months.

Figure 10 depicts the predicted annual abundances of chironomids and amphipods under treated and

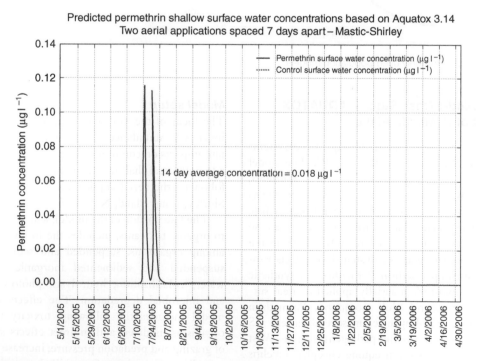

Figure 8   AQUATOX-predicted concentrations of permethrin in a shallow open surface water body.

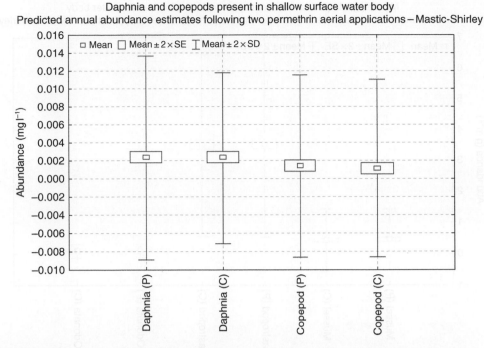

Figure 9 AQUATOX-predicted annual abundances for daphnia and copepods in treated and control simulations.

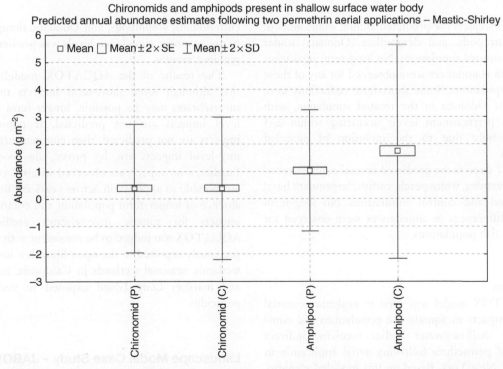

Figure 10 AQUATOX-predicted annual abundances for chironomids and amphipods in treated and control simulations.

control simulations. No significant long-term differences in abundances were observed for chironomids. Amphipods under treated conditions had a slightly lower average annual abundance than that predicted for the control (i.e., 1.0 vs. 1.7 g m^{-2}); however, this

difference was not statistically significant. Some short-term reductions were predicted for both chironomids and amphipods in the treated simulation, with recovery to pretreatment levels occurring within 2–2.5 months.

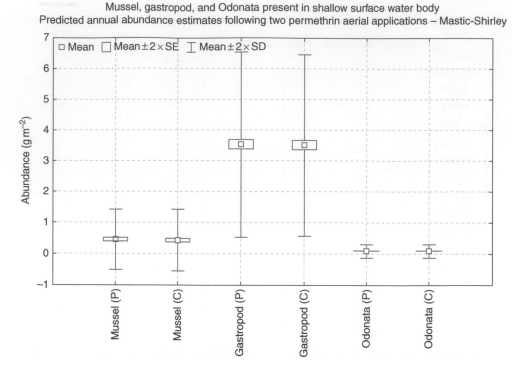

Figure 11 AQUATOX-predicted annual abundances for mussel, gastropod, and odonata in treated and control simulations.

Figure 11 depicts the predicted annual abundances of mussels, gastropods, and dragonflies (Odonata) under treated and control simulations. No long-term significant differences in abundances were observed for any of these organism populations. Some short-term reductions were predicted for Odonata in the treated simulation, with recovery to pretreatment levels occurring within 2–3 months, possibly due to the inclusion of modeled immigration.

Figure 12 depicts the predicted annual abundances of fish (i.e., silversides, white perch, catfish, largemouth bass) under treated and control simulations. No long-term significant differences in abundances were observed for any of these fish populations.

Conclusions

The AQUATOX model was used to evaluate potential long-term impacts to aquatic life populations and communities in shallow water bodies receiving indirect deposition of permethrin following aerial application in Long Island, New York. Based on this modeled scenario, no long-term impacts were predicted for a variety of aquatic life populations, including those for aquatic plants, benthic organisms (amphipods, chironomids), suspended feeders (*Daphnia*, copepods), predatory invertebrates (Odonata), mollusks (mussel), gastropods (snail), and fish. Some short-term decreases were observed for some aquatic invertebrates, such as

chironomids, amphipods, and Odonata, though recovery to pretreatment abundance levels was predicted to occur within 3 months.

The results of the AQUATOX modeling indicate that although some short-term impacts to individual invertebrates may be possible, longer-term population-level impacts are not predicted. If population-level impacts are not predicted, then commensurate community-level impacts are, by proxy, also not predicted. Further, if no impacts are observed within 1 year, impacts attributable to application across years are unlikely. The absence of longer-term population and community-level impacts for aquatic invertebrates predicted using AQUATOX was judged to be consistent with the findings previously reported in the open literature for Minnesota wetlands, seasonal wetlands in California, and wetlands and marshes Long Island exposed to vector control pesticides.

Landscape Model Case Study – JABOWA Assessment of Atmospheric CO_2 Increases on Forests

JABOWA, and its subsequent versions including JABOWA-II, is a generalized model of the reproduction, growth, and mortality of trees in mixed-species forests in response to environmental conditions. It is among the first multispecies computer simulations of terrestrial

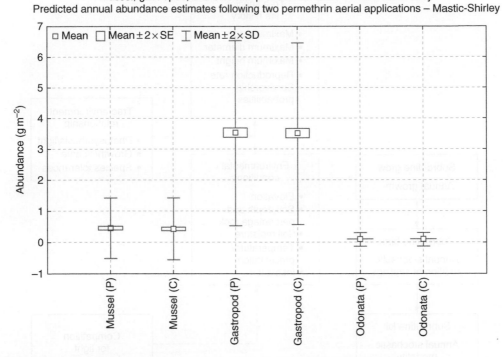

Mussel, gastropod, and Odonata present in shallow surface water body
Predicted annual abundance estimates following two permethrin aerial applications – Mastic-Shirley

Figure 12   AQUATOX-predicted annual abundances for fish in treated and control simulations.

ecosystems developed, and it has undergone extensive use and modification during the 30 years since it originated.

JABOWA was originally designed to be used for the Hubbard Brook Forest in northeastern North America; however, its underlying concepts are general, so that any nonhydrophytic tree species could be used for the simulation. The model enables the user to examine forest responses to a variety of abiotic (e.g., temperature, elevation, soil moisture-holding capacity) and biotic (e.g., competition for light) variables. The user can determine the kind and number of tree species; current versions allow for up to 45 species. Processes affecting growth and mortality in the chosen species take place independently within a series of grid areas (10 m × 10 m is the default value, which is user-adjustable in early versions of the model).

The model functions by setting a number of environmental conditions which, in turn, determine the shape of the trees, their growth, reproduction, and mortality. These outcomes define community endpoints including stand density, species diversity, and composition (**Figure 13**). The basic growth function is a species-specific growth curve (determined empirically for a tree growing under optimum conditions) with dependence on canopy-level solar radiation; the temperature index which simulates effects of monthly mean temperatures on photosynthesis rates; a soil quality index which simulates the effect of soil structure on tree growth; and, in more recent versions of the model, soil nitrogen, depth of

the water table, depth of soil, and other soil characteristics that also influence growth.

The growth function is modified by a coefficient accounting for competition between trees within the same grid plot as a function of tree density, and for the effects of the existing environment on each species. Species-specific recruitment depends on species density and a coefficient defining seed production for each species. Mortality is also modeled to affect competition for light by altering the number of tree stems on a plot. Annual tree mortality is determined for each age class and is modeled stochastically with two algorithms: the first algorithm is for healthy trees, in which it is assumed that 2% of the individuals of the species will, on average, reach the maximum known age for that species. The second algorithm is for trees that grow poorly, with a user-determined minimum growth; a second stochastic function assumes that such a tree will, on average, survive only 10 years unless growth rises above the user-set minimum.

At the landscape scale, grids are defined by state variables determining the number of saplings within a stand based on shade, elevation, soil type, soil capacity, percentage rock in soil, and monthly temperature and precipitation.

An example of applying the JABOWA landscape model was a study conducted in the United States by researchers at Indiana University and the Richard Stockton College of New Jersey to assess the response of southern Great Lakes forests to a doubling of atmospheric CO_2. For this

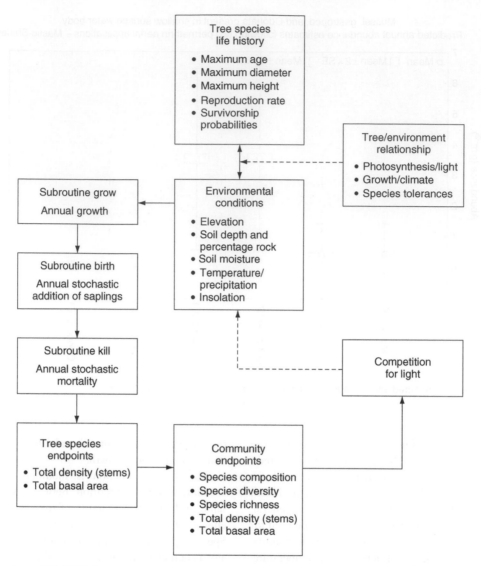

Figure 13   Structure of JABOWA forest landscape model. Copyright (© 2002) *Ecological Modeling in Risk Assessment: Chemical Effects on Populations, Ecosystems, and Landscapes,* by Pastorok RA, Bartell SM, Ferson S, and Ginzburg LR (eds.). Reproduced by permission of Routledge/Taylor & Francis Group, LLC.

study, JABOWA-II was embedded within a geographic information system (GIS) in order to examine temporal and spatial changes in forest condition in response to a changed climate relative to growth under baseline climate conditions.

Data providing the inputs for the model included:

- *type of land use*: urban, agricultural, forest, water, wetland, etc.;
- *tree species composition*: species name and diameter at breast height (dbh);
- *thermal data*: minimum- and maximum-growing degree-days used to define a thermal response curve in order to estimate tree growth increment;
- *nutrient concentrations*: available nitrogen, as a function of climate, litter quality, and soil texture (JABOWA-II

does not simulate nutrient dynamics; therefore, available nitrogen levels remained static throughout the modeling period);
- *baseline climate conditions*: average monthly temperature and precipitation across the region, spatially interpolated from 1180 weather stations.

Using these data sets, site files were generated from the average conditions existing in each $4 \, km^2$ cell for input into the JABOWA-II model. An Oregon State University general circulation model (OSUGCM) was used to generate a changed climate scenario, hypothesized as a doubling of CO_2 over an 80-year modeling period that produced linear changes in temperature and precipitation. Baseline and climate-changed conditions were iterated separately and forest growth outputs were compiled at 10-year intervals as total basal area for each species. The outputs of these two scenarios were analyzed in three ways: (1) nonspatial, specific responses to the changed climate in terms of total basal area of trees; (2) basal area-weighted population centroids were calculated to investigate changes in individual species' population distributions; and (3) maps were produced for species groups and regionally dominant species to assess spatial patterns of growth response for the larger forest community, across a landscape (e.g., **Figure 14**).

Relative to baseline forest conditions, the climate change scenario predicted large decreases in total basal area of northern conifers and northern deciduous species (>99% decrease), and population centroid shifts were primarily toward the northeast. These results were in agreement with other studies examining effects of climate change. In contrast with other studies, only slight increases in intermediate and southern species (which represent *c.* 90% of the region's basal area) were predicted from this model. The authors attributed this finding to the ability of their model to consider a high degree of spatial resolution, realism of modeled forest conditions, and accuracy in spatial configuration within the land-use matrix. Some important limitations of the model were highlighted in this application as well. These limitations included the inability to simulate nutrient dynamics, whereas changes in nutrient levels over time are a potentially important outcome of climate change; the inability to simulate non-linear responses such as variable growth responses across age classes; and the inability of JABOWA-II, which

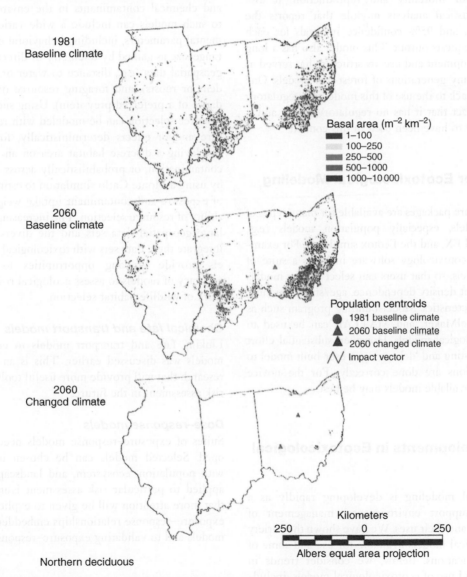

1981
Baseline climate

2060
Baseline climate

2060
Changed climate

Northern deciduous

Basal area (m^{-2} km^{-2})
■ 1–100
▨ 100–250
▨ 250–500
■ 500–1000
■ 1000–10 000

Population centroids
● 1981 baseline climate
▲ 2060 baseline climate
▲ 2060 changed climate
∧ Impact vector

Kilometers
250   0   250

Albers equal area projection

Figure 14 Results from application of JABOWA forest landscape model to assess effects of increasing atmospheric CO_2. Reprinted from Ehman JL, Fan W, Randolph JC, Southworth J, and Welch NT (2002) An integrated GIS and modeling approach for assessing the transient response of forests of the southern Great Lakes region to a doubled CO_2 climate. *Forest Ecology and Management* 155: 237–255, with permission from Elsevier.

allocates annual growth increments, to account for longer-term climatic episodes like multiyear droughts and their effects on forest communities.

JABOWA meets several criteria of a useful and relevant environmental model for terrestrial forest ecosystems. The model combines mechanistic functions and site-specific empirical relationships that accurately describe forest processes. The model can simulate a variety of ecologically relevant endpoints, including tree biomass, forest productivity, or effects of toxic chemicals. It is quite flexible; the parameters of each species are based on information for its entire range, and it has been applied to many types of forests over a range of environmental conditions in North America, Siberia, Eastern Europe, and Costa Rica. JABOWA incorporates uncertainty in the form of stochastic functions for mortality and reproduction. It also includes a statistical analysis module that reports the mean, variance, and 95% confidence intervals for each year the user requests output. This model also has a long history of development and use; its structure has served as the basis for many generations of forest-gap models. One potential drawback to the use of this model in a regulatory context is the fact that it has no regulatory status, and it does not appear to have been used in this context.

Software for Ecotoxicological Modeling

Excellent software packages are available for many types of ecological models, especially population models (e.g., RAMAS, VORTEX, and the Ecotox simulator). For example, RAMAS Ecotoxicology software includes a suite of population models, so that users can select among models with and without density dependence, age/stage structure, and other characteristics. Model-building program such as STELLA, ModelMaker, and MATLAB can be used to build ecotoxicological models. However, substantial effort is needed for testing and 'debugging' a user-built model to ensure calculations are done correctly. For the novice modeler, using available models may be best.

Future Developments in Ecotoxicological Modeling

Ecotoxicological modeling is developing rapidly as a discipline to support environmental management of toxic chemicals and their uses. We have shown the variety of ecotoxicological models available for use and some of their past applications. Below, we consider trends in development and use of ecotoxicological models, including linking other kinds of models to ecological models, spatially explicit models, probabilistic models, and ecological modeling systems.

Linking Other Models to Ecological Models

Ecotoxicological models can be linked to other kinds of models to provide integrated modeling systems. Examples are discussed below.

Wildlife resource selection models

Resource selection addresses how individuals or populations choose the spatial environment in which they conduct activities, including where and when they forage, rest, or reproduce. The spatial–temporal area over which these activities take place defines the ecological habitat of an organism or population, and these models are therefore often applied in terms of habitat selection. Habitat selection models can assist in chemical risk assessments by examining spatial–temporal intersections of receptors and chemical contaminants in the environment. Inputs to such models can include a wide variety of environmental parameters, including behavioral data (e.g., home range size as gained by telemetry or direct observations), geospatial data (e.g., distance to water or distribution of dens or roosts), and foraging resource data (e.g., abundance of a preferred prey item). Using such data, habitat resource selection can be modeled with respect to exposure and/or effects deterministically, for example, by overlaying a discrete habitat area on an area of known contamination, or probabilistically across a landscape; or by using a Monte Carlo simulation to estimate likelihood of exposure and contaminant uptake weighted by probability of resource selection. The increasing availability of large geospatial data sets and the diversity of ways to integrate these data sets with toxicological response models provide growing opportunities to increase the accuracy of models to assess ecological risk in the framework of wildlife habitat selection.

Chemical fate and transport models

Linking fate and transport models to ecotoxicological models was discussed earlier. This is an active area of research that will provide more useful tools for ecological risk assessment in the future.

Dose–response models

Suites of exposure–response models need to be developed. Selected models can be chosen in combination with population, ecosystem, and landscape models and applied to particular risk assessment issues. We expect that more attention will be given to explicit modeling of exposure–response relationships embedded in ecological models and to validating exposure–response functions.

Spatially Explicit Models

Spatially explicit models have been applied to management of toxic chemical issues primarily for fish, birds, mammals, and other vertebrate species. As discussed above, linking habitat selection models for fish and wildlife to ecotoxicological models will add realism to future risk assessments. Habitat models incorporating toxic chemical effects on plants are less common, but we expect their development and use to increase in the future. Spatially explicit models may include not only habitat models but also spatial–temporal variations in vital rates of populations, food web structure, nutrient and toxic chemical distributions, hydrologic variables, and other environmental factors.

Ecological Modeling Systems

Use of combinations of ecotoxicological models in a modeling system will likely increase in the future. The Across-Trophic-Level System Simulation (ATLSS) is a landscape model of the Everglades in the state of Florida in the US. ATLSS is a multicomponent modeling framework or set of integrated models that simulate the hierarchy of whole-system responses across all trophic levels and across spatial and temporal scales that are ecologically relevant to a large wetland system. ATLSS uses different modeling approaches tailored to each trophic level, including differential equations for process models of lower levels and age-structured and individual-based models for higher levels. Other integrated modeling systems exist that include an eutrophication model, a toxic chemical fate and transport model, a toxicity model, and a food-web model.

Probabilistic Models

Probabilistic modeling is an active area of research and many ecotoxicological models have been applied in probabilistic risk assessments. Monte Carlo analysis is a common approach to development of probabilistic risk estimates for population-, ecosystem-, and landscape-level endpoints. Because of the importance of this area, and because a discussion of probabilistic approaches is beyond the scope of this article, we urge the reader to consult the ecological risk assessment literature to learn more about this topic.

Conclusions

Ecological models are important tools for use in characterizing ecological risks of toxic chemicals because they support quantitative determinations of the ecological significance of risks expressed only in terms of organism-level endpoints. True ecotoxicological models combine steady-state or dynamic modeling of ecological systems with exposure–response models to predict the effects of toxic

chemicals on endpoints for populations, ecosystems, or landscapes. They can be applied to translate the organism-level endpoints of survival, growth, and various reproductive measures into estimates of risk or population decline, risk of extinction, risk of habitat fragmentation, or other metrics. Data are often limited on vital rates and dose–response functions needed for ecological modeling, but reasonable estimates of model parameters can often be made to meet the objectives of an ecological risk assessment. Often, a comparative assessment of risk (e.g., relative to baseline or reference) is of primary interest. Ecotoxicological models and user-friendly software for them are now widely available. Population, ecosystem, and landscape models that incorporate toxicological data (e.g., dose–response functions) are being used increasingly in environmental management of toxic chemical problems. Such models can be applied in the context of ecological risk assessments, design and evaluation of remedial actions, habitat restoration projects, and monitoring programs.

See also: Dose–Response; Exposure and Exposure Assessment.

Further Reading

Akcakaya HR and Atwood JL (1997) A habitat-based metapopulation model of the California gnatcatcher. *Conservation Biology* 11: 422–434.
Akcakaya HR, Burgman MA, and Ginzburg LR (1999) *Applied Population Ecology, Principles and Computer Exercises using RAMAS Ecolab*, 2nd edn. Sunderland, MA: Sinauer Associates.
Bartell SM, Campbell KR, Lovelock CM, Nair SK, and Shaw JL (2000) Characterizing aquatic ecological risks from pesticides using a diquat dibromide case study. Part III: Ecological process models. *Environmental Toxicology and Chemistry* 19(5): 1441–1453.
Botkin DB (1993) *Forest Dynamics: An Ecological Model*. New York: Oxford University Press.
Bridges TS and Carroll S (2000) *Application of Population Modeling to Evaluate Chronic Toxicity in the Estuarine Amphipod Leptocheirus plumulosus*. Vicksburg, MS: US Army Engineer Research and Development Center (ERDC/TN EEDP-01-44).
DeAngelis DL, Gross LJ, Huston MA, *et al.* (1998) Landscape modeling for Everglades ecosystem restoration. *Ecosystems* 1: 64–75.
Ehman JL, Fan W, Randolph JC, Southworth J, and Welch NT (2002) An integrated GIS and modeling approach for assessing the transient response of forests of the southern Great Lakes region to a doubled CO_2 climate. *Forest Ecology and Management* 155: 237–255.
Hakoyama H and Iwasa Y (1998) Ecological risk assessment: A new method of extinction risk assessment and its application to a freshwater fish (*Carassius auratus* subsp.). In: Nakanishi J (ed.) *Proceedings of First International Workshop of Risk Evaluation and Management of Chemicals*, pp. 93–110. Yokohama, Japan: Japan Science and Technology Corporation.
Jackson LJ, Trebitz AS, and Cottingham KL (2000) An introduction to the practice of ecological modeling. *BioScience* 50: 694–706.
Jørgensen SE (1994) *Fundamentals of Ecological Modelling*, 2nd edn. Amsterdam: Elsevier.
Jørgensen SE, Halling-Sorensen B, and Nielsen SN (1996) *Handbook of Environmental and Ecological Modeling*. New York: Lewis Publishers.
Munns WR, Black DE, Gleason TR, *et al.* (1997) Evaluation of the effects of dioxin and PCBs on *Fundulus heteroclitus* populations using a modeling approach. *Environmental Toxicology and Chemistry* 16: 1074–1081.

Pastorok RA, Akçakaya HR, Regan HM, Ferson S, and Bartell SM (2003) Role of ecological modeling in risk assessment. *Human and Ecological Risk Assessment* 9(4): 939–972.

Pastorok RA, Bartell SM, Ferson S, and Ginzburg LR (eds.) (2002) *Ecological Modeling in Risk Assessment: Chemical Effects on Populations, Ecosystems, and Landscapes*. Boca Raton, FL: Lewis Publishers.

Spencer M and Ferson S (1997) *RAMAS Ecotoxicology: Ecological Risk Assessment for Structured Populations*. Setauket, NY: Applied Biomathematics.

Suter GW, II (ed.) (1993) *Ecological Risk Assessment*. Boca Raton, FL: Lewis Publishers.

US Environmental Protection Agency (2004) AQUATOX (Release 2). *Modeling Environmental Fate and Ecological Effects in Aquatic Ecosystems, Vol. 1: User's Manual*, EPA-823-R-04-001. Washington, DC: United States Environmental Protection Agency.

US Environmental Protection Agency (2004) AQUATOX (Release 2). *Modeling Environmental Fate and Ecological Effects in Aquatic Ecosystems, Vol. 2: Technical Documentation*, EPA-823-R-04-002. Washington, DC: United States Environmental Protection Agency. v2.1 available at http://www.epa.gov/waterscience/models/aquatox/technical/release21-addendum.pdf (accessed November 2007).

Effects of Endocrine Disruptors in Wildlife and Laboratory Animals

F S vom Saal, University of Missouri – Columbia, Columbia, MO, USA

L J Guillette Jr., University of Florida, Gainesville, FL, USA

J P Myers, Environmental Health Sciences, White Hall, VA, USA

S H Swan, University of Rochester, Rochester, NY, USA

Introduction
Wildlife Studies
Studies with Laboratory Animals: Effects of Endocrine Disrupting Chemicals in Plastic

Summary
Further Reading

Introduction

In this article, we provide an overview of the impacts of endocrine-disrupting chemicals (EDCs) on wildlife and laboratory animals. The definition of endocrine disruption and overview of the issues in the field of endocrine disruption were covered in Endocrine Disruptor Chemicals: Overview. Here we focus primarily on the effects of chemicals with estrogenic and antiandrogenic activity in wildlife and both rats and mice in controlled studies. The chemicals discussed include those used in pesticides and plastic and other household products.

Wildlife Studies

For many decades, there has been concern about the effects of environmental contaminants on the health and persistence of wildlife populations. Prior to work over the last 10–15 years, the vast majority of these studies examined the lethal consequences of exposure, or they focused on the induction of cancer or major birth defects. Although these endpoints are still critical in the study of

toxicology, a growing collection of studies examining diverse wildlife species demonstrates that additional adverse outcomes can be produced in wildlife as a result of exposure to environmental contaminants. A number of these abnormalities have been attributed to the disruption of endocrine signaling. Below, we examine a few of the many examples of endocrine disruption in wildlife.

Fish, Vitellogenesis, and Sewage

In the 1990s, reports were published documenting that male fish living below sewage outfalls in Europe, Great Britain, North America, and Japan had elevated plasma concentrations of the yolk protein vitellogenin. Vitellogenin is normally synthesized in the liver of the female following stimulation by elevated plasma estrogens of ovarian origin. Males of many vertebrate classes, including fish, amphibians, and reptiles, have the ability to synthesize vitellogenin if stimulated by estrogen, although this does not occur normally. Intensive chemical fractionation of sewage identified two major classes of compounds capable of acting as estrogens in male fish; these included the pharmaceutical estrogen, ethinylestradiol, and the industrial chemicals, nonylphenol and octylphenol. Ethinylestradiol is a common ingredient of the human birth control pill and is excreted in the urine of females taking this pharmaceutical agent. Ethinylestradiol has been identified in the surface and reclaimed sewage waters from all continents where such studies have been performed. Similarly, nonylphenol, an alkylphenolic chemical, is widely used in industrial applications as a surfactant and is commonly released into the environment. It is persistent in the ecosystem with very large concentrations found associated with sediments and organic matter in freshwater and estuarine regions. It has been shown to be weakly estrogenic in mammalian laboratory animals, but is a potent estrogen in many fish. Laboratory-based life-cycle testing with ecologically relevant concentrations has shown that both of these compounds have adverse effects on the reproductive potential of males and females, and they also alter sex determination in developing embryos. These common pollutants have the potential to disrupt the health of individual animals and the persistence of populations; some populations have no males. It has also been suggested that endocrine disruption could be associated with the decline of commercial and sport fish populations.

Alligators and Pesticides

Alligators and crocodiles are long-lived top predator species inhabiting most subtropical and tropical wetlands. Studies begun in the late 1980s reported abnormalities in central and south Florida (USA) populations of the American alligator exposed to various contaminant mixtures associated with modern agriculture, such as insecticides, herbicides, and fertilizers. These abnormalities include altered plasma sex steroid profiles, gonadal, genital, and immune tissue anatomy, and hepatic steroid metabolism. Specifically, male alligators exposed in ovo (as embryos) to various pesticides, due to deposition in the eggs prior to being laid by the female, exhibit significantly reduced plasma testosterone concentrations, aberrant testicular morphology, and small penis size. Females from the same contaminated locations displayed significantly elevated plasma concentrations of estradiol as neonates but reduced concentrations as subadults. Subadult females also had elevated plasma concentrations of the potent androgen dihydrotestosterone. They also exhibit a high frequency of polyovular follicles, an ovarian abnormality associated with low fertility and high embryonic mortality. These contaminated populations have shown elevated embryonic mortality greater than 50%. Polyovular follicles are a documented outcome of exposure of women to the estrogenic drug diethylstilbestrol during fetal life exposed as fetuses due to their

mothers taking this drug during pregnancy, and these exposed women also suffer a decrease in fertility.

Populations displaying these abnormalities have elevated egg, tissue, or serum concentrations of a wide range of organochlorine pesticides or their metabolites, heavy metals, and other widely used agricultural chemicals, such as nitrates. Experimental exposure of developing alligator embryos to various organochlorine pesticides or their metabolites induces many of the abnormalities seen in wild populations, such as altered plasma hormone profiles and small penis size as well as altered sex determination. Concentrations required to induce these abnormalities were in the part per trillion to part per billion range, 100–1000 times lower in concentration than the reported levels in alligator eggs or serum.

Studies of mosquito fish from the same contaminated lakes indicate that the reported abnormalities are not limited to a single lake or species, as male mosquito fish have reduced tissue concentrations of testosterone, lower sperm counts, and altered reproductive behavior.

Fish and Pulp Mill Effluent

Many studies have documented the detrimental effects of pulp mill effluent on the environment over many decades. Classical ecotoxicology studies reported wide-scale disruption of populations, including the local extinction of many exposed freshwater or estuarine fish and invertebrate populations. Although modifications in the processing of pulp mill effluent have occurred over time, abnormalities persist. Studies from several Canadian locations report altered hormone profiles in fish exposed to pulp mill effluent, including alterations in hypothalamic, gonadal, and adrenal hormones. Exposed fish displayed altered stress responses and altered reproductive performance.

Masculinization of females has also been reported. For example, female mosquito fish living below effluent outfalls from paper pulp mills develop a gonopodium, a modified anal fin found in males of this species and used to transfer sperm to the female for internal fertilization. The gonopodium develops in the male following exposure to androgens, specifically testosterone. Masculinized females do reproduce but have lower production of offspring and greatly elevated levels of aromatase activity in their brain and ovary. Aromatase is an enzyme that converts the hormone testosterone to estradiol, the principle estrogen in these females. These females thus have an altered potential to produce this critical hormone that regulates reproduction.

Fish, Feedlots, and Pharmaceuticals

Modern animal production techniques in many countries involve the use of potent hormones and antibiotics. Although there has been an ongoing debate on the safety of the meat products produced from such practices, few concerns have been voiced concerning the possible ecological impacts of these practices. Guillette and colleagues have examined feral fish exposed to effluent released from animal feedlots, including urine and feces, into a natural river system. They observed that male fish exhibited many of the classical signs of exogenous androgen exposure, including reduced testicular mass, reduced plasma testosterone, and altered head morphology. With the extensive use of anabolic steroids in cattle production, the potential for wide-scale disruption of fish reproduction is possible, since the presence of ethinylestradiol in rivers after excretion by women and bacterial action in water treatment plants leads to endocrine disruption in fish. Experimental laboratory-based studies support the field observations, as low-level exposure to the commonly used anabolic steroid trenbolone (used to promote growth in cattle) alters fish development and reproduction in a manner similar to that observed in the wild fish.

These and many more observations of wildlife demonstrate that global contamination of wildlife populations has dramatic effects on the health and reproductive potential of these populations. As described in Endocrine Disruptor Chemicals: Overview, the phenotype observed in individuals is produced by the environment acting on the genotype. The abnormalities observed in wildlife are not due to classically held concepts of gene mutations. Instead, they represent alterations in the timing of gene expression and the level of gene expression. As described above, if exposure occurs during embryonic development, these alterations can be permanent.

Studies with Laboratory Animals: Effects of Endocrine Disrupting Chemicals in Plastic

Numerous studies in laboratory animals have documented profound embryonic disruption by low-level exposure to environmental chemicals including pesticides (herbicides, insecticides, and fungicides) and chemicals contained in a wide range of industrial products.

Bisphenol A: The Estrogenic Chemical Used to Make Polycarbonate Plastic

Since 1997, a large number of peer-reviewed journal articles have been published showing that bisphenol A causes harm in animals at levels to which the average human is exposed due to the widespread use of bisphenol A in plastic products and to line metal cans. Bisphenol A is one of many chemicals that have the ability to bind to estrogen receptors and initiate cellular responses similar to those caused by the endogenous estrogen, estradiol. However, bisphenol A was incorrectly initially thought to only be a very weak estrogen-mimicking chemical. Instead, recent experiments have shown that at very low doses that had previously been predicted to be safe based on models, not data, bisphenol A has dramatic adverse effects.

Recent findings include chromosomal damage in developing oocytes in mouse ovaries, and abnormalities in the entire reproductive system in male mice, including a decrease in testicular sperm production and a decrease in fertility, In addition, fetal exposure to bisphenol A increases the rate of postnatal growth and decreases the age at which females mature sexually (go through puberty). These females also have mammary gland abnormalities, and mammary glands appear precancerous by the time the females reach young adulthood. Bisphenol A also causes abnormal brain development, and changes in brain function and behavior. Bisphenol A also disrupts immune function. Bisphenol A is also an animal carcinogen, since exposure to a very low dose during early postnatal development causes prostate cancer in rats.

In addition to acting as an estrogen-mimicking chemical, bisphenol A binds to androgen receptors, but the consequence is not activation of the receptor (steroid receptors are ligand-activated transcription factors, which results either in activating or suppressing transcription of specific genes). Instead, bisphenol A acts to inhibit the ability of androgens (testosterone or dihydrotestosterone) from binding to the androgen receptors. Bisphenol A is thus also referred to as an antiandrogen. Bisphenol A at somewhat higher doses also can bind to nuclear receptors for thyroid hormone, and, again, the consequence is that bisphenol A antagonizes the action of thyroid hormone and is thus also has antithyroid hormone activity. A surprising recent finding is that bisphenol A stimulates insulin secretion but reduces insulin sensitivity in mice. This action involves activating rapid enzyme cascades via receptors associated with the cell membrane, rather than via binding to the classical estrogen receptors associated with specific regions of DNA in the nucleus. These effects are of particular concern in that while bisphenol A typically is 100–1000-fold less potent than estradiol in stimulating responses via binding to the nuclear estrogen receptors, bisphenol A and estradiol are equally potent in stimulating responses after activation of these newly discovered membrane-associated receptors.

Phthalates: Inhibitors of Testosterone Synthesis Used as Additives in Polyvinylchloride Plastic

Diesters of phthalic acid, commonly referred to as phthalates, are widely used in industry and commerce, including personal care products (such as makeup, shampoo, and soaps), plastics, medical tubing and medication coatings, paints, and some pesticide formulations. Polyvinyl chloride (PVC) plastic cannot be made without adding phthalate plasticizers, which act as softeners.

Phthalates are not chemically bonded to PVC, so the phthalates readily migrate out of the PVC (e.g., when a baby sucks on a PVC toy), resulting in human exposure to phthalates. For example, most 'rubber duckies' are about 50% PVC, and the remaining 50% is the phthalate di-isononyl phthalate (DINP), which is actually a mixture of a large number of 100 isomers.

There is consistent toxicological evidence of adverse developmental and reproductive effects of several of these phthalate esters. In particular, di-*n*-butyl phthalate (DBP), butyl benzyl phthalate (BBzP), di-2-ethylhexyl phthalate (DEHP), and DINP have been shown to disrupt reproductive tract development in male rodents. This will be discussed in more detail in Epidemiological Studies of Reproductive Effects in Humans, where similar findings are presented based on epidemiological studies on human babies. Of particular concern is DEHP, which is the most commonly used phthalate, with over 4 billion pounds produced in 2000, and its metabolites are considered to be the most potent reproductive toxicant among the many different types of phthalates. Most medical tubing contains DEHP, which has been measured in neonates in an intensive care nursery. DEHP is also used in vinyl materials in cars (producing the 'new car smell'), in flooring, toys, and a wide variety of other products. Many cosmetics contain the phthalates DEP and DBP, although manufacturers are increasingly switching to phthalate-free options. PVC also may contain bisphenol A, which, along with phthalates, will leach out of the plastic.

Summary

In this paper, we have discussed evidence from diverse sources indicating that a variety of man-made compounds can interfere with reproductive system and brain development, resulting in reduced fertility and altered brain function and behavior in wildlife and laboratory animals. Wildlife have acted as sentinels for human health for centuries. Laboratory animal studies of molecular mechanisms provide a basis for examining outcomes in humans when there is a high level of similarity between animals and humans in the mechanisms mediating responses to specific classes of chemicals, such as those with estrogenic or anti-androgenic activity. An important issue is thus whether the abnormalities reported in wildlife and laboratory animals provide a warning that human health and development are at risk. Findings relating environmental chemicals to human health are discussed in Epidemiological Studies of Reproductive Effects in Humans.

See also: Endocrine Disruptor Chemicals: An Overview; Epidemiological Studies of Reproductive Effects in Humans.

Further Reading

Ankley GT, Jensen KM, Makynen EA, *et al.* (2003) Effects of the androgenic growth promoter 17-beta-trenbolone on fecundity and reproductive endocrinology of the fathead minnow. *Environmental Toxicology and Chemistry* 22: 1350–1360.

Anway MD, Cupp AS, Uzumcu M, and Skinner MK (2005) Epigenetic transgenerational actions of endocrine disruptors and male fertility. *Science* 308: 1466–1469.

Gray LE, Jr., Wilson VS, Stoker T, *et al.* (2006) Adverse effects of environmental antiandrogens and androgens on reproductive development in mammals. *International Journal of Andrology* 29: 96–104.

Guillette LJ, Jr., Crain DA, Gunderson M, *et al.* (2000) Alligators and endocrine disrupting contaminants: A current perspective. *American Zoologist* 40: 438–452.

Guillette LJ, Jr., and Gunderson MP (2001) Alterations in the development of the reproductive and endocrine systems of wildlife exposed to endocrine disrupting contaminants. *Reproduction* 122: 857–864.

Hayes TB, Collins A, Lee M, *et al.* (2002) Hormaphroditic, demasculinized frogs after exposure to the herbicide atrazine at low ecologically relevant doses. *Proceedings of the National Academy of Sciences of the United States of America* 99: 5476–5480.

Ho S-M, Tang W-Y, Belmonte de Frausto J, and Prins GS (2006) Developmental exposure to estradiol and bisphenol A increases susceptibility to prostate carcinogenesis and epigenetically regulates phosphodiesterase type 4 variant 4. *Cancer Research* 66: 1–9.

Lilienthal H, Hack A, Roth-Harer A, Grande SW, and Talsness CE (2006) Effects of developmental exposure to 2,2,4,4,5-pentabromodiphenyl ether (PBDE-99) on sex steroids, sexual development, and sexually dimorphic behavior in rats. *Environmental Health Perspectives* 114: 194–201.

Oehlmann J, Schulte-Oehlmann U, Bachmann J, *et al.* (2006) Bisphenol A induces superfeminization in the ramshorn snail Marisa cornuarietis(Gastropoda: Prosobranchia) at environmentally relevant concentrations. *Environmental Health Perspectives* 114(supplement 1): 127–133.

Orlando EF, Davis WP, and Guillette LJ, Jr. (2002) Aromatase activity in the ovary and brain of the eastern mosquitofish (Gambusia holbrooki) exposed to paper mill effluent. *Environmental Health Perspectives* 110(supplement 3): 429–433.

Orlando EF, Kolok AS, Binzcik GA, *et al.* (2004) Endocrine-disrupting effects of cattle feedlot effluent on an aquatic sentinel species, the fathead minnow. *Environmental Health Perspectives* 112: 353–358.

Thompson RC, Olsen Y, Mitchell RP, *et al.* (2004) Lost at sea: Where is all the plastic? *Science* 304: 838.

Timms BG, Howdeshell KL, Barton L, *et al.* (2005) Estrogenic chemicals in plastic and oral contraceptives disrupt development of the mouse prostate and urethra. *Proceedings of the National Academy of Sciences of the United States of America* 102: 7014–7019.

vom Saal FS and Welshons WV (2006) Large effects from small exposures. II. The importance of positive controls in low-dose research on bisphenol A. *Environmental Research* 100: 50–76.

Vreugdenhil HJ, Slijper FM, Mulder PG, and Weisglas-Kuperus N (2002) Effects of perinatal exposure to PCBs and dioxins on play behavior in Dutch children at school age. *Environmental Health Perspectives* 110: A593–A538.

Welshons WV, Nagel SC, and vom Saal FS (2006) Large effects from small exposures. III. Endocrine mechanisms mediating effects of bisphenol A at levels of human exposure. *Endocrinology* 147: S56–S69.

Weltje L, vom Saal FS, and Oehlmann J (2005) Reproductive stimulation by low doses of xenoestrogens contrasts with the view of hormesis as an adaptive response. *Human and Experimental Toxicology* 24: 431–437.

Epidemiological Studies of Reproductive Effects in Humans

S H Swan, University of Rochester, Rochester, NY, USA

L J Guillette Jr., University of Florida, Gainesville, FL, USA

J P Myers, Environmental Health Sciences, White Hall, VA, USA

F S vom Saal, University of Missouri-Columbia, Columbia, MO, USA

Introduction

In this article we discuss the evidence for effects of endocrine disrupting chemicals on human health and the considerable challenges this new science of endocrine disruption presents to epidemiology. We then briefly describe the progress being made toward a more 'environmentally sensitive epidemiology' and illustrate these concepts with data from a recent study that is using biomarkers of low environmental levels of endocrine disrupting chemicals (EDCs) to identify some of the impacts of these chemicals on human health.

Endocrine Disrupting Chemicals with Antiandrogenic Activity

Epidemiological Studies of Phthalates

Diesters of phthalic acid, commonly referred to as phthalates, are widely used in industry and commerce, including in personal care products (such as makeup, shampoo and soaps), plastics, medical tubing and medication coatings, paints and some pesticide formulations. Polyvinyl chloride (PVC) plastic cannot be made without adding phthalate plasticizers, which act as softeners. Phthalates are not chemically bonded to PVC, so the phthalates readily migrate out of the PVC (e.g., when a baby sucks on a PVC toy), resulting in human exposure to phthalates.

Despite the growing body of literature on phthalate reproductive toxicity and data demonstrating extensive human exposure, few studies have examined the effects of these chemicals on human reproductive development. Duty and colleagues reported dose–response relationships between tertiles of mono-butyl phthalate (MBP) and sperm motility and concentration as well as between tertiles of monobenzyl phthalate (MBzP) and sperm concentration. They also reported inverse dose–response relationships between monoethyl phthalate (MEP) and increased sperm DNA damage measured using the neutral single cell gel electrophoresis (Comet) assay. In this population of men attending an infertility clinic, increased urinary concentration of MBzP was also associated with decreased follicle stimulating hormone (FSH), while increases in MBP were marginally associated with increased inhibin-B .

Over the past 5 years, increasingly detailed rodent studies have demonstrated that there is a significant reduction in anogenital distance (AGD), which is the tissue that forms into the scrotum, in Sprague-Dawley rats following prenatal exposure at high doses to BBzP, DBP, and DEHP. This demasculinizing effect is due to disruption of the androgen-signaling pathway, resulting in interference with the synthesis of testosterone in the testes of male rat fetuses. The consequence of a decrease in testosterone during the fetal period of masculinization is that normal masculinization, which occurs in response to testosterone, does not occur. Virtually every aspect of development of a masculine phenotype is disrupted by these phthalates: in addition to shortened AGD, the penile and urethral development are abnormal, resulting in an increased frequency of hypospadias (a genital malformation in which the urethral opening is improperly located), and incomplete descent of the testes (cryptorchidism), testes are abnormally small, and other reproductive organs are abnormal. Three phthalates (DEHP, DBP, and BzBP) have been shown to downregulate expression of the insl3 (insulin-like hormone 3) gene, which plays a critical role in gubernacular development and thus normal testicular descent. The cluster of morphological and functional changes induced by these phthalates has been referred to as the phthalate syndrome. Research described below has begun to relate prenatal phthalate exposure in humans to similar reproductive and developmental effects.

In a recent study, Swan and colleagues examined the concentration of urinary phthalate metabolites in mother's prenatal urine samples in relation to the AGD and other measures of male genital development. This analysis includes physical measurements on 134 boys, obtained on average at 12.6 months of age, whose mothers had participated in a multi-center pregnancy cohort study. The AGD in millimeters was measured and regression analysis used to obtain predicted AGD, controlling for age and weight at examination. Urinary phthalate metabolites were obtained in 85 prenatal urine samples provided by mothers of these boys at 28.3 weeks gestation, on average. In regression analyses the concentration of four phthalate metabolites that are found in urine (MEP, MBP, MBzP, and MiBP) were significantly, and inversely related to age-adjusted AGD. While MEHP was not related to AGD, the secondary metabolites of DEHP (MEOHP and MEHHP) were related, though not statistically significantly at $p = 0.05$. The authors considered mother's ethnicity and smoking, time of day and season of sample collection as covariates but none were retained in final models. The odds ratio for short AGD (less than 25% of expected for age and weight) were 10.2 (95% CI 2.5 to 42.2) for MBP in the highest quartile compared to the lowest quartile). For highest compared to lowest MEP, MBzP, and MiBP the odds ratios were 4.7, 3.8, and 7.3, respectively (all p values <0.05). While none of these boys exhibited frank malformations, short AGD was associated with an increased proportion of boys with incomplete testicular descent as well as with smaller penile volume.

The cut-offs defining 'high' phthalate exposure were quite similar to the 75th percentiles of phthalate distribution in a national sample (CDC 3rd Report). A comparison of these levels to doses used in most phthalate rodent studies suggests that humans are at least an order of magnitude more sensitive to these metabolites than rodents. However, until recently, phthalates have been tested singly. Recent mixture studies have shown that these exposures are at least dose-additive, so exposure to multiple phthalates at low levels, which is the typical human exposure pattern, may convey considerably greater risk than predicted by the typical single exposure rodent study. This is supported from results reported by Swan and colleagues. Boys whose mothers had high levels of multiple phthalate metabolites were at particularly high risk of short AGD. Thus, these data suggest that prenatal exposure to phthalates at current environmental levels, may produce a cluster of genital changes consistent with the phthalate syndrome previously identified in rodents.

While Swan et al. did not obtain infant hormone levels, a recent Danish study examined serum hormone levels in 96 boys at three months of age in relation phthalate metabolite levels in breast milk samples collected during postnatal months 1–3. MEP and MBP were significantly correlated with (sex hormone binding globulin) SHBG, which is the plasma-binding protein for both testosterone and estradiol ($r = 0.323$, $p = -0.002$ and $r = 0.272$, $p = 0.01$, for MEP and MBP, respectively); MMP, MEP, and MBP were positively correlated with LH/testosterone, and MBP was inversely correlated with free testosterone (all p values <0.05). (Approximately 50% of these boys were cryptorchid; these associations did not differ between normal and cryptorchid boys.) This study suggests an antiandrogenic (testosterone inhibiting) effect of perinatal phthalate exposure in humans, consistent with rodent studies.

Antiandrogenic Effects of Pesticides

Antiandrogenic effects have also been associated with exposure to pesticides. For example, the fungicide vinclozolin that is commonly sprayed on grapes in vineyards also has antiandrogenic activity. This antiandrogen action is quite different from phthalates however, in that vinclozolin binds to and thus blocks endogenous testosterone from binding to androgen receptors. Other pesticides, such as the insecticides DDT and its replacement after it was banned, the insecticide methoxychlor, have been shown to be converted in vivo to antiandrogenic metabolites; these compounds also bind to androgen receptors but, similar to vinclozolin, do not activate them and are thus antiandrogens. However, both DDT and methoxychlor are examples of products that produce complex effects by interacting with more than one aspect of the endocrine system. Commercial DDT is a mixture of a contaminant (o,p'DDT), which binds to estrogen receptors and activates responses similar to estradiol, while the active insecticide p,p'DDT is metabolized to the highly persistent compound p,p'DDE, which is an antiandrogen. Since o,p'DDT has a shorter half life than p,p'DDE, the initial consequence of DDT exposure is activation of estrogen and inhibition of androgen response systems, while the long-term consequence is persistent antiandrogenic effects. The in vivo metabolite of methoxychlor is similar to bisphenol A in that it has both estrogenic and antiandrogenic activity.

A recent finding that has generated concern is that administration of vinclozolin or methoxychlor to male rats during fetal life via administration to the pregnant mother resulted in male offspring with decreased sperm numbers and decreased fertility. This effect of exposure of F_1 generation males to these chemicals during fetal life (when genetic programming is occurring during cellular differentiation) resulted in the same phenotype occurring in F_2, F_3, and F_4 generations transmitted in each generation through the male offspring. Importantly, in each subsequent generation, the frequency of this phenotype was constant at about 90% of males. Thus, exposure of only the F_1 generation led to a 'programmed' change in the genome that was transmitted at least through 3 more generations. This phenotype was associated with epigenetic changes (covalent methylation of cytosine bases) in the sperm that was transmitted across generations. These findings raise the possibility that adverse effects caused by environmental chemical exposure during fetal life may not be just limited to the exposed generation, but through nonclassical (epigenetic) modifications of DNA, may become stably fixed into the genome of the germ line and transmitted across generations.

Issues Concerning Assessing Risks Posed by EDCs to Human Health

All of these findings suggest that it is likely that EDCs pose a significant threat to human health that classical epidemiological methods may not have the sensitivity to detect. Human studies on EDCs, which fall under the broader heading of environmental epidemiology, share many features with studies of environmental exposures, such as to radon or total suspended particulates, which are not endocrine disruptors. However, they differ from non-EDC studies in several important ways (study hypothesis, exposure(s), effect(s), model selection, analysis, and interpretation) that make detection of effects more difficult. We will consider these points and then examine a recent study that circumvents at least some of these problems.

What triggers an investigation between an EDC and a human health effect? Traditional ('classical') epidemiological studies were often designed to investigate unusual patterns of human health outcomes. Perhaps the most dramatic of these was the investigation of diethylstilbestrol (DES) in response to a cluster of seven cases of a rare vaginal cancer (clear cell adenocarcinoma) in young women. Similarly, an awareness of increasing rates of lung cancer triggered the first studies of smoking and lung cancer. Some epidemiological studies of EDCs have similar origins. Indeed, DES itself is a quintessential EDC, and current research into possible EDC involvement in breast cancer causation and fertility impairment have been provoked by observations of human trends.

Many epidemiological questions raised by EDCs have their origins, however, in observations of impacts on laboratory animals and wildlife. These include the possible role of EDCs in increases in hypospadias, the effects of phthalates on male fertility, and the impact of polybrominated diphenyl ethers (PBDEs) on neurocognitive development. In each of these cases, and many more, pronounced laboratory and field effects provoke questions about human impacts based on animal observations.

All else being equal, the ability of an epidemiological study to identify the cause of an adverse outcome decreases as the prevalence of the outcome and the number of causal factors increase. For example, the identification of DES as the cause of clear cell vaginal adenocarcinoma in young women was relatively easy because vanishingly few cases of this rare cancer had ever been documented in this age group, and no other cause had ever (before or since) been identified. Conversely, causes of breast cancer are notoriously hard to find, not only because it is a complex, multifactorial disease but because of its extremely high lifetime incidence (one in eight women). The metaphor of signal detection may be helpful in clarifying this point; high background levels of a disease contribute to background 'noise', as do alternative causes, errors in exposure identification and diagnoses) and make detection of the 'signal' (the association under investigation) difficult to identify. Epidemiology handles diseases of low incidence, and strong associations well, but multifactorial diseases of high incidence only poorly.

Consider the following thought experiment (**Figure 1**). Imagine a population of 5000 women with a significant but not unusual spontaneous miscarriage rate of 10%, normally distributed. In that hypothetical population one would expect 500 miscarriages, ± 42. In this experiment, expose 1% of the women to a contaminant that increases the risk of that abortion, on average by X-fold, with X increasing from 1 (no effect on risk) to a tenfold increase. Elevation in risk would have to be more than ninefold before the signal of exposure-induced miscarriage rose above background noise.

A crucial feature of EDCs is their 'stealth' nature. Several recent studies have demonstrated that the general population has been exposed to, and currently carries measurable levels of, tens to hundreds of EDCs. The

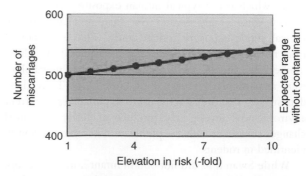

Figure 1 The expected number of miscarriages in a population of 5000 women as a function of risk elevated by exposure to a hypothetical contaminant. See text for parameters.

subject has no knowledge of these exposures; so the classical tools of epidemiologists (questionnaires, vital records, occupational histories, etc.) provide no information. These presuppose the subject's (or physician's or employer's) knowledge of exposure. Instead, it is necessary to obtain biological measures of exposure (biomarkers).

Biomarker studies require that subjects agree to provide a biological sample (e.g., blood, urine, or saliva) and give permission for its use in such a study. Obtaining subjects willing to do this, Institutional Review Boards willing to approve these protocols, and funding for such studies is becoming increasingly challenging. An increasing number of studies are taking this approach; we describe a recent example below. But for many EDCs, the analytical chemistry that would permit body burden measurements has not yet been developed, and for many for which it has, the chemical analyses are very costly, limiting sample size and thus statistical power. Moreover, the rapid metabolic degradation of some compounds means that single exposure measurements, for example from cord blood at birth, may completely miss critical exposures during pregnancy. Finally, there is now clear evidence that for some environmental chemicals, the detection limits for chemical analysis methods are far above the concentrations that are able to cause biological effects.

Whether distinguishing between exposure and non-exposure in cases and controls or estimating changes in risk as a function of increasing exposure, epidemiological studies traditionally assume monotonic dose response curves. Higher exposure levels are assumed to produce larger effects. Laboratory work with EDCs clearly shows, however, that nonmonotonic curves are commonly found. The result of use of inappropriate models, such as those that assume monotonicity of dose response, and absence of low dose effects, will result in 'false negatives'.

Epidemiology regularly compares exposed and unexposed populations. Yet the global distribution of EDCs means that finding unexposed populations is virtually impossible. Classical epidemiological studies were designed primarily to examine isolated exposures, ignoring concurrent exposures or considering them as confounding factors to be treated as 'nuisance variables'. This is inappropriate with EDCs, however, for two reasons. First, EDCs from similar and different chemical families can work through the same mechanism. They are thus substitutable. Unless the possibility of substitution is factored into the study by measuring multiple exposures and examining their joint risk, such mixtures will increase misclassification of exposure (an important source of conservative bias) and thus increase the likelihood of false negatives.

When mixtures of EDCs have been studied they have been seen to interact, often in unpredictable ways, with subadditive, additive and even synergistic effects. It is difficult, if not impossible, to isolate exposure to a single pesticide, phenol or phthalate. As a study by Thornton and colleagues showed, it is likely that all subjects are exposed to measurable amounts of large numbers of these chemicals, many of which act along common pathways. These factors pose a significant and currently unsolved challenge to epidemiology.

Long time lags between exposure and effect, which may span decades or even generations as in the case of DES, will further complicate detection of impacts. For nonpersistent compounds, all traces of the parent compound and its metabolites will likely have disappeared. With persistent contaminants, degradation of the parent compound into different metabolites, some toxic, some not, and some working via different mechanisms (e.g., o,p'DDT is estrogenic while the metabolite p,p'DDE is antiandrogenic) will further complicate interpretation, even in cases where the study has measured biomarkers of exposure.

Aside from ecological studies, epidemiology is conducted at the individual level. Effects of classical exposures are usually binary outcomes in individuals, which are well defined and severe (cancer case vs. non-case, birth with limb reduction or not). However, wildlife data suggest that changes from EDC exposure at the level of the individual are often subtle and difficult to classify (reduced fertility, poor semen quality, more feminine play behavior, genital dysmorphology). The effect of such changes at the population level, however, can be profound. As discussed above, trends in mean values of several outcomes have been reported but other changes at the population level, which may be even more profound, are increases in population variance and an increasingly non-normal (non-Gaussian) population distribution.

While EDCs manifestly present challenges to epidemiological studies, and are likely to have led to false negatives and underestimates of true risk, some progress is being made in developing approaches that acknowledge these pitfalls and employ methods explicitly designed to avoid them. One of us (Swan) has been involved in such a study, investigating reduced semen quality in relation to pesticide exposure. This study is somewhat unusual from an EDC perspective because it focuses on what appear to be adult-mediated impacts rather than developmental impacts. This avoids the problem of long time lags noted above.

A Study of Semen Quality in Relation to Pesticide Exposure

After finding that fertile men from the general population of an agrarian area (Columbia, MO) had decreased semen quality (e.g., only 58% of the number of moving sperm as men from Minneapolis, MN), pesticide exposure was examined as a cause of poor semen quality. The authors measured urinary metabolites of eight nonpersistent, current-use pesticides in two groups of men from mid-Missouri; men with all semen parameters (concentration, % normal morphology, and % motile) below median value (cases) and men in whom all semen parameters were within normal limits (controls). Pesticide metabolite levels were particularly elevated in cases compared to controls for the herbicides alachlor and atrazine, and for the insecticide diazinon (2-isopropoxy-4-methyl-pyrimidinol, or IMPY) (p values for Wilcoxon rank test $= 0.0007$, 0.012, and 0.0004, for alachlor, atrazine and IMPY, respectively). Men with higher levels of alachlor or IMPY were significantly more likely to be cases than men with low levels (OR $= 30.0$, 16.7 for alachlor and IMPY, respectively), as were men with atrazine over the LOD (OR $= 11.3$). The number of pesticides found in the urine at elevated levels was significantly related to the risk of poor semen quality (being a case rather than a control). These associations were seen in the general population, who were not occupationally exposed. The three pesticides most strongly associated with semen quality are among the five that have been measured most frequently in drinking water sources in the mid-West. These are not removed by routine water treatment. Therefore, drinking water is the most plausible route of exposure. These findings suggest that adult exposure to several widely used pesticides via drinking water is a likely cause of the reduced semen quality seen in fertile men from mid-Missouri.

Subject responses to questions about home and occupational pesticide use were not related to semen quality, suggesting that the relevant pesticide exposure was unknown to the subject. Therefore, collection of urine samples and assays for pesticide metabolites in the subject's urine using highly sensitive GC-MS were required to document exposure to the low levels of pesticides that were related to semen quality. In addition, effects were seen at the level of the individual, with likely more profound effects at the population level. The average decrease in sperm concentration in fertile men living in mid-Missouri relative to men living in Minneapolis, MO is 40 million sperm ml^{-1}. While the median sperm concentration for Missouri men (54 million ml^{-1}) was within normal limits, the sperm count for about 40% of these men fell below 40 million ml^{-1}, the point at which fertility declines significantly.

Summary

In this article we have outlined evidence from diverse sources indicating that a variety of manmade compounds can interfere with development and subsequent functioning of the reproductive system in humans. The concordance of animal and human data, where the latter are available, indicates that when human data are not available health standards should be guided by animal research on a precautionary basis. It will be decades, at best, before epidemiological science is capable of thoroughly documenting the health impacts of even a small number of the contaminants to which humans are exposed daily.

Many of the chemicals of concern were produced to improve human welfare and provide economic benefit (e.g., to increase crop production or to protect food from metal in food cans). This new science, however, is now revealing many unexpected adverse consequences, resulting from the ability of very low levels of these compounds to interfere with gene activation. Most of the chemicals now implicated were subject to little if any rigorous testing; those that were tested were found 'safe' (using criteria now known to miss important risks) and allowed to enter the marketplace. Now we are discovering their 'stealth' characteristics only long after widespread exposure has occurred.

Because of their 'stealth' nature, we are currently unprepared to detect the effects of EDCs or defend against them. Many are persistent; they cannot be removed; they are globally distributed through our atmosphere, our seas and wildlife. Others, while not persistent, should be treated as persistent because of their chronic and ubiquitous use. They act at a population level and many have the potential to (individually or cumulatively) affect future generations (e.g., by decreasing fertility, feminizing males or reducing intelligence). All these endpoints have been produced in the laboratory and many have been observed in wildlife. New data, which must be confirmed by further study, suggest that comparable changes are being produced in human populations as well. Precaution dictates that we cannot wait for "conclusive" evidence of harm to wildlife or human populations to take action.

Chemical corporations and government agencies charged with regulating chemicals in the environment (air, soil, water, and food) assure the public that these chemicals are safe. Because of absence of data concerning risk (often confused with evidence of an absence of risk) and the use of conservative models no longer supported by recent data, the public remains ignorant of the risk potential of the vast majority of chemicals. The public is routinely informed that these chemicals have been tested, that there are studies demonstrating the absence of their risk, and that regulatory agencies adequately protect public health.

Clearly significant changes are needed to bring current regulatory practices into conformity with new scientific information. We propose that testing for health effects at doses within the range of exposure of wildlife and humans (currently not done) with respect to long-latency effects of developmental exposure throughout the lifespan (currently not done) be required prior to the introduction of any chemical intended for use in commerce.

See also: Effects of Endocrine Disruptors in Wildlife and Laboratory Animals; Endocrine Disruptor Chemicals: An Overview.

Further Reading

Duty SM, Calafat AM, Silva MJ, Ryan L, and Hauser R (2005) Phthalate exposure and reproductive hormones in adult men. *Human Reproduction* 20: 604–610.

Gray LE, Jr., Wilson VS, Stoker T, *et al.* (2006) Adverse effects of environmental antiandrogens and androgens on reproductive development in mammals. *International Journal of Andrology* 29: 96–104.

Green R, Hauser R, Calafat AM, *et al.* (2005) Use of Di(2-ethylhexyl) phthalate–containing medical products and urinary levels of mono(2-ethylhexyl) phthalate in neonatal intensive care unit infants. *Environmental Health Perspectives.* 113: 1222–1225.

Jacobson JL and Jacobson SW (1996) Intellectual impairment in children exposed to polychlorinated biphenyls *in utero. The New England Journal of Medicine* 335: 783–789.

Main KM, Mortensen GK, Kaleva M, *et al.* (2006) Human breast milk contamination with phthalates and alterations of endogenous reproductive hormones in infants three months of age. *Environmental Health Perspectives* 114: 270–276.

Michaels D (2005) Doubt is their product. *Scientific American* 292: 96–101.

Parks LG, Ostby JS, Lambright CR, *et al.* (2000) The plasticizer diethylhexyl phthalate induces malformations by decreasing fetal testosterone synthesis during sexual differentiation in the male rat. *Toxicological Sciences* 58: 339–349.

Silva MJ, Barr DB, Reidy JA, *et al.* (2004) Urinary levels of seven phthalate metabolites in the US. population from the National Health and Nutrition Examination Survey (NHANES) 1999–2000. *Environmental Health Perspectives* 112: 331–338.

Skakkebaek NE, Rajpert-De Meyts E, and Main KM (2001) Testicular dysgenesis syndrome: An increasingly common developmental disorder with environmental aspects. *Human Reproduction* 16: 972–978.

Swan SH, Brazil C, Drobnis EZ, *et al.* (2003) Geographic differences in semen quality of fertile US males. *Environmental Health Perspectives* 111: 414–420.

Swan SH and Elkin EP (1999) Declining semen quality: Can the past inform the present? *BioEssays* 21: 614–621.

Swan SH, Kruse RL, Fan L, *et al.* (2003) Semen quality in relation to biomarkers of pesticide exposure. *Environmental Health Perspectives* 111: 1478–1484.

Swan S, Main K, Liu F, *et al.* (2005) Decrease in anogenital distance among male infants with prenatal phthalate exposure. *Environmental Health Perspectives* 113: 1056–1061.

Thornton JW, McCally M, and Houlihan J (2003) Biomonitoring of industrial pollutants: Health and body implications of the chemical body burden. *Public Health Reports* 117: 315–323.

Vreugdenhil HJ, Slijper FM, Mulder PG, and Weisglas-Kuperus N (2002) Effects of perinatal exposure to PCBs and dioxins on play behavior in Dutch children at school age. *Environmental Health Perspectives* 110: A593–A598.

Wilson VS, Lambright C, Furr J, *et al.* (2004) Phthalate ester-induced gubernacular lesions are associated with reduced insl3 gene expression in the fetal rat testis. *Toxicology Letters* 146: 207–215.

Exposure and Exposure Assessment

K F Gaines, Eastern Illinois University, Charleston, IL, USA

T E Chow, University of Michigan – Flint, Flint, MI, USA

S A Dyer, Washington Savannah River Company, Aiken, SC, USA

Introduction	Trophic Transfer and Exposure Estimates
Estimating Toxicant Exposure	Linking Exposure to Uptake
Modeling Exposure to Environmental Contaminants	Further Reading

Introduction

Exposure to contaminants in the environment is quantified through the ecological risk assessment (ERA) process which provides a framework for the development and implementation of environmental management decisions. The ERA uses available toxicological and ecological information to estimate the probability of occurrence for a specified undesired ecological event or endpoint. The level for these endpoints depends on the objectives and the constraints imposed upon the risk assessment process; therefore,

multiple endpoints at different scales may be necessary. ERAs often rely on the link between these undesired endpoints to a threshold of exposure to specific toxicants and toxicant mixtures. Oral reference doses (RfD), inhalation reference concentrations (RfC), and carcinogenicity assessments are the usual way these links are expressed in the ERA, and unfortunately most of these thresholds have been developed for human health assessments and not ecosystem integrity. However, since these studies often use animal models, in many cases the original empirical data can be used when trying to apply these findings to ecological consequences or to establish ecological screening values (ESVs). The ecological exposure assessment often begins by comparing constituent concentrations in media (surface water, sediment, soil) to ESVs. The ESVs are derived from ecologically relevant criteria and standards. For example, in the United States the United States Environmental Protection Agency (USEPA) Screening Values and National Ambient Water Quality Criteria (NAWQC) are often used based on 'no observed adverse effect levels' (NOAELs) or 'lowest observed adverse effect levels' (LOAELs) derived from literature to assess exposure. Radionuclide comparisons for ecological screening are typically dose-based for population level effects. In addition to the ecological threshold comparison, constituents that may bioaccumulate/bioconcentrate are identified during initial screening processes. This is done to account for toxicants that may not be present at levels exceeding ESVs, but must be considered due to trophic transfer of toxicants that may concentrate in higher-trophic-level organisms. Constituents that exceed ESV comparisons (present with means, maximums, or 95% upper confidence levels (UCLs)) are evaluated using a lines-of-evidence approach based on (1) a background evaluation, (2) a bioaccumulation/bioconcentration potential and ecotoxicity evaluation, (3) a frequency and pattern-of-exceedances evaluation based on review of exceedances to the ESVs, and (4) an evaluation of existing biological data. From this information, ecosystems can be prioritized in terms of risk and focused for proper exposure assessments. This article presents a scientific overview and review of how toxicant exposure is estimated and applied to assess ecosystem integrity.

Estimating Toxicant Exposure

The challenge in estimating exposure to toxicants is to properly model ecosystem function. Exposure modeling must consider a hierarchical scale, system dynamics, and use them to determine what the limits of predictability may be. A hierarchical approach allows modelers to characterize exposure components and their linkages among different scales of ecological organization and complexity. Therefore, exposure needs to be analyzed at multiple scales and appropriate levels of ecological organization in both space and time. Although estimating exposure in terrestrial and aquatic ecosystems is linked, there are special considerations that need to be taken for each.

Aquatic systems are considered primary integrators within a watershed because they potentially receive, through surface or subsurface drainage/discharge, toxicants from outfalls and other contaminant sources within the watershed. If these toxicants reach biologically significant levels, they would be expected to affect the numbers, types, and health of stream organisms. Often, biological sampling is conducted in a stream system to measure the cumulative

ecological effects of contaminant sources received by the aquatic system (pond, stream system, lake, etc.). Information from biological sampling is used to assess the environmental quality based on contaminant exposure and is often termed bioassessment. For aquatic assessments, evaluation areas are partitioned spatially, based on watershed boundaries and potential contaminant sources and classified in terms of the type of surface water body (streams, lakes, etc.) and their associated wetlands, including surface water, sediment, and related biota. These systems receive potential contamination discharged to surface water or migrating through groundwater from source contaminant areas, National Pollutant Discharge Elimination System outfalls, and operational facilities to points of potential receptor exposure. Ecological receptors feeding within stream-based food chains are exposed to the cumulative effects of contaminants that are released to the stream system, and their health can be considered an integrative indicator of the severity of contamination within the watershed. Because some watersheds/stream systems are large in size, the study area may need to be subdivided into subunits to facilitate the assessment process and identify areas of possible contamination with higher precision.

Exposure assessment models have primarily focused on the mobility of contaminants in the environment using vertebrates as the assessment endpoint (e.g., exposure to each contaminant in $mg\,kg^{-1}d^{-1}$ or $mg\,l^{-1}d^{-1}$). Invertebrate models are also used especially when soil screening levels (SSLs) are of interest; however, these studies usually concentrate on transfer factors associated with bioaccumulation models. As mentioned previously, the affects assessment may be expanded beyond comparison to ESVs by using existing data to conduct trophic exposure modeling (hereafter trophic modeling). Trophic modeling can be used to refine the list of toxicants to those constituents that may pose an adverse impact (significant risk) to specific ecological receptors. The trophic modeling uses toxicant exposure models for ecological receptors to calculate an exposure dose (ED) for each constituent that poses a potential risk through ingestion of contaminated media. EDs are compared with toxicity reference values (TRVs) to identify constituents with an evaluation-level hazard quotient (HQ, i.e., ED/TRV) greater than 1. Results of the HQ assessment and other weight-of-evidence criterion can be used to further refine the list of constituents of potential concern. The use of exposure assessments must be appropriate to the spatial scale across which the toxicants of interest are dispersed. The most influential factor for contaminant accumulation in wildlife in any ecosystem is how much time the individual spends exposed to the contaminant and how it utilizes the ecosystem. In areas with broad-scale contamination, this must be done at the landscape level and can be achieved by the implementation of spatially explicit models that are calibrated using data from long-term biomonitoring of large areas. Specifically, exposure assessment considers the following:

1. Chemical distribution which defines the extent of measured chemical contamination to each exposure area and the approximate acreage of each exposure group. The chemical exposures that may be experienced by ecological receptors are affected by the degree of their spatial and temporal associations with the contaminated media.

2. Receptor distribution which involves the variety of factors that may affect the extent and significance of potential exposures. Receptor exposures are affected by the degree of spatial and temporal association with the contamination. A receptors' mobility may significantly affect their potential exposures to contaminants. Many species may only inhabit the study area during the seasonal periods (e.g., breeding season, nonmigratory periods). Nonmigratory species may remain in the vicinity throughout the year. These species, particularly those with longer life spans, have the greatest potential duration of exposure. For both terrestrial and aquatic systems, some species may live their entire life cycle within the systems and others may utilize the system for forage areas, water intake, reproduction, or utilize the area for early life stages only.

3. Quantification of exposure and effects assessment defines the degree to which contaminant distributions and receptor distributions overlap and indicates which receptors are likely to have the greatest potential exposures to contaminants. This can be conducted by comparing media concentrations to ESVs or further quantify exposures by calculating an intake for each chemical in each medium (sediment, surface water, prey). The effects assessment defines and evaluates the potential ecological response to the contaminant by use of TRVs or ESVs that are the basis of the comparison. To relate these numeric comparisons to the actual receptors, biological data can be used to determine if effects are occurring in the system.

As fish and wildlife occupy different habitats within an ecosystem, they may be exposed to toxicants through three pathways: oral, dermal, and inhalation. Oral exposure occurs through the consumption of contaminants through food, water, or soil/sediment. Dermal exposure takes place when contaminants are absorbed directly through the skin. Inhalation exposure occurs when volatile compounds or fine particles are respirated to the lungs. Therefore, the total exposure experienced by an individual is the sum of exposure from all three pathways or

$$E_{\text{total}} = E_{\text{oral}} + E_{\text{dermal}} + E_{\text{inhalation}} \qquad [1]$$

where E_{total} is the total exposure from all pathways; E_{oral} is the oral exposure; E_{dermal} is the dermal exposure; and $E_{\text{inhalation}}$ is the exposure through inhalation.

In aquatic systems exposure via inhalation and dermal pathways are usually considered as one factor. This is because total uptake for free-swimming aquatic receptors is assumed to be represented by simple partitioning from surface water alone. Aquatic receptors are assumed to be in equilibrium with contaminants in the water column (this assumption in many cases is erroneous and warrants further research). Contaminant partitioning between surface water and aquatic organisms is defined by a contaminant-specific bioconcentration factor. In general, the primary mechanism of contaminant uptake for many fully aquatic species is via direct uptake across permeable membranes such as gill and gill structures (which can be addressed under dermal exposure in eqn [1]). This can occur as a passive transfer or an active biological process (osmoregulation). Prey consumption, incidental ingestion of sediment and pore water/groundwater during prey consumption, and incidental ingestion of surface water during prey consumption are usually treated as secondary uptake mechanisms since they are modeled via bioconcentration factors rather than exposure models. This potential exposure parameter should be considered spatially dynamic since contaminant concentrations change based on their distance from a source.

Dermal exposure is assumed to be negligible for birds and mammals on many hazardous waste sites relative to other routes in most cases. Feathers and fur of birds and mammals further reduce the likelihood of significant dermal exposure by limiting the contact of skin with contaminated media. However, when an exposure scenario for a receptor species is likely to result in significant dermal exposure such as through brood patches on birds, direct contact by burrowing mammals, or swimming by amphibians, this exposure pathway should be estimated using models for terrestrial wildlife listed in the 'Further reading' section. Moreover, if contaminants that have a high affinity for dermal uptake are present (e.g., organic solvents and pesticides), dermal pathways should be considered even if contact is minimal compared to the aforementioned taxa. Inhalation of contaminants is treated as negligible at many waste sites since quite often these sites are either capped or vegetated. This minimizes exposure of contaminated surface soils to winds which results in aerial suspension of contaminated dust particulates. Also, the contaminants most likely to present a risk through inhalation exposure, such as most volatile organic compounds (VOCs), will quickly volatilize from soil and surface water to air, where they are diluted and dispersed. As a result, significant exposure to VOCs through inhalation is unlikely. In circumstances where inhalation exposure of endpoint species is believed to be occurring or is expected to occur, models for vapor or particulate inhalation may be employed.

Based on these factors, most exposure models in fish and wildlife concentrate on exposure through ingestion.

The general formulas used to estimate contaminant exposure to terrestrial and aquatic wildlife via ingestion uses the ingestion rate multiplied by the concentration of the contaminant in all possible food items in relation to the body weight of the animal. Because many waste sites (contaminated areas) do not provide suitable habitats, exposure estimates are modified to be sensitive to the home-range size (total area used by an animal) or core area (areas used most often within an animal's home range) of the species as well as the habitats that are used or the probability of the species occurring in the area. These parameters are incorporated in the following equation:

$$E_j = P\left(\frac{A}{HR}\left[\sum_{i=1}^{m}\left(\frac{IR_i C_{ij}}{BW}\right)\right]\right) \qquad [2]$$

where E_j is the exposure to contaminant through ingestion (j) ($mg\,k^{-1}g\,d^{-1}$ or $mg\,l^{-1}\,d^{-1}$); P is the probability of the receptor species inhabiting a waste site or the proportion of the waste site used; A is the area (ha) of waste site; HR is the area (ha) that defines the receptor species home range or core area; m is the total number of ingested media (e.g., food, water, or soil); IR_i is the ingestion rate for media (i) ($kg\,d^{-1}$ or $l\,d^{-1}$); C_{ij} is the concentration of contaminant (j) in medium (i) ($mg\,kg^{-1}$ or $mg\,l^{-1}$); and BW is the whole body weight of endpoint species (kg).

The area is considered in two dimensions even for aquatic species since the area-to-home-range ratio is used to determine the fraction of the waste site in relation to the total area used by the animal for foraging. This could of course be modified to a volumetric parameter for aquatic species where the third dimension is necessary to determine that ratio.

Modeling Exposure to Environmental Contaminants

Advances in geographic information science (GIS) technologies such as remote sensing, spatial databases, and spatially explicit models have shown to be extremely useful in the exposure assessment process. By adopting such technologies, landscape-level exposure models can be developed by integrating the spatial parameters such as those shown in eqn [2]. These methods recognize that if a site is spatially heterogeneous with respect to either contamination or animal use, exposure models must be modified to include the dynamics imposed by those spatial factors thus improving the estimated parameters in eqn [2]. When using fish and wildlife as receptor species for mechanisms of contaminant accumulation, transport, redistribution, and as ecological endpoints, the foundations and principles of animal habitat relationships and

the interaction between spatial pattern and ecological processes must be properly modeled with particular attention to (1) spatial relationships among fish and wildlife and their habitats, (2) spatial and temporal interactions, and (3) influences of spatial heterogeneity on biotic and abiotic processes. Below, the basic elements needed to estimate the spatially explicit parameters used in most exposure models are outlined.

Data Layers for Exposure Assessment

Through various methods of data capture, such as remote sensing, global positioning system (GPS), and field survey, detailed biophysical characteristics of the landscape can be represented in a GIS. In the form of map layers, a GIS can store the spatial patterns of individual geographic phenomenon, such as habitat, land use, hydrology, population, topography, road networks, and other infrastructural information into a spatial database. The map layers are geographically referenced in a common coordinate system so that the layers are projected onto a scale-down plane surface that enables distance measurement, area calculation, and map overlay.

Historically, the map layers are often too general to make fine-resolution predictions in terms of how receptors may be utilizing contaminated areas or if the map layers were constructed with a focus on timber management and harvest rather than being designed to describe the ecosystem structure. For example, LANDSAT imagery has 30 m spatial resolution (i.e., each pixel has a ground equivalent dimension) and is commonly used for mapping the distribution of vegetation. Recently, the emergence of high-resolution remotely sensed imagery such as QuickBird, airborne visible/infrared imaging spectrometer (AVIRIS), and light detection and ranging (LIDAR) has enabled the researchers to map the three-dimensional information of the landscape with spatial resolution of <1 m and hundreds of spectral channels. Through various techniques of digital image processing, including image filtering, band ratioing, feature/pattern extraction, and spectral classification, biophysical characteristics of the landscape can be extracted from the remotely sensed imagery. The specific technology to be used to map out vegetation in the study site is dependent upon the stage of the bioaccumulation model that is being estimated; that is, the scale needed to determine transfer factors from soil to plant species to estimate bioavailability is very different from the scales needed to estimate the distributions of wildlife endpoint species. In many cases, field survey is necessary for verifying, calibrating, and validating purposes.

Once all the map layers are in digital format, the data can be compiled into a spatial database in which many spatial relationships could be explored and analyzed within and among the map layers. To assist risk assessors, the spatial database provides important information about how the focal wildlife species may use contaminated areas and how contaminants may move in the environment. Such a database is extremely useful in identifying potential data gaps and which data sources are available to assist in a risk assessment. In some cases, information on the spatial distribution of contaminants and waste units are not available, and methods of spatial interpolation can be used to generate new information based on known value at surrounding locations (see the following section).

Contaminant Distribution

For most areas it is difficult to map the distribution of contaminants in the soil or sediment. The most notable exception is gamma-ray detection for radioisotopes, which can be achieved through remote-sensing flyovers of the disturbed areas. However, when the contaminants of concern cannot be measured remotely, or the scale of such flyover data is too coarse, some sampling regime has to occur to determine their distribution. Once samples are obtained, contaminant distributions can be mapped using appropriate spatial interpolation techniques.

The first law of geography (Tobler's law) states the likelihood of things closer in distance to be more related and similar than those afar. Built upon this concept, spatial interpolation methods estimate unknown sampling points in relation to the distance of their neighbors near and far. Inverse distance weighting (IDW), local polynomial, global polynomial, spline and radial basis functions (RBSs) are deterministic interpolators that apply an established mathematical formula to the sample points. A second family of interpolation methods consists of geostatistical methods that are based on statistical models that incorporate autocorrelation (statistical relationships among the measured points). Not only do these techniques have the capability of producing prediction surfaces, but they can also provide some measure of the

accuracy of these predictions using cross-validation techniques. Kriging is the most widely used geostatistical interpolator. An important feature of geostatistical analysis is the generation of an empirical semivariogram to estimate the spatial correlation of the sampling points in space. Thus, the semivariogram quantifies how the correlation between two points in space changes as they move closer together or farther apart. This is a useful tool in its own right and defines the variance structure of the geostatistical model.

Exposure Model Classification

The development of many spatially explicit exposure models to estimate the adverse impact of specific toxicants and toxicant mixtures to the environment was in part fueled by the demand in understanding the fate of contaminants in terms of environmental risk and environmental justice. In general, an exposure model provides the framework that uses one of the many functions in combining the identified controls (i.e., factors) in assessing the ecological risks. Depending on the basis of the actual algorithms, most of the existing predictive models can be broadly classified as physical-based, statistical-based, and rule-based models. Depending on how the model treats randomness in time and space, the exposure models can further be categorized as deterministic or stochastic models (**Figure 1**). A deterministic model does not consider randomness at all; that is, a given set of input parameters always yield the same output prediction. A stochastic model allows the quantification of uncertainties in time and space, so that the same set of input parameters may have different results. In exposure modeling, uncertainties may come from the lack of input data or understanding about the physical reality, such as seasonality of the ecosystem, random behavior of individuals, etc.

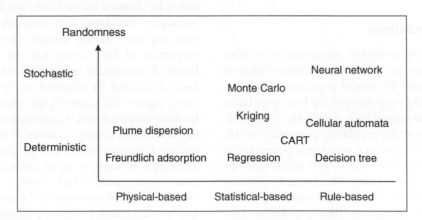

Figure 1 Classification of the spatially explicit exposure models based on the basis of algorithms and randomness.

The physical-based models adopt established laws or mathematical equations that attempt to describe the physical processes, for example, Freundlich adsorption equation, toxicokinetic model, plume dispersion model, etc. This type of model is commonly used in modeling the exposure and uptake of toxicants to the endpoints through media such as air, water, and soil. In general, a physical-based model is well established in physical laws and can be extensively applied to many endpoints. However, such models often require many physical parameters that may not be readily available (particularly in spatial data) and the accuracy of model prediction is limited by the extent of field calibration.

The statistical-based models explore the relationship (which can be attribute or spatial in nature) between identified controls and the level of exposure at endpoints with a probability distribution function, for example, generalized linear models, spatial statistical methods, etc. In most cases, this approach is based on empirical data collected from the field or in the laboratory. Many researchers utilized common statistical techniques to conduct ERA, such as logistic regression, kriging, Monte Carlo simulation, etc. The implications of the research that resulted from such models are usually restricted to the tested study areas or ecosystems with similar biophysical characteristics.

The rule-based (or agent-based) models assess the ecological impact of exposure by exploring the underlying mechanisms to simulate the process. The governing rules may be established from the literature or field experts, observed data for both training and validating (e.g., neural network, decision tree), or even arbitrary rules (e.g., cellular automata, weighted linear combination). Within this category of model, one of the most controversial components is how to determine the weight of individual controls (i.e., the impacts) in computing the exposure level. The most common weights include the population of receptor species and the space–time interaction between the endpoints and stresses.

Exposure to Populations

When conducting an ecological assessment it is often desirable to estimate the risk to a population rather to an 'at-risk individual'. To model population exposure, one must estimate the proportion of the local population exposed at levels that exceed toxic thresholds. This represents the proportion of the population potentially at risk. Specifically, the proportion of a population potentially at risk is represented by the number of individuals that may use habitat within waste unit(s). To properly estimate exposure, the movement of contaminated individuals within and between populations (metapopulation) may also be of interest especially when the proportion of

new recruits is important to estimate the effects of the contaminant on fecundity and survival.

Also, often investigators are interested in making inferences about the mean exposure to a receptor species at a waste site, but it may be erroneous to assume that the distribution of the mean is the same as that of the population. Hence, a similar procedure needs to be performed to estimate the distribution of mean exposure. By estimating the number of individuals (n) that would use the waste site(s), n home ranges for the waste site(s) can be randomly sampled, the n exposures calculated, and the average taken based on eqn [2]. This procedure is repeated (usually 1000+) times for each site creating what is commonly referred to as a Monte Carlo random sample of average exposures. Hence, the resulting simulations provide an estimate of the distribution of mean exposure using histograms and quantiles. The 2.5th and 97.5th elements in the ranked means are the estimated lower and upper bounds, respectively, of the 95% confidence interval. The mean exposures and their corresponding 95% confidence intervals provide the information necessary to conduct hypothesis testing about the mean exposure at the waste units. In practice, a researcher could test the hypothesis that the mean exposure was zero, or below (above) a given regulatory limit, by using the appropriate confidence bound (upper or lower). Another approach is to combine the results of Monte Carlo simulation of exposure with literature-derived population density data to evaluate the likelihood and magnitude of population-level effects on wildlife.

Trophic Transfer and Exposure Estimates

Although the concepts of trophic transfer and bioaccumulation are outlined elsewhere, it is worth noting here in the discussion of estimating exposure. The key to a useful exposure model is to derive a realistic link between exposure and uptake. Up to this point this article has focused on the landscape level and behavioral parameters associated with exposure models. However, most exposure models assume that the diets and the proportion of food items that the animal ingests are known. Contaminant exposure is strongly linked to the kind of material an organism occupying lower trophic levels ingest – for example, leaf versus fruit, or particular invertebrate species. Contaminant studies have relied on comparisons among a variety of target species, which confound interpretations due to dietary variations and differences in interspecific physiologies. It has been suggested that animals which show a higher diversity of food items better represent the extent of contamination and trophic transfer within a system, especially when they occupy the uppermost trophic levels. For some vertebrates, trophic level rather than body size (which is the usual parameter used) appears to be one of the most important factors.

However, accurately quantifying an animal's diet down to specific food items and quantities is extremely difficult. This is often achieved through fecal and stomach analyses, which are at best snap shots in time and do not capture the animals trophic position over a specified duration (such as weeks to seasons). These difficulties have biased exposure studies to animals with specialized diets, which is extremely unrealistic and may not adequately represent the dynamics of the ecosystem being studied.

One of the most promising techniques available that can minimize some of this variation is through analyzing tissues and food items using stable isotopic analyses. The stable isotope composition of biological materials provides insights into the life histories of fish and wildlife species. It has been demonstrated that animal tissues are enriched in ^{15}N in relation to their diet. Also, owing to differences in photosynthetic pathways, C_3, C_4, and CAM plants have different $^{13}C/^{12}C$ ratios (-32 to -22 and -23 to -9, respectively, with CAM overlapping) and can be used to identify sources of primary productivity in the diet. The differences in isotopic composition between any tissue compartment of an animal and diet is represented by a tissue–diet enrichment factor $\varepsilon_{tissue-diet}$, where $\varepsilon_{tissue-diet} \approx \delta_{tissue} - \delta_{diet}$ and 'δ' is the delta value for the isotope of interest. This technique allows researchers to determine what levels of the food chain target species within an ecosystem are occupying and potentially where an animal is foraging. This enables a more sophisticated understanding of food web structure and spatial foraging patterns thus allowing exposure estimates to be better parametrized.

Linking Exposure to Uptake

The biggest challenge to date in exposure modeling is to link exposure to uptake outside of the laboratory. A biomagnification model that was successfully applied to terrestrial biomagnification known as BIOMAG is a good example of linking exposure to uptake. The model considers target species which are top predators in significant ecosystems. The model incorporates a consideration of bioavailability (concentration in soil solution), ecology (stochastic food chain models), toxicology (simple compartment model), and a consideration of effect based on the 'no observed effect concentration' (NOEC). Movement of the contaminant through the food chain is quantified using bioaccumulation factors (BAFs); a ratio of the concentration of

contaminant in the consumer and the food that it consumed. Such approaches have been used to underpin soil standards by calculating the level of soil pollution that gives rise to the maximum tolerable risk for birds and mammals at the top of the particular food chain, and in general involves working backward through the model, starting at maximum tolerable risk. One shortfall of this and many models is controlling the sensitivity of the model especially in a situation where heterogeneity of soil contamination amplifies the variability of the model, although it is suggested that soil solution concentration should be treated as a stochastic parameter, in a similar manner to BAF. In aquatic toxicology, similar relationship can be explored by using quantitative structure–activity relationships (QSAR) in rule-based expert systems.

See also: Ecological Risk Assessment.

Further Reading

Bradbury SP (1995) Quantitative structure–activity relationships and ecological risk assessment: An overview of predictive aquatic toxicology research. *Toxicology Letters* 79: 229–237.

Cairns J, Jr. and Niederlehner BR (1996) Developing a field of landscape ecotoxicology. *Ecological Applications* 6: 790–796.

Chow TE, Gaines KF, Hodgson ME, and Wilson MA (2005) Habitat and exposure modeling of raccoon for ecological risk assessment: A case study in Savannah river site. *Ecological Modelling* 189: 151–167.

Craig H (1953) The geochemistry of the stable isotopes. *Geochimica et Cosmochimica Acta* 3: 53–92.

Cressie AC (1993) *Statistics for Spatial Data,* revised edition. New York: Wiley.

DeAngelis DL, Gross LJ, Huston MA, *et al.* (1998) Landscape modeling for everglades ecosystem restoration. *Ecosystems* 1: 64–75.

Feijtel T, Boeije G, Matthies M, *et al.* (1998) Development of a geography-referenced regional exposure assessment tool for European rivers – GREAT-ER. *Journal of Hazardous Material* 61: 59–65.

Gaines KF, Romanek CS, Boring CS, *et al.* (2002) Using raccoons as an indicator species for metal accumulation across trophic levels – A stable isotope approach. *Journal of Wildlife Management* 66: 808–818.

Gay JR and Korre A (2006) A spatially-evaluated methodology for assessing risk to a population from contaminated land. *Environmental Pollution* 142: 227–234.

Hope BK (1995) A review of models for estimating terrestrial ecological receptor exposure to chemical contaminants. *Chemosphere* 30: 2267–2287.

International Automic Energy Agency (1992) *Effects of Ionizing Radiation on Plants and Animals at Levels Implied by Current Radiation Protection Standards,* Technical Reports Series, No. 332. Vienna: International Automic Energy Agency.

Ma WC (1992) Methodological principles of using micromammals in biomonitoring and hazard assessment of chemical soil contamination. In: Donker MH, Eijsackers H, and Heimback F (eds.) *Ecotoxicology of Soil Organisms*, pp. 357–376. Boca Raton, FL: CRC Press.

Manly BFJ, McDonald LL, and Thomas DL (1993) *Resource Selection by Animals: Statistical Design and Analysis for Field Studies*. New York: Chapman and Hall.

Pesch R and Schröder W (2006) Integrative exposure assessment through classification and regression trees on bioaccumulation of metals, related sampling site characteristics and ecoregions. *Ecological Informatics* 1: 55–65.

Sample BE, Aplin MS, Efroymson RA, Suter GW, II, and Welsh CJE (1997) Methods and tools for estimation of the exposure of terrestrial wildlife to contaminants. DOE Environmental Sciences Division, Publication No. 4650.

Sample BE and Suter GW (1999) Ecological risk assessment in a large river-reservoir. Part 4: Piscivorous wildlife. *Environmental Toxicology and Chemistry* 18(4): 610–620.

Tichendorf L (2001) Can landscape indices predict ecological processes consistently? *Landscape Ecology* 16: 235–254.

Tobler WR (1970) A computer model simulating urban growth in the Detroit region. *Economic Geography* 46: 234–240.

United States Environmental Protection Agency (2005) *Guidance for Developing Ecological Soil Screening Levels OSWER Directive*, 92: 857–55. Washington, DC: United States Environmental Protection Agency.

Food-Web Bioaccumulation Models

F A P C Gobas, Simon Fraser University, Burnaby, BC, Canada

Introduction	Food-Web Model Application
Food-Web Model Philosophy	Further Reading

Introduction

Practitioners of ecotoxicology are confronted with some unique problems. For example, when assessing the impacts of chemical contamination (e.g., as result of a spill, a point source emission, or historic sediment contamination) on wildlife species, chemical concentrations at the source (e.g., effluent, water, sediments) may be known but the resulting exposure concentrations and associated risks in affected wildlife species are not. In other cases, the chemical concentration in fish or wildlife are known through monitoring programs and considered of concern, but it is unknown how the organisms acquired their chemical body burden, what the chemical concentrations at the source are, and what source concentrations should be to eliminate the concern. In both cases, it is difficult to justify remediation actions to reduce ecotoxicological risks and improve environmental health of the system as the scientific basis for the cause–effect relationship is either absent or weak. The missing link in both of these examples is knowledge of the relationship between the chemical concentrations in environmental media such as water, air, sediments, and soil, and those in the wildlife species of interest. This relationship is complex. For example, a predatory fish will take up chemical from the water via its gills, ingest contaminated sediments, and feed on a variety of prey items, each of which acquired their contaminant burden via similar mechanisms as the predator. Food-web bioaccumulation models are useful tools to handle this complexity and have been increasingly used to better characterize and understand the relationship between contaminant concentrations in abiotic environmental media and those in wildlife and humans. They have been used in risk assessments, the derivation of total maximum daily loadings (TMDLs) for impacted water bodies, the derivation of criteria for water and sediment quality criteria as well as wildlife criteria, the development of target levels for water and sediment clean-up levels and the derivation of bioconcentration factors (BCFs), bioaccumulation factors (BAFs) and biota-sediment bioaccumulation factors (BSAFs) for chemicals to support the evaluation of large numbers of commercial chemicals for bioaccumulation potential.

Several food-web models exist. The most well-used models are the ones developed by Thomann and co-workers, Mackay and co-workers, Czub and McLachlan, and Gobas and co-workers. Some of the models are readily available and can be downloaded from the Web, while others can be requested from the authors. When applying these tools to address various goals, it is important to be familiar with the main principles included in the model. It is further important to be aware of the key assumptions of the modeling approach and recognize the limitations of the model's application. It is the purpose of this article to summarize the key model principles and discuss the most important aspects of the application of the model. It is expected that this will guide the reader in further studies.

Food-Web Model Philosophy

One of the key challenges in ecotoxicology is to understand the interaction between the chemistry of the contaminant and the biology (including the physiology and ecology) of the organism. Food-web bioaccumulation models are among the important tools to document our knowledge of this interaction and to apply our knowledge to practical problems. These models incorporate a number of fundamental chemical and biological principles. In addition, to apply the model in an acceptable fashion, the model calculations have to be conducted according to a set of modeling and management principles. These fundamental principles need to be understood to recognize the application domain of the models and the strengths and limitations of the model outcomes. This section discusses these principles briefly. Current food-web bioaccumulation models show good agreement on the main principles included in the models. To avoid duplication, this section refers to the Gobas models for specific examples.

Chemical Equilibrium

One of the most important chemical principles embedded in food-web bioaccumulation is the natural tendency of chemicals to involve in net transport with the goal to achieve a chemical equilibrium. Equilibrium is thermodynamically defined as a situation where the chemical potential μ or the fugacity f (Pa) of the chemical in two or more media (e.g., water and fish) is the same, that is,

$$\mu_B = \mu_W \quad \text{or} \quad f_B = f_W \qquad [1]$$

where μ_B and μ_W and f_B and f_W are the chemical potentials and fugacities in the organism or biota and water, respectively. The fugacity of a chemical, f (in units of pascal), in a given phase is related to the molar concentration C (in $mol\,m^{-3}$) by the fugacity capacity Z (in $mol\,m^{-3}\,Pa^{-1}$) of the phase in which the chemical is solubilized:

$$f = C/Z \qquad [2]$$

The fugacity capacity Z is compound and phase specific, and it represents the capacity of that phase to sorb and retain a given chemical within its matrix. Net passive chemical transport occurs from the medium of high fugacity (e.g., water) to the medium of low fugacity (e.g., an organism) until the fugacities in both media are equal (**Figure 1**), at which the chemical concentrations in both media (e.g., in the organism, C_B, and the water, C_W ($mol\,m^{-3}$)) are related by the chemical's partition coefficient K_{BW} (unitless), that is,

$$C_B/C_W = Z_B/Z_W = K_{BW} \qquad [3]$$

The application of this principle was the foundation of some of the first bioaccumulation models for organic contaminants, in which the BCF (typically expressed in units of $l\,kg^{-1}$) of the chemical in fish was found to be related to the octanol–water partition coefficient (K_{OW}); for example, one such model estimates the BCF in fish as $0.048 K_{OW}$, where 0.048 is an estimate of the average lipid content of fish.

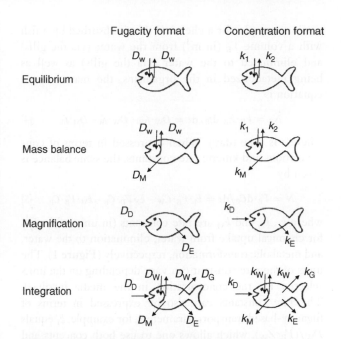

Figure 1 Conceptual diagrams illustrating key chemical principles (equilibrium, mass balance, magnification, and integration of processes) at the organism level.

Mass Balance

The second principle that is embedded in the food-web bioaccumulation models is that physical transport processes (e.g., gill ventilation flows, ingestion rates) and reaction processes (e.g., metabolic transformation) may interfere with the chemical's natural tendency to achieve a chemical equilibrium. This is captured by the mass balance equation where the net flux of chemical N_B (in units of $mol\,day^{-1}$) into an organism is the sum of the chemical fluxes into and out of the organism. When expressed in fugacity format, the flux is the product of the transport parameter, D, and the fugacity of the chemical in the medium, f, in which the transport occurs. The transport parameter can represent transport by molecular diffusion, in which case D is the product of the mass transfer coefficient S (in $m\,day^{-1}$), the area of diffusion A (m^2), and the Z in which the diffusion occurs, that is, D equals $S \cdot A \cdot Z$. D can also represent an advective transport process (e.g., gill ventilation rate or ingestion) described by a flow rate G ($m^3\,day^{-1}$) in which case D equals $G \cdot Z$. D can further describe a transformation process (e.g., metabolic transformation), often described by a transformation rate constant (day^{-1}), in which case D equals $k \cdot V \cdot Z$, where V is the volume of the medium in which transformation occurs. One of the attractive features of the fugacity approach is that it separates biological variables such as the flow rate (e.g., gill flow rate, ingestion rate, urine excretion rate) and chemical variables such as Z which control the physical–chemical partitioning of the chemical.

For example, for a chemical that is absorbed by a fish with a volume V_B (in m^3) from the water (via the gills) and eliminated to the water (via the gills) as well as being metabolized in the organisms, the mass balance equation is

$$N_B = V_B \cdot Z_B \cdot df_B/dt = D_W \cdot f_W - D_W \cdot f_B - D_M \cdot f_B \qquad [4]$$

where t is time (day). When expressed in terms of concentrations and kinetic rate constants, the same balance is given by

$$N = V_B \cdot dC_B/dt = k_1 \cdot V_B \cdot C_W - k_2 \cdot V_B \cdot C_B - k_M \cdot V_B \cdot C_B \qquad [5]$$

where k_1, k_2, and k_M are rate constants (in units of day^{-1}) for chemical uptake from water, elimination to the water, and metabolic transformation, respectively (**Figure 1**). The units of the rate constant can vary depending on the units selected for the concentration in the media involved. The rate constants can also be expressed in terms of fugacity-based transport parameters, for example, k_1 equals $D_W/(V_B \cdot Z_W)$, which allows one to use both concepts and benefit from the fugacity-based approach to distinguish between biological and chemical variables in the model.

Equations [4] and [5] illustrate that after a long-term exposure, when a steady state is reached (i.e., $N = 0$), the chemical fugacities in water and organism are no longer equal, that is, $f_B/f_W = D_W/(D_W + D_M)$, with the fugacity in the organism being smaller than that in the water. In concentration format, this equates to $C_B/C_W = k_1/(k_2 + k_M)$.

Biomagnification

One of the key principles in a food-web bioaccumulation model is the biomagnification effect, which causes the fugacity and concentration of the chemical to increase with increasing trophic level. This process can lead to food-web magnification of the chemical when this process occurs at each predator–prey interaction in the food web. Food-web biomagnification alone can produce a 10 000–100 000-fold increase in lipid-normalized concentration of a bioaccumulative substance. Biomagnification is of ecotoxicological significance because it can cause organisms at higher trophic levels to be exposed to high concentrations, which can produce toxicological effects or high risk levels. In our models and those of Mackay and co-authors, this occurs as a result of food absorption and digestion and can be described by the following mass balance equation:

$$N_B = V_B \cdot Z_B \cdot df_B/dt = D_D \cdot f_D - D_F \cdot f_B \qquad [6]$$

where f_D is the chemical fugacity in the diet, D_D is the dietary ingestion rate, which is the product of the dietary ingestion rate G_D ($m^3\ day^{-1}$) and the fugacity capacity of the diet Z_D (i.e., $G_D \cdot Z_D$), and D_F is the fecal egestion rate

of the chemical (**Figure 1**). Hence, at steady state ($N_B = 0$), it follows that

$$f_B/f_D = D_D/D_F = (G_D/G_F) \cdot (Z_D/Z_F) \qquad [7]$$

illustrating that the fugacity in the organism (f_B) exceeds that in its diet (f_D) as a result of dietary uptake because the feeding rate G_D exceeds the fecal excretion rate G_F due to food absorption and Z_D exceeds Z_F because of food digestion which leaves the feces depleted of lipids, proteins, and other food constituents that give the food its high fugacity capacity. This magnification effect is approximately 8 times in fish, but much higher in mammals, birds, and humans with a more efficient digestive system. A graphical presentation of the magnification effect is presented in **Figure 2**. However, for the gastrointestinal magnification effect to cause biomagnification, it is

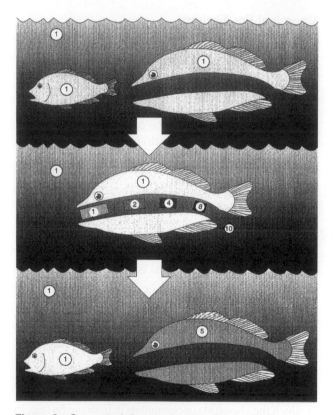

Figure 2 Conceptual diagram depicting the mechanism of biomagnification. Top panel: Predator with a chemical fugacity of 1 Pa in water with a chemical fugacity of 1 Pa consuming prey with a chemical fugacity of 1 Pa. Middle panel: Because predator and prey are at the same fugacity (i.e., no net passive uptake of chemical), food absorption (which reduces the volume of food in the gastrointestinal tract) and food digestion (which reduces the fugacity capacity of the intestinal contents) produce an increase in chemical fugacity in the gastrointestinal tract which leads to net uptake of chemical in the predator Bottom panel: Net chemical uptake will cause a fugacity in the predator that exceeds the fugacity in its prey (i.e., biomagnification) as long as the combined chemical elimination rate by metabolic transformation, elimination to water, and growth dilution are slow.

key that the combined rate of chemical elimination due to metabolic transformation and excretion in the organisms is slow. However, if the combined rate of chemical elimination is high, then a high chemical concentration in the organism cannot be maintained and the chemical will not biomagnify. Chemicals which are predominantly absorbed via the diet and subject to a high rate of chemical elimination due to metabolic transformation and other excretion rates, will exhibit concentrations in the organisms of a food chain that decline with increasing trophic level. This phenomenon is sometimes referred to as trophic dilution, which is the opposite of biomagnification.

Integration

It is important to recognize that body burdens of contaminants in animals are the combined result of a number of chemical uptake and elimination processes acting together. In water-respiring organisms (e.g., fish), the most important processes are respiratory uptake via the gills and body surface area, dietary uptake and elimination via the respiratory surface, fecal egestion, and metabolic transformation. Growth of the animal is also often viewed as an elimination process and referred to as 'growth dilution' although no chemical is actually excreted or transformed. An increase in body mass has a 'diluting' effect on the chemical mass in the organisms. Hence, growth is often treated as an elimination process in bioaccumulation models. A model for bioaccumulation in fish can therefore be formulated as:

$$
\begin{aligned}
N_B &= V_B \cdot dC_B/dt \\
&= k_1 \cdot V_B \cdot C_W + k_{D,i} \cdot V_B \cdot \sum (P_i \cdot C_{D,i}) \\
&\quad - (k_2 + k_E + k_M + k_G) \cdot V_B \cdot C_B
\end{aligned}
\qquad [8]
$$

where k_1, k_D, k_2, k_E, and k_G are the rate constants (in units of day^{-1} if concentrations are in mol m^{-3}) for chemical uptake via the respiratory area (k_1), uptake via food ingestion (k_D) and elimination via the respiratory area (k_2), excretion into egested feces (k_E), metabolic transformation (k_M) and growth dilution (k_G); P_i is the fraction of the diet consisting of prey item i, $C_{D,i}$ is the concentration of chemical (g kg^{-1}) in prey item i, k_2 is the rate constant (day^{-1}) for chemical elimination via the respiratory area (i.e., gills and skin), k_E is the rate constant (day^{-1}) for chemical elimination via excretion into egested feces, and k_M is the rate constant (day^{-1}) for metabolic transformation of the chemical (**Figure 1**). A similar approach can be followed to develop models for other species. For example, for mammals (**Figure 3**), we have used

$$
\begin{aligned}
N_B &= V_B \cdot dC_B/dt \\
&= k_A \cdot V_B \cdot C_A + k_{D,i} \cdot V_B \cdot \sum (P_i \cdot C_{D,i}) \\
&\quad - (k_O + k_E + k_U + k_G + k_P + k_L + k_M) \cdot V_B \cdot C_B
\end{aligned}
\qquad [9]
$$

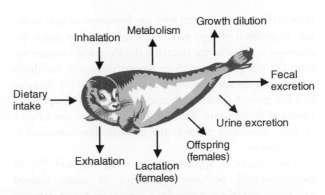

Figure 3 Chemical uptake and elimination processes included in the bioaccumulation model for organic contaminants in mammals.

where C_A is the gaseous chemical concentration in the air and k_A, k_O, k_U, k_P, and k_L are the rate constants (in units of day^{-1} if concentrations are in mol m^{-3}) for chemical uptake via the respiratory area (k_1), elimination via the respiratory area (k_O), excretion into urine (k_U), production in female animals (k_P), and lactation in female animals (k_L).

This modeling approach is based on several key assumptions. First, it is assumed that the chemical is homogeneously distributed within the organism as long as differences in tissue composition and phase partitioning are taken into account. There is considerable evidence, especially for poorly metabolizable substances after long exposure periods, that supports this assumption. However, since the model is not designed to estimate concentrations in specific organs, the model is best applied in situations where the mass or concentration of the chemical in the whole organism is of interest. Internal physiological based pharmacokinetic (PBPK) models are more suitable to estimate concentration differences between various parts of the organism. Second, it is assumed that the organism can be described as a single compartment in its exchange with its surrounding environment. The one-compartment model for an organism is best applied in situations where variations in concentration over time are relatively slow or of secondary concern. A third assumption of the model concerns chemical elimination associated with sexual reproduction and offspring production. Examples are egg deposition or sperm ejection in fish and parturition in mammals. Studies in fish have shown that lipid-normalized concentrations of many persistent organic chemicals in eggs and adult female fish are often approximately equal. This implies that while egg deposition transfers a significant fraction of the chemical body burden from the adult female fish into the eggs, the lipid-equivalent concentration within the organism remains the same. The mechanism in the model by which egg deposition can lower the internal concentration in the organism compared to fish that do not produce eggs (e.g., male fish) is through growth dilution associated with the formation of eggs in the fish. Formation of eggs produces extra tissue in which the chemical resides, hence reducing the chemical's concentration. Offspring production in female mammals and birds follow a similar mechanism. Equations [8] and [9] illustrate that this growth dilution effect is counteracted by uptake of chemical from water and the diet and that the balance of these processes controls the ultimate concentration in the organism.

The practical application of eqns [8] and [9] to environmental pollution problems is often limited by access to time-dependent model input parameter values. Hence, for the model to become useful, it is often further simplified by applying a steady-state assumption ($N_B = 0$). The steady-state assumption transforms eqn [9] into

$$C_B = \left(k_1 \cdot C_W + k_{D,i} \cdot \sum (P_i \cdot C_{D,i})\right)/(k_2 + k_E + k_M + k_G) \quad [10]$$

The steady-state assumption is reasonable for applications to field situations where organisms have been exposed to the chemical over a long period of time, often throughout their entire life. It applies best to chemicals that are subject to relatively fast exchange kinetics (e.g., lower-K_{OW} substances, small organisms), as steady state is achieved rapidly in these situations. It should be used with caution in situations where the exchange kinetics are very slow (e.g., slowly metabolizable chemicals of high K_{OW} (i.e., larger than $10^{7.5}$) in large, lipid-rich organisms), because steady state takes a long time to achieve. In cases where changes in concentrations with the age of the organism are of interest, it is possible to introduce various age classes of the species and apply the steady-state model to each age class independently. One of the implications of applying a steady-state assumption is that the growth of the organism needs to be expressed as a growth rate constant k_G, which is $dW_B/(W_B \cdot dt)$ and assumes that over the period of time that the model applies, the growth of the organism can be represented by a constant fraction of the organism's body weight W_B. The main driving forces of the kinetic bioaccumulation model are: (1) the chemical partitioning of chemical between water and the organism, represented by k_1/k_2; (2) the dietary magnification of chemical, represented by k_D/k_E; and (3) the combined rate of chemical elimination via metabolic transformation, growth dilution. It is noteworthy that any error in the estimation of the respiration rate, which affects both k_1 and k_2, has a tendency to cancel out, causing the ratio of k_1 and k_2 not to be affected. The same argument applies to the ingestion rate, which is related to the egestion rate. This gives the model some remarkable robustness. However, errors in respiration and ingestion rates do affect the relative contribution of the various uptake and elimination pathways and hence the model outcome.

Trophic Interactions

Describing the transfer of contaminants between organisms due to feeding interactions is challenging as food webs are typically complex and vary as a function of time and space. The role of food-web models is to simplify this process to a level that is understandable and can provide useful information. Since the scope of the model is typically limited to several key species, it is often sufficient to include only the most relevant trophic interactions relating to these species of interest (**Figure 4**). Also, because chemical concentrations in organism lipid tissues tend to increase significantly between trophic positions but considerably less between organisms that occupy a similar trophic level, it is often possible to 'lump' species of comparable trophic guilds. The latter should be done with caution as some species exhibit specific feeding behaviors that cannot be generalized to other organisms. In the development of a food-web structure for modeling the bioaccumulation of contaminants, some basic rules of thumb can be suggested:

1. Include species of primary management interest.
2. Include species that can be considered residents of the area of interest unless migrant species are of importance.
3. Include species representing trophic guilds that are of key relevance to the food-web transfer and accumulation of polychlorinated biphenyls (PCBs) in the species of interest. For example, phytoplankton and algae, zooplankton, filter-feeding invertebrates, benthic detritovores, juvenile and adult fish, male and female fish-eating birds, and male, female, and juvenile marine mammals.
4. Minimize the number of species included in the model by representing key trophic guilds by one or two species. This is done to simplify the model and make the calculations more transparent.
5. Include species for which empirical concentration data are available. This provides the opportunity to test and ground-truth the model's calculations.

The structure of the food web represented in food-web models is typically subject to uncertainty. As a test to check whether the trophic structure of the model is adequate for chemical bioaccumulation modeling, it is often useful to explore the relationship between the trophic level of the species of the model as assigned by the feeding relationships used in the model, and stable nitrogen isotope ratios measured in samples of the species included in the model (**Figure 5**). Stable isotope ratios provide an empirical measure of the trophic status of the organism, with stable isotope ratios increasing with increasing trophic level. A good correlation between trophic level and stable nitrogen isotope ratios provides confidence in

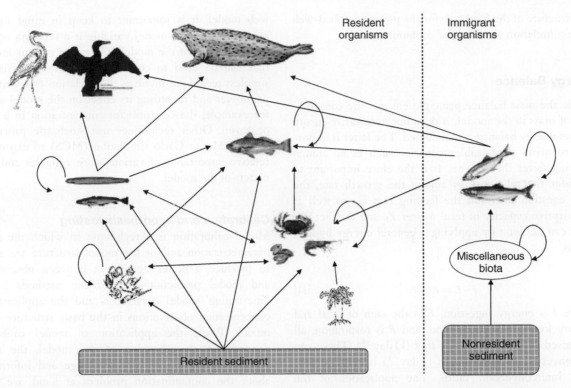

Figure 4 Illustrative example of a food-web structure used to describe the dynamics of contaminants in a West Coast marine food web that includes resident and migrant species.

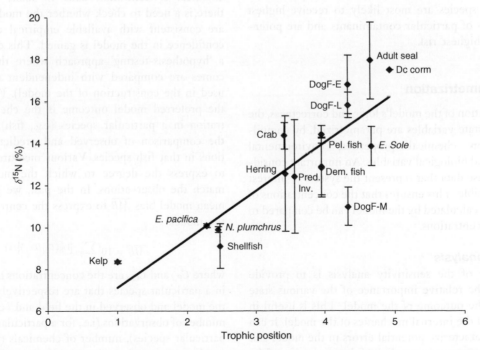

Figure 5 Illustrative example of the application of the relationship between trophic position, derived from gut content studies and stable $^{15}N/^{14}N$ isotope ratios ($\delta^{15}N$) measured in organism tissues, to test the applicability of the trophic structure of the model's food web. A high degree of correlation is indicative of an appropriate food-web structure.

the structure of the food web for the purpose of food-web bioaccumulation modeling of contaminants.

Energy Balance

While the mass balance principle ensures the conservation of mass in the model, it does not necessarily ensure that an energy balance is maintained. The latter is important to avoid implausible scenarios such as an animal growing faster than it eats. It is therefore important to consider the relationships among the growth rate, the fecal egestion rate, and the feeding rate G_D as well as the sorptive capacity of fecal matter Z_F and the diet Z_D. This can be done by applying a general energy budget, that is,

$$I - L = R + P \qquad [11]$$

where I is energy ingestion, L is the sum of fecal and urinary losses, P is production, and R is respiration, all expressed in units of energy flux (kJ day^{-1}). These can be converted to mass fluxes (g day^{-1}) by energy–biomass interconversion ratios. The application of the energy mass balance in the food-web bioaccumulation model makes it possible to include bioenergetic efficiencies in the food-web model. This provides the opportunity to apply the model to a large variety of species for which bioenergetic efficiencies are known. This is an interesting application of the model for ecotoxicological evaluations as it makes it possible to assess which species are most likely to receive highest body burdens of particular contaminants and are potentially at the highest risk.

Model Parametrization

Upon verification of the model's logic and correctness, the model's key state variables are parametrized, by compiling data on chemical properties, environmental conditions, and biological variables. An important consideration is to use data that represent the system of interest as best as possible. This ensures that the concentrations of contaminants calculated by the model can be compared to observed concentrations.

Sensitivity analysis

The purpose of the sensitivity analysis is to provide insight into the relative importance of the various state variables to the outcome of the model. This is useful in the analysis of the internal mechanics of the model. It can be used to characterize potential errors in the model and to develop a better understanding of the interaction of the processes that control the behavior of the contaminant in the food web. In multiparameter models like the food-web model, it is important to keep in mind that the sensitivity of each model variable is a function of other state variables of the model. There are various methods that can be used to conduct sensitivity analysis. The simplest technique involves the variation of a particular parameter and recording its effect on the model output; for example, the contaminant concentration in a target organism. Other techniques use stochastic procedures such as Monte Carlo simulation (MCS) to express the relative importance of various state variables and parameters of the model.

Calibration and hypothesis testing

Model calibration is a technique in which the model parametrization and/or the model structure are altered to produce a better agreement between observations and model predictions. Calibration methods include 'fine-tuning' model parameters and the application of concentration observations in the basic structure of the model. While the application of model calibration depends on the objective of the model, the model strategy, and the state of knowledge and information about the contamination problem at hand, we found that for estimating hydrophobic organic chemical concentrations in organisms of food webs, there is rarely a need for model calibration as long as the model is used within its application domain. The food-web modeling approach described in this article is a mechanistic model and can be used without including empirical chemical concentration data. In many cases though, there is a need to check whether the model calculations are consistent with available empirical data such that confidence in the model is gained. This can be done in a 'hypothesis-testing' approach where the model outcomes are compared with independent data (data not used in the construction of the model). For example, if the preferred model outcome is the chemical concentration in a particular species (e.g., fish), this involves the comparison of observed and predicted concentrations in that fish species. Various measures can be used to express the degree to which the model outcomes match the observations. In the past, we have used the mean model bias $\overline{MB}$ to express the central tendency of the model:

$$\overline{MB} = 10^{\left(\sum_{i=1}^{n} [\log(C_{P,i}/C_{O,i})]/n\right)} \qquad [12]$$

where $C_{P,i}$ and $C_{O,i}$ are the concentrations of the chemical in a particular species that are respectively predicted by the model and observed in the field and i can refer to the number of observations (i.e., for a particular chemical in a particular species), number of chemicals (in a particular species), or a number of species (for a particular chemical substance). In essence, $\overline{MB}$ is the geometric mean (assuming a log-normal distribution) of the ratio of predicted and observed concentrations. $\overline{MB}$ is a measure of the systematic over- ($MB > 1$) or underprediction ($MB < 1$) of the model. It should be stressed that in the calculation of MB, over- and underestimations have a tendency to cancel out. Hence, it describes the central tendency of the model outcome. Variability in the over- and underestimation of measured values can be represented by the 95% confidence interval of $\overline{MB}$. Due to the lognormal distribution of the ratio of predicted and observed BSAFs, this variability can be expressed as a factor (rather than a term) of the geometric mean. For example, if the 95% confidence interval of the $\overline{MB}$ is 3, it means that 95% of the predicted/observed concentration ratios are found between $\overline{MB}/3$ and $\overline{MB} \times 3$.

Uncertainty analysis

The role of the 'uncertainty analysis' is to assess the error in the model calculations. The uncertainty analysis is important because the magnitude of the model needs to be considered when interpreting the results of the model calculations for management purposes. One of the most established techniques for conducting uncertainty analysis of models is MCS, which calculates the effect of inherent error in the model state variables and parameters on the model outcome. This methodology is based on the representation of the model state variables by statistical distributions rather than point estimates. The distribution represents the uncertainty in the value of the model state variable used in the model. The distribution expresses how the state variables may vary due to geographical location, time of the year, differences in behavior among individuals of a species, and other factors. In MCS, these distributions are repeatedly sampled and the sampled values are used in the model to produce a distribution of model outcomes (e.g., chemical concentrations in fish). This distribution of model results represents the variability in the model outcome due to variability and error in the model's state variables (temperature, organic carbon content, lipid contents, etc.). It represents the model uncertainty. The uncertainties in all state variables contribute to the magnitude of the range of model outcomes; however, the contributions are not necessarily additive. The uncertainty calculated through MCS has a strong theoretical foundation. However, it is subject to difficulties associated with the characterization of errors in model parameters and it cannot include errors in model architecture.

One of the key requirements for a meaningful MCS is that the model state variables included in the MCS are independent. In a food-web bioaccumulation, this requirement can pose difficulties. For example, for a predator with three prey items, uncertainty in the proportion of prey item 1 consumed by the predator has direct

consequence for the proportion of the other two diet items in the predator's prey. Hence, diet item proportions or feeding preferences are typically not independent. Other examples of related state variables are animal size, growth rate, lipid content, and feeding rate. Also, regression coefficients in regression equations introduce co-dependency. Before attempting to apply MCS to food-web bioaccumulation models, it is important to ensure that in the model structure chosen the model state variables are indeed independent since lack of dependence creates biologically implausible model outcomes that should not be considered. A second requirement for MCS is a realistic characterization of the uncertainty in each state variable. The latter is not always available for all model state variables and parameters. For example, feeding preferences derived from gut content studies contain considerable uncertainty that is often not characterized.

An alternative method for determining model uncertainty is to use calculated differences between observed and predicted chemical concentrations. This method applies the mean model bias and its 95% confidence interval to the model outcome (e.g., concentration) to predict a distribution of concentrations that includes 95% of the observed data. This method requires that observed concentration data are available. If this is not the case, it is sometimes possible to use uncertainty calculations from application of the model to other systems as a measure of model uncertainty. The application of available site-specific concentration data to characterize model uncertainty has the advantage that estimates of model uncertainty are grounded in empirical observations. However, this method is subject to the limitations of the sampling programs used to obtain the contaminant concentrations. For example, in larger systems, monitoring programs may only have collected samples from a subpopulation of the larger population of a particular species. In that case, it is possible that the distribution of the concentrations in the sampled organisms does not accurately represent the actual distribution of chemical concentrations in the population of the system. In such cases, the model uncertainty may be underestimated.

Food-Web Model Application

Food-web bioaccumulation models have been applied in a number of different ways. However, in terms of assessing ecotoxicological risks, two main methods of application should be emphasized. The first method, referred to as the 'forward calculation', uses observed distributions in measured chemical concentrations in the water and sediments as the starting point of the model (i.e., the external variable or forcing function) to calculate the anticipated corresponding concentration in the wildlife species of interest. The resulting distribution can then be compared to tissue residue guidelines or toxicological threshold values to derive the fraction of the affected wildlife population that contains chemical concentrations greater than or below the reference value of interest. This is illustrated in **Figure 6**, which illustrates the application of the model to the calculation of PCBs concentrations in harbor seal pups as a result of exposure to PCB concentration distributions in water and sediments. In this particular example, the distribution of PCB concentrations in water and sediments of the affected water body can be expected to produce PCB concentrations that vary substantially as described by the distribution with a large percentage of the concentrations exceeding the threshold effects concentration.

The second method, referred to as the 'backward calculation', applies the model to back-calculate what

distribution of PCB concentrations in water and sediments can be expected to produce a particular distribution of concentrations in the target species of ecotoxicological concern or interest. The application of the model is illustrated in **Figure 7** and refers to a management goal to ensure that 95% of the population of a particular target species (e.g., seal pups) contains chemical (e.g., PCBs) concentrations less than a threshold effects concentration (e.g., 5000 ng/g ww). The food-web bioaccumulation model is then used to calculate what distribution in PCB concentrations in the sediments of the system can be expected to produce this distribution. The relationship between PCB concentrations in water and sediments, determined from monitoring programs or estimated using models, needs to be known for this purpose. The calculated distribution can then serve as a remediation or pollution control objective or a sediment quality criterion for the protection of wildlife species.

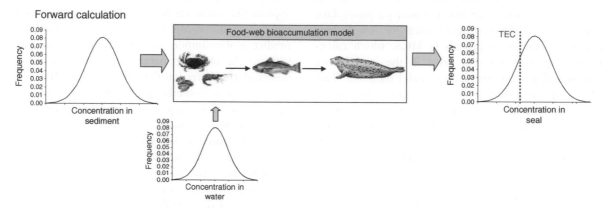

Figure 6 Illustration of the application of the food-web model to assess the ecotoxicological risk of a contaminant in a higher-trophic-level organism (seal). Observed chemical concentrations in sediment and water (presented as statistical distributions) are entered in the model to derive the chemical concentration distribution in a resident seal population and the incidence of concentrations greater than the toxicological threshold effect concentration (TEC).

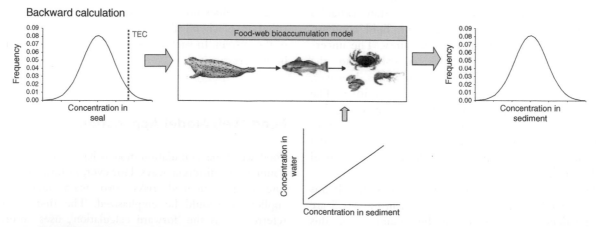

Figure 7 Illustration of the application of the food-web model to derive the system-wide sediment concentration distribution that can be expected to meet an acceptable risk level, set at 5% of the target population of seals exceeding the toxicological threshold effect concentration (TEC).

Examples of the application of the food-web bioaccumulation for ecotoxicological risk assessment, including the use of forward and backward calculations, include the San Francisco Bay food-web bioaccumulation model. The model is documented in a Gobas and Arnot report listed on the website of the San Francisco Bay Clean Estuary Partnership (CEP), and can be downloaded in the form of a Microsoft EXCEL® workbook from http://www.rem.sfu.ca. The purpose of this model is to estimate concentrations of PCBs in a set of key species that reside in the Bay, including double-crested cormorants, the Forster's tern, the harbor seal, and three fish species that are frequently caught by fishermen in the bay, as a result of PCB concentrations in sediments and water in the bay. The model can be used to determine what concentrations of PCBs in the water and sediments of the bay need to be reached to achieve an adequate margin of safety in wildlife and humans exposed to PCBs in the bay area. This information can be used as part of a TMDL characterization to formulate remedial actions to achieve desired water quality goals. The management module includes a simple worksheet to conduct two types of calculations, viz. 'forward' calculations to estimate the concentrations of PCBs in biota of the bay from PCB concentrations in the sediments of the bay and 'backward' calculations to calculate the PCB concentrations in the sediments of the bay that are required to meet PCB concentration based criteria in fish and wildlife for the bay. The backward calculation is designed to determine target PCB concentrations in sediments that meet ecological and/or human health criteria.

Other applications of the food-web bioaccumulation include the estimation of the BAF and BCF for fish species in lower, middle, and upper trophic levels of aquatic food webs. The model predictions can include the effect of metabolic transformation and trophic dilution on the BAF if a reliable estimate of the chemical's metabolic transformation rate in fish is available. The model is named BAF-QSAR v1.1 and is coded in a Microsoft EXCEL® workbook, is freely available for download, and can be run for a large number of chemicals.

Food-web bioaccumulation models have also been used for the derivation of water quality guidelines. For example, the Gobas 1993 model, which was originally published for application to the Lake Ontario ecosystem and has since been applied to many other ecosystems by several authors, has been reviewed and adopted by the US Environmental Protection Agency for developing water quality criteria and waste load allocations in the US under the Great Lakes Water Quality Initiative (EPA-822-R-94-002). This model has been updated and is now referred to as AQUAWEB v1.1. It provides site-specific estimates of chemical concentrations in organisms of aquatic food webs from chemical concentrations in the water and the sediment. Key revisions included new equations for the partitioning of chemicals into organisms, new kinetic models for predicting chemical concentrations in algae, phytoplankton, and zooplankton, new allometric relationships for predicting gill ventilation rates in a wide range of aquatic species, and a novel mechanistic model for predicting gastrointestinal magnification of organic chemicals in a range of species. The model has been evaluated using empirical data from three different fresh water ecosystems involving 1019 observations for 35 species and 64 chemicals. Both models are coded in one Microsoft EXCEL® workbook and can be downloaded from http://www.rem.sfu.catoxicology.

See also: Bioaccumulation; Biomagnification.

Further Reading

Arnot JA and Gobas FAPC (2003) A generic QSAR for assessing the bioaccumulation potential of organic chemicals in aquatic food-webs. *QSAR and Combinatorial Science* 22: 337–345.

Arnot JA and Gobas FAPC (2004) A food web bioaccumulation model for organic chemicals in aquatic ecosystems. *Environmental Toxicology and Chemistry* 23: 2343–2355.

Campfens J and Mackay D (1997) Fugacity based model of PCB bioaccumulation in complex aquatic food webs. *Environmental Science and Technology* 31: 577–583.

Connolly JP and Pedersen CJ (1988) A thermodynamically based evaluation of organic chemical bioaccumulation in aquatic organisms. *Environmental Science and Technology* 22: 99–103.

de Bruyn AMH and Gobas FAPC (2006) A bioenergetic biomagnification model for the animal kingdom. *Environmental Science and Technology* 40: 1581–1587.

Gobas FAPC (1993) A model for predicting the bioaccumulation of hydrophobic organic chemicals in aquatic food webs: Application to Lake Ontario. *Ecological Modelling* 69: 1–17.

Gobas FAPC, Kelly BC, and Arnot JA (2003) Quantitative structure activity relationships for predicting the bioaccumulation of POPs in terrestrial food-webs. *QSAR and Combinatorial Science* 22: 346–351.

Gobas FAPC and Morrison HA (1999) Bioconcentration and bioaccumulation in the aquatic environment. In: Boethling R and Mackay D (eds.) *Handbook of Property Estimation Methods for Chemicals: Environmental and Health Sciences,* pp. 139-232 Boca Raton, FL: CRC Press (ISBN 1-56670-456-1).

Gobas FAPC, Wilcockson JWB, Russell RW, and Haffner GD (1999) Mechanism of biomagnification in fish under laboratory and field conditions. *Environmental Science and Technology* 33: 133–141.

Gobas FAPC, Zhang X, and Wells RJ (1993) Gastro-intestinal magnification: The mechanism of biomagnification and food-chain accumulation of organic chemicals. *Environmental Science and Technology* 27: 2855–2863.

Kelly BC and Gobas FAPC (2003) An arctic terrestrial food-chain bioaccumulation model for persistent organic pollutants. *Environmental Science and Technology* 37: 2966–2974.

Kelly BC, McLachlan MS, and Gobas FAPC (2004) Intestinal absorption and biomagnification of organic contaminants in fish, wildlife and humans. *Environmental Toxicology and Chemistry* 23: 2324–2336.

Mackay D (2001) *Multimedia Environmental Models: The Fugacity Approach,* 2nd edn., 261pp. Boca Raton, FL: Lewis Publishers.

Nichols JW, McKim JM, Andersen ME, *et al.* (1990) A physiology based toxicokinetic model for the uptake and disposition of waterborne organic chemicals in fish. *Toxicology and Applied Pharmacology* 106: 433–447.

Thomann RV (1989) Bioaccumulation model of organic chemical distribution in aquatic food chains. *Environmental Science and Technology* 23: 699–707.

Relevant Website

http://www.rem.sfu.ca – The Environmental Toxicology Research Group, School of Resource and Environmental Management (REM).

Mutagenesis

C W Theodorakis, Southern Illinois University Edwardsville, Edwardsville, IL, USA

Introduction
DNA Damage
Mutations

Effects on Fitness and Ecological Parameters
Further Reading

Introduction

Mutagenesis is the formation of mutations in DNA molecules. There are a variety of mutations that can occur in DNA, such as changes in the DNA sequence or rearrangement of the chromosomes. Such mutations may occur spontaneously, as a result of 'mistakes' that occur during DNA replication or mitosis. Spontaneous mutations are essential to produce genetic variation necessary for natural selection. Mutations may also occur as a result of environmental exposure to genotoxins (chemicals that alter the structure of DNA). Mutagenesis is of concern because it may lead to irreversible effects that can affect fitness of organisms, which in turn may affect population-level processes.

There are potentially thousands of mutagenic and genotoxic agents to which organisms are exposed. Examples of the classes of mutagenic compounds, the DNA damage they elicit, and their sources in the environment are listed in **Table 1**. Each genotoxin may elicit many different types of DNA damage.

DNA Damage

Types of DNA Damage

Because most environmentally induced mutations originate as DNA damage, any discussion on mutagenesis must begin with a discourse on this subject. For the sake of clarity, the structure of DNA bases is given in **Figure 1**. Many classes of DNA damage can lead to mutations, as illustrated in **Figure 2**. Such DNA lesions include damage to DNA bases or to the deoxyribose sugar, base loss, strand breakage, and DNA cross-links (**Figure 2**).

Adducts

Numerous mutagens can form DNA adducts, which are molecules that form covalent bonds with DNA. Some chemicals transfer a methyl or ethyl group to a nucleotide base. **Figure 2b**, 1, illustrates a generalized structure of such a methyl-adducted base. Other chemicals form bulky adducts, so called because they are composed of relatively large and bulky molecules. A number of chemicals are not mutagenic in their native state, but require metabolic oxidation to convert them to mutagenic intermediates. These include polycyclic aromatic hydrocarbons (PAHs). **Figure 2b**, 2, is a schematic representation of a benzo[*a*]pyrene (a PAH that is a common environmental contaminant) adduct. Another type of adduct is lipid aldehyde adducts (**Figure 2b**, 6), which are formed as a result of oxidative damage to lipids, and are discussed in the next section.

Oxidative damage

Oxidative damage occurs as a result of interaction of free radicals or singlet oxygen (molecular oxygen in an excited state) with DNA. The most common oxyradicals include hydroxide radicals ($OH^{\cdot}$) and the superoxide anion O_2^-. Oxyradicals and singlet oxygen are potential mutagenic chemicals known as reactive oxygen species (ROSs). These ROSs are produced to some extent by endogenous metabolic processes, for example, during mitochondrial respiration, metabolism of natural and man-made hydrocarbons, and metabolism of fats. However, some chemicals may stimulate cells to over-produce ROSs metabolically. Besides metabolic processes, some hydrocarbons and heavy metals may convert molecular oxygen to superoxide.

ROSs can damage DNA in two ways. First, the ROSs themselves can form chemical bonds to nucleotide bases

Table 1 Examples of common mutagenic and genotoxic chemicals, their sources in the environment, and the mechanism of formation of DNA damage

Agent	Environmental source	Damage caused	Mechanism
Polycyclic aromatic hydrocarbons	Combustion of organic matter and fossil fuels, crude oil and coal spills and leaching, copier toner cartridges, coal coking, creosote, used oil and lubricants, asphalt	Adducts[a] Oxidative damage[a,b]	Metabolic activation Induction of cytochrome P450 Formation and redox cycling of quinones
Alkylating agents, nitrosamines	Rubber industry, dyes	Methylated or ethylated bases	Metabolic activation
Halogenated organics (PCBs, dioxins, chlorinated solvents, perfluorocarbons, brominated aromatic hydrocarbons)	Industrial manufacturing, paper processing, electrical insulators, cleaning and degreasing agents, solvents, chemical industry, combustion and manufacture of plastics, flame retardants, stain repellents	Oxidative damage[a,b] Adducts	Induction of cytochrome P450 Interference with mitochondrial function Modification of peroxisome function
Pesticides[c]	Agricultural, commercial, and residential applications	Oxidative damage[a,b] Methylated or ethylated bases (some)	Induction of cytochrome P450 Redox cycling (diquat) Interference with mitochondrial function Modification of peroxisome function Metabolic activation
Transition metals, heavy metals, and arsenic	Industrial manufacturing, agricultural chemicals, ore mining and smelting, steel manufacture, building materials and paints, gasoline additives, fossil fuel extraction, combustion of coal, battery manufacture and disposal, metal plating, photographic emulsions, paper manufacture	Oxidative damage[a,b] Adducts, cross-links (As, Cr, Pt)	Reduction of O_2 to form superoxide[d] Reduction of hydrogen peroxide[d] Catalysis of quinone redox cycling[d] Interference with mitochondrial metabolism Inhibition of DNA repair Inhibition of antioxidant enzymes Glutathione depletion Direct DNA binding
Ionizing radiation	Uranium ore mining and fuel processing, nuclear energy, nuclear weapons, combustion of coal	Oxidative damage,[a,b] base loss and fragmentation, DNA–DNA cross-links	Formation of oxyradicals from water and oxygen Excitation of oxygen to singlet oxygen Direct interaction of radioactive particle with DNA sugars and bases
UV light	Sun	Oxidative damage[a,b] Pyrimidine dimers, 6-4 photoproducts	Excitation of oxygen to singlet oxygen Interaction of UV light with bases

[a]Adducts and oxidized bases may lead to production of abasic sites via destabilization of the glycoside (sugar base) linkage.
[b]Oxidative damage includes oxidized bases, change in chemical structure of bases (e.g., open rings), strand breaks, base loss, DNA–protein adducts, and lipid aldehyde adducts.
[c]Includes insecticides (organochlorines, organophosphates, carbamates, pyrethroids), herbicides, and fungicides.
[d]Transition metals only.

(**Figure 2b**, 3 and 4). Second, the oxyradicals may cause internal rearrangement of the DNA to form fragmented bases or open-ring structures (**Figure 2b**, 6). These ROSs may also react with cellular lipids or phospholipids, which leads to formation of lipid adducts (**Figure 2b**, 6). Finally, oxyradicals may oxidize proteins, creating protein radicals, which can form covalent attachments to DNA in the form of DNA–protein cross-links (**Figure 2b**, 8).

Base loss, sugar damage, and strand breakage
Base loss, sugar damage, and strand breakage may occur in several ways. For example, a base may be hydrolyzed from the deoxyribose sugar (**Figure 2c**, 9) enzymatically – during DNA repair (see below) – as a result of oxyradical attack, or as a result of bulky adducts or oxidized bases. This site is called an abasic site. Base loss can also occur as a result of free radical attack on the sugar, resulting in sugar

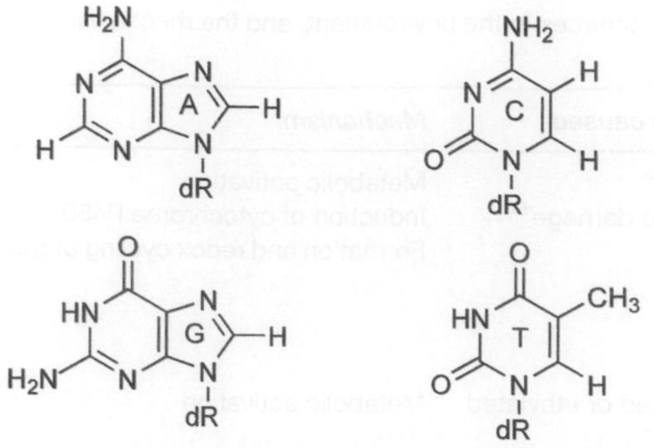

Figure 1 Schematic diagram representing the structure of DNA bases. A, adenine; C, cytosine; G, guanine; T, thymine.

damage (**Figure 2c**, 10). In addition, sugar damage may result in DNA strand breakage (**Figure 2c**, 11). Strand breaks may also be formed by hydrolysis of the sugar–phosphate bond (**Figure 2c**, 12). These types of strand breaks may be produced transiently during the DNA repair process. However, some chemicals might inhibit repair enzymes, resulting in persistent strand breaks.

Because DNA is a double-stranded molecule, strand breaks may occur in one (single-strand breaks, SSBs) or both of the DNA strands (double-strand breaks, DSBs; **Figure 3**). DSBs are less easily repaired and more persistent than SSBs, and are more effective in producing deleterious cellular effects. There are basically three ways in which DSBs may be formed. First, there may be two SSBs directly across from each other or in close proximity (**Figure 3**). Second, if an SSB is unrepaired and the cell tries to replicate a DNA molecule with an SSB, this may result in a DSB. Third, some types of enzymes can produce DSBs. If left unrepaired, DSBs can lead to chromosomal mutations, as discussed below, or may lead to cell death.

DNA cross-links

An additional class of chemically induced DNA lesions includes DNA cross-links. These are formed when some chemical agents such as *cis*-platinum (a chemotherapeutic agent), arsenic, or chromate can form adducts to two or more bases simultaneously. DNA cross-linking agents may covalently cross-link adjacent nucleotide bases on the same strand (intrastrand cross-links; **Figure 2c**, 15) or on opposite strands bases or (interstrand cross-links; **Figure 2c**, 16). Alternatively, cross-linking agents may link proteins to the DNA bases (DNA–protein cross-links).

Radiation-induced DNA damage

Chemicals are not the only environmental agents that can cause mutations. Radiation, a type of electromagnetic energy, may also be mutagenic. In general, there are two primary categories of mutagenic ionizing radiation and ultraviolet (UV) radiation. Although there have been

claims that other types of electromagnetic energy – such as magnetic fields, microwaves, and radiowaves – are mutagenic or carcinogenic, to date, the evidence for this remains equivocal.

Ionizing radiation includes alpha particles (two protons and two neutrons, that is, a helium nucleus), beta particles (high-energy electrons), and gamma particles (high-energy photons). The sources of ionizing radiation in the environment may be natural or man-made. Natural sources include cosmic radiation – originating from the sun, stars, or other celestial bodies – and naturally occurring radioisotopes. Man-made sources are listed in **Table 1**. Ionizing radiation could produce base or sugar radicals, which are unstable and rapidly react with other macromolecules or undergo internal molecular rearrangements. This results in strand breakage, base loss, fragmented bases, DNA–DNA cross-links (**Figure 2b**, 5), or DNA–protein cross-links (**Figure 2b**, 8). Alternatively, the radioactive particles can interact with water or oxygen, which produces ROSs and singlet oxygen, which leads to oxidative DNA damage.

Another type of radiation is UV radiation. Because the source of UV radiation is the sun, environmental sources of UV radiation are entirely natural. However, anthropogenic activities may result in increased exposure or susceptibility to UV-induced mutagenesis. For example, chlorofluorocarbons (CFCs) may react with ozone in the upper atmosphere to convert it to molecular oxygen. Because ozone strongly absorbs solar UV light, this may result in increased UV reaching the Earth. Also, changes in climate (e.g., due to buildup of atmospheric CO_2) or draining of wetlands may lower water levels and expose aquatic organisms to more UV. Furthermore, some chemicals may inhibit an organism's natural ability to repair or prevent UV-induced DNA damage, or may react with UV to produce ROS.

UV can cause DNA damage in two mechanisms. First, UV can convert molecular oxygen into singlet oxygen (an energized, highly reactive form of oxygen). This may lead to increase oxidative DNA damage. Second, UV radiation can directly interact with DNA bases to produce so-called dimers and photoproducts (**Figure 2c**, 13 and 14, respectively).

Repair of DNA Damage

There are several different pathways involved in the repair of modified DNA bases. One such pathway is termed nucleotide excision repair, which repairs bulky adducts, lipid aldehyde adducts, and UV photoproducts (**Figure 4a**). A second type of DNA repair is termed base excision repair. This type of repair is used on oxidized bases, AP sites, methylated bases, and some SSBs, and is illustrated in **Figure 4b**. Another type of DNA repair is DSB repair, which may involve homologous

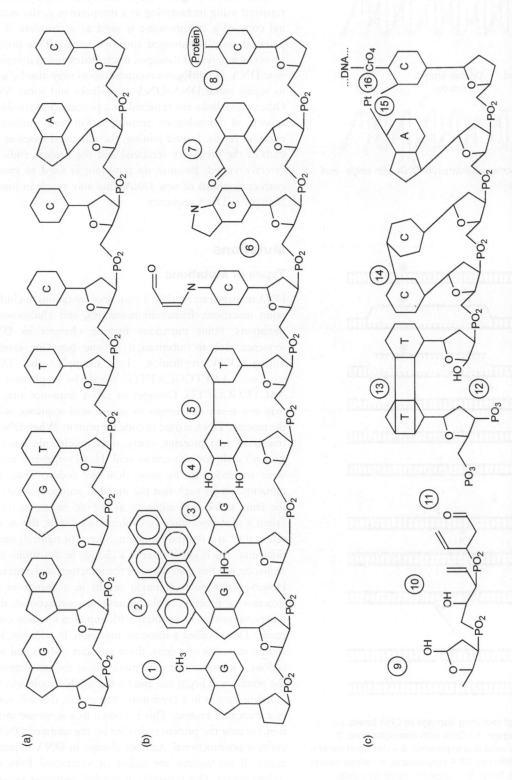

Figure 2 Diagram of the types of DNA damage that can occur as a result of exposure to genotoxic agents. (a) Undamaged DNA; (b) 1 – Methylated guanine, 2 – benzo[a]pyrene adduct, 3 and 4 – oxidized bases, 5 – DNA cross-link 6 – two examples of lipid aldehyde adducts, 7 – open ring base, 8 – DNA–protein cross-link; (c) 9 – abasic site (hydrolysis of glycosidic linkage), 10 – sugar damage leading to base loss, 11 – sugar damage leading to strand break, 12 – hydrolysis of sugar–phosphate bond, 13 – thymine dimer, 14 – cytosine 6-4 photoproduct, 15 – DNA–DNA cross-link (interstrand), in this case mediated by *cis*-platinum (complete structure of *cis*-platinum not shown), 16 – DNA–DNA cross-link (interstrand), in this case mediated by chromate.

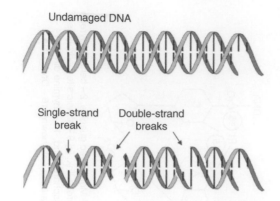

Figure 3 Schematic representation of DNA with single- and double-strand breaks.

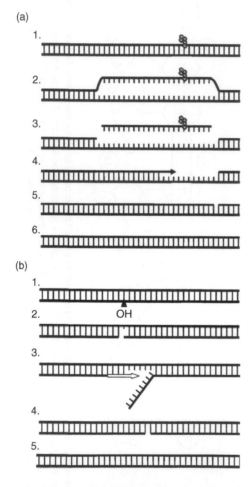

Figure 4 Methods of repairing damage to DNA bases. (a) Nucleotide excision repair: 1 – DNA with damaged base, 2 – damaged DNA is unwound and separated, 3 – damaged section is excised, 4 – gap is filled by DNA polymerase, 5 – single-strand gap remains after gap filling, 6 – ligase connects free ends.
(b) Base excision repair: 1 – DNA with damaged base, 2 – damaged base is removed and nick is made in DNA, 3 – DNA polymerase simultaneously displaces damaged section (producing a 'flap') and synthesizes new DNA, 4 – flap is cut, leaving a single-strand nick, 5 – DNA ligase connects the two free ends of the nick.

recombination or direct end rejoining. In homologous recombination (**Figure 5**), a damaged DNA strand is repaired using its homolog as a template (e.g., the maternal copy of a chromosome is used as a template if the paternal copy is damaged and vice versa). This process involves removal of damaged nucleotides and synthesis of new DNA. Homologous recombination may also be used to repair some DNA–DNA cross-links and some SSBs. Other cross-links are repaired in a process that combines aspects of homologous recombination and nucleotide excision repair. In end joining, the damaged bases at the ends of the break are removed and the broken ends are directly joined. Because no template is used to ensure correct synthesis of new DNA, this may result in loss or changes in DNA sequence.

Mutations

Types of Mutations

DNA damage can result in a variety of mutations, including point mutations, frameshift mutations, and chromosomal mutations. Point mutations include changes in DNA sequence due to substitution of one base for another during DNA replication. For example, the DNA sequence AATTCGCATTG could be replicated as AACTCGCCTTG. Changes in DNA sequence may or may not result in changes in amino acid sequence when the mutated DNA is used to code for protein. When DNA is translated into proteins, every three nucleotide bases (a 'codon') code for one amino acid. However, many amino acids are coded for by more than one codon. Thus, if a mutation occurs such that the mutated sequence codes for the same amino acid sequence as the old sequence, this is called a silent mutation. In evolutionary terms, this is also referred to as a neutral mutation. Silent (or neutral) mutations may also occur if there is a change in the amino acid sequence, but this does not alter the structure of the protein. However, if a point mutation results in a change in the structure or function of the protein, a nonfunctional, dysfunctional protein or a protein with impaired function could result. This is called a missense mutation. In addition, in a coding sequence of a gene, there are start codons and stop codons – locations that determine where the translation of the protein will begin and end on the mRNA molecule. If a mutation results in a premature stop codon, this will result in a truncated protein. This is known as a nonsense mutation, because the protein coded for by the mutated DNA is entirely nonfunctional. Another change in DNA sequence occurs if nucleotides are added or subtracted from the coding region. This is called a frameshift mutation, because it changes the reading frame and leads to a complete change in the amino acid sequence coded by the DNA.

Chromosomal mutations (also known as cytogenetic mutations) are changes in the structure or number of

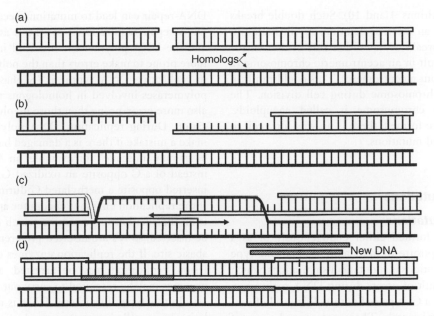

Figure 5 Homologous recombination repair of DNA DSB. (a) Damaged and undamaged homologs pair up; (b) damaged sections are removed by nucleases; (c) damaged and undamaged homologs cross over, polymerases use undamaged homolog to synthesize new DNA in damaged homolog; (d) DNA is cut at crossovers and ligated.

chromosomes. Chromosomal mutations are alternatively called chromosomal aberrations, chromosomal rearrangements, cytogenetic effects, cytogenetic aberrations, or clastogenic effects. The process of producing such effects is referred to as clastogenesis. Chromosomes can be visualized when they condense during mitosis or meiosis, and can be stained with various dyes. Because some regions stain darker than others, this produces a banding pattern when the chromosome is observed under a microscope. An unreplicated chromosome with a representative banding pattern is schematically illustrated in **Figure 6a**. The circle at the center represents the centromere: the place where the mitotic spindle attaches during cell division. The numbers refer to various positions on the undamaged chromosome. A DSB may lead to a break in the chromosome, as illustrated in **Figure 6b**.

If this break is unrepaired, it may lead to loss of a portion of the chromosome, called a deletion. If a piece of chromosome is deleted from the end, as illustrated in **Figure 6c**, this is called a terminal deletion. If there are breaks in two different chromosomes, they may exchange the ends distal to the breaks, a process known as translocation (**Figure 6d**). In some cases, a piece of a chromosome may be inserted in the interior of a different chromosome, a process known as insertion (**Figure 6e**). If there are two or more breaks in the same chromosome, a number of things can also happen. For example, an inversion may take place, where the piece of chromosome between the two breaks is 'flipped' (**Figure 6f**; this takes

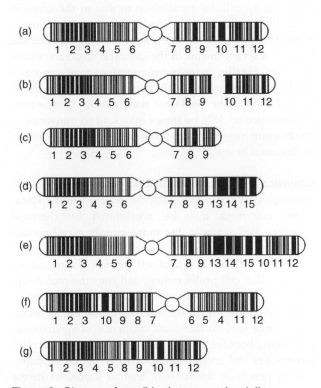

Figure 6 Diagram of possible chromosomal mutations. (a) Undamaged chromosome; (b) chromosome with break, (c) terminal deletion of positions 10–12; (d) translocation of a section containing positions 13–15 from another chromosome, (e) insertion of section 13–15 in between positions 9 and 10; (f) inversion of section from positions 4–10; (g) internal deletion of section 6–7, which contains the centromere.

place between positions 4 and 10). Such double breaks may also result in an internal deletion. As illustrated in **Figure 6g**, if the breaks are on either side of the centromere, this may result in an acentromeric chromosome (a chromosome without a centromere). This could result in loss of an entire chromosome during cell division. The loss of an entire chromosome is called aneuploidy. Unfortunately, once formed, there is no way for a cell to repair chromosomal mutations.

Formation of Mutations

Spontaneous mutations

Mutations can be formed either endogenously or as a consequence of exposure to mutagenic agents. One type of endogenous mutations is spontaneous mutations, which may occur as a result of errors during DNA replication, for example, when a G is paired with a T instead of a C. This results in a mismatch. There are several types of enzymes, called mismatch repair enzymes, which can correct such mistakes. However, sometimes this mistake is not repaired and this results in a mutation. Strand breaks and abasic sites can also form spontaneously, for example, due to thermal energy arising from the heat produced by cellular metabolism or due to the inherent instability in the chemical bonds. Other types of spontaneous DNA damage include loss of amino groups on the bases or rearrangements in the chemical structure within the bases. Finally, endogenous mutations may occur as a result of endogenous DNA damage caused by ROSs formed by routine oxidative metabolism. Spontaneous and endogenous DNA damage may lead to mutations in mechanisms similar to mutagen-induced DNA damage, as discussed below.

Chemically induced mutations

Although they occur naturally, the occurrence of spontaneous mutations may be accelerated by chemical exposure. For example, the more rapidly a cell divides the greater the chance of a spontaneous mutation. Some chemicals increase the rate of cell division in some tissues (this is called cell proliferation), and thus the probability that a spontaneous mutation will occur. In addition, inhibition of DNA repair by arsenic, cadmium, or other metals may lead to increased incidence of spontaneous mutations, because of the reduced rate of removal of mismatches and endogenous DNA damage. Also, exposure to genotoxic agents may lead to mutations in the mismatch repair or other repair genes, leading to decreased rates of repair.

Mutations can also be induced by exposure to mutagenic compounds, which certainly applies to point mutations. Point mutations can be induced when damaged DNA is repaired or undergoes replication.

DNA repair can lead to mutations because most types of DNA repair require DNA synthesis as an essential step, and the DNA polymerases involved in DNA repair are more prone to make errors than the polymerases involved in replicative DNA synthesis (S phase synthesis). The polymerases involved in homologous recombination are also more error-prone than those involved in DNA replication. During replication, DNA polymerases may also make a mistake if there is a damaged base. Such damaged bases may 'miscode'; for example, an A may be inserted instead of a C opposite an oxidized G, and a T may be inserted opposite a methylated G during DNA synthesis. If the replication enzymes encounter an abasic site, there is no information to determine which nucleotide should be inserted, so A's are inserted preferentially opposite an abasic site. If the replication enzymes encounter a bulky lesion, replication may be arrested, and a new set of enzymes may be recruited to carry out translesion synthesis. In this type of synthesis, DNA is replicated past the lesion by so-called error-prone polymerases. These polymerases may induce a mutation because (1) the damaged bases may miscode, for example, an A may be inserted instead of a C opposite an adducted G, or (2) these polymerases are inherently error-prone, so they may make a mistake even at a site where there is no damage. Similar events may occur to produce frameshift mutations.

Frameshift mutations may occur in one of two ways. The first method involves replication of damaged bases. Deletion of one or more damaged bases (**Figure 7a**) may occur if there is a sequence with two or more of the same bases side by side, in this case two G's, one of which is damaged. During synthesis, the damaged G may 'bulge out' of the DNA strand, and the C on the opposite strand may then bind with the next, undamaged, G (**Figure 7a**). When DNA replication resumes, the new strand has a one-base deletion. Alternatively, if a DNA strand with a damaged G is replicated, a C may be inserted opposite the damaged G, but then it may be displaced by an A (some chemically modified G's may bind with A just as well as, or even better than, C). In this case, the C may bulge out, resulting in the newly synthesized strand having an extra base inserted. A second method of frameshift mutations may occur as a result of intercolating agents. These are chemicals that can intercolate or 'slip' between DNA bases, and may mimic a DNA base during DNA replication. represents a damaged guanine. Dotted line represents newly synthesized DNA.

Chromosomal mutations may occur as a result of DSBs. If such breaks are unrepaired, this may result in chromosomal deletions. Errors in repair of DSBs may lead to inversions, translocations, or insertions.

Finally, some organisms may undergo a phenomenon known as adaptive mutagenesis. In this process,

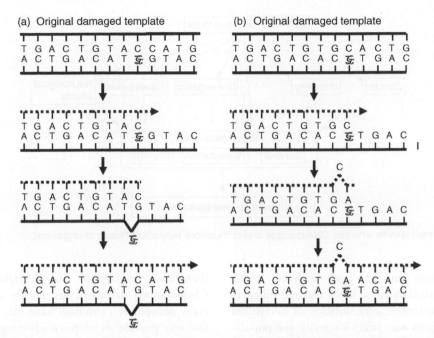

(a) Original damaged template   (b) Original damaged template

Figure 7 Hypothetical mechanisms for DNA damage-induced frameshift mutation formation. (a) Deletion in newly synthesized DNA; (b) insertion in newly synthesized DNA.

environmental stressors cause an increase in endogenous or spontaneous mutations, presumably by endogenous inhibition of repair and mismatch detection. This is thought to be an adaptive mechanism whereby bacteria create *de novo* genetic variation, because some of the new variants may survive the stress better than others. It is not known if adaptive mutation occurs in eukaryotes, or genotoxic stressors can also induce adaptive mutations. However, a similar process occurs in cancer cells, which gradually accumulate more and more mutations after initiation of the tumor – a process called genomic instability. Latent genomic instability can also occur in radiation-exposed cells, which may spontaneously develop high numbers of mutations long after radiation exposure and initial repair of the damage to DNA.

Modulators of Mutagenesis

There are variety of endogenous and environmental factors that can modulate genotoxic responses and mutagenesis. For example, in some species, development of neoplasia ('cancer') is sex dependent, so that mutagenesis is perhaps modulated by estrogen or other hormones. Because cell division in embryonic, larval, and juvenile organisms is more rapid than in adults, the adults may be less susceptible to such damage. Also, persistent DNA lesions (mutations or chromosomal abnormalities) may accumulate over time, so that older individuals are more likely to exhibit neoplasia or other mutagenic effects. Additionally, variation between individuals may be due to different exposure histories or genetic variability in

cellular uptake, excretion, xenobiotic metabolism, or DNA repair. Environmental factors that modulate DNA damage and mutagenesis include temperature (which may mediate carcinogen metabolism or DNA repair rates in these ectotherms), dissolved oxygen concentration in water (which may mediate oxidative stress), salinity or ionic composition of water, or food availability and chemical composition. Also, the amount of DNA damage in fish may vary with season, perhaps due to temperature or bioenergetic or hormonal status. Furthermore, concomitant exposure to other chemicals may promote DNA damage or promote mutagenesis. Thus, the amount of DNA damage induced by complex mixtures may be much more than that predicted by single-chemical genotoxic effects. Finally the degree of genotoxic or mutagenic effects may be mediated by intra- and interspecific interactions such as competition, predation, parasitism, trophic structure and complexity of the ecosystem, and population density of affected organisms.

Effects on Fitness and Ecological Parameters

Environmentally induced DNA damage and resultant mutations may be pertinent for ecologically relevant organisms because they may affect organismal-level fitness components. This may be translated into effects on populations, and eventually communities and ecosystems. This is illustrated in **Figure 8**. First, because DNA

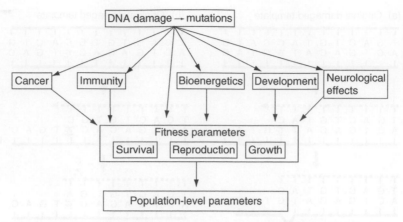

Figure 8 Possible mechanisms whereby DNA damage and/or mutations may affect fitness of organisms.

damage and mutations can lead to cell death and cancer, this may affect survival. Because DNA damage enhances the rate of cell senescence, accumulation of unrepaired damage and mutations may affect longevity and population age structure. DNA damage and mutations have their greatest deleterious effect on rapidly dividing cells. Because gonadal germ cells are rapidly dividing, they are particularly susceptible to the effects of DNA damage and mutations. Growth may also be affected because of induced cell death, interference with DNA replication, or induced delay of cell division (DNA damage induces cell cycle delay, a phenomenon that halts the cell cycle to allow time for repair before DNA replication or mitosis). Immune cells, both mature white blood cells and white blood cell stem cells (which divide rapidly), are also particularly susceptible to the effects of genotoxins and mutagens. An inhibited immune system may in turn affect fitness of affected organisms. Genotoxic effects may also affect bioenergetics or organisms for two reasons. First, DNA repair is an energetically expensive process. Second, mitochondria contain their own DNA, and damage and mutations in mitochondrial genes may affect mitochondrial function and ATP production. Furthermore, DNA damage and mutagenesis may affect development, both by inducing teratogenic (causing developmental defects) mutations and by delaying development because of cell cycle delay and interference with DNA replication. Finally, DNA damage in nerve cells may result in acute neurological effects, neurodegeneration, or neurodevelopmental effects.

Mutations in germ cells may result in heritable mutations that can be passed on to future generations (transgenerational effects) or can spread through the populations. Spontaneous and endogenous germ-line mutations are relevant in evolutionary terms. Missense, nonsense, and frameshift mutations are almost always harmful, or deleterious, but in some rare instances may produce a phenotype with an adaptive advantage

(beneficial mutations) in certain environments. Chromosomal mutations are mostly deleterious (especially aneuploidy), although some may be neutral, and a few may provide an adaptive advantage. If they do provide such an advantage, they may increase in frequency in the population due to natural selection. If dominant mutations in germ cells (i.e., gamete stem cells) result in the death of the offspring, these are called dominant lethal mutations. Deleterious mutations may eventually be removed from the population, but this may take many generations, especially if they are only mildly deleterious. If the deleterious mutation is recessive rather than dominant, it may persist indefinitely in the population. Neutral, beneficial, recessive, and mildly deleterious mutations may persist in the populations and increase population genetic diversity over time.

Environmentally induced mutations (i.e., due to chemical exposure, ionizing radiation, or UV) can affect survival, metabolism, growth, reproduction, propensity to develop cancer, or behavior in offspring or other descendants at any life stage. If these mutations are expressed in a dominant fashion, their effects may be always apparent when a mutant allele is present. If they are recessive mutations, the mutant phenotype is apparent only in the homozygous state. Exposure to mutagenic agents may increase the mutation rate of populations, that is, the number of new mutations per generation.

The relative number of persistent deleterious mutations in the population is called the mutational load. The deleterious effects of mutational load may depend on population size, because smaller populations have a higher level of inbreeding (the mating of genetically similar individuals). This leads to increased number of homozygous loci in the population, which increases the chance that deleterious recessive mutations are expressed – a process known as inbreeding depression. Small populations may experience inbreeding depression, which leads to further reduction in population size due to decreased average fitness, which leads to further inbreeding depression, etc., such that these populations may spiral toward extinction in a phenomenon called mutational meltdown. Exposure to mutagenic agents may hasten this process. Thus, although exposure to mutagenic contaminants may increase population genetic diversity and thus hasten the rate of evolution, loss of fitness may also result in population bottlenecks and reduction of genetic diversity. If the relative loss of fitness is genotype dependent, this may lead to evolution of more mutagen-resistant populations.

Further Reading

Loechler EL (1996) The role of adduct site-specific mutagenesis in understanding how carcinogen–DNA adducts cause mutations:

Perspective, prospects and problems. *Carcinogenesis* 17: 895–902.
Norbury CJ and Hickson ID (2001) Cellular responses to DNA damage. *Annual Reviews of Pharmacology and Toxicology* 41: 367–401.
Rosenberg SM (2001) Evolving responsively: Adaptive mutation. *Nature Reviews Genetics* 2: 504–515.
Sarasin A (2003) An overview of the mechanisms of mutagenesis and carcinogenesis. *Mutation Research* 544: 99–106.
Shugart LR, Theodorakis CW, Bickham AM, and Bickham J (2002) Genetic effects of contaminant exposure and potential impacts on animal populations. In: Calow P (ed.) *Handbook of Ecotoxicology*, pp. 1129–1148. Oxford: Blackwell Science.
Wang Z (2001) DNA damage-induced mutagenesis: A novel target for cancer prevention. *Molecular Interventions* 1: 269–281.
Wirgin I and Theodorakis CW (2002) Molecular biomarkers in aquatic organisms: DNA- and RNA-based endpoints. In: Adams SM (ed.) *Biological Indicators of Aquatic Ecosystem Health*, pp. 73–82. New York: American Fisheries Society.

Reproductive Toxicity

L V Tannenbaum, US Army Center for Health Promotion and Preventive Medicine, Aberdeen, MD, USA

Published by Elsevier B.V., 2010.

Introduction
Scope of the Issue
Standard Approach to Reproductive Toxicity
  Assessment – Hazard Quotients
Other Approaches to Reproductive Toxicity
  Assessment

Looking to the Future
Does Reproductive Toxicity Really Need to Be
  Evaluated?
Summary
Further Reading

Introduction

Although chemicals released to the environment have the potential to tax any number of biological processes in nonhuman species, the case can be made that reproductive effects are the ones that need to be most seriously addressed. Whereas the public's perception of reproductive toxicity may primarily or only reflect matters of societal relevance (e.g., hunting or fishing as recreational pastimes may all of a sudden be jeopardized; the desire to have in abundant number, a species whose image is on a state emblem), a stark biological reality should heighten the attention given to reproductive toxicity, more so than any other chemically caused ill-effect – if reproduction is being compromised in a species, extinction, in a specific locality or beyond, may ensue. The argument can be made that the primacy of reproductive toxicity as an ecotoxicological endpoint of concern is reflected in

this encyclopedia's arrangement. Of nearly 70 subtopical entries under the larger heading of 'ecotoxicology', just four involve specific types of illness or disease; aside from 'reproductive toxicity', there are only 'carcinogenic effects', 'sublethal effects', and 'teratogenesis' that are discussed.

Scope of the Issue

A proper review of reproductive toxicity within an ecotoxicological context should draw out a distinction between concerns over large geographical expanses and those that pertain to discrete and relatively small area locations. Examples of the former would be the US's Gulf States region, or the Eastern deciduous forest; an example of the latter would be a 5 ac hazardous waste site that is managed under a program such as the US Environmental Protection Agency's (EPA) Superfund Program. For the large expanses, the chemical or chemicals responsible for reproductive effects typically have numerous release points, and have had wide-area distribution of the toxicants achieved by dynamic global processes, such as aerial transfer via wind entrainment. The very real problem of eggshell thinning in numerous bird species that came to our attention in the 1960s and 1970s, and that led to severe bird population declines, exemplifies this pattern. The extensive use of chlorinated organic pesticides for two decades prior, led to substantial accumulations of DDT and its associated metabolites in soil, which were conveyed to lakes and streams through overland flow events. Through bioaccumulation, the pesticides entered the birds' aquatic diet items, and interfered with shell deposition during development. This situation led to incubating eggs having their shells crack under the weight of nesting parents. Reproductive toxicity in this scenario entered the picture when it became apparent that eggshells were breaking during development, and when population declines were being noted. Thus, there was a critical need to assign toxicological causation to the discovered effects, and for a variety of reasons. First, with the reproductive toxin identified, man could intervene and ensure that further releases of the offending agents could be curbed. Second, toxicological principles could possibly be applied to minimize further impacts from the already-released organochloride pesticides. Finally, protocols could be developed to test other pesticides for their potential to induce reproductive effects prior to their commercial availability and use.

Within a context of discrete and relatively small contaminated properties (such as hazardous waste sites), the role of reproductive toxicity is entirely different. What triggers the development and application of such science is rarely the discovery of a population-impacting reproductive effect, but rather the anticipation that there could be one. The addition of the Superfund Program to the EPA's mission in 1980 required health risk assessments to be conducted for ecological receptors as well as for humans. The outcomes of the largely reproduction-based ecological assessments intend to either express an acceptable or unacceptable risk. In the latter case, the recommendation is made to study the issue further or to remediate the site outright, because it is assumed that continued chemical exposures will eventually trigger population-leveling effects. It is chiefly the ecological risk assessment (ERA) process of the Superfund Program and other similarly aligned initiatives that has fueled the interest in reproductive toxicity for ecological (i.e., nonhuman) species, and the generation of toxicological data over the last two decades or more to support such concerns.

Standard Approach to Reproductive Toxicity Assessment – Hazard Quotients

Although not expressly stated in ERA guidance, reproduction is rather clearly the toxicological endpoint of greatest concern in ERA work. Aside from reproduction being an essential biological function that allows for perpetuation of a species, reproduction's popularity may reflect the myriad ways in which this function can be compromised. A partial list of reproductive biology elements that can be impaired to some degree in chemically exposed terrestrial receptors includes: behavior (e.g., mate recognition, courtship displays), spermatogenesis and oogenesis, litter/clutch/brood number, litter/clutch/brood size, mating frequency, birth/neonatal weight, spontaneous abortion rate, and dam weight. In the interest of having available tools with which to assess the potential for reproductive effects cropping up at contaminated sites, toxicologists scour the peer-reviewed scientific literature for toxicity studies where one or more reproductive effects are the endpoints. If the toxicity databases are lacking, new studies are conducted to furnish the required data. From these studies, that are almost always conducted with laboratory species, safe- and/or effect-level chemical-specific 'doses' (in units of milligrams of the chemical ingested/kilograms of body weight of the test species/day) are derived. The doses, termed toxicity reference values (TRVs), serve as the denominators of the simple ratios of desktop ERAs, termed hazard quotients (HQs), where a receptor's supposed potential for risk is calculated:

$$HQ = \frac{\text{Estimated daily chemical intake through the diet (mg/kg/day)}}{\text{Toxicity reference value (mg/kg/day)}}$$

The numerator of the HQ, in the same units as the denominator, is the receptor's estimated chemical intake. A simplified example will illustrate the calculation. If one wanted to calculate the HQ for mercury for red fox at a contaminated site, one would first determine how many milligrams of mercury a fox consumes in a given day. If we approximate a fox's daily diet to consist of field mice only (in actuality, a red fox has an omnivorous diet), the mercury concentration in one field mouse would first be determined either through modeling or by actual measure. This figure would then be multiplied by the number of mice a fox likely consumes in a given day. The total milligrams of mercury consumed per day would then be normalized to the fox's body weight, to render an estimated daily mercury dose in mg/kg/day – the HQ's numerator. This figure would then be divided by the reproduction-based mercury TRV, and the resultant unitless ratio is the HQ. For all of their popularity and ease in construction and use, the commonly computed reproduction-based HQs are crude measures of reproductive toxicity. As unitless metrics, they do not express risk, the probability of there being a negative effect, in this case, the probability of a receptor developing a compromised reproductive condition. One other problem (of many) with HQs, is that the values generated almost always exceed 1.0, indicating that receptors are consuming more than a safe chemical dose, and suggesting that the receptors are likely exhibiting reproductive effects.

Much of the inexactitude of the above reproductive toxicity assessment scheme stems from the manner in which the laboratory exposures of the test species radically differ from the actual exposures of receptors in the wild. By way of example, nearly all the laboratory studies with mammals are conducted with mice or rats, but the mammalian species that are of concern in ERAs are the larger, higher-trophic-level, and wider-ranging species, such as fox, deer, raccoon, and coyote. Aside from this key species difference, the form of the chemical tested with is rarely the one that the receptor in the wild encounters. Even in the rare case of the chemical form being the same, the chemical in the outdoors has been subjected to numerous environmental factors over the decades since the site became contaminated (temperature extremes, precipitation, photoincident light, ionizing radiation, etc.), and these have likely served to significantly alter the chemical's toxicity. The laboratory studies also occur under fixed temperature and lighting conditions, quite unlike the variable environments of the site receptors. Several other significant departures from the site condition may throw into question the utility of the chemical concentrations or doses that are deemed to be safe or harmful to ecological receptors. The laboratory studies almost always test a singular chemical's effect, but at contaminated sites, there are commonly a dozen or more chemicals of concern. Although a well-controlled lab study can essentially prove causation (i.e., that the administered chemical, alone, produced an effect), the utility of the study information is compromised because of the likely operating synergistic or antagonistic properties of the collective chemical mixture presented to the receptor in the wild. One other key difference is that almost all laboratory toxicity studies are of single-generation exposures (i.e., the exposure occurred during a portion of the lifetime of a cohort of test animals), while the receptors at contaminated sites have been exposed for tens of generations.

Other Approaches to Reproductive Toxicity Assessment

The food ingestion/food chain model that generates HQs is one approach of several used to assess reproductive toxicity in nonhuman species. At present the HQ approach,

essentially a dose comparison, is not even workable for two terrestrial animal groups, amphibians and reptiles, nor is it used for fish and other aquatic species. In the case of the amphibians and reptiles, there is a dearth of toxicity information of the dose–response type, but a greater difficulty to surmount is that of identifying the dominant mode of chemical uptake for these receptors. It is recognized that terrestrial ecological receptors, like man, have three operating routes of uptake – ingestion, inhalation, and dermal contact. With reasonable supports, it is assumed that the predominant route for birds and mammals is ingestion, and consequently ERAs do not even attempt to quantify for birds and mammals, the potential for reproductive effects (or any other systemic effects for that matter) that may stem from the other two routes. For amphibians and reptiles, where the integument is often quite moist, and where the animal lies closely appressed to the substrate, dermal contact and respiration through the skin may constitute the predominant route of chemical exposure and subsequent uptake. Should transdermal exposures constitute the primary concern, the requisite empirical toxicity studies to support reproductive assessments would first have to be conducted for there to be a useful and reliable assessment tool. Presently, amphibians and reptiles are not evaluated in ERAs altogether (i.e., for any toxicological endpoint), except in an occasional crude qualitative manner. This is somewhat surprising and unfortunate, in light of these receptors being understood to be sensitive bioindicators, and where many have argued that in response to man-induced environmental changes (pollution, primary among them), population declines are widespread, and species are vanishing. In the absence of an appropriate ingestion model, ecotoxicologists may turn to evaluating tissue concentrations of bioaccumulated chemicals (in specific organs, whole-body measures, bird eggs, or perhaps in the case of amphibians, in the gelatinous egg masses that undergo external development). The difficulty with such an approach is that it assumes that a higher tissue concentration is necessarily unhealthful, and there is little evidence to support such an assumption. With virtually the entirety of ecotoxicological databases being of the administered dose genre, little or no attention has been given to assigning effects, reproductive ones included, to tissue burden. The prospects are somewhat better for fish and other aquatic species, where efforts are underway to compile data sets that identify the principal organs that load chemicals, and to establish linkages of tissue concentrations and reproductive impairments. Despite efforts to do so, there appear to be many glaring examples of highly contaminated fish that although unquestionably unhealthful for the would-be human consumer, demonstrate no apparent reduction in fish health (as in fecundity). Examples would be alewife, shad, perch, and bass in the waters of the 40-mi-long Hudson River polychlorinated biphenyl (PCB) Superfund site, in upstate New York.

One common approach to assessing reproductive toxicity involves the use of 'toxicity tests' (also termed bioassays). Here, a standardized test species (e.g., fathead minnow, manure worm, the amphipod crustacean, *Hyalella azteca*) is exposed to a contaminated medium (such as site topsoil, whole effluent from a water treatment facility, or a specified dilution of the effluent) while under highly controlled laboratory conditions. Often the test endpoint, or one of several, is a reproductive one. Extreme care must be taken to ensure that the site-specific media samples satisfy the requirements to rear and maintain the test species. Certain invertebrates, for example, only fare well when species-specific sediment grain size specifications are met. Other species may only be able to tolerate a very narrow salinity range. Should essential life-supporting features of the contaminated site's media not well match those of the commercially available test species to be used, the situation may be ripe for drawing errant conclusions. A noteworthy and statistically significant negative change in a reproductive measure, such as reduced cocoon production in earthworms placed in jars of contaminated site soil (relative to the cocoon production rate of worms placed in jars of reference location soils), could have nothing to do with soil contaminant levels. Such toxicity testing is not without its share of drawbacks, and many of these reflect the dissimilarity of the imposed chemical exposures of the test organisms and the natural environmental exposures of receptors in the real world. Consider the case of a waterbody with several known contaminants in the shallow sediment's bioactive zone (the top several inches). Stakeholders might collectively agree to conduct a chronic freshwater toxicity test using the freshwater waterflea, *Ceriodaphnia dubia*. Since the established test for this species is one that monitors survival and the number of offspring produced in a water-column species, the sediment is first agitated in the laboratory to liberate to the column water that lies above the sediment, the contaminants that are bound to the sediment matrix. The 7-day static-renewal test to be run, where newly prepared water (the elutriate) can circulate through the test chambers each day, will require that ample site sediments have been brought to the laboratory beforehand. Although the test can detect statistically significant reductions in reproductive success, the following cannot be overlooked:

- the test species (*Ceriodaphnia dubia* in this hypothetical example) may not occur in the contaminated site's sediment;
- a 'column water' test species was used to speak to sediment-dwelling/exposed species;
- the commercially bred *Ceriodaphnia* sp. used in the test had no prior history of living amid contaminated water or sediment, unlike the macroinvertebrates that reside today in the waterbody's contaminated sediments (in all likelihood, the site sediments have been contaminated

for several decades, and have consequently allowed vast opportunities for the site biota to adapt);
- a statistically significant difference, as in a measured reduction in the number of *Ceriodaphnia* sp. offspring produced, is not necessarily a 'biologically' significant difference for the test species;
- identification of valid, reproducible, and statistically significant reproductive impairment in the test organisms, does not necessarily mean that there is impairment in the actual site-exposed sediment-dwelling invertebrate species;
- there is enormous potential for error when extrapolating from 'failed' test responses; the actual waterbody-associated aquatic fauna, for whose protection a site cleanup could reasonably proceed (such as larger fish; cleanups do not occur for water fleas), may bear no ill-health effects, reproductive or otherwise (standard toxicity tests, like the *Ceriodaphnia* one discussed here, have no ability to speak to other, and especially higher-food-chain, aquatic species, e.g., fish).

To counterbalance the many uncertainties associated with toxicity testing, and in particular when evaluating water, it is recommended that the testing involve at least two species, as one fish, and one invertebrate. Still there are often problems associated with interpreting and applying toxicity test outcome information, and much of this reflects the desire to extrapolate from toxicity test species to higher trophic level species, that is, those for whose protection a site cleanup could realistically proceed. A stakeholder could argue that with only the pollution-tolerant invertebrates in a streambed being capable of reproducing normally, the fish that feed on these invertebrates are receiving a nutritionally compromised diet, and are consequently at risk of not receiving enough dietary energy such that they themselves can reproduce adequately. The only way to verify such an argument would be to conduct the empirical research to support the contention. In this case, not only would the nutritional value of the pollution-tolerant invertebrates need to be measured and possibly shown to be inferior, but it should be ascertained if in fact there is a depauperate resident fish population. Focused study might reveal that although the food base is predominantly or only comprised of pollution-tolerant forms, the biomass in the contaminated stream is actually greater than what was present prior to the contaminant releases. One of the more common misapplications of toxicity test outcome information concerns work with earthworms. Where the earthworms that have been exposed to a site's contaminated soil show reduced reproduction, there are those that would like to use the test results to estimate the corresponding reduction in the local songbird population. Such expressed wishful intentions overlook at least three realities: that the test was never designed to be used to make a population assessment for birds; that there is no way to relate an earthworm toxicity test outcome, such as reduced reproduction, to a corresponding impact in birds; and that with so much time having already elapsed at a site since it became contaminated, the local birds have rather assuredly adapted as necessary.

A review of toxicity testing for reproductive endpoints would be less than complete if it did not acknowledge the phenomenon of test subjects having imposed unnatural contaminant exposures. FETAX (frog embryo teratogenesis assay – *Xenopus*) is a 96 h whole embryo assay for detecting teratogenic (developmental) effects, and thereby an indirect reproduction assessment method. Frog embryos of the species *Xenopus laevis* are placed into aquaria with contaminated water, but just prior to this, the jelly coat of the embryos is carefully removed. This procedure, although well intended, vastly increases the likelihood that one or more contaminants in water will cause various malformations in the developing frogs. On the one hand, the philosophy behind the test is admirable – there is a keen interest in uncovering early on, the slightest potential for developmental malformations to occur. The counter-argument though is equally appreciated – what utility is there in evaluating the malformations, when in the real-world exposure, the jelly coat is not removed, but is rather intact? Should there be striking differences in response, as in the case where malformations are only observed in the embryos exposed to the contaminated water (and not in the control-aquarium embryos), such information comes only at the expense of having artificially tampered with nature. Given the prior discussion, it should be clear that toxicity testing for the purpose of enhancing our understanding of reproductive toxicity, whether in a contaminated site context or not, should only be applied in a weight-of-evidence, or lines-of-evidence, context.

For certain animal groups, there is the prospect of collecting somatic measurements that bear on reproductive capability, and then endeavoring to interpret the degree of reproductive well-being from the gathered information. Birds and smaller mammals at contaminated locations lend themselves to this work, where brood patches and placental scars are evaluated, respectively. Simplistically, fewer brood patches and fewer placental scars can indicate a reduction in the number of offspring produced. Caution must be exercised when reviewing the data though, because linking somatic differences such as these to particular site chemicals is not a straightforward process, and especially when most contaminated sites have multiple toxicants present. There is also the added possible complication of wrongly ascribing altered measures to specific contaminated sites when they in fact stem from chemical exposures that occurred tens and hundreds of miles away from the animal's point of capture, as in the case of migratory birds. A related but yet

different approach in ecotoxicology's quest to identify clear somatic markers of chemically caused reproductive impairments, involves assessing the configuration and deployment of reproductive organs. Irregularities in the shape of a female rodent's uterine horns for example, can signify reduced reproductive capability.

Looking to the Future

One new approach that is only first beginning to be explored, is that of directly assessing the field-exposed animal for its own reproductive health condition. Although researchers may be disinclined to venture outside of the laboratory and into highly variable environmental settings to conduct unusual and possibly labor-intensive work, the potential gains can far outstrip any complications. Direct health assessment for reproductive effects is predicated upon an entirely different understanding of the potential for contaminated areas to pose health risks. The new approach recognizes that at virtually all contaminated sites where reproductive and other assessments are to be made, multiple decades have elapsed since the contamination was released, and consequently tens, and in some cases hundreds, of generations of ecological species (e.g., small rodents) have already cycled through by the present day. Thus, it is really too late to be endeavoring to project or forecast the likelihood of reproductive effects first arising; effects, if they were ever to occur, should have already been expressed. Additionally, with almost no documented cases of ill-health ever having been reported for terrestrial receptors at contaminated sites, the anticipation is that such a direct assessment method would reveal the same. Finding this supposition to be true would have vast ramifications for environmental management; we would then know that although chemicals remain in environmental media, cleanups are not necessary because the chemicals are not impinging on an animal's ability to survive and perpetuate its own.

Evaluating the sperm parameters of count, motility (the percentage of properly swimming sperm), and morphology (the percentage of normally/abnormally shaped sperm) in adult male small rodents collected at Superfund-type terrestrial sites (for a comparison with the parameters of rodents from nearby noncontaminated sites) has substantially advanced the field of reproductive toxicity for ecological receptors. This approach keenly recognizes that each of the sperm parameters is a barometer of reproductive success, a situation much akin to the case in humans, thereby explaining today's rather routine clinical studies investigating the causes of infertility. For the small rodent grouping, which is for all intents and purposes the maximally exposed terrestrial receptor (given the group's nonmigratory nature and high degree of site contact), it is known rather precisely how much of a reduction in either count or motility, and how much of an increase in morphology, is needed to compromise reproduction. For nearly all other biological measures that can be collected in exposed and nonexposed animals, whether in the laboratory or the field, it is not known how much of a difference equates with a demonstration of impact, and consequently, only absolute measurement differences can ordinarily be reported. Where none of the sperm parameter thresholds are exceeded, it is logical to conclude that the larger, higher-trophic-level and wider-ranging receptors are also not experiencing reproductive impairment. Although this testing scheme would appear to be somewhat lacking, in that females are not evaluated, the US EPA, among others, find that there is abundant information to reliably conclude that the sexes respond similarly. Thus where demonstrated reproductive effects occur in one sex of a species, the response (or lack thereof) in the other sex should be the same.

There is promise too for the development of a female-based homolog of the above-described sperm parameter reproductive assessment scheme. From laboratory-based studies with mice and rats, it is clear that certain chemical exposures (e.g., the pesticide methoxychlor) can arrest the normal development of ovarian follicles. In theory, ongoing empirical research will lead to a chemical-specific understanding of the degree to which follicle development needs to be reduced or arrested such that fecundity is offset.

Given the ever-increasing contaminant releases to the environment from our highly industrialized society, both in volume and in the number of constituents involved, the argument can be made that our ability to assess the potential for reproductive effects by employing calculations that involve dose estimation, uptake, biotransformation, and metabolic rates, is being far outstripped. Potentially, the direct assessment of reproductive capability provides a way around the problem. Care must be taken though to conduct the empirical research necessary for the accurate identification of reproductive impacts should they exist. By way of example, one may detect distinct reductions in the gonadal mass of largemouth bass that have been exposed to dietary mercury under controlled laboratory conditions, but the extent of the reduction needed to compromise reproduction must be known. Conceivably, a gonadal mass reduction of some order could coincide with improved reproduction, and it could be also that reproductive impacts only occur when there are drastic reductions in gonadal mass. In the case of the rodent sperm counts, this is precisely the case; an 80–90% reduction is first needed for there to be compromised reproduction, a figure very much contrary to what would be expected. In a similar way, the brood patches of birds and the placental scars on mammals could be routinely examined for an assessment

of reproductive health, but needed are rather iron-clad thresholds for effect, as in "how many fewer placental scars in a certain age mammal is indicative of a certain percentage reduction in fecundity?"

Does Reproductive Toxicity Really Need to Be Evaluated?

An honest assessment of the potential for there to be chemically caused reproductive impacts to ecological receptors in the wild, reveals rather limited opportunities for such occurrences when patterns of animal movement, and other spatial considerations are taken into account. In the common case, there is often an insufficient degree of exposure to create a situation where reproductive effects could take hold. This situation is often born of contaminated properties routinely being particularly small (perhaps 5 or 10 ac), and species having either relatively huge home ranges, naturally sparse distributions, or both. An example will demonstrate the point. The smallest home range of mink (*Mustela vison*), reputedly the mammal that is the most sensitive to PCBs in the environment, is about 600 ac, and the highest density for this species is 0.04 animals per acre. In the hypothetical, albeit unlikely case of a 40 ac PCB-contaminated parcel (most contaminated Superfund National Priority List sites are 20 ac in size or smaller), there would be an anticipation of only two animals present, a situation that alone would not suggest that reproductive effects be monitored. Additionally, any one mink would be only expected to be present about 7% of the time at this hypothetical site, a situation unlikely to trigger reproductive impacts. Consideration of the mink's specific habits, as in it having a linear home range, would further reduce opportunities for contact with the PCB-affected media, and the chances of there being reproductive impacts.

Summary

Mankind's use of and reliance on chemicals, unfortunately, ensures that there will be always be unavoidable chemical releases to the environment. Consequently, there will be ongoing opportunities for nonhuman receptors to be exposed to these, and also for the essential biological function of reproduction to be altered in some way. Although there are established methods for evaluating the reproductive toxicity of chemicals, these are plagued by uncertainties. It would seem that laboratory-based methods can only fall short of the mark, given the growing number of chemicals in use today, and the reality that chemical mixtures will nearly always be at play. Field-based (direct) assessments of reproductive toxicity within an ecological context may hold the best promise for improving our understanding of this challenging and vital subject area.

See also: Bioaccumulation; Ecological Risk Assessment.

Further Reading

Chapin RE, Sloane RA, and Haseman JK (1997) The relationships among reproductive endpoints in Swiss mice, using the reproductive assessment by Continuous Breeding Database. *Fundamental and Applied Toxicology* 38: 129–142.

Perrault SD (1998) Gamete toxicology: The impact of new technologies. In: Korach KS (ed.) *Reproductive and Developmental Toxicology.* New York: Dekker.

Sample BE, Opresko DM, and Suter GW, II (1996) *Toxicological Benchmarks for Wildlife: 1996 Revision.* Oak Ridge, TN: Lockheed Martin Energy Systems.

Tannenbaum LV, Thran BH, and Williams KJ (2007) Demonstrating ecological receptor health at contaminated sites with wild rodent sperm parameters. *Archives of Environmental Contamination and Toxicology* 53 (in press).

US Environmental Protection Agency (1993) *Wildlife Exposure Factors Handbook, Vol. I,* 2 vols. Office of Research and Development. EPA/600/R-93/187a.

US Environmental Protection Agency (1994) Eco Update, Intermittent Bulletin Volume 2, Number 1, Using Toxicity Tests in Ecological Risk Assessment. Office of Solid Waste and Emergency Response, EPA 540-F-94-012.

US Environmental Protection Agency (1995) Eco Update, Intermittent Bulletin Volume 2, Number 2, Catalogue of Standard Toxicity Tests for Ecological Risk Assessment. Office of Solid Waste and Emergency Response. EPA 540-F-94-013.

US Environmental Protection Agency (1996) Guidelines for Reproductive Toxicity Risk Assessment. Office of Research and Development, EPA/630/R-96/009.

Zenick H, Clegg ED, Perrault SD, Klinefelter GR, and Gray LE (1994) Assessment of male reproductive toxicity: A risk assessment approach. In: Hayes AW (ed.) *Principles and Methods of Toxicology,* 3rd edn., pp. 937–988. New York: Raven Press.

Risk Management Safety Factor

A Fairbrother, Parametrix, Inc., Bellevue, WA, USA

Published by Elsevier B.V., 2010.

Introduction	Precaution
Safety Factors	Summary
Alternative Methods for Incorporating Uncertainty	Further Reading

Introduction

Ecological risk assessments are management decision tools that organize and integrate information to estimate the likelihood and magnitude of an undesirable environmental response. Because ecological systems are highly complex, their responses to contaminants are inherently uncertain. This uncertainty results from a combination of factors, including ignorance of the true nature of ecological responses, intrinsic variability in biological and physical phenomena, and imprecision in measurement capabilities. As the consequences of underestimating risk are likely to be more severe than overestimates (e.g., irreversible harm to the environment vs. increased costs of less risky alternatives), risk managers prefer to err on the side of conservatism. This can be done in several ways, including using lower-end estimates of the range of potential effects and higher-end measurements of exposure. Alternatively, adding a safety factor to the analysis provides a simple way of ensuring that unknown differences (such as differing sensitivities between responses observed in the laboratory and those expected

in the natural environment) are taken into account. This article summarizes the safety factors in use today in various countries (e.g., the United States, Canada, Britain, and the European Union), discusses the basis for their selection, and examines some of the biases and pitfalls in their use.

Safety Factors

The term safety factor is synonymous with the term uncertainty factor and grew out of what was known as application factors. For the purposes of this discussion, the term safety factor will be used to encompass all these terms. In human health and ecological risk assessments, safety factor is defined as a number by which an observed or estimated no observed adverse effect concentration (NOAEC) or dose is divided to arrive at a criterion or standard that is considered safe (or has an acceptable level of risk). Safety factors are, of course, used in engineering as well, and depend on the structure that is being supported. Most codes require a safety factor of 2, although bridges and large structures generally apply safety factors of 5–10. In all cases, the safety factor will vary, depending on such things as the circumstances under which it is to be used and the amount of data available for making the decision.

Safety factors were first proposed in 1945 as a way to calculate presumed harmless concentrations in water of toxic substances based on dose–response relationships of standard aquatic test organisms such as fish, algae, or daphnia (water fleas). The initial approach was to divide the lowest acute toxicity value by 3. This eventually was increased to dividing by multiples of 10, to take into account more of the uncertainty in extrapolating data from a few species tested for a short exposure period to all species that may be chronically exposed in different types of aquatic systems. About 10 years later, in 1954, human health risk assessors proposed the use of uncertainty factors when making regulatory decisions regarding chemical use. A 100-fold uncertainty factor was proposed, and used for many years to account for uncertainties in extrapolating animal data for the protection of humans. This has since been expanded to include margins of safety for sensitive subgroups (e.g., children), different endpoints, and potential effects. However, as the discussion of safety factors in this article will be limited to ecological assessments, their use in human health protection will not be discussed further.

The use of safety factors is now an accepted part of the hazard analysis in ecological risk assessments. They are used to account for the unknowns when predicting the threshold for effects of potentially hazardous substances to fish, wildlife, plants, and invertebrates. Toxicity tests can be conducted on only a few selected species held in the artificial environment of the laboratory, yet the results must be used to make predictions about potential risks to all species in any natural environments. Some margin of safety is needed to ensure protection for species or environments that are more sensitive than those that are tested. It is important to emphasize that safety factors are used only for effects assessments. Exposure estimates do not include safety factors but rather are based on either worst-case scenarios or use various statistical methods to be inclusive of the potential range(s) of exposures that might occur. Safety factors (**Table 1**) are used to account for:

- intraspecies heterogeneity: accounts for differences between individuals of the same species, due to genetics, testing differences, or laboratory factors;
- interspecies extrapolation: accounts for differences between species due to variable physiology and response mechanisms;
- acute-to-chronic comparisons: estimates effects from long-term exposures that may span the full life cycle of organisms, from data generated from short experimental exposure times;
- LOAEC-to-NOAEC extrapolations: estimates the no effects threshold from test data that predict only the lowest effect level where responses are first observed;
- laboratory-to-field extrapolations: accounts for differences in field conditions (e.g., additional stresses such as predators, climate, or poor nutrition) that may alter the response of organisms to toxicants when compared to studies conducted under controlled conditions in laboratory settings.

Table 1 Safety factors used in ecological risk assessments

Type of extrapolation	Assumptions	Safety factor
Intraspecific	Includes individual variation, test-to-test variation, and differences between laboratories	2–10
Interspecific	Wildlife species, acute data	10–20
	Wildlife species, chronic data	5
	Aquatic species	2–4
	Plants	2–15
	Amphibians and reptiles	10–1000
Acute to chronic	Primarily for aquatic species; based on some measured values. Size of safety factor depends upon number of species tested	1–40
LOAEC to NOAEC	Depends upon the number of studies of a particular chemical and the range of species tested	10–1000
Laboratory to field	Accounts for differences between laboratory test conditions and variable real world environments	10–100
Other	Freshwater to saltwater	10 000
	Aquatic to terrestrial	10
	Water column to sediment organisms	10
	Individuals to populations	1–10

See the European TGD listed in 'Further reading' for detailed tables for each type of environmental media (soil, water, sediments).

In general, the more the data that are available, the smaller the safety factor that is required. For example, if extrapolations across species are made on the basis of a single species, then a safety factor of 100 or 1000 may be used. Adding additional species that represent different taxonomic groups (e.g., an algae, a fish, and an invertebrate) will reduce the safety factor to 10. If more than 10 species have been tested, then a statistical distribution can be developed to predict the sensitivity of the most affected 5% of the species (see below for more detailed discussion of this approach), in which case a safety factor of 5 is applied or, in some cases, no safety factor is used at all. Occasionally, safety factors as high as 10 000 have been used, for example, when setting toxicity thresholds for marine organisms under European risk assessment guidelines. The following sections provide additional details on the use of safety factors for each of the extrapolation categories listed above.

Intraspecies Heterogeneity

Because no two organisms are exactly alike, variation in response between individuals within the same species is a common occurrence. If all organisms were exactly the same, there would simply be a threshold dose (or concentration) where all of the organisms responded and below which none responded to the chemical exposure. This variation among individuals is described by the dose–response curve (**Figure 1**), which shows the proportion of the animals that is affected by the chemical at a given concentration. As with most statistically generated

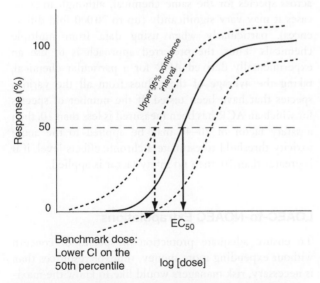

Figure 1 Dose–response curve showing the percent of test organisms that respond at each chemical concentration. The dotted lines on either side of the solid curve represent the 95% confidence interval. The EC_{50} (or LC_{50}, if the response measured is mortality), is determined by reading across from where 50% of the organisms respond, and then down to the corresponding chemical concentration that caused this level of response.

functions, there is uncertainty around this curve which can be quantified by the 95% confidence intervals on either side of the mean (the dotted curves in **Figure 1**). Conservatism in the risk estimates is then included through the decision of which point on the curve to use as an effect metric. In acute (short-term) studies, this often is the LC_{50} which is the concentration (or dose) that kills 50% of the test animals. Using the upper confidence interval from the dose–response curve at this value adds further conservatism to the estimate (i.e., results in a benchmark dose which is the lower confidence interval around the LC_{50} concentration; see **Figure 1**). For chronic (long-term) studies, a nonlethal endpoint generally is used, often selecting the EC_{10} or EC_{20} (effects threshold for 10% or 20% of the tested organisms, respectively). Again, the lower confidence interval of the EC_{10} will provide additional protection and is used as the benchmark dose. An alternative to using confidence intervals about the $LC(EC)_x$ value is to apply a safety factor to the value from the calculated curve (i.e., the solid line in **Figure 1**). Several studies have examined the intraspecific variability in both aquatic and terrestrial organisms, and conclude that a safety factor of 10 is appropriate for bounding the variability of within species responses.

Intraspecific variability also includes test-to-test variation or differences between laboratories. This results from variations in the strain of organisms used for testing, deviations from specifications in the physical or chemical properties of the test media (water or soil), and other operator-dependent effects. For standard aquatic test organisms, there is a two- to fivefold difference between laboratories in acute toxicity (LC_{50}) values when the same chemical is tested in the same species. Intralab replication of the same test by the same researchers usually results in only a twofold variation. Generally, this variability is taken into account as part of the intraspecies safety factor, and no additional safety factor is applied.

Interspecies Extrapolation

The most common use of safety factors is to extrapolate data from tested species to other, nontested organisms to provide protection for at least 95% of all species. Unfortunately, while there is a general pattern to relative sensitivity of species (particularly within particular classes of chemicals), the same species will not be the most (or least) sensitive for all chemicals. Therefore, it generally is not possible to *a priori* pick the most sensitive species for testing. Examination of data from pesticide studies where a large number of chemicals have been tested in the same animal species suggests that the maximum difference in species sensitivity following dietary (chronic) exposure is about sevenfold for birds and fourfold for mammals, and acute lethal doses vary no more than tenfold (birds) or 20-fold (mammals). These data provide support for use of an interspecies wildlife safety factor for pesticide risk assessments of 10–20 when using acute data or 5 when using chronic (dietary) exposures. However, the range in response to nonpesticide chemicals is greater (likely due to the greater range of modes of action of these substances), with 95% of the results being within a factor of 50. Bayesian statistical analyses currently are being investigated as a possible way of using this prior knowledge of relative sensitivities to make predictions about interspecific responses to new chemicals that may result in more reliable safety factors.

For terrestrial plants, a safety factor of 2 will capture 80% of the total potential variability among genera within a single family. However, most extrapolations are done across families or orders which appear to require a safety factor of at least 15 to capture 80% of the variability. Acute toxicity for freshwater aquatic organisms, on the other hand, can vary over 5 orders of magnitude (100 000-fold). This larger variability likely is due to inclusion of multiple classes of organisms in the database, including plants, vertebrate fishes, and invertebrates. Variability within a single order ranges from one- to fourfold differences. As with other taxonomic groups, no single species is always the most sensitive to all chemicals.

For taxonomic groups where the range of relative sensitivities is unknown (e.g., reptiles, amphibians, soil invertebrates, or saltwater species), the lowest LC_{50} (or LD_{50}) value (or other measured endpoint) would be used and a larger safety factor applied to that value. In these instances, the safety factor generally ranges from 10 to 1000, depending upon how many species have been tested, whether there are at least three different trophic levels represented (a plant, a primary consumer, and a carnivore), and if the studies are acute (i.e., very short) or chronic (i.e., long-term) exposures.

If more than eight species are tested, then a statistical distribution of their LC_{50}s (or another benchmark toxicity value) is derived and the point on the curve that represents the LC_{50} of the most sensitive 5% of the species is determined (**Figure 2**). This is discussed in more detail below (see the section titled 'Alternative methods for incorporating uncertainty').

Acute-to-Chronic Comparisons

Many toxicity studies are conducted over a very short time period (generally a few hours); these are known as acute toxicity studies. However, in nature, organisms frequently are chronically (i.e., continuously) exposed to pollutants. For aquatic organisms, it is frequently possible to establish the relationship between the toxicity thresholds following acute exposure and those following long-term (days to months), chronic exposures. An acute-to-chronic ratio (ACR) can be derived by dividing the acute LC_{50} by the chronic NOAEC. Because many

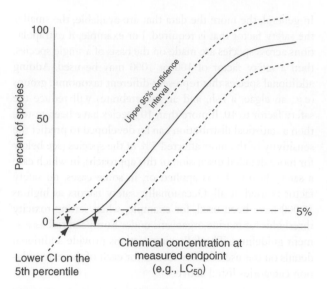

Figure 2 Species sensitivity distribution, showing the cumulative density function of the response of all tested species. The dotted lines on either side of the solid curve represent the 95% confidence interval. The toxicity threshold value is taken as either the 5th percentile on the mean (solid line) curve, or the 5th percentile from the upper confidence interval (dotted line) curve.

more species are subjected to acute toxicity studies than to chronic ones, the ACR derived from a few tested species may be used to estimate the chronic no effect level for many other (nontested) species. It is important to note that this extrapolation is done almost exclusively for aquatic organisms and that ACRs have not been generated for plants or wildlife. This ACR often is similar across species for the same chemical, although in some cases it may vary significantly (up to 20 000-fold differences), particularly when using data from multiple chemicals. Thus, the preferred approach is to use an experimentally derived ACR for a particular chemical, taking the average of the values from all the various species that have been tested. If the number of species for which an ACR has been measured is less than 10, then a safety factor of 20–40 may be applied to the acute toxicity threshold to estimate a chronic effects level; if it is greater than 10, then no safety factor is applied.

LOAEC-to-NOAEC Extrapolations

To ensure adequate protection of species of concern without expending more money, time, or resources than is necessary, risk managers would like to know the maximum amount of a chemical to which organisms can be exposed without showing any significant adverse effects. If this threshold is incorrectly estimated to be higher than what really occurs, then organisms may continue to be affected even after corrective or preventive actions are put in place whereas if the estimate is too low then unnecessary mitigation expenses might be incurred. Therefore, it is important to estimate the toxicity threshold as accurately as possible and to apply a safety factor that will err on the side of ecosystem protection. It is now commonly accepted that nearly all contaminants have a threshold of effect, although thresholds for cancer initiations may be so low as to be considered nonexistent. For the purposes of ecological risk assessments, however, cancer endpoints generally are not considered and all chemicals are assumed to have measurable thresholds.

Most risk assessments are based on the chronic NOAEC. This is the highest tested concentration of a chemical that causes no statistically significant response of the test organisms under a particular study design. However, this piece of information frequently is not available, and only the lowest concentration that did cause a statistically significant effect (the LOAEC) is reported. It is difficult to extrapolate NOAECs from the LOAECs for most chemicals, as often there is an insufficient amount of information to reach any definitive conclusion. Many studies report unbounded values, that is, either all test organisms showed some response (an unbounded LOAEC) or no test organisms responded (an unbounded NOAEC). Furthermore, even studies that result in calculation of both an NOAEC and an LOAEC cannot definitively define the true threshold value. The NOAEC and LOAEC values are generated using an analysis of variance statistic that is dependent upon study design factors such as number of organisms in each test concentration, exposure levels tested, and variability of the organism responses. Given these design dependencies, different NOAEC or LOAEC values may be generated for the same organism–chemical combinations simply by redesigning the study. Even assuming that the resulting values are true representations of no or low effects, they frequently are at least an order of magnitude apart and the actual concentration where at least some organisms begin to show measurable responses to chemical exposure lies at some unknown distance between the two. Therefore, the convention has evolved to estimate the no effect level (NEL) as the geometric mean of the NOAEC and LOAEC values. Again, this is most frequently applied in aquatic risk assessments, but there have been some similar applications to soil organism tests (e.g., soil invertebrate or plant toxicity testing).

Because of the inherent biases in derivation of the NOAEC and LOAEC data, there are no reliable statistics to estimate what the usual ratio of these two values might be for a variety of species across chemical classes. Therefore, an alternative approach has been suggested to use dose–response functions (**Figure 1**) rather than the analysis of variance approach, and develop benchmark doses rather than NOAEC–LOAEC values. The NEL threshold is then estimated by applying a 100-fold safety factor to the acute LC_{50} (or LD_{50}) value (i.e., dividing it by 100) or a tenfold safety factor if the benchmark dose is based on a chronic endpoint. While this has begun to be accepted in some risk assessment applications, especially for terrestrial wildlife, the NOAEC–LOAEC approach remains standard practice in many others. Safety factors may be applied to the NEL and often are based on professional judgment. Depending upon the number of studies of a particular chemical and the range of species tested, safety factors may range from 10 to 1000.

Laboratory-to-Field Extrapolations

Organisms tested under laboratory conditions may not respond the same as their wild counterparts. Laboratory tests frequently are conducted using highly controlled conditions that are known to be nonstressful to the organism (e.g., most appropriate temperatures, constant and sufficient feeding of nutritionally adequate diets, absence of predators, etc.), although it is possible to purposely simulate environmental stress (e.g., temperature regime changes) or unintentionally introduce novel stresses (e.g., isolation housing of gregarious species or fluorescent lighting wavelengths and flicker rates). In nature, organisms are exposed to multiple stressors, simultaneously and often continuously. Field studies generally are less controlled than laboratory studies, although use of caged animals, potted plants, or careful application of chemicals can standardize many of the study variables. Such experiments generally benefit from less variable and more defined exposure but do not reliably control other environmental stressors. Stress of any kind, whether in laboratory of field, may alter organisms' physiology and therefore change their response(s) to a particular stressor, such as a chemical pollutant. Many studies have been conducted to examine the differences between responses under these two sets of conditions (laboratory and field) with mixed results. About half the time, laboratory studies yielded lower toxicity thresholds and the other 50% of the time organisms in the field were more sensitive. Obviously, this remains an area of high uncertainty, and safety factors of 10 to 100 may be applied to laboratory data to compensate.

Other Safety Factors

Other safety factors are used occasionally by some jurisdictions, but not others. For example, the United States Environmental Protection Agency (US EPA) will sometimes estimate effects to terrestrial organisms by dividing an aquatic toxicity threshold by 10 (based on professional judgment only, with no empirical basis for this particular value). In Canada, toxicity thresholds for sediment-dwelling (i.e., benthic) organisms frequently include a tenfold safety factor, even if information from toxicity tests conducted with sediment organisms is available. This is to account for the potential of ingested chemicals to add to toxicity resulting from gill uptake from water. As mentioned previously, effects thresholds for saltwater organisms may be estimated using data from freshwater organisms and the application of a very large (up to 10 000) safety factor. Britain and other European countries also apply safety factors when estimating changes in population growth rates as a consequence of measured effects to individual organisms. This seems counterintuitive as populations have many compensatory factors that allow some effects to occur to individuals before changes in population growth rates are manifest (as exemplified by hunting or harvest mortality of ubiquitous animals such as deer).

It is common practice to use multiple safety factors when developing a regulatory threshold that is protective of organisms of concern. For example, if toxicity data are available from only two aquatic species tested in short-term studies (e.g., an LC_{50}), then a tenfold safety factor is applied to account for interspecific difference, another tenfold factor (or, in some cases, a 100-fold factor) is used for acute-to-chronic conversion, and yet another tenfold factor may be applied for laboratory-to-field extrapolations. The end result is a total safety factor of 1000 to 10 000 being applied to the measured toxicity data. This drives the acceptable risk threshold very low. While this is appropriate for humans, we need to ask if it is equally appropriate for valued ecological resources (i.e., plants, fish, and wildlife). Toxicity data are based on studies of the effects of chemicals on particular attributes of single organisms, and risk assessments generally provide estimates of potential effects on organism health (including longevity and reproductive ability). However, naturally occurring species depend upon relative fitness of adults and their potential to contribute to the gene pool and successfully raise offspring. Populations can sustain significant mortality or reduced reproductive rates before entering into an inevitable decline. Therefore, it is likely that use of multiple safety factors results in overly protective threshold values. However, how chemical exposures affect species interactions or whether adverse affects to populations of one species (e.g., predators) benefit another (e.g., prey species) remain largely unknown. Competing environmental management goals also add significant debate to required level of protection. For all these reasons, generation of additional data to more accurately estimate the variability in organism responses is highly encouraged as the use of large, and often arbitrary, safety factors can then be avoided.

Alternative Methods for Incorporating Uncertainty

Alternatives to the use of safety factors have been proposed, although no method is completely free from uncertainty. The use of data to estimate ACRs for a particular chemical has been discussed above, and is the oldest example of attempting to develop a robust data set for estimating beyond the tested species. More recently, the use of species sensitivity distributions (SSDs) has been gaining favor as a more accurate estimate of differences in sensitivity among species. However, no consensus exists on the number of species required to develop a sufficiently robust distribution, and what the minimum degree of taxonomic relationship should be among the species tested (i.e., is it sufficient to be of different genera or should they represent different orders?). The US EPA requires a minimum of eight genera representing at least three trophic levels when using SSDs for derivation of water-quality criteria or mitigation goals. In the European Union, eight to ten genera are required if an SSD is to be used instead of an interspecies extrapolation factor, and other jurisdictions are considering as many as 30 species. There is general agreement, however, that aquatic and terrestrial species should be analyzed separately, and within terrestrial systems animals and plants should have separate estimates (note that aquatic toxicity threshold estimates generally use algae as a surrogate for all aquatic plants, although it is likely that rooted macrophytes (larger plants) will respond differently to pollutants in the water column, particularly those that are emergent above the water line).

Another source of uncertainty when using distributions of species responses (**Figure 2**) is the point on the curve that is chosen as the toxicity threshold. There is general agreement that the value that defines the concentration where only 5% of the species would show a response is appropriate for management purposes (i.e., 95% of the species would be protected). However, there is disagreement on how the uncertainty in this estimate should be incorporated. It is recognized that the species toxicity endpoints that are used to derive the SSD contain the variability discussed in the above sections on intraspecies extrapolations and, therefore, that a curve based on single points for each species will contain significant uncertainty. Ideally, more than one value would be generated for each species and the species (or genus) mean value used to derive the distribution curve. The US EPA uses the concentration where 5% of the species are protected (the 5th percentile of the distribution) for setting water quality criteria, while other decisions are based on the lower confidence limit of this value. The size of the confidence interval around the 5% value depends on the number of data points (i.e., species) used to generate the SSD and the inherent variability in their responses. The difference between using eight, ten, or 20 species appears to be a factor of 3. Some jurisdictions may consider using 3 as a safety factor on the 5% estimate if less than ten species are tested.

Other methods of uncertainty analysis that are applied to exposure estimates cannot be used in effects analysis. For example, uncertainty bounds could be applied, providing estimates of risk using the most sensitive and least sensitive organism. This bounds the possibilities of where the true risk lies, and allows the risk manager to determine a level of conservatism based on degree of risk aversion rather than assuming scientific certainty. The difficulty with this approach is that there still is no empirical way of determining what is the most or least sensitive organism and what exposure concentration would represent the threshold where these organisms would begin to show effects.

Precaution

When using the safety factor approach, confidence intervals are not given and the degree of protection is usually unknown. Because of the somewhat arbitrary nature of safety factors, their application must be appropriate to the particular purpose and documentation describing the logic used in selecting a particular number should always be provided. The use of the same safety factor (e.g., 10) in all cases has been discouraged by several regulatory and scientific bodies in favor of case-specific values. In particular, the scale, frequency, severity, and potential for long-term consequences of the environmental insult must be taken into account. Providing protection against an acute spill that is easily remediated may require less precaution (and therefore smaller safety factors) than the permitted continual release of a pollutant discharge. The decision to use a safety factor approach, and the eventual selection of the appropriate number, is more a management decision than a scientific methodology. The desired level of protection (which is inversely proportional to the degree of risk aversion of the regulatory body) will play a large role in determining how much of a margin of safety to build into a management decision.

There are those who propose that a severe lack of data (e.g., only one or two acute toxicity values) should preclude the use of a risk assessment in favor of a precautionary approach. That is, rather than applying a large safety factor and allowing the action to proceed, no action should be taken until sufficient data are generated to have at least a reasonable estimate of the potential range of sensitivities. While this may have merit within the context of new chemical releases or discharges, it provides no means for establishing remediation or cleanup levels for environments with historic pollution issues, nor does it allow for any means of an assessment of the comparative risk of current practice with proposed new technologies.

Summary

Until more information becomes available on the relative sensitivity of different species and the effect of natural conditions on their response to novel stresses, some approach is needed to ensure that risk management decisions are made in a manner that is protective under most of the likely scenarios. The use of safety factors ensures that the risk estimates will be conservative; the challenge, of course, is to keep from being significantly overprotective or defining regulatory values that are below natural background levels or levels of contamination that are known to have no discernible ecological effects. Selection of an appropriate safety factor most often is based on experience and best professional judgment and can be highly variable among regulatory agencies or between risk assessors. Documentation of reasoning for why particular safety factors were selected often is lacking in risk assessment reports but is critical if reviewers and the general public are to understand how risk management decisions are made and the degree of conservatism or bias incorporated into the supporting assessment. Fortunately, an effort is now being made to develop a scientific basis for the use of safety factors to at least narrow the selection range and make more accurate extrapolations. Additionally, other methods of estimating uncertainty are being incorporated more routinely into ecological risk assessments with the goal of moving completely away from the use of judgment-based safety factors. More sophisticated methods of toxicity estimation (e.g., quantitative structure–activity relationships based on genomic response patterns or physiologically based toxicokinetics models) may also provide a science-based estimate of comparative toxicity responses. Although all these methods show promise, it is likely that safety factors will continue to be used for at least the foreseeable future to ensure that risk management decisions are appropriately protective of the most sensitive valued ecological resources.

See also: Acute and Chronic Toxicity; Biogeochemical Approaches to Environmental Risk Assessment; Dose–Response; Ecotoxicology: The Focal Topics; Ecotoxicology Nomenclature; Ecotoxicological Model of Populations, Ecosystems, and Landscapes; Ecological Risk Assessment.

Further Reading

Chapman PM, Fairbrother A, and Brown D (1998) A critical evaluation of safety (uncertainty) factors for ecological risk assessment. *Environmental Toxicology and Chemistry* 17: 99–108.

Duke LD and Taggart M (2000) Uncertainty factors in screening ecological risk assessments. *Environmental Toxicology and Chemistry* 19: 1668–1680.

European Commission (2003) *Technical Guidance Document on Risk Assessment in Support of Commission Directive 93/67/EEC on Risk Assessment for New Notified Substances Commission Regulation*

(EC) No. 1488/94 on Risk Assessment for Existing Substances Part II, 337pp. Luxembourg: Office for Official Publications of the European Communities L – 2985, http://ecb.jrc.it/Documents/TECHNICAL_GUIDANCE_DOCUMENT/EDITION_2/tgd part2_2ed.pdf (accessed November 2007).

Posthuma L, Suter GW, II, and Traas TP (eds.) (2001) Species Sensitivity Distributions in Ecotoxicology, 616pp. Boca Raton, FL: CRC Press.
Warren-Hicks W and Moore D (eds.) (1998) Uncertainty Analysis in Ecological Risk Assessment, 315pp. Pensacola, FL: SETAC Press.

Hill's Postulates

P C Chrostowski, CPF Associates, Inc., Takoma Park, MD, USA

Introduction
Evolution of Causation Theory and Criteria
Applications of Causation Criteria to Ecological Science

Summary and Conclusions
Further Reading

Introduction

The ability of a stressor to evoke a response in an individual, community, or population is often easy to observe. It is substantially more difficult, however, to demonstrate that the stressor caused the response. Philosophers have been interested in the subject of causation since the time of Aristotle; however, scientific interest has been more recent. Ecotoxicological associations in the natural environment are rarely unambiguous and scientists often depend on weight of evidence approaches such as ecological risk assessments to resolve the ambiguities. These weight of evidence approaches often depend on determinations that exposure of an organism to a chemical results in a toxicological response. To cite just one possible example, we consider the scenario where there is an observed reduced species diversity or abundance in a population of fish in a stream flowing through an urban area compared to a control population. Meaningful mitigation of this situation requires that the cause or causes of the biological conditions associated with diversity and abundance be identified. These causes may be toxicological, related to nutrients, habitat, predator–prey relationships, competition, and other factors. Application of causation criteria such as Bradford Hill's postulates may assist the investigator in identifying unambiguous causal elements and developing a mitigation strategy.

This article opens with a discussion of the historical evolution and theory of causation criteria focusing on Bradford Hill's postulates and related systems. This is followed by a discussion of applications of causation criteria to ecotoxicological problem solving.

Evolution of Causation Theory and Criteria

Although philosophers had discussed the concept of causation since ancient times, the first exposition of causation using scientific principles was proposed by Robert Koch in 1884. Koch developed a series of four postulates to aid microbiologists in determining if exposure to pathogenic bacteria caused a disease. The postulates include the following.

- The microorganisms should be found in all cases of the disease in question and the distribution in the body should be in accordance with the lesions observed.
- The microorganisms should be grown in pure culture in vitro for several generations.
- When the pure culture is inoculated into susceptible animal species, the typical disease must result.
- The microorganism must again be isolated from the lesions of such experimentally produced disease.

In general, Koch's postulates have been satisfied for most known pathogens although some pathogens such as *Mycobacterium leprae* (leprosy) cannot be grown in culture and others such as *Neisseria gonorrhoeae* (gonorrhea) have no adequate animal model of infection. Recently, the microbiological community has adopted molecular techniques for identification and experimentation; thus, molecular analogs have been proposed to augment Koch's original postulates.

As long as the focus of environmental health was on infectious disease, Koch's postulates were an adequate basis for determinations of disease causation. In the mid-twentieth century, however, concerns began to arise regarding potential associations between chemical exposure and disease. Rather than taking a strict experimental approach, environmental health scientists began to rely increasingly on epidemiology to make inferences between chemical exposure and disease. For example, the 1964 report of the US Surgeon General on smoking and health concluded that cigarette smoking caused lung cancer. This conclusion was based on historical associations that met tests of strength, consistency, specificity, temporal relationship, and coherence.

In 1965, British epidemiologist Sir Austin Bradford Hill presented a formal system for assisting researchers in determining if an association between chemical exposure and a disease could be considered causal. The factors presented by Hill are often known as the Bradford Hill postulates or Bradford Hill criteria, although it must be noted that Hill never intended these factors to be used in a vacuum or as a checklist in determinations of causation. Hill considered the situation in which observation had revealed an "association between two variables perfectly clear-cut and beyond what we would care to attribute to the play of chance" and listed nine factors that should be considered.

Strength of the association. In epidemiology, the strength of the association is usually measured by the relative risk (or odds ratio). A strong association (high numerical value of a relative risk) supports a hypothesis of causality of the association. A weak relative risk fails to support the hypothesis. Hill used well-known examples from historic epidemiology such as Percival Pott's observation of the association between exposure to soot and testicular cancer in chimney sweeps and John Snow's observations about cholera and polluted water to illustrate the concept of strength of the association. Some investigators, and indeed some courts and regulatory bodies, have codified numerical values of relative risk that demark 'strong' or 'weak' in this context; however, none of these limits have been generally accepted. A hypothetical causal relationship is also thought to be more credible if it is precise. Precision in this sense being interpreted as narrow confidence intervals around the relative risk.

Consistency. Consistency normally refers to reproducibility of results. In epidemiology, consistency is demonstrated by repeated findings of associations in different population groups or in different settings. Hill used the example of and association between tobacco smoking and cancer that had been observed in 29 retrospective and seven prospective epidemiologic studies to illustrate consistency.

Specificity. Specificity often applies to the ability of a particular toxicant to cause a response. In epidemiology, the issue is expressed in the elimination of confounders; in medicine the issue is expressed through the concept of differential diagnosis.

Temporality. The exposure must precede the disease. This is a threshold requirement for causation. If a latency period is associated with the development of a particular disease, the duration of the latency should be taken into account.

Biological gradient. The biological gradient often is the dose–response relationship in toxicology. An observed dose–response or concentration–response relationship supports a hypothesis of causation. Locational gradients and variable durations of exposure may also be used in causation analysis.

Plausibility. The assertion should be biologically plausible. Most investigators use information from field observations, laboratory bioassays, chemical measurements, structure–activity relationships, pharmacokinetics, and *in vitro* testing to reach as judgment regarding causation. When considered as a whole, these various sources of information should be compatible with each other. This factor depends on the state of biological knowledge. For example, a mode of action of a particular toxicant may be unknown at one point in time but discoveries in a related area of research may result in an understanding of the mode of action at a later date. In the first instance, the association may be considered implausible, whereas in the second it may be deemed plausible.

Coherence. Hill believed that an interpretation of causality should not conflict with generally known facts of the natural history or biology of the response.

Experiment. Experimental evidence results when investigators are able to adequately control the variables in a test. With the exception of a randomized clinical trial, human experimental data are generally not available. With species other than humans, experimental data may be the foundation upon which a determination of causation is made.

Analogy. Hill believed that, in some cases, it was possible to draw judgments about causation on the basis of analogy. His examples were birth defects associated with thalidomide and rubella arguing that a similar drug or viral disease could be presumed to have similar effects.

These postulates have been widely applied in human epidemiology and, to a lesser extent, in cognate fields. Although mostly used to analyze observational data, they may also be applied to experimental data. Bradford Hill, in fact, was the originator of the randomized clinical trial in pharmaceutical testing which relies on carefully controlled experimental protocols.

In 1976, A. S. Evans updated Koch's postulates integrating concepts proposed by Hill into a unified concept of disease causation that could be applied to both infectious and noninfectious diseases. Evans postulates included the following.

- The presence of the response should be significantly higher in those exposed than in controls.
- Exposure to the hypothesized cause should be more frequent among those with the response than in controls when all other risk factors are held constant.

- Incidence of the response should be significantly higher in those exposed to the cause than those not so exposed, as shown by prospective studies.
- Temporally, the response should follow exposure to the hypothesized cause.
- A spectrum of host responses should follow exposure to the hypothesized agent along a logical biological gradient.
- A measurable response following exposure to the cause should have a high probability of appearing in those lacking the response prior to exposure.
- Experimental reproduction of the response should occur more frequently in animals or humans exposed to the hypothetical cause compared to an unexposed group.
- Elimination or modification of the hypothetical cause or of the vector carrying it should decrease incidence of the response.
- Prevention or modification of the host's response on exposure to the hypothetical cause should decrease or eliminate the response.
- All of the relationships and findings should be biologically plausible.

In 1991, Mervyn Susser critically analyzed scientific thinking concerning causation and developed a refined group of criteria that he felt were most useful and least tautological:

1. Strength is the size of the estimated risk given the constraints of probability levels, confidence intervals, or other measure of likelihood.
2. Specificity is the precision with which one variable, to the exclusion of others, will predict the occurrence of another:
   (a) specificity in the cause implies that a given effect has a unique cause, and
   (b) specificity in the effect implies that a given cause has a unique effect.
3. Consistency is the persistence of an association upon repeated testing:
   (a) survivability is the number, rigor, and severity of tests of association; and
   (b) replicability is the number and diversity of tests of association.
4. Predictive performance is the ability of a causal hypothesis drawn from an observed association to predict an unknown fact that is consequent on the initial association.
5. Coherence is the extent to which a hypothesized causal association is compatible with existing knowledge:
   (a) theoretical coherence describes compatibility with existing scientific theory;
   (b) factual coherence is compatibility with existing knowledge;

(c) biologic coherence is compatibility with knowledge drawn from a species other than the species from which the causal hypothesis was drawn; and
(d) statistical coherence is compatibility with a conceivable model of the distribution of cause and effect.

The application of sets of criteria or postulates such as those proposed by Bradford Hill, Evans, or Susser has not been without its detractors. Some of the criticism stems from the use of the Bradford Hill postulates as checklists, although this was not intended by its author. Even if the postulates are used correctly, as a broad framework for analysis, many investigators have pointed out deficiencies in the individual postulates or caveats that should be observed in their application. For example, cases where there strong associations but no causation are well known (e.g., the association between water ingestion and mortality). Theoretical objections have also been raised to application of postulates for determination of causation. The school of the twentieth-century philosopher Karl Popper believes that science progresses by rejecting or modifying causal hypotheses rather than demonstrating causation. Because causation is often associated with litigation or regulatory decisions regarding chemical safety, testifying experts or advocates for or against a particular decision have voiced objections to particular postulates particularly when application of the postulate contradicts the authoritative position of the expert. For example, while acknowledging that a strong association is more likely to be causal than a weak one, an authoritative attack on this postulate may note that this does not preclude the possibility of a weak association being causal. Similarly, the authoritative position may argue that biological plausibility may be a mere reflection of the state of biological knowledge and that supporting experimental evidence, while valuable, may be precluded by practical or ethical considerations.

More recently, the concept of evidence-based toxicology has been developed as a comprehensive framework for causation. Drawing on the emerging field of evidence-based medicine, Philip Guzelian and co-workers have attempted to formalize a process for determination of causation that substitutes reliance on a comprehensive scientific database analysis for authoritative statements. In general, Guzelian and co-workers have proposed three main stages of evaluation consisting of a total of 12 individual steps:

1. collecting and evaluating the relevant data,
   (a) source,
   (b) exposure,
   (c) dose,
   (d) diagnosis,
2. collecting and evaluating the relevant knowledge,
   (a) frame the question,

(b) assemble the relevant literature,

(c) assess and critique the literature, and

3. joining data with knowledge to arrive at a conclusion,

    (a) general causation (answer to the framed question)

    (b) dose–response

    (c) timing

    (d) alternative cause

    (e) coherence

In this system, the first stage is problem specific and corresponds most closely to the problem formulation and exposure assessment stages in risk assessment. Source evaluation includes identification of the chemicals of potential concern and elucidation of exposure-related properties such as bioavailability and persistence. Exposure and dose evaluations are both qualitative and quantitative and include such elements as pathway analysis and dose calculation. Diagnosis presupposes that damage or harm has occurred.

Field biology, laboratory testing, and pathology in addition to other specialties are used to diagnose ecological problems. The second stage corresponds loosely to the toxicological evaluation (i.e., hazard identification and dose–response quantification) stage of a risk assessment. Framing the question involves developing a conceptual model that poses specific questions that may be addressed by recourse to the scientific literature. Guzelian and co-workers list toxicological delimiters that may be used as aids in framing questions (modified slightly here to reflect current risk assessment usage):

1. effect delimiters,

    (a) time frame (acute, subchronic, chronic),

    (b) site of action (local or systemic),

    (c) possibility of toxicological interactions,

    (d) certainty of diagnosis,

2. route of exposure delimiters,

    (a) source (medium, matrix, single or multiple chemicals),

    (b) duration and frequency,

    (c) continuous or intermittent,

    (d) constant or variable magnitude,

3. dose delimiters,

    (a) time frame (acute, chronic),

    (b) rate,

    (c) frequency and duration,

    (d) variable or constant,

    (e) dose metric (applied, absorbed, target tissue), and

4. extrapolations

    (a) interspecies,

    (b) *in vitro* or *in vivo*,

    (c) experimental or observational.

The last stage corresponds most closely to the application of classical causation criteria such as those postulated by Bradford Hill and others. Although relatively new and

designed for specific application to causation of human disease, evidence-based toxicology bears significant promise for wider application in environmental risk assessment.

Scientific and philosophical debates regarding the nature of causation and applications of criteria to determine causation continue to the present day and are unlikely to be resolved in the immediate future. Rothman and Greenland have recently taken the position that causal inference is most appropriately viewed as an exercise in the measurement of an effect rather than a criteria-driven process for deciding if the effect is present. They present counter-evidence for each one of Bradford Hill's criteria and imply that criteria are only valid to generate hypotheses that may be subsequently tested. Kundi has proposed a dialog approach to evaluate causal inference relying on concepts involving multiple causes and addressing issues of temporality, association, environmental, and population equivalence. Many of these debates are more theoretical and academic rather than pragmatic in nature. Regardless of the ultimate outcome of the debates, causation criteria such as those proposed by Bradford Hill and Susser, especially when used in the context of evidence-based toxicology, are a valuable tool for risk assessors, toxicologists, and epidemiologists to use in evaluating accumulated weight of evidence in complex situations involving multiple stressor stimuli and responses.

Applications of Causation Criteria to Ecological Science

Ecological risk assessors are often faced with problems of large scale and involving multiple impacted species and multiple stressors. The United States Environmental Protection Agency (USEPA) has devoted a substantial amount of effort to stressor identification including causation analysis derived from human epidemiology. Among other things, USEPA proposed a generalization of Koch's original postulates to include chemical toxicants:

- The injury, dysfunction, or other potential effect of the toxicant must be regularly associated with exposure to the toxicant in association with any contributory causal factors.
- The toxicant or a specific indicator of exposure must be found in the affected organisms.
- The effects must be seen when healthy organisms are exposed to the toxicant under controlled conditions.
- The toxicant or specific indicator of exposure must be found in the experimentally affected organisms.

USEPA also developed a general scheme for analyzing causation derived from the Bradford Hill postulates. In this scheme, USEPA proposes the use of four criteria that are based on associations derived primarily from the problem under consideration:

- Co-occurrence interpreted as the spatial co-location of the candidate cause and effect.
- Temporality.
- Biological gradient often in the form of a dose–response relationship.
- Complete exposure pathway or the pathway that a stressor takes to reach a receptor.

In addition, USEPA proposes the use of five additional criteria that are based on either the problem under consideration or other situations:

- Consistency of association or the repeated observation of the effect and candidate cause in different times or places. The more observations and the more diverse the situations, the greater the case for causation.
- Experiment referring to manipulation of a hypothetical cause by removing the exposure rather than a laboratory test.
- Plausibility or the degree to which the cause and effect relationship could be expected in light of what is known about mechanism and stressor response.
- Analogy.
- Predictive performance – whether the hypothetical cause has any initially unobserved properties that were predicted to occur.

Finally, USEPA proposes two criteria that are integrative in nature:

- Consistency of evidence – whether the hypothetical relationship is consistent with all lines of evidence.
- Coherence of evidence – whether a conceptual or mathematical model can explain any apparent inconsistencies among lines of evidence.

USEPA recommends considering all individual lines of evidence using a semi-quantitative scoring system and evaluating the total weight of evidence. USEPA also stresses the necessity for uncertainty and confidence analyses as components of the causation evaluation.

A paradigmatic ecotoxicologic problem is the North American Laurentian Great Lakes which have substantial ecological value as well as significance to Canada and the United States as fisheries and sources of drinking water. Over the past 50 years, environmental impacts to the Great Lakes have included population declines, reproductive failure, and deformities in fish, reptiles, and pisciverous birds, among other impacts. Multiple stressors associated with these problems include the presence of many persistent pollutants, discharges of nonpersistent toxic substances, overfishing, eutrophication, depredations of introduced exotic species, and degradation in habitat. For intervention in the form of regulation or remediation to be successful, it is desirable to be able to differentiate among the various stressors and determine which are causative agents associated with ecological impacts.

In 1991, Glen Fox was one of the first to propose the application of causation criteria to impacts in the Great Lakes. Fox recognized the parallel between human epidemiology and epizoology or ecoepidemiology which he took to be the study of ecological causes and effects in specific localities or among populations, communities, and ecosystems. Borrowing from pathobiology, Fox defined 'disease' to be any failure of normal homeostatic processes at any level of biological organization. With this broad definition in hand, Fox discussed the Bradford Hill and Susser criteria in the ecological context. For example, he recognized that criteria regarding specificity should be viewed in light of current biological thinking that many diseases have multiple causes and that a single substance may have a number of different biological effects with different underlying mechanisms.

The actual application of causation criteria to Great Lakes problems was undertaken by scientists from the International Joint Commission who evaluated associations related to chemical exposure and avian chick mortality and deformity and reproductive success of lake trout among other topics. In a paper published in 1997, Michael Gilbertson presented a case study applying epidemiologic causation criteria to the Great Lakes mortality, edema, and deformities syndrome (GLEMEDS). GLEMEDS is a syndrome observed in common terns and herring gulls in which there is a high incidence of chick mortality, edema, and deformity. No bacterial or viral pathogens were found on culture; however, eggs showed a high level of various persistent organochlorine compounds including polychlorinated dibenzo dioxins and furans (PCDD/PCDFs) and polychlorinated biphenyls (PCBs). The criterion of specificity was evaluated by the ability of specific PCDD/PCDF and PCB congeners to cause the syndrome in laboratory animals. The strength of the association was evaluated using correlations between PCDD/PCDF toxicity equivalents and egg mortality and deformity in nesting colonies of double-crested cormorants and Caspian terns. Time order was evident in the observation that bald eagles exhibited similar reproductive failure subsequent to the introduction of DDT in the 1940s in addition to other, more anecdotal, evidence. The observation that numerous species (including herring gulls, roseate, common, Caspian, and Forster's tern, and double-crested cormorants) showed similar effects was used to demonstrate consistency. Coherence was demonstrated by recourse to the scientific literature and determination that there was a biologically plausible mechanism of action. Gilbertson was able to conclude that the association between exposure to persistent organochlorine compounds and

GLEMENDS was a causal one and recommended legislative and regulatory courses of action to mitigate the problem.

Another application of causation criteria is in the use of weight of evidence approaches for evaluating contaminated sediment. The causal hypothesis of chemical contamination impairment of benthic communities is typically evaluated using such an approach. The weight of evidence is developed from individual lines of evidence including:

- sediment contaminant chemistry and geochemistry,
- benthic invertebrate community structure as revealed by field observation,
- sediment toxicity testing, and
- consideration of bioaccumulation or biomagnification.

The first three of these lines of evidence constitute the well-known and widely used sediment triad. Wenning and co-workers have proposed seven criteria, derived from epidemiologic criteria, for using line of evidence data to evaluate causation in the context of contaminated sediment and impacted benthic communities including:

- co-occurrence or spatial correlation in which the impacted community co-occurs with the chemical contamination,
- temporal correlation,
- effect magnitude (corresponding to strength of association),
- consistency of the association at multiple sites,
- experimental confirmation in the field or in the laboratory,
- biological plausibility of the stressor-effect linkage, and
- specificity or consideration that the stressor causes a unique effect.

Adams and Collier conducted a test of the utility of causation criteria in a variety of ecotoxicological contexts. In this test, eight investigators were requested to use seven causal criteria to evaluate their individual studies. The criteria proposed included strength of association, consistency of association, specificity of association, time order/temporality, biological gradient, experimental evidence, and biological plausibility. Organisms involved in the individual case studies included fish, benthos, clams, and birds. Chemicals of potential concern included radionuclides, PCBs, polycyclic aromatic hydrocarbons (PAHs), silver, insecticides, metals, and complex mixtures. Toxicological endpoints included genetic and biochemical alterations, liver histopathology, reproduction, and population impacts. The utility of the criteria was judged on the basis of a semiquantitative scale similar to that used by USEPA using the following categories:

- convincing evidence presented that the criterion was met,
- strong evidence presented yet some questions raised,
- more likely than not or only little evidence presented,
- evidence presented for both accepting or rejecting the criterion or no available evidence for a particular criterion, and
- evidence presented that argued against accepting the criterion.

Application of this scoring system to the criteria for the case studies showed that specificity of association was the least useful of the criteria followed by temporality. The presence of multiple stressors was hypothesized as the basis of the lack of utility for specificity. This test also found that causation was more difficult to demonstrate at the community level of organization compare to lower levels of biological organization.

Summary and Conclusions

The desirability of criteria to assess cause and effect relationships between exposure to environmental agents and disease has been obvious for over 200 years. Scientists, primarily epidemiologists, and philosophers have debated the development and use of causation criteria and the debate continues to the present. Regardless of the ultimate outcome of the debate, the use of criteria derived from or resembling Bradford Hill's postulates is a pragmatic approach that can serve to formalize determinations of causation and ensure some degree of consistency among practitioners. Although epidemiologists have been applying causation criteria to questions regarding chemical exposure and human disease for well over half a century, only recently has this technique been employed in the ecotoxicological context. The application of causation criteria to ecotoxicological problem solving has great promise; however, there is a great need for ecotoxicologists and ecological risk assessors to continue to integrate new concepts such as evidence-based toxicology into their causation thinking.

See also: Dose–Response; Ecological Risk Assessment; Exposure and Exposure Assessment.

Further Reading

Adams SM (2003) Establishing causality between environmental stressors and effects on aquatic ecosystems. Human Ecological Risk Assessment 9: 17–35.
Brooks GF, Butel JS, and Morse SA (2001) Jawetz, Melnick, & Adelbergs Medical Microbiology. New York: Lange.
Collier TK (2003) Forensic ecotoxicology: Establishing causality between contaminants and biological effects in field studies. Human Ecological Risk Assessment 9: 259–266.
Evans AS (1976) Causation and disease: The Henle–Koch postulates revisited. Yale Journal Biology Medicine 49: 175–195.

Gilbertson M (1997) Advances in forensic toxicology for establishing causality between Great Lakes epizootics and specific persistent toxic chemicals. *Environmental Toxicology and Chemistry* 16: 1771–1778.

Guzelian PS, Victoroff MS, Halmes NC, *et al.* (2005) Evidence-based toxicology: A comprehensive framework for causation. *Human & Experimental Toxicology* 24: 161–201.

Hill AB (1965) The environment and disease: Association or causation? *Proceedings of the Royal Society of Medicine* 58: 295–300.

Kundi M (2006) Causality and the interpretation of epidemiologic evidence. *Environmental Health Perspectives* 114: 969–974.

Rothman KJ and Greenland S (2005) Causation and causal inference in epidemiology. *American Journal of Public Health*, 95(supplement 1): S144-S150.

Susser M (1991) What is a cause and how do we know one? A grammar for pragmatic epidemiology. *American Journal of Epidemiology* 133: 635–648.

United States Environmental Protection Agency (USEPA) (2000) Stressor identification guidance document. EPA-822-B-00-025. Washington, DC.

Wenning RJ, Batley GE, Ingersoll CG, and Moore DW (eds.) (2005) *Use of Sediment Quality Guidelines and Related Tools for the Assessment of Contaminated Sediments*. Pensacola, FL: SETAC.

Teratogenesis

J M Conley and S M Richards, University of Tennessee at Chattanooga, Chattanooga, TN, USA

Introduction	Wildlife Teratogenesis
Types of Environmental Teratogens	Summary
Teratogenic Effects	Further Reading
Stage of Development and Species Susceptibility	

Introduction

Many agents, whether man-made or naturally occurring, are teratogenic to a wide range of vertebrate and invertebrate taxa. A teratogen is any agent that physically or chemically alters developmental processes and produces structural deformities in an organism. These deformities are commonly known as birth defects. Indeed, teratogenesis literally translates to 'monster creation' from the Greek words *teras* (monster) and *genesis* (creation, formation). Traditional teratologists considered only gross anatomical abnormalities to be a teratogenic defect. Current definitions of teratogenesis include all morphological abnormalities along with subtle defects in behavior, biochemistry, and learning. Morphological defects are the most commonly found and studied abnormality in an ecological context. Thus, this article focuses on morphological defects occurring in ecological species from a range of taxa and the teratogens to which the defects are associated.

Teratogenesis, and the resultant deformity, is dependent upon the timing and progression of developmental pathways. Developmental pathways vary significantly between species, yet there are common fundamental processes that mark the beginning stages of development in all multicellular organisms. All sexually reproducing organisms fuse sperm and egg to form a zygote during fertilization. The zygote develops into an embryo through steps such as cell proliferation, differentiation, and migration in a process known as embryogenesis. Organogenesis occurs during embryogenesis and involves steps that form the organs and limbs of the embryo. At the end of embryogenesis the conceptus is considered a fetus or larva (depending on the species). Further development is characterized by growth and physiologic maturation of the fetus giving rise to a neonate, or metamorphosis of the larva into adult form. These events are extremely sensitive to perturbation and transpire in a specific amount of time and order unique to each taxon.

Teratogenesis is the creation of an abnormal organism due to a disruption in the structural formation of the organism during one of the developmental stages. Teratogens disrupt somatic cells, not germ cells, and are therefore not heritable by future generations. Conversely, mutagens alter germ cells causing genetic disruptions that are heritable by future generations. In addition, mutagens can exert an effect at any point in the organism's life, whereas teratogens only have effects at an organism's developmental stage.

Types of Environmental Teratogens

The three main categories of teratogens suspected of affecting wildlife are mechanical disruptors, environmental factors, and chemical contaminants. These teratogens are introduced to the environment through natural processes and/or anthropogenic sources. The nature of the teratogen and the timing during which it insults the developing embryo or fetus is critical to the type of abnormal effect it will produce (**Table 1**).

Mechanical Disruptors

Mechanical disruptors physically alter the development of tissue through scarring or impeded growth. Many mechanical disruptions are due to biotic interactions in nature and can include ecologic predators and parasitic implantations. For example, a predatory fish attempting to consume a tadpole (larval frog) could inadvertently amputate one of the developing hind limbs without killing the organism. Then, depending on the ability of the organism to regenerate the tissue, a malformed limb could be produced. Parasitic implantations include cysts created during the life cycle of parasites, which can embed in sensitive areas at critical times and alter the development of the host organism (see the section titled 'Amphibians'). Mechanical disruptors are also abiotic. For example, a

Table 1 Examples of environmental teratogens and their associated ecological receptors. Not all of the compounds are teratogenic at environmentally relevant concentrations

Suspected teratogen	Affected organism	Teratogenic effect	Reference
Acid mine drainage (combination of Cd, Cu, Zn)	Chironomidea (Diptera)[a]	Mouthpart deformity	Swansburg et al. (2002)
Atrazine	African clawed frog[b]	Hermaphroditism/reduced larynx	Hayes et al. (2002)
Barium	California mussel[c]	Abnormal morphology/developmental delay	Spangenberg and Cherr (1996)
Chlorpyrifos (organophosphate insecticide)	African clawed frog[b]	Spinal/muscular deformity	Richards and Kendall (2003)
Coal combustion waste (combination of Cd, Cu, Se)	Bullfrog[d]	Spinal curvature	Hopkins et al. (2000)
DDE	American alligator[e]	Reduced phallus and cuff	Milnes et al. (2005)
Hypoxia	Zebrafish[f]	Spinal curvature/developmental delay	Shang and Wu (2004)
Polyaromatic hydrocarbons (PAHs)	Pacific herring[g]	Reduced jaw/spinal curvature	Carls et al. (1999)
Polychlorinated biphenyls (PCBs)	American kestrel[h]	Bill and limb deformities	Fernie et al. (2003)
	House wren[i]	Heart deformities	De Witt et al. (2006)
	Tree swallow[j]	Heart deformities	De Witt et al. (2006)
	Double-crested cormorant[k]	Crossed bill	Larson et al. (1996)
PCBs, PCDDs, PCDFs	Snapping turtle[l]	Tail, limb, and shell deformities	Bishop et al. (1998)
Selenium	Rainbow trout[m]	Skeletal, facial deformities/reduced fin	Holm et al. (2005)
	American coot[n]	Missing appendages	Ohlendorf et al. (1986)
	Mallard[o]	Skeletal anomalies	Ohlendorf et al. (1986)
Trematode metacercariae (multiple species)	Pacific tree frog[p]	Limb malformation/absence	Johnson et al. (1999)
	Axolotl salamander[q]	Limb malformation/absence	Sessions and Ruth (1990)
UVB radiation	Northern leopard frog[r]	Limb malformation/absence	Ankley et al. (1998)
Vitamin D_3 deficiency	Double-crested cormorant[k]	Crossed bill	Kuiken et al. (1999)

[a]Multiple species.
[b]Xenopus laevis.
[c]Mytilus californianus.
[d]Rana catesbeiana.
[e]Alligator mississippiensis.
[f]Danio rerio.
[g]Clupea pallasi.
[h]Falco sparverius.
[i]Troglodytes aedon.
[j]Tachycineta bicolor.
[k]Phalacrocorax auritus.
[l]Chelydra serpentina serpentina.
[m]Oncorhynchus mykiss.
[n]Fulica americana.
[o]Anas platyrhynchos.
[p]Hyla regilla.
[q]Ambystoma mexicanum.
[r]Rana pipiens.

larval fish may be abraded by sand or gravel in a stream, disrupting the development of fin tissue, creating a malformed fin.

Chemical Contaminants

Many chemical contaminants are xenobiotic, meaning they have no intrinsic function in living organisms, and may be teratogenic. Certain chemicals (metals and nutrients) have critical biotic functions in proper amounts but may be teratogenic at excessive concentrations. Agricultural chemicals and industrial pollutants are commonly suspected chemical teratogens. These contaminants can enter the environment from a variety of sources including direct application, soil runoff, effluent discharge, and atmospheric deposition. Agricultural chemicals include fertilizers and a variety of pesticides such as herbicides, insecticides, and fungicides. Industrial pollutants include a wide array of chemicals such as chlorinated hydrocarbons, metals, and aromatic hydrocarbons. Many of these chemicals are by-products of industrial processes that are either discharged into the aquatic environment or are discharged into the atmosphere and subsequently deposit on land or water. Chemical pollutants receive a lot of scrutiny and government regulation due to the large-scale production, application, and potential impact on nontarget species. Thus, the vast majority of ecotoxicological teratogenesis research has focused on chemical contaminants.

Endocrine disrupting compounds

Recently, endocrine disrupting compounds (EDCs) have been implicated as teratogens. The EDCs are xenobiotic chemicals that interfere with an organism's normal physiology by mimicking the action of hormone agonists or antagonists leading to changes in the synthesis, action, and/or metabolism of hormones. As a result of their effect on hormones, structural deformities may result. Due to the hormonal mode of action and subsequent reproductive organ alteration, EDCs are often associated with reproductive abnormalities in wildlife. It should be noted that EDC is a descriptive category within the broader category of chemical pollutants. Many chlorinated hydrocarbons (e.g., polychlorinated biphenyls (PCBs) and organochlorines (OCs)) are recognized as having EDC characteristics in addition to their effects on other tissues.

Environmental Factors

Naturally occurring and anthropogenic environmental factors may directly or indirectly lead to teratogenesis. Examples include ultraviolet (UV) radiation, eutrophication, temperature, and pH. Solar radiation in the UVB range can disrupt tissue development. Multiple wavelengths of UV radiation can also alter chemical contaminants through direct or indirect photolysis to produce chemicals that are more potent teratogens than the original compound. Eutrophication is the increase in nutrient content of a water body commonly leading to poor water quality (anoxic or hypoxic conditions), which can be teratogenic. Eutrophication has also been linked to an increase in the population of intermediate hosts of amphibian parasites. Alterations in the environment that affect temperature of nest sites or larval habitat can produce abnormalities in developing organisms. Virtually all oviparous organisms are highly temperature sensitive during development in the egg and abnormally hot or cold conditions can be teratogenic. Temperature variation affects the growth and migration of cells during embryogenesis leading to limb and organ deformities. Similarly, variation in pH can be teratogenic. Highly acidic conditions, especially in aquatic oviparous species, directly affect developing organisms, as well as increase the toxicity of xenobiotic contaminants.

Teratogenic Effects

Four potential results of teratogenic insult include death, malformation, functional disorder, and growth retardation. Exposure to any teratogen at a high enough concentration or long enough duration will directly kill the organism or produce such severe abnormalities that eventually the organism dies. At lesser concentrations or shorter durations, the teratogen will disrupt the natural processes of development and produce a malformation to a body structure such as the palate, eyes, axial skeleton, or internal organs. With a functional disorder, all body parts (e.g., organs, limbs) develop structurally correct, but a teratogenic insult alters the proper action of the structure. For example, functional disorders can create an imbalance in the biochemistry or reduce the efficiency of an organ or organ system. Growth retardation is simply a stunting of the optimal growth of the organism, typically resulting in reduced morphometrics and body mass.

The most commonly described teratogenic effects occurring in ecosystems are external malformations. This is because malformation of an exterior body structure is readily noticeable whereas death, functional disorder, and growth retardation are not as obvious and require more investigation to diagnose. Malformations can be manifest as defects in internal structures such as organs but this requires necropsy to identify. Teratogens leading to death may go largely unnoticed because dying organisms may be inactive and/or consumed by predators or scavengers. Functional disorder and growth retardation are less noticeable because all external structures may appear normal and further investigation is required to ascertain whether organs are functioning properly and that growth has transpired unimpeded.

Stage of Development and Species Susceptibility

Common Stages of Development

In general, across taxa, the type and degree of defect are determined by the time and duration that the developing organism is exposed to a teratogen. The earliest developmental stages are prone to the most dramatic effects. When teratogenic insult occurs very early in development, it is often too much for the organism to overcome and death is often the result. Organogenesis, a later developmental stage, is more resilient to teratogenic insult; interferences at this stage tend to result in structural malformations, as opposed to death. During organogenesis, different organ systems have specific times in which they are rapidly developing and most sensitive to teratogenic insult. The processes within organogenesis vary between species as to their timing and sequence of events (see the section titled 'Species susceptibility').

The fetus is a later developmental stage than the embryo and is unique to organisms that undergo direct development. The fetus is generally more resilient than the embryo, but is still highly susceptible to teratogenic insult. Fetal teratogenesis is typically manifested as growth retardation and functional disorders. A larva is also a later developmental stage than the embryo, but is unique to organisms that undergo complex (metamorphic) development. The larva is also more resilient than the embryo but still very susceptible to teratogenic insult.

Species Susceptibility

One of the most important factors in ecological teratogenesis is the wide variability in the timing of developmental processes between species. In the rat, for example, embryogenesis takes place during the first 17.5 days of gestation, whereas frogs undergo embryogenesis in the first day of gestation. Thus, the postconception time point at which both of these organisms are subject to the same birth defect is very different. For example, in rat organogenesis, the developing eyes are most susceptible to teratogenic insult between gestation days 8 and 10, whereas the palate is most susceptible between days 12 and 14. In the frog, however, the susceptibility of structures is separated by hours, not days. Within organogenesis, development of some organs takes place concurrently. This overlap in developmental processes adds further complexity to the

issue and introduces the potential for a spectrum of malformations.

The type of gestation is also important and highly variable in nature. Viviparous organisms (all mammals, many reptiles, some amphibians and fishes) carry and directly nourish the developing organism in the womb until birth. Oviparous organisms (all birds, monotremes; many fishes, amphibians, reptiles, and invertebrates) develop within an egg that may or may not be protected by an eggshell. The oviparous embryo develops in the egg and receives nourishment from the yolk until mature enough to exit the egg. Ovoviviparous organisms (many fishes, reptiles, and invertebrates; some amphibians) develop inside an egg, which remain inside the mother until hatching or just before hatching. Viviparous, and to an extent ovoviviparous, embryos are exposed to teratogens based on the mother's exposures throughout gestation. Oviparous embryos have a much shorter maternal exposure to teratogens.

The type of egg and the environment where the eggs are laid is critical to the teratogenic exposure of oviparous embryos. Most aquatic species lay eggs that are more fluid and do not contain an eggshell, whereas terrestrial species tend to lay eggs that do contain an eggshell. Eggs that do not contain an eggshell tend to be more susceptible to teratogenic insult.

The exposure of the developing organism is also dependent on the overall life cycle of the species. Organisms that undergo metamorphosis (many invertebrates, amphibians, and fishes) develop from embryo to larva to adult without a fetal stage. In many situations the larva is a free-living organism that develops critical structures without the protection of the maternal womb or an eggshell. A good example is the frog. In the larval stage, frogs are developing limbs and overall adult body structure. Exposure to a teratogen at this time is critical and largely without protection. Organisms that undergo direct development (many mammals, birds, and reptiles) have the protection of the womb (and associated maternal, physical, and enzymatic defenses) or an eggshell.

Breeding season and birthing location is critical to the types and duration of offspring teratogen exposure. Factors such as seasonal variation in agricultural pesticide use and UV radiation intensity can coincide with the egg-laying season of oviparous organisms in a way that makes the offspring more prone to teratogenic insult. Whether the organisms are aquatic or terrestrial plays a major role in the types of potential teratogens. For example, toxicants that adhere to soil are typically hydrophobic while compounds in the water column are hydrophilic. This factor is very important when considering aquatic organisms because they are captive to their environment throughout their life and across generations.

Wildlife Teratogenesis

Birth defects in wild organisms are not a new phenomenon. Wild organisms have always been subject to teratogens of various natures. Unfortunately, anthropogenic influences have increased the number, intensity, and/or volume of teratogenic insults. Due to the reasons mentioned above, ecotoxicologists interested in teratogenesis first focused on waterborne pollutants and their effects on fish. Historically, birds and mammals have also received considerable investigation, with amphibians and reptiles receiving the least attention. Within the last decade, however, the focus has changed. Studies of amphibian teratogenesis have seemed to surpass that of the other vertebrates. Multiple man-made compounds (pharmaceuticals, food additives, industrial chemicals, radioactive isotopes, etc.) have been shown in laboratory tests to be teratogenic. The following section addresses only those teratogens that have been shown to occur in the environment and have a scientific correlation to a specific deformity.

Invertebrates

Research on effects of contaminants on invertebrates has primarily focused on whether a chemical has the potential to be toxic to natural invertebrate assemblages. Subtle effects such as teratogenesis have not received much attention. Reports of invertebrate teratogenesis outside of the laboratory setting are rare and infrequently attributable to a single compound. However, there are a few accounts of invertebrate teratogenesis. Chironomid larvae inhabiting streams receiving metal mine drainage displayed mentum deformities that correlated with elevated tissue concentrations of copper, cadmium, and zinc. Barium has been found to produce developmental abnormalities in the marine bivalve, *Mytilus californianus*, including abnormal shell calcification and deformed embryo morphology.

Fishes

Fish, much like amphibian larvae, are completely confined to the aquatic environment and therefore relatively prone to adverse effects from waterborne contamination. As a result, the literature is replete with studies investigating the toxicity of environmental contaminants on wild fish populations. The most common toxicants of concern include organochlorines, PAHs (polycyclic aromatic hydrocarbons), metals, and pesticides. Constituents from all of these classes have been shown to alter the development of embryonic fish. However, many natural stress factors including variation in temperature, oxygen levels,

and salinity are also identified as having impact on developmental processes.

Numerous studies of natural fish populations in close proximity to industrialized areas have found fish with abnormalities such as intestinal defects, bent or spiral spinal axis, eye defects, and craniofacial defects. The specific teratogens are not easily identified but could potentially be any of the above-named environmental contaminants. PAHs have been identified as significant teratogens in wild Pacific herring (*Clupea pallasi*). Embryonic PAH exposure produced skeletal and craniofacial defects including reduced jaw size, spinal curvature, and failure to develop pectoral fins. Selenium has also been identified as a fish teratogen. For example, rainbow trout (*Oncorhynchus mykiss*) collected from watersheds of coal and uranium mines have structural abnormalities (e.g., curvature of the spine and craniofacial defects of the jaw and eye) correlated with selenium concentrations. Numerous such accounts implicate selenium as a fish teratogen.

Pollution-induced environmental conditions have also been determined to produce abnormal development in wild fish. Hypoxia (low dissolved oxygen) commonly occurs in natural water bodies as a result of pollution-induced algal blooms and thermal pollution, for example. Sublethal levels of hypoxia are responsible for teratogenesis in developing wild fish. Fish developing in a hypoxic environment display spinal curvatures, increases in testosterone, and decreases in estradiol.

Amphibians

In 1995, a middle-school group in Minnesota was investigating wildlife in a nearby wetland and discovered a large number of northern leopard frogs (*Rana pipiens*) with hind limb malformations. This discovery sparked great interest in the scientific community and resulted in a dramatic increase in amphibian teratogenesis research, specifically anuran (frog) research. Since that time there have been many accounts of free-living frog populations with incidence of teratogenesis as high as 60%. Many scientists feel that this is a much greater incidence of deformity as compared to historical rates or normal rates of malformation, which range from 2% to 6%. Whether there is a real difference in historical deformity incidence versus current incidence is debatable. Some scientists state that the increased incidence of malformation is merely a result of increased investigation and that the actual rate of deformities in frogs has not changed.

The common limb deformities observed in wild frogs have been divided into four main categories: no limb, extra limbs or elements, reduced limbs or elements, and complete but malformed limbs or elements. No limbs (**Figure 1a**), or amelia, is the absence of one or both of the hind limbs as well as the socket where the limb bones attach to the hip.

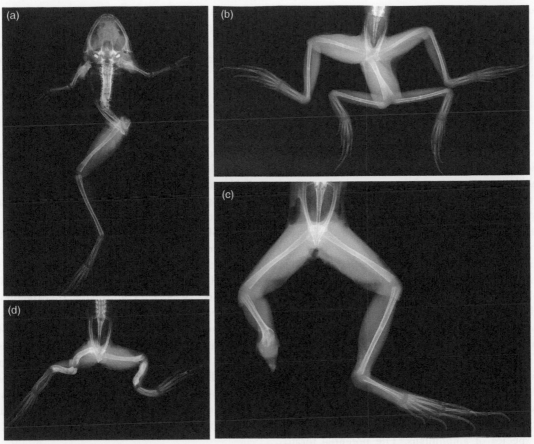

Figure 1 Hind limb deformities in northern leopard frogs (*Rana pipiens*). (a) Frog displays amelia (missing limb). (b) Frog displays polymelia (extra limbs) with supernumerary pelvic elements. (c) Frog displays ectromelia (truncated limb) of the tibiofibula. (d) Frog displays rotational malformation of the hind limbs and skin webbing (not visible). Adapted from Meteyer CU, Loeffler IK, Fallon JF, *et al.* (2000) Hind limb malformations in free-living northern leopard frogs (*Rana pipiens*) from Maine, Minnesota, and Vermont suggest multiple etiologies. *Teratology* 62: 151–171, Copyright 2000 Wiley InterScience. Reprinted with permission of Wiley-Liss, Inc., a subsidiary of John Wiley & Sons, Inc.

Extra limbs or elements can be manifest as polymelia (extra limbs; **Figure 1b**), polydactyly (extra digit and metatarsal bone), or polyphalangy (extra digit). Reduced limbs or elements are classified as phocomelia (abnormally structured, leg-like appendage), ectromelia (reduced/truncated limb; **Figure 1c**), ectrodactyly (missing digit or part of digit), or brachydactyly (shorter digit). Complete but malformed limbs have all bones necessary but they are either rotated (**Figure 1d**), have a skin web, bone bridge, or micromelia (stunted limb compared to complement).

Several etiologies have emerged as to the cause of the malformations and there is much contention with this issue as well. The predominant suspects are the parasitic trematode (*Ribeiroia*), UVB radiation, and various chemical contaminants (**Table 1**). *Ribeiroia* utilize fish and amphibians as secondary hosts and form cysts known as metacercariae to house the parasite until it reaches its final host, a bird or mammal. Metacercariae have been found to infest the areas around the developing limb buds and have been directly correlated with abnormalities of the hind limbs due solely to their presence in the tissue (mechanical disruption). Indeed, the simple act of metacercariae implantation has been found to physically disrupt developing hind limb tissue. Other cysts have also been shown to mechanically disrupt developing hind limb tissue as well, but *Ribeiroia* is the dominant antagonist.

UVB radiation is often cited as a reason for increased incidence of amphibian malformation. This stems from several factors that seem to correlate with increased deformity incidence, namely increased UVB from ozone depletion, seasonal variations in ozone intensity, and amphibian decline, a global issue with UVB intensity increasing worldwide. While these are all plausible theories, they are difficult to adequately test. Thus, the UVB theory of amphibian malformation has received a lot of scrutiny. Laboratory studies, however, have confirmed that developing amphibians exposed to intense UVB radiation do produce malformed hind limbs; but the malformations typically develop in a symmetrical fashion, which is not common in the environment.

Agricultural pesticides have received considerable attention as possible culprits of amphibian limb deformation. Retinoids are responsible for controlling the function of many developmental processes in amphibians. Amphibians develop hind limb malformations when exposed to retinoic acid (vitamin A derivative). Some environmental contaminants and their derivatives, such as the insecticide methoprene, may mimic retinoic acid and alter amphibian development. However, the agricultural chemical/retinoic acid link to amphibian limb deformities has not been clearly established. Similarly, the herbicide atrazine (one of the most widely utilized agricultural chemicals in the world) has also been suspected as an amphibian teratogen. While atrazine has not been found to induce limb malformations in the African clawed frog (*Xenopus laevis*), it is recognized as a potential EDC. In some studies, *Xenopus* larvae exposed to atrazine have exhibited hermaphrodism, reduced larynx size, and altered blood hormone concentrations. In other studies, atrazine had no such effects. At this time, the issue of whether or not atrazine is a teratogen or EDC is highly controversial and yet to be resolved.

It is apparent that many compounds in the environment can be teratogenic to amphibians under the right conditions. At this point, researchers have not found a single causative agent that can be attributed to all of the varying types of hind limb malformations as seen in the multiple species of amphibians. The potential teratogenic agent for producing these malformations is probably not one universal entity, but more likely a combination of multiple factors. The sum of these teratogens and their spectrum of teratogenic malformations may be a contributor to the global decline in amphibian populations.

Reptiles

While there are accounts of reptiles with multiple limbs, missing limbs, and other birth defects, the incidence of malformation has not been documented as with amphibians.

This may be due to an actually lower rate or a reduced sampling effort as compared to the amphibians. The recent focus has been on the effects of chlorinated hydrocarbons and metals on the endocrine function and development of reptiles. The American alligator (*Alligator mississippiensis*) has received considerable attention in this context. The examination of wild alligators from Lake Apopka, Florida, is one of the most prominent studies of reptiles and teratogenesis of EDCs. Organochlorine pesticides such as dichlorodiphenyltrichloroethane (DDT), its metabolites 1,1-dichloro-2,2-bis(*p*-chlorophenyl)ethylene (DDE) and dichlorodiphenyldichlorethane (DDD), and their isoforms, as well as toxaphene, dieldrin, chlordane, and nonachlor are suspected of producing a feminizing effect in male alligators from Lake Apopka. These feminized male alligators are characterized by a reduced phallus size and altered blood hormone concentrations.

Turtle malformations are linked to chemical contaminants as well. Chlorinated hydrocarbons such as polychlorinated dibenzodioxins (PCDDs), polychlorinated dibenzofurans (PCDFs), and PCBs are correlated with malformation of common snapping turtles (*Chelydra serpentina serpentina*). Sites containing PCBs, PCDDs, and PCDFs produced abnormalities including tail deformities, amelia, ectrodactyly, micromelia, carpace (shell) deformities, eye deformities, and jaw deformities.

Birds

The hallmark of avian teratogenesis is the malformation in bills resulting in crossing of the maxilla and mandible (**Figure 2**). Such beak malformations in wild birds are correlated with PCB concentrations. The Great Lakes region in particular has received considerable attention due to the relatively high frequencies of avian beak malformations in addition to dwarfed limbs, club feet, skeletal deformities, and edema giving rise to the term GLEMEDS (Great Lakes embryo mortality, edema, and deformity syndrome). This syndrome is attributed to the

Figure 2   (a) and (b) Bill malformations of double-crested cormorants (*Phalacrocorax auritus*) associated with vitamin D$_3$ deficiency. Copyright © (1999) Society of Environmental Toxicology & Chemistry. From *Environmental Toxicology and Chemistry*, by Kuiken T, Fox GA, and Danesik KL. Reprinted by permission of Alliance Communications Group, a division of Allen Press, Inc.

high contamination rate of the Great Lakes region with industrial pollutants such as the organochlorines (PCB, PCDD, and PCDF). Similar effects are also observed in relation to selenium contamination through trophic biomagnification. The most commonly afflicted birds are fish-eating (piscivorous) species such as bald eagles (*Haliaeetus leucocephalus*), double-crested cormorants (*Phalacrocorax auritus*), and kestrels (*Falco sparverius*).

Environmental contaminants may transfer from the mother to the chick via yolk proteins and fat during egg development or through direct application to the exterior of the egg. Many environmental contaminants are lipophilic and therefore have the potential to permeate the eggshell membrane and impact the embryo or fetus. This exposure may be through atmospheric deposition when a parent is not tending the nest or from the contaminated plumage of a parent.

As with amphibian deformities, there are naturally occurring incidences of avian malformation. Crossed bills are also associated with deficiencies in vitamin D_3, which is critical for maintaining proper levels of calcium in all vertebrates and subsequent bill formation in birds. For example, *P. auritus* purposely not exposed to significant concentrations of PCBs produced severely deformed bills due to a deficiency in vitamin D_3. Another seemingly natural event that could lead to developmental defects is that of irregular parental incubation behavior. It is possible that subsequent temperature extremes (high or low) could lead to altered development.

Mammals

While the medical literature is replete with mammalian teratogenesis data resulting from research on pharmaceuticals, illicit and licit compounds, and work-related chemical exposures, ecological information regarding wild mammal teratogenesis is very limited. The majority of teratogenic data are from laboratory studies on standard, homogeneous test species, typically rats or mice. These studies have predominantly been conducted to assess the potential effects of contaminants on human health with less regard for ecological implications.

Upper-trophic-level mammals have the highest risk of bioaccumulating toxic environmental contaminants and potentially are under the greatest threat of teratogenic effects. Some polar bear (*Ursus maritimus*) cubs in Norway displayed hermaphroditic characteristics. While the animals were genetically female they displayed both male and female genitalia. The cause is undetermined but EDCs that may accumulate in the fatty tissue of polar bear prey is a likely suspect. A similar scenario has recently been observed in beluga whales inhabiting the Gulf of St. Lawrence.

One reason wild mammal teratogenesis is less common may be due to the viviparity of mammals, whereby the developing embryo and fetus are more protected from

external factors such as chemical contamination. The developing organism is only exposed to the compounds that enter the mother and then pass through multiple physical barriers (i.e., multiple lipid bilayer membranes and the placental barrier). An additional level of protection is provided by the mother's enzymatic and immune system, both of which can remove or lessen the impact of teratogens that may otherwise contact the embryo or fetus.

Summary

The underlying factors that drive teratogenesis and the type of resultant morphological, behavioral, biochemical, and learning effects are quite complex. Most of that mechanistic information is beyond the scope of this article. In general, naturally and artificially occurring teratogenic defects happen in all organisms with the greatest variable being the type of the teratogen and the timing of exposure. Teratogens are broadly defined because any physical or chemical agent that produces a birth defect is a teratogen. The majority of ecotoxicological teratogenic research and concern, however, has been on chemical contaminants in the environment. The primary contaminants of concern are organochlorine pesticides, metals, PAHs, and chlorinated hydrocarbons (PCBs, PCDDs, PCDFs).

Amphibian teratogenesis has received the most attention in recent years due to the indication of global amphibian population decline and the visual impact of a deformed frog. Bird teratogenesis has also been relatively well researched historically, perhaps because birds are highly conspicuous and abnormality tends to draw public attention. It is important to remember, however, that animals in all ecosystems are subject to and affected by not only chemical, but also mechanical and environmental teratogens.

See also: Ecotoxicology: The History and Present Directions; Polychlorinated Biphenyls; Polycyclic Aromatic Hydrocarbons.

Further Reading

Ankley GT, Tietge JE, DeFoe DL, *et al.* (1998) Effects of ultraviolet light and methoprene on survival and development of *Rana pipiens*. *Environmental Toxicology and Chemistry* 17: 2530–2542.

Bishop CA, Ng P, Pettit KE, *et al.* (1998) Environmental contamination and developmental abnormalities in eggs and hatchlings of the common snapping turtle (*Chelydra serpentina serpentina*) from the Great Lakes–St. Lawrence River basin (1989–91). *Environmental Pollution* 101: 143–156.

Carls MG, Rice SD, and Hose JE (1999) Sensitivity of fish embryos to weathered crude oil. Part I: Low-level exposure during incubation causes malformations, genetic damage, and mortality in larval Pacific herring (*Clupea pallasi*). *Environmental Toxicology and Chemistry* 18(3): 481–493.

DeWitt JC, Millsap DS, Yeager RL, *et al.* (2006) External heart deformities in passerine birds exposed to environmental mixtures of polychlorinated biphenyls during development. *Environmental Toxicology and Chemistry* 25(2): 541–551.

Fernie K, Bortolotti G, and Smits J (2003) Reproductive abnormalities, teratogenicity, and developmental problems in American kestrels (*Falco sparverius*) exposed to polychlorinated biphenyls. *Journal of Toxicology and Environmental Health, Part A* 66: 2089–2103.

Finnell RH, Gelineau-van Waes J, Eudy JD, and Rosenquist TH (2002) Molecular basis of environmentally induced birth defects. *Annual Review of Pharmacology and Toxicology* 42: 181–208.

Fry DM (1995) Reproductive effects in birds exposed to pesticides and industrial chemicals. *Environmental Health Perspectives* 103(supplement 7): 165–171.

Guillette LJ, Jr. (2000) Contaminant-induced endocrine disruption in wildlife. *Growth Hormone and IGF Research* 10(supplement B): S45–S50.

Hayes TB, Collins A, Lee M, *et al.* (2002) Hermaphroditic, demasculinized frogs after exposure to the herbicide atrazine at low ecologically relevant doses. *Proceedings of the National Academy of Sciences of the United States of America* 99(8): 5476–5480.

Holm J, Palace V, Siwik P, *et al.* (2005) Developmental effects of bioaccumulated selenium in eggs and larvae of two salmonid species. *Environmental Toxicology and Chemistry* 24(9): 2373–2381.

Hopkins WA, Congdon J, and Ray JK (2000) Incidence and impact of axial malformations in larval bullfrogs (*Rana catesbeiana*) developing in sites polluted by a coal-burning power plant. *Environmental Toxicology and Chemistry* 19(4): 862–868.

Johnson PT, Lunde KB, Ritchie EG, and Launer AE (1999) The effect of trematode infection on amphibian limb development and survivorship. *Science* 284: 802–804.

Kuiken T, Fox GA, and Danesik KL (1999) Bill malformations in double-crested cormorants with low exposure to organochlorines. *Environmental Toxicology and Chemistry* 18(12): 2908–2913.

Larson JM, Karasov WH, Sileo L, *et al.* (1996) Reproductive success, developmental anomalies, and environmental contaminants in double-crested cormorants (*Phalacrocorax auritus*). *Environmental Toxicology and Chemistry* 15(4): 553–559.

Loeffler IK, Stocum DL, Fallon JF, and Meteyer CU (2001) Leaping lopsided: A review of the current hypotheses regarding etiologies of limb malformations in frogs. *Anatomical Record (New Anatomy)* 265: 228–245.

Meteyer CU, Loeffler IK, Fallon JF, *et al.* (2000) Hind limb malformations in free-living northern leopard frogs (*Rana pipiens*) from Maine, Minnesota, and Vermont suggest multiple etiologies. *Teratology* 62: 151–171.

Milnes MR, Bermudez DS, Bryan TA, Gunderson MP, and Guillette LJ, Jr. (2005) Altered neonatal development and endocrine function in *Alligator mississippiensis* associated with a contaminated environment. *Biology of Reproduction* 73: 1004–1010.

Ohlendorf HM, Hoffman DJ, Saiki MK, and Aldrich TW (1986) Embryonic mortality and abnormalities of aquatic birds: Apparent impacts of selenium from irrigation drainwater. *Science of the Total Environment* 52: 49–63.

Richards SM and Kendall RJ (2003) Physical effects of chlorpyrifos on two stages of *Xenopus laevis*. *Journal of Toxicology and Environmental Health, Part A* 66: 75–91.

Sessions SK and Ruth SB (1990) Explanation for naturally occurring supernumerary limbs in amphibians. *Journal of Experimental Zoology* 254: 38–47.

Shang EH and Wu RS (2004) Aquatic hypoxia is a teratogen and affects fish embryonic development. *Environmental Science and Technology* 38: 4763–4767.

Spangenberg JV and Cherr GN (1996) Developmental effects of barium exposure in a marine bivalve (*Mytilus californianus*). *Environmental Toxicology and Chemistry* 15: 1769–1774.

Swansburg EO, Fairchild WL, Fryer BJ, and Ciborowski JJ (2002) Mouthpart deformities and community composition of Chironomidae (Diptera) larvae downstream of metal mines in New Brunswick, Canada. *Environmental Toxicology and Chemistry* 21(12): 2675–2684.

PART C

Chemicals with Ecotoxicological Effects

Antagonistic and Synergistic Effects Antifouling Chemicals in Mixture

S Nagata, X Zhou, and H Okamura, Kobe University, Kobe, Japan

Introduction
Materials and Methods
Results and Discussion

Conclusion
Further Reading

Introduction

Since the aquatic environments such as coastal regions of marine area, lakes, rivers, etc., are burdened with thousands of pollutants, it is important to pay attention to their toxicities against the organisms in those ecosystems. Among these pollutants, we cannot turn a blind eye to the role of antifouling chemicals, which have been developed and used widely in the world. Fundamentally, most of them are highly toxic to organisms from both chronic and acute aspects, and thereby they function to control the adhesion and growth of various types of organisms on submerged structures of ships' hulls. Through human activities such as agriculture, aquaculture, or sailing of vessels around the world, these chemicals are discharged into the natural environment.

Under the direction of International Maritime Organization since 2003, organotin-based antifoulants, which have been most widely used mainly as antifouling paints for many years, have been strictly regulated and their use was prohibited because of their severe negative effects on organisms. The ecotoxicological behaviors of new antifoulants in place of organotin compounds, however, have been insignificantly understood. Until recently, the reports on the toxicity of chemicals have been limited to those of each single chemical, but actually we frequently encounter the mixed state of chemicals rather than the single form – for instance, paint products are manufactured by mixing different kinds of antifoulants. Thus, it seems to be indispensable to estimate and evaluate the toxicities in the mixed state of these chemicals as fast and as quantitatively as possible.

Numerous types of organisms have been used in the different bioassay systems for the detection of toxicities. Among these assays, the usage of bioluminescent bacterium was proposed as a relatively simple method based on the reduction of bioluminescence intensity (BLI) by toxic compounds. Symbiotic bacterium *Vibrio fischeri*, which lives in the light organs of fish of the family Monocentridae as well as in the cephalopods *Sepiola* and *Euprymna*, is highly sensitive to environmental changes. In concrete terms, BLI changes of cells are strictly related to the cellular activities which ensure to reflect the certain level of contamination of the surrounding environment. An assay system using the fresh cells of *V. fischeri* was proposed in the previous studies of the authors, which showed fairly good response to the toxicity level of the samples tested.

The difference of toxicities between single chemicals and their mixture has been attributed to the interaction effects, which leads to the classification into five typical patterns as follows: antagonism, no addition, partial addition, concentration addition, and supra-addition. In this regard, the highest toxicity and sum of all toxicities (ST) are the important critical indices to classify the interaction effects. In general, toxicity of each chemical has been expressed in terms of toxicity unit (TU), which was defined as the ratio of the concentration of each chemical to the respective EC_{50} value. Actual toxicity unit (ATU) for a mixture was the percentage after division by 50% for the data such as cell viability or growth inhibition. To clarify the interaction effects, total theoretical toxicity unit (TTTU) corresponding to ST was calculated from each chemical in the mixture and compared with ATU. If ATU is less than TTTU, the interaction effect was defined as antagonism. Using the same way for analyses, each combination is reasonably categorized as follows: if ATU is equal to or higher than TTTU, they are designated to the concentration addition or synergism, respectively. Based on the methodology mentioned above, Könemann proposed mixture toxicity indices (MTIs) as a quantitative indicator for the toxicity in mixed chemicals. Calculations of MTI values enable us to estimate the extent of toxicity enhancement or reduction by mixing the single chemicals together. In the case of samples showing the synergistic effect (MTI > 1), the relatively higher the values of MTI are, the more serious is the enhancement of toxicity. In the antagonistic effect, on the contrary, the relatively lower the values of MTI are, the more serious is the reduction of toxicity.

In this section, the authors demonstrate the evaluation data obtained by their assay system for 11 different kinds of single antifouling chemicals as well as 45 combinations composed of two of them. In addition, the interactions of these chemicals were examined based on the EC_{50} values as well as the percentage of inhibition efficiency (INH, %) for single and mixed chemicals, the latter of which were also analyzed through MTI calculations to classify the effects by mixing the chemicals.

Materials and Methods

Culture Conditions

Luminescent bacterium *V. fischeri* DSM 7151 was grown in a luminescence (LM) medium, in which 0.5% yeast extract (Difco Laboratories, Detroit, MI, USA), 0.5% tryptone (Difco), 0.1% $CaCO_3$, and 0.3% glycerol were present in artificial seawater (ASW, JIS K-2510). pH of the medium was adjusted to 7.0 with NaOH. Cell growth was initiated by adding 1% (v/v) of preculture and incubated with a rotary shaker (120 rpm) at 30 °C for 18 h.

Antifouling Chemicals Tested

As the antifoulants tested, the authors used $CuSO_4$ and ten different types of antifouling chemicals shown in **Figure 1**, the purities of which were >95 %. Among them, 2-methylthio-4-*tert*-butylamino-6-cyclopropylamino-*s*-triazine (Irgarol 1051) and 3-(3,4-dichlorophenyl)-1,1-dimethylurea (Diuron) possess the effect of algaecide and inhibit photosystem II by interfering with the electron-transport chain of photosynthesis in chloroplasts. *N*-Dichlorofluoromethylthio-*N'*,*N'*-dimethyl-*N*-phenylsulfamide (DCF) has been used as a fungicide in agriculture for its actions against a wide range of organisms. High activity was observed for SEA-NINE 211 against a wide spectrum of bacteria, diatoms, fungi, and algae. 3-Iodo-2-propynyl butylcarbamate (IPBC) is not only an inhibitor of acetylcholinesterase in animals but also a highly effective fungicide and bacteriocide. Zinc 2-pyridinethiol 1-oxide (Zn-pt) and copper 2-pyridinethiol 1-oxide (Cu-pt) are

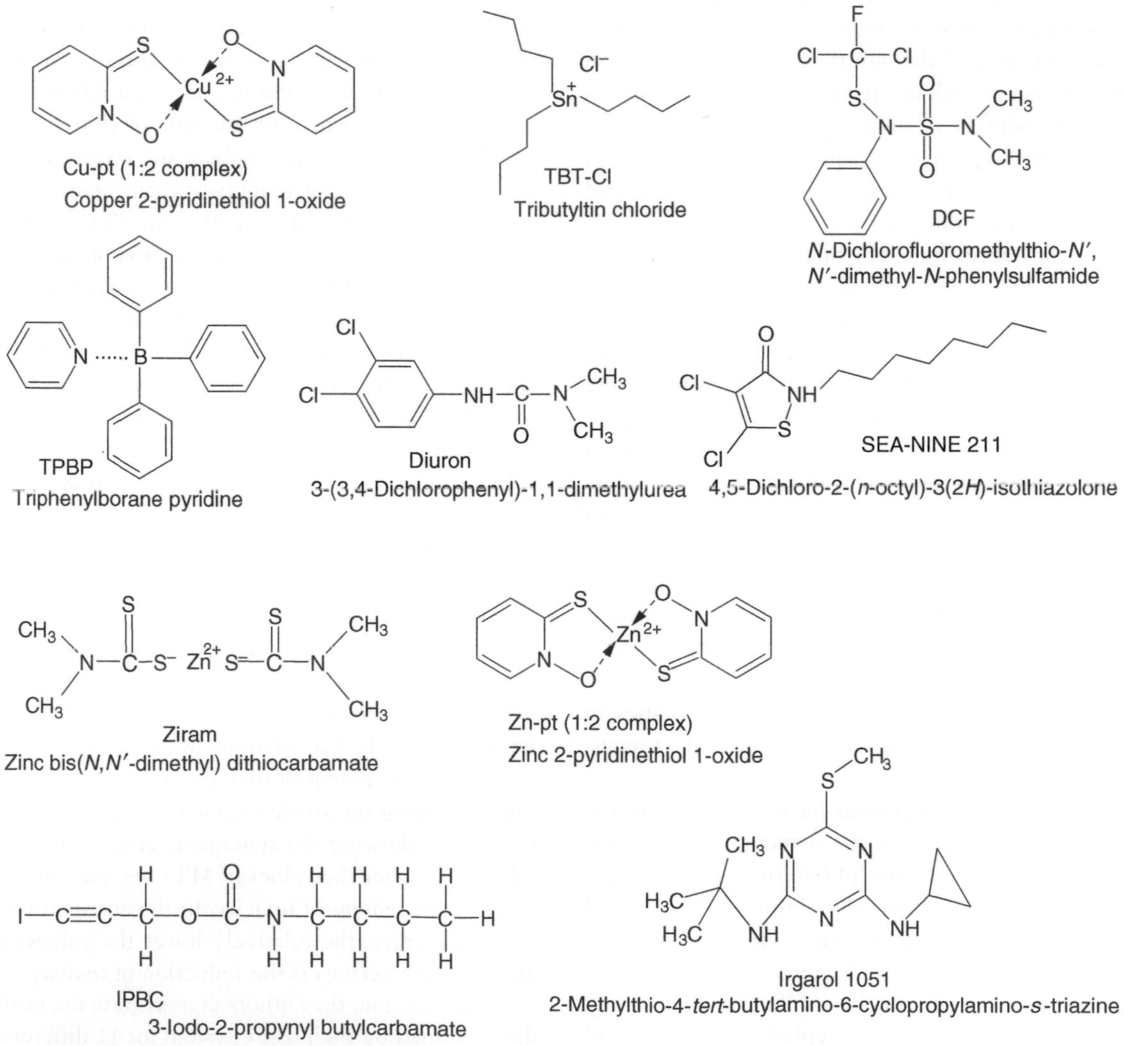

Figure 1 Molecular structures of antifouling chemicals examined.

well known as effective biocidal agents and have been widely used in personal-care products such as antidandruff shampoos and a particularly desirable biocide against soft fouling, respectively. Zinc bis(N,N'-dimethyl)dithiocarbamate (Ziram) and triphenylborane pyridine (TPBP) have been frequently used as fungicide and antifouling biocide, respectively. Each chemical dissolved in dimethyl sulfoxide (DMSO) was added into ASW at a concentration of 1% (v/v). To determine the EC_{50} values of each single chemical as well as its combinations precisely, concentrations of each chemical were increased by 1.5-fold between adjacent concentrations. The concentration ratios of two kinds of chemicals in a mixture were manipulated by using each EC_{50} value in a single form. The controls for all samples were ASW containing 1% (v/v) DMSO.

Measurements of BLI

As described in the previous papers of the authors, the cells of *V. fischeri* DSM 7151 were grown in the LM medium, centrifuged at $10\,000 \times g$ for 1 min at $4\,^{\circ}C$, and then suspended in ASW. Aliquot 0.2 ml cell suspension of *V. fischeri* was added into each well of a 96-well microplate. Then, BLI changes in each well containing 0.2 ml of sample or control were followed using a multidetection microplate reader (Powerscan HT, Dainippon Pharmaceutical, Osaka, Japan) during 30 min of incubation.

Calculations on Percentage of INH and EC_{50}

Values of INH (%) were used to determine the toxicity of chemicals against *V. fischeri* DSM 7151. They were calculated from BLI changes of both samples and controls, as follows:

$$INH(\%) = \left[1 - \left(\frac{IT_t \times IC_0}{IT_0 \times IC_t} \right) \right] \times 100 \qquad [1]$$

where IC_0 and IT_0 are the initial BLI values of controls and samples, respectively, and IC_t and IT_t are the BLI values of controls and samples at exposure time (t) to cell suspension, respectively. The INH(%) values presented here are the averages of at least three independent measurements. The EC_{50} values are the concentrations corresponding to INH(%) = 50 in the relationship between INH(%) and concentration of chemicals.

Calculation of MTIs

MTI values were calculated by the following equation of Könemann:

$$MTI = 1 - \left(\frac{\log M}{\log M_0} \right) \qquad [2]$$

where $M_0 = M/f_{Max}$ and $M = \sum f(i) = \sum C(i)/LC_{50}(i)$. f_{Max} is defined as the largest $f(i)$ value in the mixture. $C(i)$ and $LC_{50}(i)$ are the concentrations of ith component in the mixture and its LC_{50} value, respectively. EC_{50} values are used instead of LC_{50} in this article. When MTI values were lower than zero, the mixture potency was defined as antagonism. If they are equal to zero, there was no additive effect. The mixture potency was defined as partially additive when the values were higher than zero but less than 1. When they are equivalent to or greater than 1, the mixture potency was defined as additive or synergistic, respectively.

Results and Discussion

Toxicity Evaluation of Single Chemicals

As a typical example, BLI changes of *V. fischeri* in the presence of $CuSO_4$ or Ziram were shown in **Figures 2a** or **2b**, respectively. Their reductive effects on the BLI values of *V. fischeri* were enhanced with an increase of incubation time. Toxic effects by different concentrations of $CuSO_4$ and Ziram were recognized after 20–30 min of incubations in the presence of 0.20 and $0.27\,mg\,l^{-1}$, respectively, from which the authors selected 30 min of incubation to judge the BLI changes of *V. fischeri* in the later sections due to the clearer difference of toxicity than that of 20 min or less. From the different reduction rates of BLI between samples and controls, they calculated the EC_{50} values for each single chemical (**Table 1**). Based on the EC_{50} values of respective antifouling chemicals, the authors classified them into three groups. The first group is highly toxic with low EC_{50} values ($<0.75\,mg\,l^{-1}$), and is as follows: tributyltin chloride (TBT-Cl), Zn-pt, Cu-pt, $CuSO_4$, Ziram, SEA-NINE 211, and TPBP. The second one shows medium toxicity, with EC_{50} values ranging from 8.5 to $12.8\,mg\,l^{-1}$ for IPBC and Diuron. The third group consists of DCF and Irgarol 1051 with low toxicity ($>30\,mg\,l^{-1}$ of EC_{50} values). Among highly toxic chemicals, it is interesting to note that most of them contain the metal elements in their respective molecular structures. On the other hand, toxicity ranking of these chemicals detected in the present assay is consistent with the previous data reported through the analyses of various kinds of bioassay systems, which certifies that the present assay system possesses high reliability and sufficient sensitivity to detect the toxicity of organic chemicals.

(a)

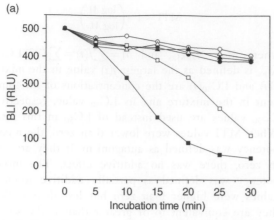

(b)

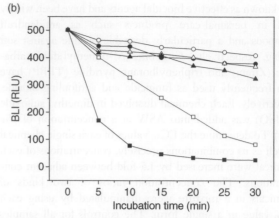

Figure 2   Typical inhibition patterns of the BLI of *V. fischeri* in the presence of CuSO₄ or Ziram. BLI changes in *V. fischeri* were followed in the presence of (a) CuSO₄ and (b) Ziram as a function of the incubation time. Different concentrations of CuSO₄ were added by 0.06 (closed circles), 0.09 (open triangles), 0.13 (closed triangles), 0.20 (open squares), and 0.30 (closed squares) mg l^{-1}. For Ziram, the following concentrations were added: 0.08 (closed circles), 0.12 (open triangles), 0.18 (closed triangles), 0.27 (open squares), and 0.42 (closed squares) mg l^{-1}. Control samples containing ASW with 1% (v/v) DMSO are shown as open circles. All the data were obtained from triplicate experiments, averages of which were described. RLU, Relative Luminescence Unit.

Table 1   EC$_{50}$ values of single chemicals examined (mg l^{-1})[a]

Chemical	EC$_{50}$ value
TBT-Cl	0.02 ± 0.00
Zn-pt	0.08 ± 0.01
Cu-pt	0.12 ± 0.01
CuSO₄	0.22 ± 0.02
Ziram	0.31 ± 0.02
SEA-NINE 211	0.35 ± 0.02
TPBP	0.75 ± 0.05
IPBC	8.49 ± 0.79
Diuron	12.74 ± 1.21
DCF	39.02 ± 2.18
Irgarol 1051	>40.00[b]

[a]The EC$_{50}$ values were calculated at 30 min of incubation from triplicate experiments.
[b]The EC$_{50}$ value of Irgarol 1051 was undetectable due to its solubility limitation (about 40 mg l^{-1}).

Toxicity Evaluations of Mixed Chemicals

Detection of EC$_{50}$ values of mixed chemicals

To clarify whether the toxicities of mixed chemicals were enhanced or reduced in comparison with the cases of single chemicals, each combination of two chemicals was evaluated in the present system. As shown in **Figure 3a**, BLI changes of *V. fischeri* were followed for the mixture of SEA-NINE 211 and TPBP, where they were reduced when the concentrations of two kinds of chemicals as well as incubation times increased. The inhibitory effects became clear at 20–30 min of incubations when the mixture concentrations of SEA-NINE 211 and TPBP were higher than 0.21 and 0.56 mg l^{-1}, respectively. The values of INH (%) were enhanced by an increase in the

concentration of both SEA-NINE 211 and TPBP in combination, as shown in **Figure 3b**. From the crossing point of 50% INH value, the authors determined the EC$_{50}$ value for the mixture of SEA-NINE 211 and TPBP to be 0.19 and 0.47 mg l^{-1}, respectively. EC$_{50}$ values for other mixtures (*A*, *B*) were also calculated in the same way described above, which are summarized in **Figure 4**.

The toxicities of samples can be shown based on the EC$_{50}$ values, in which lower values reflect the higher toxicity, since 50% BLI values in the cell suspension of *V. fischeri* are inhibited by relatively low concentration of test samples. The EC$_{50}$ values in **Table 1** and **Figure 4** show the toxicities of single and their mixed chemicals, respectively. To determine the interaction effects in each combination of two chemicals, their toxicities should be compared with those of respective single chemicals.

Comparison of EC$_{50}$ values of single and mixed chemicals

BLI changes in cell suspension of *V. fischeri* were followed as a function of the incubation time and different concentrations consisting of CuSO₄ and Ziram. As shown in **Figure 5**, the mixture consisting of 0.05 mg l^{-1} of CuSO₄ and 0.06 mg l^{-1} of Ziram resulted in a significant reduction of BLI value in comparison with the control. In the presence of the same level of concentrations in CuSO₄ or Ziram of single form, however, BLI values were almost the same as control, as shown in **Figures 2a** and **2b**. The EC$_{50}$ values of this mixture were calculated from the BLI changes in the presence of various concentrations of CuSO₄ and Ziram (**Figure 5**). The EC$_{50}$ values in the single form of CuSO₄ and Ziram were sufficiently high,

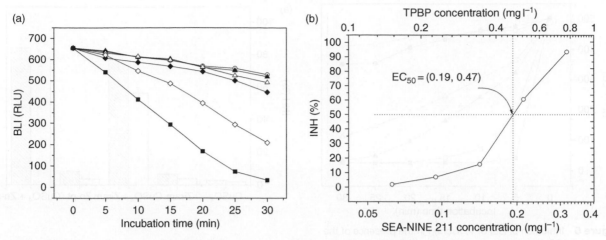

Figure 3 Inhibition of BLI of *V. fischeri* in the presence of mixed chemicals, SEA-NINE 211 and TPBP. (a) BLI changes in *V. fischeri* were followed in the presence of the mixture composed of SEA-NINE 211 and TPBP as a function of the incubation time. Different concentrations of SEA-NINE 211 and TPBP were added by 0.06 and 0.16 (closed triangles), 0.09 and 0.25 (open triangles), 0.14 and 0.38 (closed diamonds), 0.21 and 0.56 (open diamonds), and 0.32 and 0.84 mg l^{-1} (closed squares), respectively. Control samples containing ASW with 1% (v/v) DMSO are shown by open circles. (b) The INH (%) values of BLI for this mixture were calculated at 30 min of incubation. The EC$_{50}$ values of mixed chemicals were obtained as the concentration corresponding to 50% of INH value. All the data were obtained from triplicate experiments, averages of which were described.

Irgarol 1051									
Diuron	UD^a								
Ziram	0.268, 19.05^b	0.281, 6.434							
Cu-pt	0.097, 14.86	0.123, 6.027	0.071, 0.153						
Zn-pt	0.063, 17.16	0.071, 4.094	0.042, 0.106	0.058, 0.068					
TPBP	UD^a	0.714, 5.976	0.589, 0.216	0.553, 0.094	0.443, 0.064				
SEA-NINE 211	0.247, 19.29	0.216, 5.442	0.190, 0.209	0.180, 0.092	0.137, 0.060	0.185, 0.465			
DCF	22.80, 13.44	32.18, 6.103	25.75, 0.213	21.86, 0.084	18.50, 0.061	21.05, 0.477	26.42, 0.199		
CuSO$_4$	0.099, 16.08	0.122, 4.221	0.048, 0.072	0.045, 0.032	0.047, 0.028	0.110, 0.452	0.093, 0.127	0.118, 21.59	
IPBC	UD^a	4.393, 6.434	3.472, 0.222	3.160, 0.094	2.674, 0.068	2.899, 0.507	3.004, 0.175	4.393, 33.92	4.393, 0.186
	Irgarol 1051	Diuron	Ziram	Cu-pt	Zn-pt	TPBP	SEA-NINE 211	DCF	CuSO$_4$	IPBC

Figure 4 EC$_{50}$ values for the mixed chemicals examined (mg l^{-1}). The EC$_{50}$ values are calculated at 30 min incubations from triplicate experiments. ^{a}UD: undetectable due to the low toxicity of Irgarol 1051. ^{b}The values in (A, B) of each cell indicate that 50% inhibition for BLI of *V. fischeri* was afforded by the mixture composed from the chemicals in row and column titles, the concentrations of which are A and B, respectively.

0.22 and 0.31 mg l^{-1}, respectively, but they became markedly low when they were mixed, as shown in **Figure 4**. That is to say, the mixture became highly toxic, although the respective single chemical showed low level of toxicity. Thus, the concentration to reveal the same level of toxicity in the mixture as cases of respective single chemicals resulted to less than one-fifth. It might be of value

to point out that the EC$_{50}$ values for the combinations of CuSO$_4$ together with Cu-pt, Zn-pt, or Irgarol 1051 became lower in comparison with those of single chemicals, as shown in **Table 1** and **Figure 4**. The same concentrations of their combinations containing CuSO$_4$ showed much higher inhibitory activities than respective single chemicals, probably owing to the synergistic effect

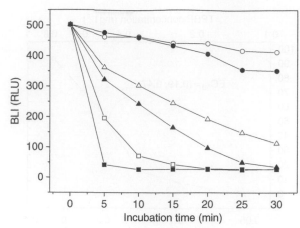

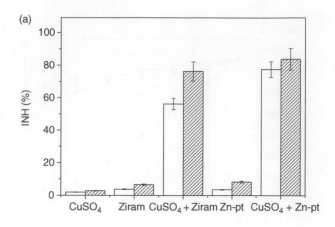

Figure 5 Inhibition of BLI of *V. fischeri* in the presence of the mixture, CuSO₄ and Ziram. BLI changes in *V. fischeri* were followed in the presence of the mixture composed from CuSO₄ and Ziram as a function of the incubation time. Different concentrations of CuSO₄ and Ziram were added by 0.03 and 0.04 (closed circles), 0.05 and 0.06 (open triangles), 0.07 and 0.09 (closed triangles), 0.10 and 0.14 (open squares), and 0.15 and 0.21 mg l⁻¹ (closed squares), respectively. Control samples containing ASW with 1 % (v/v) DMSO are shown as open circles. All the data were obtained from triplicate experiments, averages of which were described.

in combinations with CuSO₄. All of the EC₅₀ values in CuSO₄-containing combinations mentioned above were reduced to one-fourth or one-fifth of those of each single chemical. The EC₅₀ values for the other combinations were half as those of their single chemicals.

Typical patterns of interaction of mixed chemicals
Further evidence of toxicity enhancement by mixing the chemicals was recognized for the combinations such as Ziram + CuSO₄, Zn-pt + CuSO₄, SEA-NINE 211 + TPBP, and Cu-pt + Diuron. As shown in **Figure 6a**, INH(%) values of both Ziram and CuSO₄ were quite low when they were present as single forms. By mixing them, however, they increased to as high as 57% and 76% at the incubation times of 15 and 30 min, respectively. Similar toxicity enhancement with mixing was observed for the combination of Zn-pt and CuSO₄.

The relationship between the toxicities of single chemicals and their combinations shown in **Figures 6b** and **5c**, however, is different from that of **Figure 6a**. Inhibitory reduction of BLI due to the action of SEA-NINE 211 (0.19 mg l⁻¹) or TPBP (0.47 mg l⁻¹) presented as single form showed approximately 25% value at 30 min of incubation, as shown in **Figure 6b**, while it increased to 44% by mixing them. In the case of Diuron and Cu-pt, however, the toxicity did not change remarkably by mixing them, that is, the toxicity level of their mixture was almost same as that of Cu-pt alone (**Figure 6c**). In comparison with two examples of combination shown in **Figures 6b** and **6c**, the mixtures of

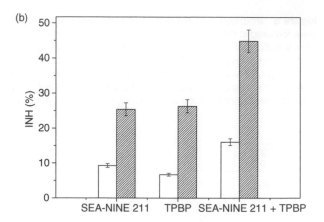

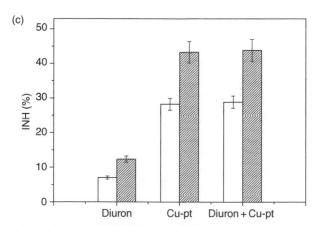

Figure 6 Typical patterns of interactions in some combinations of antifouling chemicals. The INH (%) values were calculated based on the reductions of BLI for single chemicals as well as their combinations at 15 and 30 min of incubations, which are shown by open and shaded bars, respectively. As typical patterns, the combinations of CuSO₄ with Zn-pt or with Ziram, SEA-NINE 211 with TPBP, and Diuron with Cu-pt are shown in (a), (b), and (c), respectively. For both single and mixed forms, the concentrations of CuSO₄, Zn-pt, Ziram, SEA-NINE 211, TPBP, Diuron, and Cu-pt were fixed to be 0.06, 0.04, 0.08, 0.19, 0.45, 5.00, and 0.10 mg l⁻¹, respectively. All the data were obtained from triplicate experiments, averages of which were described with standard deviations.

CuSO$_4$ with either Ziram or Zn-pt showed a significant increase of toxicity in combination (**Figure 6a**), from which it is concluded that these combinations are classified as synergistic.

In the same way, toxicity analyses for CuSO$_4$ in combination with Cu-pt or Irgarol 1051 were performed, and then marked synergistic effects were also recognized although data was not shown. The interactions of the combinations shown in **Figures 6b** and **6c** are obviously different from the synergism, the details of which are discussed in the next section.

MTI Analyses of Interactions

To classify the types of all interactions in the combinations examined, MTI values were calculated from the EC$_{50}$ values of single chemicals and their combinations at 30 min of incubation. As shown in **Figure 7**, MTI values of the mixed chemicals examined were positive except for the combination of Cu-pt and Diuron, −0.006. As realized from the data shown in **Figure 6c**, the toxicity of Cu-pt and Diuron in combination was just the same as that of single Cu-pt, suggesting that the addition of Diuron did not bring about significant contribution. This MTI value, however, was almost zero, and thus we can conclude that the interaction of this combination has little additive effect. Other combinations were positive and thus regarded as having no antagonistic effects.

MTI values of the mixtures consisting of CuSO$_4$ with Irgarol 1051, Zn-pt, Ziram, and Cu-pt resulted in >1.73,

2.20, 2.21, and 2.33, respectively (**Figure 7**). Their MTI values were in good agreement with their synergistic interactions, as indicated by the toxicity comparisons of single and combined chemicals described above. MTI values of 1.2–1.3 were obtained for the following mixtures: DCF with Irgarol 1051, CuSO$_4$ with Diuron, Zn-pt with Ziram, and CuSO$_4$ with SEA-NINE 211. Since MTI values of these combinations were larger than 1, they were regarded as synergistic. The levels of toxicity enhancement, however, were lower than those in the cases of CuSO$_4$-containing combinations as mentioned above. On the other hand, the interactions of IPBC-containing mixtures with Diuron, TPBP, or SEA-NINE 211 were classified as concentration additive, since their MTI values were nearly 1. MTI values of other mixtures were 0–1, partially additive, which form the majority of the data shown in **Figure 7**. A typical example of partially additive is shown in **Figure 6b**, where the toxicity of combinations became larger than that of either SEA-NINE 211 or TPBP alone, but it was smaller than the sum of their theoretical toxicities of single chemicals.

Among the four chemicals showing marked synergistic effects in the mixture with CuSO$_4$, they commonly contain metal, sulfur, and nitrogen atoms in their molecular structures, except for Irgarol 1051 (see **Figure 1**). In addition, the metals in three of the chemicals are loosely bounded with sulfide linkage through electron donation. These molecular characteristics should be taken into consideration to clarify the mechanism of these interactions in future investigations.

Irgarol 1051	Diuron	Ziram	Cu-pt	Zn-pt	TPBP	SEA-NINE 211	DCF	CuSO$_4$	IPBC	
	UD^a	≥0.333^b	≥0.557^b	≥0.785^b	UD^a	≥0.291^b	≥1.183^b	≥1.729^b	UD^a	Irgarol 1051
		0.223	−0.006	0.399	0.125	0.591	0.424	1.267	1.000	Diuron
			0.858	1.295	0.381	0.562	0.556	2.210	0.758	Ziram
				0.568	0.366	0.435	0.602	2.326	0.370	Cu-pt
					0.399	0.577	0.570	2.198	0.544	Zn-pt
						0.751	0.737	0.839	0.985	TPBP
							0.510	1.201	1.078	SEA-NINE 211
								0.871	0.308	DCF
									0.360	CuSO$_4$
										IPBC

Figure 7   MTI values for the mixed chemicals examined. The MTI values are calculated from EC$_{50}$ values of single chemicals and their mixture at 30 min incubations. ^{a}UD: undetectable due to the low toxicity of the mixture containing Irgarol 1051. ^{b}Since the EC$_{50}$ value of Irgarol 1051 was ≥40 mg l^{-1}, the MTI values were larger than those shown here.

According to the MTI analyses, there was no antagonistic effect among the present combinations examined. Additive effects were observed in the mixtures containing IPBC with Diuron, TPBP, or SEA-NINE 211. Certain synergistic effects were found in the combinations such as DCF with Irgarol 1051, $CuSO_4$ with Diuron, Zn-pt with Ziram, and $CuSO_4$ with SEA-NINE 211. Significant stimulation of toxicity occurred when Irgarol 1051, Ziram, Zn-pt, or Cu-pt were combined with $CuSO_4$, which is classified as synergistic effects. The other combinations were regarded as partially additive.

Toxicity Enhancement with Cu²⁺

It is of value to note that the mixture of $CuSO_4$ with other chemicals brought about the marked synergistic effects. In this regard, it seems to be of necessity to examine what kind of roles copper ion plays in the mixtures. Copper ions have generally been used in the form of CuO or Cu_2O in the antifouling industry owing to their poor solubility in water, which leads to a constant release of copper ions from the surface of a ship hull coated with paints. Since their poor solubility makes it difficult to prepare the samples with appropriate concentrations in the present study, $CuSO_4$ was used in place of CuO or Cu_2O. Single chemical toxicities of Irgarol 1051, Ziram,

Cu-pt, or Zn-pt were evaluated in the presence of high concentration of SO_4^{2-}, since ASW used for the sample preparation contains a high concentration of SO_4^{2-} ($2700\,mg\,l^{-1}$), but no Cu^{2+}. Their toxicities, however, were much lower in comparison with those detected in the presence of Cu^{2+}. Therefore, the increased toxicities of these chemicals in combination with $CuSO_4$ would be conceivably attributed to the presence of Cu^{2+} rather than SO_4^{2-}.

To confirm the assumption mentioned above, $CuCl_2$ was also used in place of $CuSO_4$ and mixed with Zn-pt, Irgarol 1051, Ziram, and Cu-pt. As shown in **Figure 8**, INH (%) values in the presence of $0.05\,mg\,l^{-1}$ of $CuCl_2$ were found to be less than 5% at 30 min of incubation. The inhibitory activities of Zn-pt, Cu-pt, Ziram, and Irgarol 1051 were less than 10% at the concentrations of 0.04, 0.05, 0.08, and $9.0\,mg\,l^{-1}$, respectively. When each of them was combined with $0.05\,mg\,l^{-1}$ of $CuCl_2$, all of the INH (%) values became higher than 60%. The degrees of toxicity enhancement were in good agreement with the values of both MTI and INH (%) calculated for their combinations with $CuSO_4$, as shown in the table inserted in **Figure 8**. These results strongly suggest that the marked synergistic effects of these chemicals in combination with $CuSO_4$ or $CuCl_2$ are ascribed to the presence of Cu^{2+}.

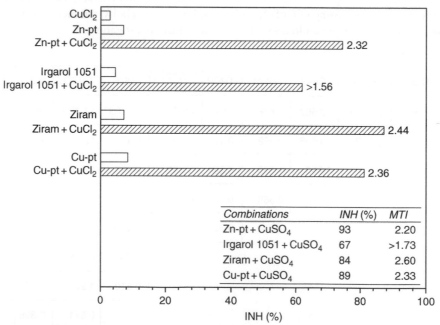

Combinations	INH (%)	MTI
Zn-pt + $CuSO_4$	93	2.20
Irgarol 1051 + $CuSO_4$	67	>1.73
Ziram + $CuSO_4$	84	2.60
Cu-pt + $CuSO_4$	89	2.33

Figure 8 Effects of Cu^{2+} on the toxicity enhancement. BLI reductions were obtained for the single chemicals (open symbols) and their combinations (shaded symbols) in the presence of 0.3 M NaCl, from which the INH (%) values were calculated at 30 min of incubation. For both single chemicals and their mixtures, the concentrations of $CuCl_2$, Zn-pt, Irgarol 1051, Ziram, and Cu-pt were fixed to be 0.05, 0.04, 9.0, 0.08, and $0.05\,mg\,l^{-1}$, respectively. The numbers shown at the right-hand side of bars are MTI values for each chemical mixed with $CuCl_2$ at 30 min of incubation. The numbers in the table inserted are the values of INH (%) and MTI which were calculated based on the BLI changes for the mixtures of $0.04\,mg\,l^{-1}$ of $CuSO_4$ with the same concentrations of Zn-pt, Irgarol 1051, Ziram, and Cu-pt as the case of $CuCl_2$ at 30 min of incubation. All the data were obtained from triplicate experiments, averages of which were described.

Copper ions are widely distributed in the natural environments and serve as metal element prerequisite for the growth of most of plants and animals. In marine environments, however, the level of concentrations of Cu^{2+} have been detected as a complex with various kinds of chemicals. Cu^{2+} in most of organisms is little accumulated due to its nonlipophilicity, but suppression of mitosis through glutathione reduction and breakage of the cellular defense against oxygen-free radicals might be brought about, probably due to the passive diffusion of Cu^{2+} into the cells. The most bioavailable and toxic form of unbounded Cu^{2+} is thought to be the free hydrated ion form, $Cu(H_2O)_6^{2+}$. Chemical form of ionic copper is governed by external pH, salinity, and concentration levels of dissolved organic matter. Thus, the toxicity of Cu^{2+} to bacteria also depends on not only the individual species but also the physiological and environmental conditions. The mechanism of toxicity enhancement might be ascribed to the formation of lipophilic organic copper complexes between Cu^{2+} and some antifouling chemicals, which are able to diffuse across the plasma membrane of the cells more easily in comparison with the inorganic state of Cu^{2+}. Organic copper complexes in the cell cytosol conceivably dissociate and exchange the transport ligands with intracellular complexes. Similar mechanism on the toxicity enhancement against microalgae was observed for the combination of dithiocarbamate and some heavy metals. On the basis of the present results showing that synergistic effects occurred in combinations with Cu^{2+}, the authors suggest that these antifouling chemicals promote the uptake of a variety of toxic heavy metals into the cell cytoplasm through the formation of complexes in the aquatic environments.

Until recently, much attention has been paid to the toxicity of single chemicals rather than mixed systems. Interactions of mixed chemicals have rarely been studied, probably due to the large amounts of analytical work to determine the actual status of chemicals present in natural environments. There are some data of interaction effects consisting of several antifouling chemicals combined with methyl-*tert*-butyl ether (MTBE), a fuel oxygenate that is added to gasoline. Thus, MTBE frequently contaminates the aquatic systems and possibly forms some complexes with a variety of chemicals. Hernando *et al.* examined the interactions between MTBE and some of the antifouling chemicals using two kinds of microorganisms, *V. fischeri* and *Daphnia magna*. As shown in **Table 2**, the mixtures examined resulted in toxicity enhancement, except for SEA-NINE 211 and MTBE. Therefore, it is necessary to pay attention to the fate of some general ionic and organic materials, since they possibly enhance their toxicity by forming complexes with each other.

Table 2 MTI values of the mixtures consisting of methyl-*tert*-butyl ether (MTBE) and some antifouling chemicals for *V. fischeri* and *D. magna*

Mixture	V. fischeri	D. magna
Diuron–MTBE	6.7	6.3
DCF–MTBE	−0.7	3.4
Irgarol 1051–MTBE	6.1	6.0
SEA-NINE 211–MTBE	−5.0	−4.6

Conclusion

The present assay system using fresh cells of *V. fischeri* proves to be enough sensitive and reliable to evaluate the toxicities of antifouling chemicals, irrespective of whether they are present as single or mixed form. The results obtained by the present system show that the toxicities of these antifoulants markedly increase on combination when compared with their toxicity as single agents. On the whole, most of the EC_{50} values obtained for single chemicals were reduced to one-half to one-fifth on mixing with other chemicals, suggesting that toxicity enhancement has occurred. Remarkable synergetic effect was observed in the mixtures containing $CuSO_4$ from the analyses of both EC_{50} and MTI values. The mechanism of toxicity enhancement by Cu^{2+} in the mixture might be attributed to the formation of lipophilic complex with organic molecules, which easily enter into the cell cytoplasm and function negatively for cellular activities.

According to the present data, the antifouling chemicals examined have a more serious impact on both environments and organisms than previously recognized. In particular, we have to keep in mind the marked stimulation of toxicity when organic chemicals or ions are mixed, as seen in natural environment. Thus, related studies of toxicity enhancement by various combinations of chemicals as well as their detailed mechanisms of actions are needed for a complete understanding of the environmental effects of these chemicals.

See also: Acute and Chronic Toxicity; Copper.

Further Reading

Bruland KW (1980) Oceanographic distributions of cadmium, zinc, nickel and copper in North Pacific. *Earth Planetary Science Letters* 47: 176–198.

Carlos AG, Beatriz DA, and Vicente JDR (1988) The use of calcium resinate in the formulation of soluble matrix antifouling paints based on cuprous oxide. *Progress in Organic Coatings* 16: 165–176.

Codina JC, Munoz MA, Cazorla FM, et al. (1998) The inhibition of methanogenic activity from anaerobic domestic sludges as a simple toxicity bioassay. *Water Research* 32: 1338–1342.

Ermolayeva E and Sanders D (1995) Mechanism of pyrithione-induced membrane depolarization in *Neurospora crassa*. *Applied and Environmental Microbiology* 61: 3385–3390.

Fernández-Alba AR, Hernando MD, Piedra L, and Chisti Y (2002) Toxicity evaluation of single and mixed antifouling biocides measured with acute toxicity bioassays. *Analytica Chimica Acta* 456: 303–312.

Gatidou G, Kotrikla A, Thomaidis N, and Lekkas T (2004) Determination of two antifouling booster biocides and their degradation products in marine sediments by high performance liquid chromatography–diode array detection. *Analytica Chimica Acta* 505: 153–159.

Hernando MD, Ejerhoon M, Fernández-Alba AR, and Chisti Y (2002) Combined toxicity effects of MTBE and pesticides measured with *Vibrio fischeri* and *Daphnia magna* bioassays. *Water Research* 37: 4091–4098.

Ince NH, Dirilgen N, Apikyan IG, Tezcanli G, and Üstün B U (1999) Assessment of toxic interactions of heavy metals in binary mixtures: A statistical approach. *Archives of Environmental Contamination and Toxicology* 36: 365–372.

Jonathan TP and Kenneth WB (1997) Trace metal exchange in solution by the fungicides Ziram and Maneb (dithiocarbamates) and subsequent uptake of the lipophilic organic Zn, Cu and Pb complexes into phytoplankton cells. *Environmental Toxicology and Chemistry* 16: 2046–2053.

Könemann H (1981) Fish toxicity tests with mixtures of more than two chemicals: A proposal for a quantitative approach and experimental results. *Toxicology* 19: 229–238.

Koutsaftis A and Aoyama I (2006) The interactive effects of binary mixtures of three antifouling biocides and three heavy metals against the marine algae *Chaetoceros gracilis*. *Environmental Toxicology* 21: 432–439.

Nagata S and Zhou XJ (2006) Analyses of factors to affect the bioassay system using luminescent bacterium *Vibrio fischeri*. *Journal of Health Science* 52: 9–16.

Okamura H and Mieno H (2006) Present status of the antifouling systems in Japan: TBT substitutes in Japan. In: Konstantinou IK (ed.) *The Handbook of Environmental Chemistry, Vol. 5: Antifouling Paint Biocides*, part O, pp. 201–212. New York: Springer.

Ren S and Frymier PD (2005) Toxicity of metals and organic chemicals evaluated with bioluminescence assays. *Chemosphere* 58: 543–550.

Ruby EG (1996) Lessons from a cooperative, bacterial–animal association: The *Vibrio fischeri–Euprymna scolopes* light organ symbiosis. *Annual Review of Microbiology* 50: 591–624.

Stauber JL and Florence TM (1987) Mechanism of toxicity of ionic copper and copper complexes to algae. *Marine Biology* 94: 511–519.

Voulvoulis N, Scrimshaw MD, and Lester JN (2002) Comparative environmental assessment of biocides used in antifouling paints. *Chemosphere* 47: 789–795.

Wegrzyn G and Czyz A (2002) How do marine bacteria produce light, why are they luminescent, and can we employ bacterial bioluminescence in aquatic biotechnology? *Oceanologia* 44: 291–305.

Zhou XJ, Okamura H, and Nagata S (2006) Remarkable synergistic effect in antifouling chemicals against *Vibrio fischeri*. *Journal of Health Science* 52: 243–251.

Zhou XJ, Okamura H, and Nagata S (2006) Applicability of luminescent assay using fresh cells of *Vibrio fischeri* for toxicity evaluation. *Journal of Health Science* 52: 811–816.

Zhou XJ, Okamura H, and Nagata S (2007) Abiotic degradation of triphenylborane pyridine (TPBP) antifouling agent in water. *Chemosphere* 67: 1904–1910.

Antibiotics in Aquatic and Terrestrial Ecosystems

B W Brooks, Baylor University, Waco, TX, USA

J D Maul, Texas Tech University, Lubbock, TX, USA

J B Belden, Baylor University, Waco, TX, USA

Introduction

Pharmaceuticals and Personal Care Products

In recent years, emerging contaminants, including pharmaceuticals and personal care products (PPCPs), have received unprecedented attention from the scientific and regulatory communities and media at the global level. These compounds are often referred to as emerging contaminants because of the limited understanding of environmental occurrence, disposition, or fate, and responses of aquatic and terrestrial ecosystems to realistic exposure scenarios. Such limited information, and the presumption that concentrations to which organisms are routinely exposed are relatively small (e.g., $<100\,\mu g\,kg^{-1}$ in soil, $<1\,\mu g\,l^{-1}$ in water), have generally hindered the development of sediment/soil screening guidelines and ambient water quality criteria in the United States, or similar metrics in other parts of the world.

PPCPs primarily enter surface waters in discharge from municipal wastewater treatment plants (WWTPs) because plant processes do not completely remove them from inflowing waste. However, PPCPs may also enter the environment from terrestrial applications of biosolids or wastewater, biosolid-derived products, leachates from landfills, discharges from on-site septic or aerobic treatment systems, aquaculture practices (e.g., net pens), or transport from intensively reared agricultural areas (e.g., confined animal feeding operations) to aquatic systems. Human and veterinary pharmaceuticals are excreted as a combination of metabolites and parent compounds or are directly discarded into wastewater as unused medications. Although many pharmaceuticals are primarily excreted as metabolites, microbial activity in a WWTP may cleave conjugated metabolites, potentially resulting in reactivation to parent compounds prior to discharge to a receiving ecosystem. Mixtures and metabolites present challenges to understanding exposure and effects, as a number of PPCP classes (e.g., antibiotics) also are expected to co-occur in complex mixtures distributed between solid and aqueous components of the environment.

Unlike non-point aquatic pesticide runoff that generally tends to occur over short periods of time following rain events, potentially resulting in acute toxicity to aquatic organisms, lower-level PPCP exposures at environmentally relevant levels over longer time periods may subtly modulate and alter biochemical, physiological, reproductive, and ecological processes. Subsequently, current approaches for assessing risk to ecosystems that rely on short-term toxicity tests may not be appropriate for PPCPs. The continuous release of PPCPs from WWTPs has resulted in use of the term 'pseudopersistent' to describe the environmental fate of these compounds, since rates of effluent introduction often exceed PPCP half-lives in a receiving ecosystem. Thus, PPCPs present unique challenges to scientists, ecosystem managers, and policymakers, and have stimulated the global scientific community to work toward development of new ecological risk assessment and water quality paradigms for these compounds.

Antibiotics and Antimicrobial Agents

Of the classes of PPCPs in the environment that have received a relatively high level of study are antibiotic therapeutics and antibacterial agents. Similar to other pharmaceuticals, antibiotics are specifically used to target various biomolecules in animals. Generally classified as personal care products, antimicrobial agents including triclosan and triclocarban are heavily utilized in products ranging from soaps to toothpaste. Such molecules that implicitly target microorganisms are of particular concern for 'nontarget' species residing in aquatic and terrestrial ecosystems, as well as the critical ecosystem functions (e.g., nutrient cycling and decomposition) that are mediated by lower-trophic-level organisms. **Table 1** provides a summary of the major antibiotic classes and select representatives that are used as human or veterinary medicines.

Occurrence and Fate of Antibiotics in the Environment

Occurrence

Representatives from a large number of antibiotic classes including the macrolides, quinolones, sulfonamides, and tetracyclines have been reported in WWTP effluents, surface waters, sediments, and soils, indicating potential

exposure via multiple routes for aquatic and terrestrial organisms. In most cases, and particularly within WWTP effluents, these compounds typically occur as mixtures in the environment. While no standardized analytical techniques are available for antibiotics in various matrices, liquid chromatography and gas chromatography tandem mass spectrometry (LC–MS/MS and GC–MS/MS) approaches are most commonly used to achieve low-level detection limits and high qualitative identification. Antibiotics are routinely detected in WWTP effluents and surface waters in the $ng\,l^{-1}$ range, with the highest levels detected in surface waters at $15\,\mu g\,l^{-1}$ for sulfadimethoxine, a sulfonamide.

In addition to contaminating surface water, antibiotics from WWTP effluents may also contaminate related sediment and soil. For instance, in a recent study from the United States Geological Survey (USGS), the average concentrations of trimethoprin and erythromycin in WWTP effluents were relatively low, $111\,ng\,l^{-1}$ and below detection limits, respectively. However, both compounds were also detected in downstream sediments, 1.22 and $5.47\,\mu g\,kg^{-1}$, soils treated with municipal wastewater, 25.7 and $2660\,\mu g\,kg^{-1}$, and in biosolid-derived components, 11.8 and $5\,\mu g\,kg^{-1}$, respectively.

As discussed further below, the relatively low lipophilicities of many antibiotics suggest that uptake into organisms is limited. However, previous studies identified oxytetracycline residues in fish following aquaculture application, and an even more recent study detected trimethoprim and erythromycin in tissues of fish residing in an effluent-dominated stream, corresponding with the USGS report of relatively high levels of these compounds in sediments downstream from a WWTP discharge. Thus, the occurrence of antibiotics and their potential to affect terrestrial and aquatic organisms via various exposure routes requires further investigation.

Environmental Fate

Of the antibiotic classes reported to occur in the environment, the fate of quinolones has been the most studied due to extensive usage in veterinary (e.g., aquaculture, poultry) and human medicine. The quinolones along with most classes of antibiotics are highly water soluble with characteristically low octanol–water partitioning coefficients. Despite high water solubility, several antibiotic classes, notably tetracyclines, fluoroquinolones, and macrolides, are likely to strongly adsorb to soil, sediment, and organic material (including manure). Specifically, compounds that are zwitterionic can adsorb to clay and organic matter through cation exchange and the degree of adsorption is pH dependent. Soil-to-water partitioning coefficients have been reported for tetracycline and oxytetracycline between 400 and 2000 (based on K_d values, $l\,kg^{-1}$), fluoroquinolones have reported K_d values of $250–6000\,l\,kg^{-1}$, and tylosin, a

Table 1   Primary classes of antibiotics used in human and veterinary medicines

Antibiotic class	Example therapeutic
Aminoglycosides	Lincomycin
β-Lactams	Amoxicillin
2,4-Diaminopyrimidines	Trimethoprim
Macrolides	Erythromycin
Pleuromutilins	Tiamulin
Quinolones	Ciprofloxacin
Sulfonamides	Sulfamethoxazole
Tetracyclines	Oxytetracycline

common macrolide used in both human and veterinary medicine, has reported K_d values between 8 and $1301 kg^{-1}$. Such high sorption to soils indicates that leaching through a soil column is unlikely and that contamination via runoff is most likely through movement attached to particles. Thus, these groups are less likely to contaminate surface and groundwater unless direct input from wastewater effluent or agricultural lagoons occurs. Other antibiotics with low K_d values (e.g., sulfamethazine) are more likely to be transported via water.

Once bound to sediment, soil, or manure, many antibiotics appear to be relatively persistent in the environment. An environmental half-life of greater than 30 days has been reported for chlortetracycline in chicken manure and soil; an oxytetracycline half-life has been reported as high as 150 days in marine sediment. Most fluoroquinolones have reported half-lives of greater than 30 days in sediment, while sulfa antibiotics including sulfadiazine, sulfamethoxazole, and sulfadimethoxine have reported half-lives in marine sediment of greater than 40 days, with some studies reporting no degradation after 180 days. In loam and clay-loam soils, a recent study reported that sarafloxacin degraded less than 1% after 65 days and ciprofloxacin has been reported to have minimal degradation after 40 days. Macrolides may be less persistent than other antibiotic classes with reported half-lives of less than 10 days.

Photodegradation is a major degradation pathway for several antibiotic groups including the fluoroquinolones and tetracyclines where half-lives may be only a few hours. Generally, photodegradation is expected to result in loss of antibacterial activity, although studies of comparative metabolite toxicity are only beginning to appear in the peer-reviewed literature. It is important to note that photodegradation is unlikely to occur once the compound is bound to sediment, prolonging the half-life of a number of antibiotics, particularly in highly turbid ecosystems. However, it is unknown whether antibiotics bound to organic matter and sediment remain biologically active. Thus, aqueous exposure of aquatic organisms to these compounds may be limited to effluent-dominated ecosystems receiving continuous loadings of antibiotics, leading to the potential 'pseudopersistent' scenarios described above. Sediment and organic material within effluent-dominated ecosystems may present continuous exposure at higher levels than measured in the overlaying water, highlighting the importance of characterizing exposure contributions via aqueous and dietary routes.

Ecological Effects of Antibiotics

Effects Analysis Considerations

Because antibiotics used as human and veterinary medicines specifically target undesired microorganisms, the potential impacts of these biologically active compounds are of particular concern for beneficial microbial communities and other organisms in aquatic and terrestrial ecosystems. For example, fluoroquinolones exert their toxicity to bacteria through inhibition of DNA gyrase, while β-lactams target transpeptidase and transglycosylase, ultimately affecting cell walls. Microorganisms are critically important to the functioning of aquatic and terrestrial systems. Because the most probable route of antibiotics into the aquatic environment is via wastewater effluents, it is reasonable that effects may be greatest in aquatic systems receiving these inputs (e.g., streams, lakes). Within a stream ecosystem, it is possible that antibiotic impacts could affect processes such as microbial decomposition of organic matter, macroinvertebrate nutrition including either bacteria directly ingested or those acting as symbionts within the invertebrate gut, and/or nitrification and denitrification processes.

The goal of an ecological risk assessment is protection of populations, communities, and ecosystems and related functions. Measures of effect are selected to support an assessment endpoint, which is a clearly defined ecological value that is to be protected such as ecosystem structure (e.g., organisms, populations, communities) or function (e.g., leaf decomposition, nutrient cycling). However, functional responses to stressors are routinely not observed before structural changes occur in an ecosystem. Previous authors have noted that the critical functions of ecosystems regulated by microorganisms may be more subtly affected over longer time periods by antibiotics released to the environment.

Thus, it is critical that measures of effect be selected carefully during risk assessment if the potential hazards of antibiotics in the environment are to be appropriately characterized. For example, a major concern is that antibiotics found in the environment may cause increased resistance in natural bacterial populations.

Single-Species Ecotoxicity

Characterization of antibiotic effects on ecosystems has largely been limited to standardized single-species ecotoxicology studies with survival, growth, or reproduction as the primary endpoints. Among the most sensitive aquatic primary producers to antibiotics are cyanobacteria (blue-green algae), which have been identified as more sensitive than green algae for a number of compounds. For example, in a series of toxicity tests with seven fluoroquinolones, EC_{50} for growth rates ranged from 7.9 to $1960 \mu g\, l^{-1}$ for the cyanobacterium, *Microcystis aeruginosa*, with levofloxacin and flumequine the most and least toxic, respectively. Between two different laboratories, ciprofloxacin EC_{50} for *M. aeruginosa* ranged from 5 to $17 \mu g\, l^{-1}$. For the green algae, *Pseudokirchneriella subcapitata*, growth rate EC_{50}s were at least 10 times greater, ranging from 1.1 to $23.0 mg\, l^{-1}$ with clinafloxacin and lomefloxacin the most and least toxic, respectively. In a separate study, the ciprofloxacin

EC_{50} for *P. subcapitata* was reported to be $3 \, mg \, l^{-1}$. The marine cryptophycean, *Rhodomonas salina*, was less sensitive with growth rate EC_{50} of 10, 18, and $24 \, mg \, l^{-1}$ for oxolinic acid, flumequine, and sarafloxacin, respectively. However, *P. subcapitata* may be more sensitive than *R. salina* to trimethoprim with EC_{50} values reported at 16 and $80.3 \, mg \, l^{-1}$, respectively.

Phytotoxicity of antibiotics warrants study because of the prokaryotic-like nature of the plant cell chloroplast. Most macrophyte toxicity data have focused on tetracyclines and fluoroquinolones with standardized toxicity tests using *Lemna gibba* or *L. minor*, in which the more sensitive response variables appear related to growth (e.g., frond number, wet weight). For *L. minor*, EC_{50} for reproduction (i.e., new frond growth) ranged from $51 \, \mu g \, l^{-1}$ for levofloxacin to $203 \, \mu g \, l^{-1}$ for ciprofloxacin among six fluroquinolones examined. Effects on other responses, such as wet weight, frond number, chlorophyll *a*, chlorophyll *b*, and carotenoids, have been studied in *L. gibba*. Of these, the most sensitive endpoints of exposure to five fluroquinolones were wet weight and frond number, and ranking of toxicity among compounds was generally the same for both endpoints. Wet weight EC_{50} ranged from 97 to $913 \, \mu g \, l^{-1}$ and EC_{50} for frond number ranged from 116 to $1146 \, \mu g \, l^{-1}$ for lomefloxacin, levofloxacin, ciprofloxacin, ofloxacin, and norfloxacin (in order from most to least toxic). Similar EC_{50} values have been observed for tetracyclines. Specifically, chlortetracycline, doxycycline, tetracycline, and oxytetracycline EC_{50}s for wet weight and frond number were reported as 219, 316, 723, 1010 and 318, 473, 1114, and $1401 \, \mu g \, l^{-1}$, respectively. A recent review paper on antibiotic effects on plants identified that triclosan presented the greatest risk to primary producers if traditional endpoints (e.g., growth) and existing regulatory exposure models are used.

Tetracyclines have been shown to affect environmentally relevant microorganisms such as those within activated sludge from WWTPs. For aerobic sludge, growth inhibition EC_{50}s were 0.08, 0.03, and $0.08 \, mg \, l^{-1}$ for tetracycline, chlortetracycline, and oxytetracycline, respectively. One of the tetracycline degradation products, although not one of the primary degradation products, was 2.7 times more potent than the parent compound. It has also been shown that bacterial genera associated with stream water and several bacterial genera (i.e., *Flavobacterium* spp. and *Pseudomonas* spp.) isolated from the gut of detritivorous invertebrates are all within the spectrum of antimicrobial activity of at least one fluoroquinolone, ciprofloxacin.

Antibiotics generally appear to be far less toxic to invertebrates and vertebrates than to aquatic microorganisms or plants. Cladocerans may be most sensitive to tetracyclines. For the cladoceran *Daphnia magna*, the no observed effect concentration (NOEC) for survival during a 48 h tetracycline exposure period was $340 \, \mu g \, l^{-1}$ while the EC_{50} for reproduction was $44.8 \, \mu g \, l^{-1}$. Cladocerans appear to be much less sensitive to erythromycin with reported acute

EC_{50}s for immobilization or mortality ranging from 200 to $400 \, mg \, l^{-1}$. In addition, a series of toxicity tests involving ciprofloxacin, levofloxacin, lomefloxacin, ofloxacin, enrofloxacin, flumequine, and clinafloxacin, all indicated very limited mortality for the cladoceran *D. magna* and the fathead minnow, *Pimephales promelas*, at concentrations as high as $10 \, mg \, l^{-1}$. In another study, the reported ciprofloxacin NOECs for *D. magna* and the zebrafish, *Brachydanio rerio*, were 60 and $100 \, mg \, l^{-1}$, respectively. For many aquatic vertebrates, such as fish, $mg \, l^{-1}$ exposure concentrations can act as growth promoters and are used readily for this purpose in many aquaculture practices. However, potential ecological consequences (e.g., fitness) associated with accelerated growth promotion have not been explored.

Although antibiotics can be introduced to the terrestrial environment via dispersion of sewage treatment sludge on agricultural fields, to date very little research with limited number of compounds (e.g., tetracyclines and sulfonamides) has been conducted on terrestrial impacts. It is also feasible that terrestrial systems that are closely linked to aquatic systems receiving antibiotic inputs could be indirectly affected by impacts to aquatic system structure and function. In agricultural soil, MIC_{50}s (minimum inhibitory concentrations) for sensitive *Pseudomonas* spp. were 2.0, 0.5, and $1.0 \, mg \, l^{-1}$ for tetracycline, chlortetracycline, and oxytetracycline, respectively, and $0.25 \, mg \, l^{-1}$ for other sensitive strains of bacteria for each of these compounds. It is important to note that the presence of divalent metals may reduce the antibacterial effects of tetracyclines via chelation reactions. Such bioavailability and effects relationships need further investigation.

Population, Community, and Ecosystem Responses

While single-species toxicity tests provide valuable information, these short-term laboratory tests may not be representative of the most sensitive species and are not intended to assess structural or functional responses to chemical stressors. Few studies have examined the effects of antibiotics on aquatic microbial communities and populations. One study, using microcosms consisting of stream sediment, water, and leaf material, reported an effect of 12 day ciprofloxacin exposures on function of aerobic microbial communities, specifically communitylevel physiological profiles. Shifts in microbial community function were observed at concentrations of $10 \, \mu g \, l^{-1}$ and above, whereas at $1 \, \mu g \, l^{-1}$, function of communities in terms of carbon substrate utilization was similar to controls. Similarly, in ternary combination with two other pharmaceutical compounds, $10 \, \mu g \, l^{-1}$ ciprofloxacin exposures did not alter bacterial abundance in aquatic microcosms. Another freshwater lentic microcosms study assessed ecological responses to a mixture of tetracycline, oxytetracycline, doxycycline, and chlortetracycline. While both structural and functional effects were observed in this study, effect levels were determined at higher levels than those reported in aquatic systems.

Ecosystem-level responses such as leaf litter breakdown by detritivores and microbes as well as detritivore growth and condition indices (i.e., responses that can be directly related to bacterial populations) were not affected by ciprofloxacin exposure concentrations of 10, 1, and 0.1 $\mu g\,l^{-1}$ in laboratory microcosm experiments. In the multiple pharmaceutical microcosm exposure involving ciprofloxacin (as described above), community-level effects on zooplankton and phytoplankton were only observed at the greatest ciprofloxacin exposure concentration $(1000\,\mu g\,l^{-1})$. For both phytoplankton and zooplankton, overall abundance increased at this concentration while diversity declined, indicating disruption of community structure and suggesting possible release from competition or predation processes for some taxa.

While the potential for greatest effects of antibiotics may be observed in stream ecosystems, no studies have assessed antibiotic effects on structural or functional response variables in model stream systems. Lotic systems involve physical, chemical, and biotic characteristics that vary greatly from lentic systems and variation in these characteristics could significantly influence occurrence, fate, and effects of antibiotics. Thus, it is likely that a robust understanding of cause and effect relationships between environmentally realistic antibiotic exposures and stream ecosystem functional responses will only be possible if sophisticated lotic mesocosms are employed in future studies. In addition, it is likely that most cases of antibiotic contamination in streams will be accompanied by excessive nutrients and particulate matter due to co-contamination of either sewage or animal waste. Here again, appropriately designed stream mesocosm experiments coupled with other lines of evidence (e.g., biomarkers of exposure or effect) can be used to appropriately characterize antibiotic and other stressor effects on aquatic ecosystems.

Ecological Risk Assessment of Antibiotics

Prospective Risk Assessments

Ecological risk assessments are generally considered either prospective or retrospective in nature. Prospective risk assessments are performed for products prior to introduction to the environment; ecological risk assessment for antibiotics has largely focused on these regulatory assessment paradigms. Currently, ecological risk of pharmaceuticals to aquatic organisms is characterized deterministically with a hazard quotient (HQ), which describes the relationship between a predicted environmental concentration (PEC) and a predicted no effect concentration (PNEC), and traditional endpoints

(e.g., survival, growth). If an HQ (PEC/PNEC) is >1, then a risk to ecosystems is considered possible.

While tiered approaches are employed during prospective assessments of antibiotics, several criticisms of current regulatory risk assessment paradigms have recently surfaced in the peer-reviewed literature. Specifically, environmental introduction concentration (EIC) modeling, trigger values for tiered decisions, and selection of single-species toxicity tests and response variables for human and veterinary medicines have all been questioned. For example, several government agencies employ $EIC_{aquatic}$ and EIC_{soil} trigger values for assessment of human and select veterinary medicines of $1\,\mu g\,l^{-1}$ and $100\,\mu g\,kg^{-1}$, respectively, below which ecological effects are believed to be minimal. In addition, a tenfold dilution factor is applied to an $EIC_{aquatic}$ to estimate PECs of antibiotics released from municipal effluent discharges. This dilution factor is not appropriate for effluent-dominated ecosystems.

In many cases, standardized single-species acute toxicity tests are used to establish PNEC values for antibiotics if a trigger value is exceeded. After data from these studies are collected, default assessment factors (AFs) of 100 or 1000 are applied to EC_{50} values to account for potential chronic effects. However, it is not transparent how default AFs were derived from empirical data. While an AF approach is appropriate for some classes of compounds, pharmaceuticals may have acute to chronic effects ratios (ACRs) higher than the default AFs currently used in regulatory risk assessments. If chronic testing is required, standardized single-species toxicity tests with response variables such as algal or juvenile fish growth or *D. magna* reproduction are assessed. However, alternative endpoints or response variables such as those described in the effects section above may be more appropriate to assess chronic toxicity for antibiotics and other pharmaceuticals.

Hot Spots for Retrospective Risk Assessments

Retrospective assessments are performed for products once they enter the environment, or for site-specific locations believed to have experienced degraded ecological integrity. However, studies of antibiotic impacts on ecosystem structure and function are lacking, particularly for streams and terrestrial systems. It is plausible that ecosystems associated with intensive antibiotic use in agricultural practices or rapidly urbanizing regions may be particularly susceptible to antibiotic impacts based on relatively high introduction concentrations.

Of the various aquatic systems receiving point-source municipal discharges, effluent-dominated systems are most likely to experience water quality degradation by antibiotics. Instream flows of many historically ephemeral streams are perennially dominated by effluent discharges, especially in urbanizing areas located within arid and semiarid regions. For example, the majority of dischargers in the states of

Texas, New Mexico, Oklahoma, Arkansas, and Louisiana are defined as effluent-dominated streams under low-flow conditions. Stream flows in the Trinity River basin below the Dallas/Ft. Worth metropolitan area are often >90% return flows from effluent discharges (**Figure 1**).

Watersheds influenced by confined animal feeding operations (CAFOs) or intensively reared animal farming (**Figure 1**) also represent potential study sites for antibiotics and other veterinary medicines. Efforts by researchers in the United States, Canada, and Europe are increasingly characterizing occurrence and fate of antibiotics in systems adjacent to CAFOs. However, occurrence, fate and transport, and effects in potential worse-case scenarios have not been investigated. As an example, the North Bosque River watershed located in central Texas, USA, contains a high density of dairy CAFOs in its headwater streams (**Figure 1**). Although this watershed is one of the most heavily studied in the United States for nutrients and pathogen loadings from dairy CAFOs to surface water, an understanding of

antibiotic impacts on terrestrial and aquatic biota is not available.

Conclusions and Recommendations

Although antibiotics have received increased study relative to other PPCPs in ecosystems, it remains clear that a greater understanding of environmental fate and effects is required. Specific areas of study should include the partitioning of antibiotics from water to soil, sediment, and other environmental compartments, the mechanistic ecotoxicological effects (e.g., target-specific responses that can be related to ecologically relevant effects) of antibiotics on organisms residing in terrestrial and aquatic ecosystems, and the individual, mixture, and degradate effects of antibiotics on structural and functional response variables, particularly in stream ecosystems. Whereas ecological risk-assessment procedures for antibiotics and other PPCPs employ deterministic hazard quotients to characterize risk, probabilistic approaches are increasingly applied to

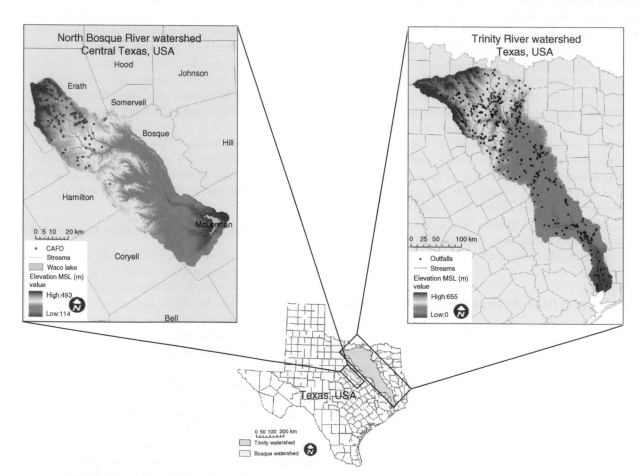

Figure 1 Examples of potential worse-case scenarios for studying watersheds impacted by antibiotics used in veterinary (North Bosque River dominated by confined animal feeding operations (CAFOs)) and human medicines (Trinity River dominated by wastewater effluent outfalls).

other chemical stressors because variability associated with exposure and effect estimates and various uncertainties may be quantified. Increased information for occurrence, fate, and effects will support more definitive probabilistic assessments of antibiotics in the environment. Such assessments will ultimately reduce uncertainty associated with current application of default application factors, dilution scenarios, and potentially affected ecosystem components (e.g., microorganisms and higher trophic levels). Lotic mesocosms provide valuable model study systems for antibiotics because appropriately designed systems integrate physical, chemical, and biological factors within exposure/threshold response experiments using model systems that are the most likely to receive antibiotic contaminants.

Empirical models of contaminant transport and fate in streams have not been linked to physically based, high-fidelity hydrologic models to date. However, cutting-edge watershed models such as the Soil Water Assessment Tool (SWAT) that respond to changes in both external and internal processes that affect stream flow, turbidity, nutrients, etc., provide an approach with increased environmental realism and are recommended for antibiotics and other PPCPs. Coupling such state-of-the-art spatial analysis, loading estimates similar to those used in the Pharmaceutical Assessment and Transport Evaluation (PhATE) model, and SWAT watershed modeling techniques with an enhanced understanding of loading rates, fate pathways, and the magnitude, frequency, and duration of various exposure routes will lead to reduced uncertainty associated with predictions of emerging water quality issues (e.g., antibiotics, other PPCPs) and reduced incertitude associated with appropriate water resource management decision making.

Regions with intensive CAFO and municipal wastewater irrigation to terrestrial systems likely represent ideal study sites for terrestrial ecosystems. Similarly, potential worse-case scenarios such as CAFO-influenced and rapidly urbanizing, effluent-dominated streams may provide useful model systems for defining the occurrence, fate, and effects of antibiotics in aquatic systems and are likely candidates for retrospective risk assessments. Water resource management decisions associated with antibiotics in such watersheds, particularly those located in arid regions, will continue to receive attention from the scientific, regulatory, and industry sectors as an ecotoxicology paradigm shift continues from the dilution (e.g., the solution to pollution is dilution) to the boomerang paradigm (e.g., what one discards into the environment may come back to be harmful).

See also: Ecological Risk Assessment.

Further Reading

Ankley GL, Brooks BW, Huggett DB, and Sumpter J (2007) Repeating history: Pharmaceuticals in the environment. *Environmental Science and Technology* 41: 8211–8217.

Boxall ABA, Fogg LA, Blackwell PA, *et al.* (2004) Veterinary medicines in the environment. *Reviews of Environmental Contamination and Toxicology* 180: 1–91.

Brain RA, Hanson M, Solomon KR, and Brooks BW (2007) Aquatic plants exposed to pharmaceuticals: Effects and risks. *Reviews of Environmental Contamination and Toxicology* 192: 67–115.

Brain RA, Johnson DJ, Richards SM, *et al.* (2004) Effects of 25 pharmaceutical compounds to *Lemna gibba* using a seven-day static-renewal test. *Environmental Toxicology and Chemistry* 23: 1035–1042.

Brain RA, Wilson CJ, Johnson DJ, *et al.* (2004) Effects of a mixture of tetracylines to *Lemna gibba* and *Myriophyllum sibiricum* evaluated in aquatic microcosms. *Environmental Pollution* 138: 425–442.

Brooks BW, Ankley GL, Hobson J, *et al.* (in press) Aquatic hazard assessment of veterinary medicines. In: Crane M, Barrett K, and Boxall A (eds.) *Veterinary Medicines in the Environment*. Pensacola, FL: SETAC Press.

Brooks BW, Riley T, and Taylor RD (2006) Water quality of effluent-dominated stream ecosystems: Ecotoxicological, hydrological, and management considerations. *Hydrobiologia* 556: 365–379.

Daughton CG and Ternes TA (1999) Pharmaceuticals and personal care products in the environment: Agents of subtle change? *Environmental Health Perspectives* 107: 907–938.

Dietrich D, Webb SF, and Petry T (eds.) (2005) *Hot Spot Pollutants: Pharmaceuticals in the Environment*. Burlington, MA: Academic Press.

Halling-Sørenson B, Nors Nielsen S, Lanzky PF, *et al.* (1998) Occurrence, fate and effect of pharmaceutical substances in the environment – A review. *Chemosphere* 36: 357–393.

Hirsch R, Ternes T, Haberer K, and Kratz KL (1999) Occurrence of antibiotics in the aquatic environment. *Science of the Total Environment* 225: 109–118.

Jørgensen SE and Halling-Sørensen B (2000) *Special Issue: Drugs in the Environment*. *Chemosphere* 40: 691–793.

Kolpin DW, Furlong ET, Meyer MT, *et al.* (2002) Pharmaceuticals, hormones, and other organic wastewater contaminants in US streams, 1999–2000: A national reconnaissance. *Environmental Science and Technology* 36: 1202–1211.

Kümmerer K (ed.) (2004) *Pharmaceuticals in the Environment: Sources, Fate, Effects and Risks*, 2nd edn. Berlin: Springer.

Maul JD, Schuler LJ, Belden JB, Whiles MR, and Lydy MJ (2006) Effects of the antibiotic ciprofloxacin on stream microbial communities and detritivorous macroinvertebrates. *Environmental Toxicology and Chemistry* 25(6): 1598–1606.

Robinson AA, Belden JB, and Lydy MJ (2005) Toxicity of fluoroquinolone antibiotics to aquatic organisms. *Environmental Toxicology and Chemistry* 24: 423–430.

Tolls J (2001) Sorption of veterinary pharmaceuticals in soils: A review. *Environmental Science and Technology* 35: 3397–3406.

Williams RT (ed.) (2005) *Human Pharmaceuticals: Assessing the Impacts on Aquatic Ecosystems*. Pensacola, FL: SETAC Press.

Wilson CJ, Brain RA, Sanderson H, *et al.* (2004) Structural and functional responses of plankton to a mixture of four tetracyclines in aquatic microcosms. *Environmental Science and Technology* 38: 6430–6439.

Benzene

J R Kuykendall, ChemRisk, Inc., Boulder, CO, USA

Published by Elsevier B.V., 2010.

Introduction

This article is intended to summarize the current knowledge of the detrimental effects of benzene in the ecosystem, with particular emphasis on potential toxicity to the aquatic and terrestrial organisms. The chemical and physical properties of benzene are discussed as they relate to chemical transit and cycling through the ecosystem. Since toxicity of benzene is known to be mediated by its metabolic activation, a brief discussion of the formation of oxidative metabolites is included. Concern for chemical contamination of habitats must be weighed in the proper context of risk versus concentration. A careful analysis of the benzene concentrations necessary to induce harmful biological effects is included in this discussion.

Chemistry and Physical Properties of Benzene

The chemical structure of benzene was determined in 1834 by the German chemist Eilhardt Mitscherlich from the University of Berlin. Benzene was found to be a cyclic hydrocarbon with a molecular formula of C_6H_6, making it a highly unsaturated compound with an index of hydrogen deficiency equal to four. Benzene is rather stable and reacts preferentially by substitution of hydrogen for another group, such as a hydroxyl, rather than by addition reactions. Substituted benzene derivatives maintain their aromatic character, presumably due to the retention of their resonance distribution of electrons between the ring carbons. Benzene derivatives were originally termed aromatic hydrocarbons, due to the fragrant nature of some of these compounds. In physical appearance, benzene is a colorless liquid which has a characteristic sweet odor. Most humans can smell benzene at a threshold of 1.5–5 ppm in air and taste the compound at concentrations as low as 0.5–4.5 ppm in water. Benzene has a high vapor pressure of 95.2 mmHg at 25 °C and volatilizes at room temperature. Despite being nonpolar, benzene is moderately soluble in water (1780 mg l^{-1} at 25 °C). The high Henry's law coefficient (5.5×10^{-3} atm m^3 mol^{-1} at 25 °C) for the compound would explain why benzene released to the environment partitions from surface water to the atmosphere.

The ability to substitute various functional groups onto the ring structure makes benzene a useful intermediate in the synthesis of many industrial aromatic compounds. Because of its symmetric, unsaturated hydrocarbon structure, benzene has been widely used as a nonpolar industrial solvent. The widespread use of benzene for industrial purposes has made it ubiquitous in the atmosphere and aquatic habitats, causing concern for the health and safety of humans, plants, fish, and wildlife.

Anthropogenic Sources of Benzene

Benzene is commonly used as a solvent and a synthetic intermediate in the chemical and pharmaceutical industries. Over 90% of benzene produced annually is used in chemical synthesis applications, including ethylbenzene (styrene production), cumene (production of phenol and acetone), cyclohexane, and nitrobenzene (production of aniline and resins). Benzene can prevent engine knocking and is added to gasoline (at 1–2% by volume) as a replacement for alkyllead compounds. Combustion of fossil fuels accounts for approximately 82% of benzene released into the atmosphere as automotive exhaust, 14% from industrial activities, 3% from human individual activities, and 0.1% from cigarette smoke. This is in comparison to human exposure sources, which would include 45% from cigarettes, 34% from individual activities, 18% from automobile exhaust, and 3% from industrial activities.

Benzene has been produced commercially from coal since 1849, although petroleum has been used since 1951 and is currently the major source of its production. Benzene is consistently one of the top ten organic chemicals produced annually in the US, with peak benzene production in the US having occurred in 2000, with 8.1 million metric tons (17.8 billion pounds) produced. Estimates of annual benzene releases to the environment is found in the Toxics Release Inventory (TRI) data and listed according to each state in the US. In 2002, it was estimated that 2700 metric tons (6 million pounds) of benzene were released to the atmosphere, suggested to account for 88% of the total released from facilities required to report to TRI. These should be considered underestimates, because not all facilities are required to report releases. Of the prior estimates, approximately 8.6 metric tons (19 000 pounds) of benzene were released to surface water from domestic manufacturing and processing facilities during 2002, accounting for about 0.3% of the estimated total environmental releases. In addition, approximately 4.5 metric tons (100 000 pounds) of benzene were released to soils from these same domestic facilities, representing 0.17% of estimated total releases. Another 314 metric tons (692 000 pounds) were released by underground injection, comprising another 10% of estimated environmental releases.

Environmental Persistence of Benzene

Natural sources of benzene include gas emissions from wild fires and volcanoes, yet benzene is most notable as a constituent of crude oil. Benzene quickly dissipates in the environment, and is generally only found in high concentrations after an industrial accident or chemical spill. Benzene can readily pass from soil or water into the air, and can then reenter the soil or aquatic ecosystems as a residue in snow or rainwater. In the absence of water, benzene volatilization from soil is not rapid. When uniformly distributed at 1 and 10 cm depths, volatilization half-lives of benzene from dry soil were 7 and 38 days, respectively.

The soil organic carbon sorption coefficient (K_{oc}) ranges from 60 to 83, which helps to explain why benzene is so mobile in hydrated soil, as the compound will rapidly leach into groundwater. Conversely, benzene released into water would not be expected to adsorb to sediment or suspended solids to any significant degree. When deposited directly onto soil, the composition of soil can have a direct impact on benzene sorption, as soils with increasing organic matter possess increasing affinity for the compound. The low density of benzene ($0.878\,7\,g\,cm^{-2}$ at $15\,°C$) would facilitate its migration to the air interface with water, enhancing volatilization at the surface. The half-life of benzene in surface water ($25\,°C$ at 1 m depth) is estimated to be 4.8 h based on evaporative loss, while half-lives in groundwater can reach one year. When examined in a model river system that is 1 m deep, with a flow rate of $1\,m\,s^{-1}$ and a

wind velocity of $3\,m\,s^{-1}$, the half-life of benzene due to volatilization at $25\,°C$ was reduced to 2.7 h. Other estimates of volatilization half-lives for a model river and model lake are 1 h and 3.5 days, respectively.

Benzene is chemically stable in soil and water, yet oxidative degradation of benzene occurs within a few hours to days after volatilization into the air. Atmospheric hydroxyl radical attack represents the most significant process involved in environmental degradation of airborne benzene. In addition, oxidation by nitrate radicals and ozone are known to occur at a low level. Since benzene does not absorb light of wavelengths above 290 nm, direct photooxidation of benzene is unlikely. However, photochemically produced hydroxyl radicals will react with benzene, giving a calculated atmospheric half-life of approximately 13 days. Acceleration of benzene degradation can occur in polluted air by interaction with nitrogen oxides and sulfur dioxide (both present in smog), dramatically reducing the half-life of benzene to 4–6 h. Atmospheric degradation of benzene can lead to the formation of phenol, nitrophenol, nitrobenzene, formic acid, and peroxyacetyl nitrate.

Although benzene is quite stable in purified water, a slow process of benzene degradation, known as indirect photolysis, will occur at water–soil interfaces. This process involves energy transfer from activated humic and fulvic acids, which are ubiquitous, primary constituents of soils in aquatic areas. These acids act as photosensitizers by indirectly generated singlet oxygen or hydroxyl radicals which react with benzene. Benzene that remains in soil and is not photochemically oxidized can be available for biodegradation, a process which is maximum at low benzene concentrations (near 1 ppm).

Of particular ecological significance is the microbial degradation of low levels of benzene by several species of aerobic bacteria, through nitrogen- and oxygen-enhanced oxidative processes. Environmental factors which affect biodegradation of benzene include the presence of specific microbial populations and their nutritional sources, temperature, pH, and levels of dissolved oxygen. In addition, the concentration of benzene itself can affect its biodegradation. At concentrations above 2 ppm, biodegradation of benzene is not observed, presumably due to toxicity to the microbes themselves. The highest rates of biodegradation occur at concentrations below 1 ppm, where decay rates of 20–50% per day have been reported. Microbial metabolism of benzene occurs through formation of cis-dihydrodiols, and further transformation to catechols, which are susceptible to ring opening. Soil bacteria which have been reported to degrade benzene under aerobic conditions include *Pseudomonas* species, *Nocardia* species, and *Nitrosomonas europaea*. Benzene is not readily biodegraded under anaerobic conditions, since low oxygen levels make it necessary to use an alternative electron acceptor (which could include nitrate, carbonate, or ferric iron). Sufficient quantities of the alternative acceptor species would be necessary for substantial anaerobic oxidation. Microbial degradation of benzene would also depend on the metabolic capacity of the bacterial community. Those bacteria containing monohydroxylases will cause ring hydroxylation of benzene to phenol, while dioxygenases will catalyze formation of pyrocatechol and hydroquinone. Controlled experiments have demonstrated that aqueous (soil-free) cultures of bacteria can begin to degrade benzene within 12 h, with a half-life of approximately 60 h and almost complete degradation by 90 h.

In most environmental contaminations, several aromatic hydrocarbons are present. Benzene metabolism can be altered by the presence of other aromatic hydrocarbons. For example, low concentrations of some aromatic compounds, such as xylene, may induce the expression of mixed function oxidases in some organisms and increase benzene metabolism. Other aromatic compounds, such as toluene, may compete for metabolic processes and reduce benzene metabolism. Since metabolic activation of benzene is necessary for its toxicity, compounds which compete for oxidative metabolism, such as toluene, may inadvertently reduce the toxicity of benzene by reducing formation of reactive metabolites.

Uptake and Metabolism of Benzene

Although benzene readily penetrates biological membranes, it does not seem to accumulate appreciably in plants, fish, and birds. With a bioconcentration factor of 1.1–2.0, it is unlikely that benzene will accumulate in the food chain, such that most concern for oral exposure would be from drinking of contaminated water. Benzene may become incorporated into plants, the majority of which is believed to occur by air-to-leaf transfer rather than root uptake. Vegetative contamination of exposed food crops used for human consumption and animal forage has been estimated to be 587 ng kg^{-1} (~0.6 ppb). When benzene enters an animal by ingestion of contaminated food or water, it can pass directly through the lining of the gastrointestinal tract into the bloodstream. Due to its lipid solubility, benzene localizes in the liver and bone marrow of animals, where it can be stored temporarily. Benzene is metabolized and ultimately eliminated as conjugated metabolites in the urine. Harmful effects of benzene are associated with production of toxic intermediary metabolites, making organs with oxidative metabolic capacity (such as hepatic mixed function oxidases) especially sensitive to reactive metabolites and targets for the toxic effects. Inhalation or ingestion of benzene can cause acute toxicities including irritation to mucous membranes, as well as neurotoxic effects including restlessness, convulsions, and death from respiratory failure.

Following ingestion, benzene is well absorbed in rodents, with over 90% and 97% absorption in rabbits and rats, respectively. Humans absorb 70–80% of inhaled benzene, which represents the most significant exposure route. Following inhalation, ingestion, or dermal absorption, benzene rapidly distributes throughout the body where it can be metabolized in selected tissues or can be excreted unchanged in exhaled air. Benzene can enter the fetal bloodstream through the placenta, yet it is generally not considered to be teratogenic to experimental animals or humans. Some embryotoxic and fetotoxic effects have been reported in rat studies, such as low birth weight and increased skeletal variations, but exposures less than 10 ppm are not associated with adverse fetal effects. A limited number of studies suggest that benzene can partition into breast milk, but the consequences to neonatal health are not yet known.

Although benzene is classified as a known carcinogen, it is generally thought to require metabolic activation by mixed function oxidases to form one of several DNA damaging nucleophiles. Experiments have shown that 'knockout mice' deficient in the cytochrome P450 2E1 enzyme (CYP2E1) are resistant to the carcinogenic effects of benzene, verifying the requirement for metabolic activation. The predominant benzene metabolite is benzene oxide, a very reactive epoxide which can rearrange nonenzymatically to form phenol (**Figure 1**). Phenol is the predominant urinary metabolite of benzene in mammals and is a known hematotoxin. Alternatively, benzene oxide can be further acted upon by CYP2E1 to form benzene oxepin (**Figure 1**), which can undergo iron-catalyzed ring opening to form reactive *trans,trans*-muconaldehyde. A third possibility involves epoxide hydratase action on benzene oxide to generate catechol (**Figure 1**).

Phenol can be further acted upon by CYP2E1 to generate hydroxyquinone or catechol. A third round of CYP2E1 oxidation of either compound would yield 1,2,4-benzenetriol (**Figure 1**), which can form a semiquinone radical and an active oxygen which are believed to play a role in benzene carcinogenesis. Metabolic detoxification of any of the four phenolic metabolites (phenol, catechol, hydroquinone, and 1,2,4-benzenetriol) can occur by sulfonic or glucuronic conjugation and allowing urinary excretion. Alternatively, benzene oxide can be inactivated by reaction with glutathione to form S-phenylmercapturic acid (**Figure 1**), which is also excreted in the urine.

Benzene oxepin is found in chemical equilibrium with benzene oxide, and can be acted upon by iron-generated hydroxyls to generation of *trans,trans*-muconaldehyde (**Figure 1**). This dialdehyde has the potential to form adducts or cross-links with biomolecules, including proteins and/or nucleic acids. Muconaldehyde is detoxified by oxidation to muconic acid by sequential action of alcohol and aldehyde dehydrogenases (**Figure 1**).

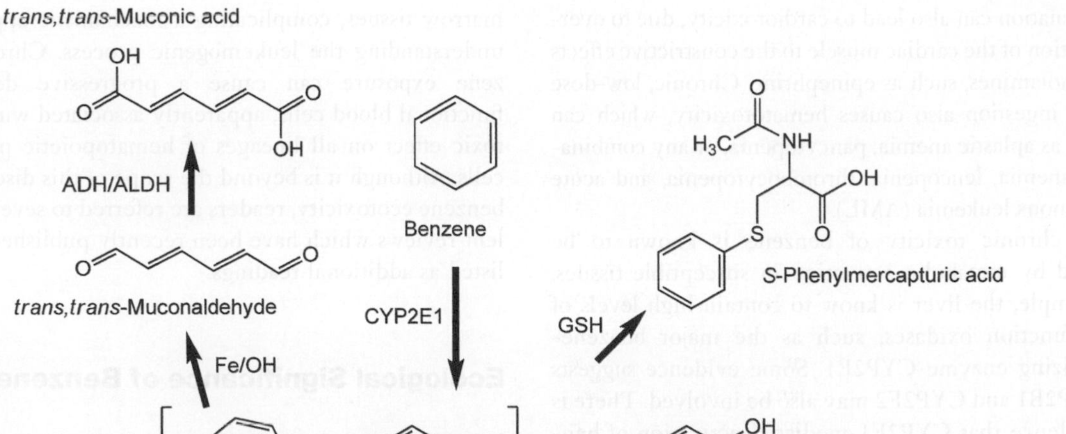

trans,trans-Muconic acid

OH

O

OH

ADH/ALDH

Benzene

CYP2E1

H_3C NH

OH

O

S

S-Phenylmercapturic acid

GSH

trans,trans-Muconaldehyde

O

O

Fe/OH

OH

EH

OH

Benzene oxepin

O

Benzene oxide

Benzene dihydrodiol

Spontaneous rearrangement

DHDD

OH

OH

OH

Phenol

CYP2E1

Catechol

CYP2E1

1,2-Benzoquinone

O

O

CYP2E1

CYP2E1

MPO

O

O

OH

OH

OH

OH

NQ01

HO

HO

OH

1,2-Benzoquinone

Hydroquinone

1,2,4-Benzoquinone

Figure 1 Metabolic activation and detoxification of benzene. Metabolic activation of benzene proceeds through a series of oxidative reactions catalyzed by mixed function oxidases, such as the 2EI isozyme of cytochrome P450 (CYP2EI) and myeloperoxidase (MPO), as well as dihydrodiol dehydrogenase (DHHD) and NAD(P)H:quinone oxidoreductase (NQ01).

Benzene metabolites of concern include benzene oxide, phenol, hydroquinone, and muconaldehyde. Being an epoxide, benzene oxide will readily react with nucleophilic centers in amino acids of proteins or bases of DNA. Benzoquinones and hydroquinones are also known to bind to proteins and nucleic acids to form covalent products. These reactions can cause inactivation of protein function and DNA damage that may lead to mutation and possibly chromosome breakage. Lymphocytes from humans and animals exposed to benzene are known to harbor damaged chromosomes, a condition which can persist for months to years following exposure. The link between specific types of chromosomal damage and subsequent development of leukemia in humans is under intense investigation. An oral reference dose in humans (allowable tolerances in food) is

0.017 mg/kg/day, while the US Environmental Protection Agency (USEPA) has determined the slope factor for carcinogenic risk from oral exposure to be 0.029 mg/kg/day. USEPA estimates of a lifetime daily exposure to 70, 7, and $0.7 \mu g l^{-1}$ benzene in drinking water would equate to a 1:10 000, a 1:100 000, and a 1:1 000 000 elevated cancer risk in humans, respectively.

Organ-Specific Toxicity

In addition to gastrointestinal irritation, acute high-dose ingestion of benzene causes neurological toxicity, most likely due to the rapid uptake of benzene into tissues with high lipid content, such as the nervous system. Acute high-dose inhalation can also lead to cardiotoxicity, due to oversensitization of the cardiac muscle to the constrictive effects of catecholamines, such as epinephrine. Chronic, low-dose benzene ingestion also causes hematotoxicity, which can manifest as aplastic anemia, pancytopenia, or any combination of anemia, leucopenia, thrombocytopenia, and acute myelogenous leukemia (AML).

The chronic toxicity of benzene is known to be mediated by metabolic activation in susceptible tissues. For example, the liver is know to contain high levels of mixed function oxidases, such as the major benzene-metabolizing enzyme CYP2E1. Some evidence suggests that CYP2B1 and CYP2F2 may also be involved. There is also evidence that CYP2E1-mediated activation of benzene occurs in the bone marrow, a major target of its toxicity and carcinogenicity. Bone marrow stroma cells of rats contain higher levels of glutathione and quinine reductase than their counterparts in mice. These represent major detoxification systems and their lower levels in mice could explain their higher susceptibility to benzene-induced hematoxicity than rats.

Benzene is able to stimulate its own metabolism in liver tissues, presumably by inducing synthesis of elevated levels of CYP2E1. Metabolites of benzene, such as phenol, hydroquinone, benzoquinone, and catechol, have also been shown to induce cytochrome levels in human hematopoietic tissues. It is thought that exposure to chemicals able to simulate these enzymes may inadvertently increase the toxic potential of benzene to these tissues. Benzene can be metabolized by a variety of major and minor pathways. As explained previously, predominate routes of metabolism are hepatic oxidative pathways leading to formation of phenol, catechol, and quinol (dihydroxybenzene). Phenol and the other hydroxybenzenes are further conjugated with inorganic sulfate or phenyl sulfate and excreted in the urine. However, saturation of the major pathways of benzene metabolism can occur, so that lesser pathways can become important, such as conjugation of phenol with cysteine to form phenylmercapturic acid or further oxidation of phenol to trihydroxybenzene. It has been suggested that more toxic metabolites are formed by the high-affinity, low-capacity pathways, such as those which become most readily saturated. This may explain why some studies have found that slightly higher doses of benzene do not lead to formation of corresponding higher levels of toxic metabolites. This may inadvertently provide a protective response from metabolic activation of benzene.

The predominant health hazard of benzene in humans is an association with AML. Metabolites assumed to be involved in the hematotoxic and leukemogenic effects of benzene include benzene oxide and the reactive products of the phenol pathway such as phenol, catechol, hydroquinone, and 1,4-benzoquinone. Metabolic activation of benzene does not occur to an appreciable degree in bone marrow tissues, complicating a mechanistic approach to understanding the leukemogenic process. Chronic benzene exposure can cause a progressive decline in functional blood cells, apparently associated with a cytotoxic effect on all lineages of hematopoietic progenitor cells. Although it is beyond the scope of this discussion of benzene ecotoxicity, readers are referred to several excellent reviews which have been recently published and are listed as additional readings.

Ecological Significance of Benzene

With the exception of accidental spillage of petroleum products, the routine levels of environmental benzene exposure are not generally associated with risk to fish and wildlife. Reasons for a reduced concern of the environmental risk of benzene include: (1) the lack of evidence for bioaccumulation of benzene, (2) the low potential for persistence due to its high volatility from surface waters and soil, and (3) the rapid photooxidation of airborne benzene and its biodegradation in soil and water. Studies have shown that high levels of benzene are toxic to terrestrial and aquatic life under controlled conditions.

The levels of benzene in unpolluted air and surface waters are often below the current analytical detection limits. Although benzene does occur naturally, its primary source of production is known to be petroleum products (and the exhaust from their combustion). The median benzene concentration of ambient air samples from urban areas in the US from 1984 to 1986 was 2.1 ppb (detection limit 0.007 ppb), as determined by the USEPA. Drinking water in the US typically contains less than 0.1 ppb, yet some concern is justified for exposure from consumption of contaminated water drawn from wells near landfills, gasoline storage tanks, and industrial areas. As discussed previously, benzene is rapidly disseminated and degraded in the environment. As a result, environmental benzene toxicity is generally associated with exposure from some concentrated source, such as a leaking storage tank or a petrochemical spill. In order to determine the magnitude of benzene exposures which would be tolerated by aquatic and terrestrial life, controlled laboratory experiments have been conducted with several species exposed for short durations to increasing concentrations of benzene. Oak Ridge National Laboratory (ORNL) has also established no observable adverse effect levels (NOAELs) for several terrestrial mammals, as well as benchmark water levels which should be expected to be generally regarded as safe (**Table 1**). Feeding experiments determined NOAELs that ranged from 2.2 mg/kg/day for whitetail deer up to 33.1 mg/kg/day for the short-tailed shrew. It appears, therefore, that larger animals may be less tolerant to the

Table 1 Benzene levels considered to be safe to terrestrial wildlife (ORNL, 1994)

Species	NOAEL (mg/kg/day)	Benchmark (water) ppm
Short-tailed shrew	33.135	150.613
Little brown bat	41.65	260.318
White-footed mouse	29.201	97.336
Meadow vole	23.23	170.355
Cottontail rabbit	7.803	80.722
Mink	8.287	83.708
Red fox	5.045	59.741
Whitetail deer	2.189	33.426

chronic toxic effects of benzene. A similar effect is seen with benzene in drinking water, benchmark water concentrations (generally regarded as safe) which ranged from *c.* 30 ppm for deer, up to 150 and 260 ppm for shrews and bats, respectively.

Effects of waterborne benzene on aquatic wildlife have also been studied, but concentrations believed to be safe are somewhat lower. This is probably due to the ability of benzene to cross respiratory membranes and enter the circulatory system of aquatic animals. A similar situation exists for airborne toxicity in mammals. Benchmark work

at ORNL has also been used to determine field concentrations which are unlikely to represent an ecological risk, such that water concentrations below these levels for each species present at a given site would generally be considered safe. The ORNL benchmark for sediment concentration unlikely to present an ecological risk is 0.052 ppm (dry weight) at 1% organic carbon. ORNL estimates that chronic exposure of most freshwater fish to 8.25 ppm is tolerated with no minimal adverse effects, while *Daphnia* species can tolerate *c.* 98 ppm under chronic exposure conditions. Low bioconcentration factors have been reported for aquatic organisms such as fish, algae, plants, bacteria, and macroplankton. In fact, there is no evidence available to support biomagnification of benzene in aquatic ecosystems. This may be a result of its rapid metabolic transformation in most species, as well as its rapid volatilization from aquatic habitats. The most sensitive aquatic animal identified by short-term continuous exposure tests is the leopard frog, with an LD50 of 3.7 ppm benzene during its 9-day embryo–larval stages (**Table 2**). Rainbow trout and coho salmon represent two of the more sensitive fish species, with 96-h LD50s ranging from 5 to 10 ppm benzene.

The growth of freshwater algae (*Selenastrum capricornutum*) was reduced by 50% following an 8-day exposure to 41 ppm benzene.

Table 2 Acute benzene LD50 values (ppm) for several representative aquatic species after continuous exposure for up to 4 days (Toxic Substances Data Base, 2006)

	24 h	48 h	96 h
Grass shrimp (*Palaemonetes pugio*)			27
Crab larvae, stage 1 (*Cancer magister*)			1108
Shrimp (*Cragon franciscorum*)			20
Brine shrimp (*Artemia*)	66	21	
Pacific herring (*Clupea harengus pallasi*)		20–25	40–45
Bluegill (*Lepomis macrochirus*)	100	22.5	20
Coho salmon (*Onchorhynchus kisutch*)	542	14.1	9.8
Rainbow trout (*Onchorhynchus mykiss*)		56	5.3–9.2
Fathead minnow (*Pimephales promelas*)			34–35 (soft water); 12–32 (hard water)
Mexican axoltl salamander (*Ambystoma mexicanum*)		370	
Leopard frog (*Rana pipiens*)			3.7
Clawed toad (*Xenopus laevis*)		190	

Summary

Doses of benzene in drinking water necessary to induce exposure-related health effects in terrestrial animals following acute ingestions would typically be in the range of 60–150 ppm under most environmental contamination situations. Threshold levels of acute lethal effect to aquatic animals generally range from 3 to 20 ppm. To put this into environmental perspective, if all of the benzene produced in the US in 2000 were spilled and evenly distributed into Lake Michigan or Lake Erie, water concentrations would be c. 1.6 and 16.7 ppm, respectively. Although benzene may represent a significant health risk to humans under high exposure or occupational scenarios, concern surrounding benzene ecotoxicity is generally limited to situations of chemical spillage or leakage from storage vessels near aquatic ecosystems. Concern is further diminished by the ecological disposition of benzene, which causes this compound to readily migrate from contaminated soils and water into the atmosphere. Atmospheric degradation and microbial biogradation of benzene further reduce the toxic potential of this widely distributed industrial compound.

See also: Bioaccumulation.

Further Reading

Amdur MO, Doull J, and Klaassen CD (eds.) (1991), *Casarett and Doull's Toxicology. The Basic Science of Poisons*, 4th edn. New York: McGraw-Hill, Inc.
Chakraborty R and Coates JD (2004) Anaerobic degradation of monoaromatic hydrocarbons. *Applied Microbiology and Biotechnology* 64: 437–446.
Diaz E (2004) Bacterial degradation of aromatic pollutants: A paradigm of metabolic versatility. *International Microbiology* 7: 173–180.
Dragun J (1998) *The Soil Chemistry of Hazardous Materials*, 2nd edn. Amherst, MA: Amherst Scientific Publishers.
Gibson JS and Harwood C (2002) Metabolic diversity in aromatic compound utilization by anaerobic microbes. *Annual Review of Microbiology* 56: 345–369.
Holeckova B, Piesova E, Sivikova K, and Dianovsky J (2004) Chromosomal aberrations in humans induced by benzene. *Annals of Agricultural and Environmental Medicine* 11: 175–179.
Irwin RJ, Van Mouwerik M, Stevens L, Seese MD, and Basham W (1997) *Environmental Contaminants Encyclopedia*. Fort Collins, CO: National Park Service, Water Resources Division (distributed within the Federal Government as an electronic document).
Morgan GJ and Alvares CL (2005) Benzene and the hemopoietic stem cell. *Chemical Biological Interactions* 153–154: 217–222.
Pyatt D (2004) Benzene and hematopoietic malignancies. *Clinical Occupational and Environmental Medicine* 4: 529–555.
Rana SV and Verma Y (2005) Biochemical toxicity of benzene. *Journal of Environmental Biology* 26: 157–168.
Toxicological profile for benzene (2005) United States Public Health Service.
Verschueren K (1996) *Handbook of Environmental Data on Organic Chemicals*, 3rd edn. New York: Van Nostrand Reinhold.

Copper

G F Riedel, Smithsonian Institution, Edgewater, MD, USA

Introduction
Chemistry of Copper
Occurrence and Mining of Copper
Copper Production and Usage
Biochemistry
Copper as a Pollutant
Further Reading

Introduction

This article presents a view of the occurrence, usage, and ecological effects of the element copper in terrestrial and aquatic ecosystems. The first section will discuss the chemical properties of copper which control its biochemical and geochemical properties. Subsequent sections will describe its natural sources, its human uses, its biochemistry, its geochemistry, its distribution, and effects in natural systems.

Chemistry of Copper

Copper (Cu), is the 29th element in the periodic table, on the first row of transition elements. It comprises about 0.006% (60 mg kg^{-1} or parts per million, ppm) of the Earth's crust. Natural copper is composed of two isotopes, ^{63}Cu at 69.17% and ^{65}Cu at 30.83% abundance. Several radioisotopes of Cu can be prepared; the most common is ^{64}Cu, with a half-life of 12.7 h.

Copper is found naturally as divalent Cu^{2+} and monovalent Cu^+, and sometimes as elemental Cu (Cu^0). Cu^{3+} can be produced in the laboratory, but is highly reactive and not commonly found in nature. Cu^{2+} is the most common form in solution (especially in oxic solution), but Cu^{1+} is predominant in minerals of copper, Cu_2O, Cu_2S, and $CuFeS_2$. Cu^{2+} forms strong square planar complexes with a variety of inorganic and organic substances.

Occurrence and Mining of Copper

Copper is one of the few metals found in its native elemental state. Elemental copper is found associated with copper ores around the globe, including North America, Africa, Asia, Europe, and South America. Copper and copper ores are commonly found in basaltic rocks, with its most common ore being chalcopyrite ($CuFeS_2$). It was probably the first metal used by man; its use for tools dates to *c.* 10 000 BC, and copper jewelry has been found that date back to 8700 BC. It was also most likely the first metal to be smelted (refined from a non-metallic ore); copper smelting operations dating to about 5000 BC have been found. Artisans soon discovered the art of alloying copper, adding a small amount of tin to produce bronze, which is harder and more suitable for tool and weapon making than pure copper. Later, copper was alloyed with zinc to make brass.

Copper Production and Usage

Today, copper is mined on every continent except Antarctica. The world's major sources of copper are in the countries of Chile, the United States, Peru, China, Russia, Poland, and Zambia. These mines are often mixed metal-sulfide ores, from which several metals can be extracted simultaneously. Copper is separated from ore by two principal methods, pyrometallurgically (smelting; roasting the ore to drive off the oxides/and or sulfur) or hydrometallurgically (treating the ore with chemicals to leach the copper out of it).

World production of copper in 2003 was approximately 13.6 million metric tons and was projected to grow to 18 million by the end of 2006. Copper's value as an electrical conductor in wire and cables accounts for approximately 75% of its usage, with structural and building uses of copper and it alloys (tubes, sheets, and rods) accounting for the majority (22%) of the remainder. All other uses (chemicals, pesticides, etc) constitute a mere 3%.

Biochemistry

Copper Requirements and Deficiencies

Copper is both a necessary micronutrient for most living things, and a potential toxicant. Cu ions are important constituents in the active sites of a number of important enzymes. These include amine oxidase, ammonia monooxygenase, Cu, Zn superoxide dismutase, cytochrome *c* oxidase, dopamine β hydroxylase, methane monooxygenase, N_2O reductase, nitrite reductase, tyrosinase, ubiquinone oxidase, and phytocyanin. Copper even forms the basis of a respiratory oxygen carrier protein, hemocyanin, in crustaceans, analogous to hemoglobin in vertebrates. As is apparent from this list, copper is an important mediator in a wide variety of biological functions, including energy production, protection from free radicals, and tissue building. As a consequence organisms have evolved mechanisms for accumulating and regulating copper. A protein called Ctr1 (copper permease) that helps transport copper into cells has been identified in a wide variety of organisms, from single-celled yeasts to land plants and vertebrates.

As a required micronutrient, inability of cells to get required copper can lead to limitation and even death. The recommended level of Cu consumption for humans is 1.5–3.0 mg d^{-1}. In humans (and other mammals) insufficient dietary copper can lead to anemia, impaired wound healing, and artery and heart disease. Copper is accumulated in the liver and kidneys, and circulated in the blood in the protein ceruloplasmin. Excessive zinc and iron consumption can reduce the absorption of Cu (and vice versa). One serious syndrome in infants, 'Menkes' kinky hair disease' is linked to a defect in the copper transport system. This syndrome is an X-linked recessive trait, and so occurs almost exclusively in male children. Kinky, brittle hair has also been observed in sheep from regions with copper deficient soil.

In plants, copper is important in nitrogen metabolism, protein synthesis, and chlorophyll production. Signs of copper deficiency in plants can include chlorosis (lack of chlorophyll) either in leaf tips, or mottling, and deformed new growth. Copper deficiency in plants can be caused by low copper concentrations in particular soils, or soil chemistry conditions that inhibit copper uptake (e.g., alkaline or highly organic soils). As copper is removed from soil by the successive removal of plant biomass containing the copper, soils can become copper deficient. Some crops (e.g., flax) are very susceptible to Cu deficiency, while others (e.g., canola) rarely if ever show signs of copper deficiency. Chronic Cu deficiency can reduce some grain yields up to 20% without showing visible symptoms. Copper (Cu sulfate or Cu-EDTA) treatment to the leaves or soil is used to alleviate Cu deficiency. Cu deficiency in soils, and thus plants, can be reflected in Cu deficiency in livestock (sheep, goats, and cattle) grown on local graze. Again, as with plants, Cu absorption by livestock may be suppressed by excessive concentrations of other chemicals (Zn, Fe, Mo, sulfates).

Very low concentrations of copper in many open ocean surface waters have suggested that copper may be one of several trace elements that potentially limits growth of phytoplankton in the ocean. For phytoplankton, copper has been shown to be important in the uptake of iron, which is also a limiting element.

Copper Toxicity

Excessive concentrations of copper are toxic, and even lethal. Ingestion, inhalation, or skin absorption of large amounts of copper can cause a metallic taste in the mouth, abdominal pain, nausea and vomiting, diarrhea, headache, and shock. As little as $11\,\mathrm{mg\,kg^{-1}}$ is regarded as a minimally toxic dose to humans. Large quantities have been lethal. Damage to brain, kidney, liver, and digestive tract may also result from acute toxicity. However, absorption of copper by mouth and skin is fairly low and vomiting frequently further limits absorption, so much of the damage from acute Cu exposure is a chemical 'burn' caused by its caustic nature and local protein damage.

Chronic toxicity to Cu is rarely observed in humans, but liver disease has been observed in vineyard workers after years of applying copper-based fungicides to grapes. A genetic defect causing excessive Cu accumulation is known as 'Wilson's disease'. This defect causes accumulation of Cu in the liver and kidneys up to 30 times normal concentrations. Early signs include a brown ring around the cornea of the eye (Kaiser-Fleisher ring), anemia, jaundice, and swelling. If untreated it can be fatal; however, it is readily treated using chelating agents to remove copper, dietary zinc supplementation to reduce copper absorption, and limiting copper in the diet. Wilson's disease is caused by an autosomal recessive. This means that both parents must carry at least one copy of the gene for the child to inherit it, and males and females are equally at risk.

While copper toxicity can be a problem with livestock (sheep in particular), it is largely caused by excessive concentrations added to feed as a supplement, rather than occurring naturally as a result of accumulation in forage from soil. Given the complex interactions between copper and other trace elements for absorption and effects, it is often difficult to regulate the supplementation correctly. Acute or chronic copper toxicity is rare in crop plants, and is usually traced to repeated or excessive use of copper based fungicides or livestock supplementation.

Copper is extremely toxic to many aquatic and marine organisms. Algae and phytoplankton are inhibited by copper in the water at part per billion $(\mu g\,l^{-1})$ concentrations; copper is widely used as an algaecide. The toxicity of dissolved copper is determined in large part by its chemical speciation. Inorganic and most organic complexes reduce the uptake and toxicity of copper. Some selected results of toxicity tests with a wide variety of freshwater and marine organisms is shown in **Table 1**. Given the wide variation in methods used and species tested it is difficult to make direct comparisons but some of the more sensitive species are strongly inhibited or killed at concentrations that occur in polluted environments. Even within the same species, the results

Table 1   Selected results of copper toxicity tests on aquatic organisms

Target organism	Reported effect	Toxicity threshold $(\mu g\,l^{-1})$
Freshwater plants		
Chlorella pyrenoidosa	Growth inhibition	50
Anabaena flos-aquae	75% Growth inhibition	200
Navicula incerta	4 day EC 50 (growth rate)	10 450
Lemna minor	7 day EC 50 (growth rate)	119
Nitzschia palea	100% Growth inhibition	5
Marine plants		
Prorocentrum micans	5 day EC 50 (growth rate)	10
Asterionella japonica	72 h EC 50 (growth rate)	12.7
Nitzschia closterium	96 h EC 50 (growth rate)	33
Thalassiosira pseudonana	72 h EC 50 (growth rate)	5
Macrocystis pyrifera	96 h EC 50 (photosynthesis rate)	100
Freshwater animals		
Daphnia magna	LC 50	5–200
Limnodrilus hoffmeisterii	LC 50	53
Campeloma decisum	LC 50	1700
Oronectes rusticus	LC 50	3000
Oncorynchys gardeneri	LC 50	20–298
Lepomis macrochirus	LC 50	200–10 000
Saltwater animals		
Acartia tonsa	LC 50	17–55
Crassostrea virginica (embryo)	LC 50	18–128
Rangia cuneata	LC 50	7400–8000
Homarus americana	LC 50	100
Pseudopleuronectes americanus	LC 50	52–271

EC 50, estimated concentration of copper causing 50% of reported effect; LC 50, estimated concentration of copper causing 50% mortality in the experimental population in 96 h.

vary substantially, depending upon the water chemistry, the testing protocol, and the life stage of the test organisms.

Copper as a Pollutant

Sources of Copper Pollution

As one of the great metals of commerce, it is not surprising that copper released by humans into the environment is in significant excess over what might be found naturally. Copper (and other) pollution has occurred in the vicinity of copper mines and smelting operations since mankind began the activity several millennia ago. The excavation of Cu-containing earth at open pit copper mines can produce copper rich dusts which are spread in the wind around the mine site. Most of these ores are sulfide minerals, and oxidize in the air to sulfates, producing sulfuric acid, which renders the copper in the mineral highly soluble. The same process occurs when mine tailings (the low-grade material removed from the mine but not sent for smelting) sit in the open. Freshly exposed to air and water, sulfide minerals oxidize, producing sulfuric acid, resulting in metal rich acid mine drainage. This acidic, contaminated water can drain into surface streams or into the ground, contaminating aquifers. Since the Clean Air and Clean Water Acts have been implemented in the US, pollution from these sources has been substantially reduced, but not eliminated. Similar processes occur at mines other than copper mines where Cu may not be the element of primary interest, but occurs in elevated concentrations (Zn, As, Ni, Pb etc.). Pollution control methods include dust abatement, lining tailing dumps with impervious liners, chemically treating the waters discharged from the mine, and recycling the captured materials into the product stream.

Refining of copper (both pyro- and electrometallurgically) can be a significant source of pollution. In smelting, the heating of the ore to drive off the sulfides and reduce the copper to its elemental state releases a variety of compounds, including copper, sulfur dioxide, arsenic, cadmium, lead, mercury, and zinc to the air and waste water. In modern operations in the US, much of the pollution is captured by scrubbers and water treatment plants. For example, much of the sulfur dioxide can be captured as sulfuric acid and sold as a by-product. However, before the invention and implementation of the controls on air pollution, many smelting sites produced wide swaths of polluted land with both sulfur and metals. Outside the more developed countries, this is still a common practice.

While leaching and electrorefining doesn't have the potential to produce the large amounts of air pollution that smelting does, the large amounts of caustic and toxic chemicals used in the process present ample opportunity for water and soil pollution. Again, in developed countries, these wastes are largely, but not entirely, controlled.

There are many other significant sources of copper pollution. These include burning of fossil fuels, particularly coal, which releases copper in both fly and bottom ash, industrial incinerators, used motor oils, city water treatment sludge, sewage, and sewage sludge and so on. The copper in copper plumbing dissolves slowly and adds copper to the water supply and waste stream, copper roofs leach copper when exposed to acidic rain, and brake pads of modern cars contain copper which wears away over the course of use. The condensers in some power plants use Cu/Ni alloys as heat exchangers; fouling resistance is a bonus.

Copper has a long history of being used as a pesticide or fungicide, and is often released to the environment for that purpose. Paris green, (copper acetoarsenate), originally used as a wall paper pigment, was adopted as a pesticide in the 1860s. 'Bordeaux mix' (copper sulfate and lime) is still used to control fungus on grapes and other crops. It was originally adopted in France in the 1880s to prevent grape theft by making consumers of unwashed grapes ill, but proved effective against mildew. The hulls of some wooden ships were clad with copper sheeting after the mid-1700s to inhibit fouling and invasion by ship worms, *Teredo* sp. (actually wood burrowing clams). Some vessels are currently built with Cu/Ni alloy hulls, which save considerable maintenance compared to a steel hull due to inhibition of fouling. Copper-based anti-fouling paints have been used to inhibit fouling on boats for several centuries, and are still in wide use. Copper in several forms (including copper napthenate, chromated copper arsenate, copper chromium fluoride, ammoniacal copper zinc arsenate, ammoniacal copper quat, and copper-8-hydroxyquinolinate) has long been used to preserve wood from insects and decay. Copper sulfate is sometimes used to inhibit algae growth in lakes and ponds. In short, many potential sources of copper pollution exist, and have existed since mankind began to mine and refine it.

Copper in the Environment and Copper Pollution

As noted previously, the mean crustal abundance of Cu is approximately $60 \, mg \, kg^{-1}$, but surface soil tends to be somewhat lower, with a mean of *c.* $30 \, mg \, kg^{-1}$. Depending on the source rock of the soil, this could be highly variable, but it serves as a baseline for comparison for enrichment of Cu in soils and sediment. Some Cu in surface waters is natural, from rainfall and weathering of soils. Copper concentrations in freshwater can vary widely depending on the hydrology and pollution level, but concentrations of low to subpart per billion ($\mu g \, l^{-1}$) levels would be typical in a region with little or no anthropogenic contribution. Copper concentrations in open ocean seawater are relatively low and probably not substantially influenced by anthropogenic sources. In the central gyres of the major oceans, surface values for copper may be as low as $0.05 \, \mu g \, l^{-1}$. In open ocean depth profiles, Cu exhibits what is often referred to as a 'nutrient-like' profile, with low concentrations in the surface waters, and higher concentrations at depth. These result from scavenging of dissolved copper in the surface waters by algae and other particles, and its subsequent downward transport by sinking particles. The particles are dissolved at depth by various processes, and the Cu is returned to the dissolved phase. Deep ocean bottom water copper concentrations reach approximately $0.5 \, \mu g \, l^{-1}$.

Not surprisingly, given its widespread mining, refining, and use as a structural material and biocide, the occurrence of copper at concentrations above 'natural' concentrations is ubiquitous near human development.

Mining and refining of copper and other metals can be a significant source of Cu pollution. One particularly well-studied site is near Sudbury, Canada, a site where a large mixed sulfide deposit bearing Cu, Ni, and platinum group metals have been mined and smelted since the late 1800s. Concentrations of Cu in soils as high as $2800 \, mg \, kg^{-1}$ have been recorded nearby. The contamination, the result of air deposition from the mining and refining operations, was largely restricted to the surface 20 cm. Vegetation from the area was also examined and had concentrations up to $360 \, mg \, kg^{-1}$ in quaking aspen and $220 \, mg \, kg^{-1}$ in forage grass. Typical 'background values' were $10–20 \, mg \, kg^{-1}$ and $5–10 \, mg \, kg^{-1}$, respectively.

Surface waters in the region are also significantly contaminated with metals, with concentrations reaching single to tens of $\mu g \, l^{-1}$. The acidic precipitation resulting from sulfur dioxide from smelting operations results in a substantial pulse of acidic water and metals in spring runoff. Since 1978, when emission controls were implemented and sulfur emissions began to decline, the acidity of lakes and streams and their metal concentrations have declined.

Sudbury is by no means unique; it is simply one example of the effects of one of many large-scale metal mining and refining operations throughout the world. Clarks Fork River, in Montana, a tributary of the Blackfoot River, has been heavily impacted by acid and metals pollution from the Anaconda copper mine. The site is being cleaned up under the 'Superfund' law, costing an estimated $100 million dollars, entailing 10 years of effort and cleanup of 120 miles of river, stabilizing 56 miles of river bank to prevent erosion of contaminated material, treatment of 700 acres of soil, disposal of contaminated soil into ponds, and reestablishment of river bank vegetation.

While mining and smelting operations are a classic case of 'point source' pollution, much copper pollution comes from numerous ill-defined sources, 'nonpoint source' pollution. For example, in the typical urban environment, Cu is emitted as airborne particles from fossil fuel combustion and brake wear, into surface waters from runoff from streets, copper roofs and gutters, agrochemicals, and from copper piping into the water waste stream.

Elevated Cu concentrations have been widely observed in urban soils and dusts, storm runoff from streets, and discharges from sewage treatment plants.

Overuse of copper supplementation or pesticides and its deliberate use as an algaecide in water can lead to accumulation of concentrations on land and waters receiving their runoff. This, however, tends to be a localized problem.

Coastal and estuarine sediments are one of the primary sinks for waterborne pollution. Estuaries and harbors, in particular, are the sites of many industrial activities, and traditionally, receiving bodies for many waste streams. Consequently, the sediments of harbors and estuaries are often highly contaminated with Cu (as well as large number of other contaminants). The United States National Oceanic and Atmospheric Administration (NOAA) Program, the National Status and Trends Program, also known as 'the Mussel Watch' monitors contaminants including Cu in coastal sediments and organism tissues. Concentrations of Cu in sediments from some of the more contaminated sites in the US exceed $500 \, mg \, kg^{-1}$.

Surface waters may receive Cu pollution from numerous sources, direct atmospheric deposition, urban or agricultural runoff, sewage discharge, acid mine drainage, or other industrial wastes. Similarly, copper concentrations in some polluted estuaries may reach a few parts per billion.

Cu has a strong tendency to adsorb to organic or inorganic particles. This is often measured by a distribution coefficient:

$$k_d = C_s / C_w$$

Where C_s is the concentration of contaminant in the solid phase and C_w is the concentration of the contaminant dissolved in the water. For a given water chemistry and sediment type, k_d is approximately constant regardless of total Cu concentrations; thus dissolved and particulate Cu concentrations tend to rise and fall together. Typical values of k_d for Cu are of the order of $10^3 \, l \, g^{-1}$.

Despite widespread contamination, it is often difficult to demonstrate that pollution by copper is causing environmental problems. One problem with demonstrating harm from elevated concentrations is that pollution rarely consists of a single toxicant. In most cases, pollution results in several to many different toxicants being released. For example, usually several toxic elements are present in the same ores, and are released in same process. In copper mining and refining, As, Cu, Pb, Cd, Zn, and SO_2 are common co-contaminants. In such a complex mixture the cause of environmental damage is not necessarily clear-cut and may ultimately be due to complex interactions among the contaminants. Similarly, coastal regions (particular harbors near large urban areas) are usually contaminated with a number of metals and organic contaminants. Careful testing is required to determine which contaminants are exerting the strongest negative effects in such cases.

In aquatic systems, the toxicity of copper is controlled largely by the complexation of copper by inorganic and organic ligands. For the simplest biota such as algae and bacteria, that do not ingest food, but extract nutrients from solution, the uptake and toxicity of copper is controlled almost exclusively by the fraction of copper remaining unbound or 'free' (there are exceptions; a few organic ligands form lipid soluble complexes which are more readily taken up than inorganic copper). For example, in seawater, inorganic complexation of copper can reduce the free copper to approximately 5% of the total copper, and organic complexation can further reduce the fraction of free copper to as little as 0.1% of the total. As a result of inorganic complexation, copper standards to protect aquatic life in freshwater systems are often based on water hardness, a rough measure of the amount of inorganic ions. Free copper ion concentrations of as little as 10^{-13} M can be inhibitory to marine algae, bacteria, or small zooplankton. Given the extent of complexation, these can be produced by total copper concentrations of c. 10^{-9} M, or approximately 0.1 $\mu g\,l^{-1}$. However, there are some copper complexes which are more available and more toxic than uncomplexed copper. These include complexes of 8-hydroxyquinoline and dithiocarbamate pesticides. Concentrations of copper which inhibit some, but not all species of algae can produce dramatic shifts in phytoplankton community composition with little change in total biomass.

In larger aquatic organisms, which may take in copper both from diet and solution, the issue of metal uptake and toxicity is more complex. The relative importance of dissolved copper and copper in the diet may depend on many things, including the availability of forms of copper in the food and water, the rate at which Cu is lost from the organism, and whether that depends on its source.

Some noted cases of high concentrations of Cu have been found, one of the earliest was accumulations of massive levels of Cu in oysters in the estuarine portion of Thames River in England in the mid-1800s. Copper concentrations in oysters were so high, the flesh turned green and produced a bad taste and even illness in people who ate the oysters. This has also happened more recently in Chesapeake Bay and Taiwan, where the sources of copper were linked to local industries. However, it is not known whether the oysters suffered any ill effects from the elevated tissue copper concentrations. Oysters have a remarkable ability to sequester copper and some other heavy metal ions in sulfide granules within their tissues.

One of the factors which control Cu and other metal toxicity in sediments is reduction and complexation by sulfide. Most sediment is anaerobic below a certain sediment depth, when oxygen is used up by the respiration of organic materials. Below that depth, sulfate may be reduced

by bacteria to sulfide, a compound which has very strong complexation coefficients for many metals. In the presence of S^{2-}, Cu^{2+} is reduced to Cu^+, and precipitated out as solid Cu_2S, or more complicated mixed solids such as FeCuS. Provided that the sum of Cu and other metals complexed by sulfide more strongly than Fe and Mn (e.g., Zn, Cd, Ag) are in lower molar concentration than the total sulfide, it has been shown that there is unlikely to be overt metal toxicity. However, sulfide concentrations are highly variable; depending upon temperature, organic loads, and sediment irrigation; thus the toxicity of such sediment may vary over the course of time independent of metal concentrations. In oxygenated surface sediments, Cu toxicity can be similarly controlled by complexation with sediment organic matter and adsorption on to mineral surfaces.

See also: Acute and Chronic Toxicity; Bioaccumulation; Bioavailability; Ecotoxicological Model of Populations, Ecosystems, and Landscapes.

Further Reading

Ankley GT (1996) Evaluation of metal/acid-volatile sulfide relationships in the prediction of metal bioaccumulation by benthic macroinvertebrates. *Environmental Toxicology and Chemistry* 15: 2138–2146.

Boyce R and Herdman WA (1897) On a green leucocytosis in oysters associated with the presence of copper in the leucocytes. *Proceedings of the Royal Academy of London* 62: 30–38.

Boyle EA and Huested S (1983) Aspects of the surface distributions of copper, nickel, cadmium, and lead in the North Atlantic and North Pacific. In: Wong CS, Boyle EA, Burton JD, and Goldberg ED (eds.) *Trace Metals in Seawater*, pp. 379–394. New York: Plenum.

Coale KH (1991) Effects of iron, manganese, copper and zinc enrichments on productivity and biomass in the subarctic Pacific. *Limnology and Oceanography* 36: 1851–1864.

Danks DM, Cartwright E, Stevens BJ, and Townley RR (1973) Menkes' kinky hair disease: Further definition of the defect in copper transport. *Science* 179: 1140–1142.

Dhawan A, Ferenci P, Geubel A, et al. (2005) Genes and metals: A deadly combination. *Acta Gastroenterologica Belgica* 68: 26–32.

Epstein E and Bloom A (2004) *Plant Nutrition*. Sunderland, MA: Sinauer Associates.

Hudson R (2005) Trace metal uptake, natural organic matter and the free ion model. *Journal of Phycology* 41: 1–4.

Millardot PM (1885) *The Discovery of the Bordeaux Mixture: Three Papers. Translated from the French by F.J. Schneiderhan (1933)*, 25pp. St. Paul, MN: APS Press. American Phytopathological Society.

Monastersky R (1996) Ancient metal mines sullied global skies. *Science News* 149: 230.

Moore JN and Luoma SN (1990) Hazardous wastes from large-scale metal extractions: A case study. *Environmental Science & Technology* 24: 1278–1285.

Peers G, Qusnel S, and Price NM (2005) Copper requirements for iron acquisition and growth of coastal and oceanic diatoms. *Limnology and Oceanography* 50: 1149–1158.

Reinfelder JR, Wang W-X, Luoma SN, and Fisher NS (1997) Assimilation efficiencies and turnover rates of trace elements in marine bivalves: A comparison of oysters, clams and mussels. *Marine Biology* 129: 443–452.

Riedel GF, Abbe GR, and Sanders JG (1998) Temporal and spatial variations of trace metal concentrations in oysters from the Patuxent River, Maryland. *Estuaries* 21: 423–434.

Winterhalder K (1996) Environmental degradation and rehabilitation of the landscape around Sudbury, a major mining and smelting area. *Environmental Reviews* 4: 185–224.

Relevant Websites

http://epa.gov – Ambient Water Quality Criteria for Copper - 1984, US Environmental Protection Agency.

http://entweb.clemson.edu – Chronological History of the Development of Insecticides and Control Equipment from 1854 through 1954, Clemson Entamology.

http://www.dartmouth.edu – Copper: An Ancient Metal, Dartmouth.

http://metallo.scripps.edu – Copper Proteins, Metalloprotein Database and Browser.

http://www.ene.gov.on.ca – Metals in Soil and Vegetation in the Sudbury Area(Survey 2000 and Additional Historic Data), Ministry of the Environment.

http://minerals.usgs.gov – Minerals Information, Mineral Resources Program.

http://extoxnet.orst.edu – Pesticide Information Profiles, The Extension Toxicology Network.

http://www.sudburysoilsstudy.com – Sudbury Soils. Study.

http://ccma.nos.noaa.gov – The Center for Coastal Monitoring and Assessment.

http://www-naweb.iaea.org – Trace and macro-elements – Cu Values Assigned in Anthropogenic Pollution Materials, International Atomic Energy Agency.

http://digestive.niddk.nih.gov – Wilson's Disease, National Digestive Diseases Information Clearinghouse (NDDIC).

Crude Oil, Oil, Gasoline and Petrol

C Y Lin and R S Tjeerdema, University of California, Davis, CA, USA

Introduction

Composition and Chemistry of Crude Oil and Gasoline

Fate of Oil

Toxic Effects of Oils on Marine Organisms and Ecosystems

Summary and Conclusions

Further Reading

Introduction

Petroleum is a naturally occurring substance thought to be formed from decaying plants and animals under pressure and heat over a long period of time. Petrogenic and biogenic hydrocarbons coexist in the environment and it is often impossible to distinguish their origins. The complexity of an oil spill in the ocean is much greater than in other places – such as on land. Petroleum, or its various products, enter the ocean by accidental release from subsurface blowouts during offshore exploration or production, the grounding or sinking of a tanker, the rupture of an underwater oil pipeline, human activity on land or offshore, or discharge of oil over an extended period of time.

Once an oil spill occurs, responding agencies need to first decide whether to treat it. If treated, the best type of remediation measure should be considered. The natural dispersion of spilled oil determines what species are potentially influenced and their level of exposure, while the geological environment and weather conditions greatly influence the fate. If water and air temperatures are high, volatile fractions evaporate quickly from the surface, decreasing the toxic components that can enter the water column and impact

aquatic organisms. Conversely, marine animals that inhale the volatile fractions may develop brain lesions and disorientation. If wind and currents are actively transporting spilled oil further offshore, reduction in the bioavailability of toxic components along the coast can result in reduced impacts to near-shore organisms. However, if such components are persistent, then transfer offshore can influence populations of other species. Therefore, when considering an oil spill event, a broad picture of the tradeoffs between species and effects should be evaluated.

Since petroleum and its products are complex mixtures, it is very difficult to estimate their impacts on different organisms – and even more so on an entire ecosystem. There is currently little knowledge on the synergistic and antagonistic interactions of petroleum hydrocarbons. Moreover, as the various components are dependent on their origin, the exact composition of petroleum at each spill is different. The most notable toxic effect is short-term mortality. However, chronic exposure to sublethal concentrations is also important to consider – and whether an ecosystem can recover over a long period of time. In this article, the composition of crude oils, their transfer and fate in the ocean, and their toxicity to marine organisms and ecosystems are discussed.

Composition and Chemistry of Crude Oil and Gasoline

Crude Oil

Petroleum is a complex mixture of hydrocarbons in gaseous, liquid, or solid form. The composition and properties of crude oil are dependent on the origin, age, and conditions of the source geologic formation. Geochemists use the composition of light hydrocarbons (C_1–C_9), which constitute about 50% of the carbon in crude oil, to identify its origin and determine how, why, and where petroleum exists as well as its migration pattern. The carbon number and structure of hydrocarbons are both highly variable in different crude oils. Usually very small hydrocarbons (C_1–C_4) are gaseous at ambient temperatures and can be used as fuel (e.g., methane, propane, butane) or for making specific products such as polymeric materials and plastics. Larger hydrocarbons (C_5–C_{12}) are liquids which can be easily heated to become gaseous and are thus used as fuels and solvents. Still larger hydrocarbons (C_{13}–C_{17}) are liquids and are used as fuels and lubricants. Hydrocarbons with carbon numbers larger than 17 are usually solids. The compositional differences explain why some crude oils contain more low-boiling components, while others contain more high-boiling components.

Petroleum hydrocarbons (e.g., aliphatic and aromatic hydrocarbons) can also be classified according to their structures; they usually are divided into three classes: paraffins, naphthenes, and aromatics (**Figure 1**). Paraffins are saturated hydrocarbons (alkanes) of straight or branched chains and without ring structure (a saturated hydrocarbon means no additional hydrogen atoms can be added to the carbons). Examples of paraffins with straight chains are methane, ethane, propane, and butane (all gaseous), and pentane and hexane (which are liquids). Branched-chain paraffins are usually present in the heavier fraction, and

they possess a higher octane rating. Paraffins account for 2–50% of the composition of crude oils.

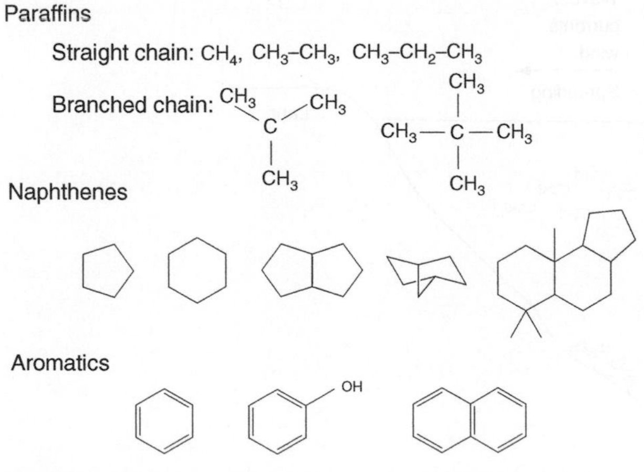

Naphthenes are saturated hydrocarbons possessing ring structure; they may feature a paraffinic side chain. Naphthenic hydrocarbons are the most abundant class present in most crude oils (accounting for 25–75%). As their amount increases, the boiling point of the fraction proportionally increases – with the exception of heavy oil fractions. Monocyclic naphthenes are distributed in the light fractions of crude oil, while polycyclic naphthenes are mainly in the heavier oil fractions. So far no evidence exists for the structures of napthenes with ring numbers in excess of five. However, polycyclic naphthenes with ring numbers of seven or eight have been suggested by current mass spectrometry techniques.

Aromatic hydrocarbons contain aromatic nuclei such as benzene, naphthalene, or phenanthrene which may link with paraffin and/or naphthene ring side chains. The amount of aromatics usually varies from 15% to 50% in crude oils, and it tends to be concentrated in the high-boiling fractions. The most abundant compounds in this group are benzols, naphthalenes, phenanthrenes, pyrenes, and anthracenes, and their properties are very different from those of the aliphatic hydrocarbons. Aromatic hydrocarbons are believed to provide the largest contribution to the toxicity of crude oil. The sum total concentration of benzene, toluene, ethylbenzene, and the xylenes (*o*, *m*, and *p*, referred to as BTEX) is commonly used to represent the total amount of volatile monocyclic aromatic hydrocarbons (MAHs) in petroleum.

Besides hydrocarbons, petroleum contains small amounts of sulfur-, nitrogen-, and oxygen-containing compounds, as well as trace amounts of metallic constituents. These heteroatomic constituents heavily influence the properties of crude oil, even though their concentrations are relatively low. Asphalthenes (high molecular mass hydrocarbons) and resins are generally soluble in aromatic solvents but not in paraffinic solvents. Asphalthenes containing numerous heteroatomic components are the most complex known substances in crude oils, and both asphalthenes and resins (as emulsification agents) stabilize water-in-oil emulsification.

Sulfur-containing hydrocarbons are the most important of the heteroatomic constituents in petroleum (they account for *c.* 0.1–5% of the total composition). Each crude oil has its own distinctive type and proportion of sulfurous hydrocarbons, depending upon geological environment and time. Compounds may include H_2S, mercaptans, thiophenols, thiophenes, and benzothiophenes. In general, the higher the density of crude oil, the higher its sulfur content. Most sulfur-containing compounds are highly toxic to marine organisms.

The amount of nitrogen in different crude oils can account for 0.02–1.5% of the total content. There are

Paraffins

Straight chain: CH_4, CH_3–CH_3, CH_3–CH_2–CH_3

Branched chain:

Naphthenes

Aromatics

Figure 1 Three classes of petroleum hydrocarbons (paraffins, naphthenes, and aromatics).

many nitrogen-containing compounds in oil, including pyridines, hinolines, acredines, indols, and carbozoles. Nitrogen-containing hydrocarbons significantly contribute to the properties of natural surfactants; they also greatly influence the physical–chemical properties and toxicity of crude oils.

In most crude oils, the amount of oxygen-containing hydrocarbons varies from 0.1% to 3% of the total composition; it is also directly proportional to boiling point. Over 20% of these compounds are concentrated in the asphalthene and resin fractions, and may include neutral compounds such as ethers, anhydrides, furans, and acidic oxygens.

The proportions of both hydrocarbons and metals in crude oils can also vary greatly. Therefore, volatility, specific gravity, and viscosity can also vary just as greatly. These variable components help to determine physical–chemical characteristics and biodegradation processes, and further determine the bioavailability to, and toxicity in, marine organisms.

Gasoline

Crude oil can be separated into different fractions for diverse uses via fractional distillation. In general, gasoline represents the fraction with a boiling range of −1 to 180 °C; it may include about 500 compounds in the C_3–C_{12} range. The gasoline fraction mainly contains saturated hydrocarbons or alkanes, unsaturated hydrocarbons, naphthene or cyclic hydrocarbons, aromatics, oxygenates, and other heteroatom-containing compounds. While saturated hydrocarbons are thermally

and chemically stable, alkenes or unsaturated hydrocarbons are unstable, toxic, and have high octane ratings. Aromatic hydrocarbons are still more toxic and possess the highest octane numbers. Small polycyclic aromatic hydrocarbons (PAHs) such as naphthalene are at relatively low levels in gasoline. No large, highly toxic, multiringed PAHs are present in gasoline at significant concentrations. Oxygenates, which do not contribute energy content, provide reasonable antiknock value. Therefore, oxygenates are a reasonably good substitute for the aromatics. Several other additives are also used to prevent knocking and improve octane number. The components of gasoline depend on location and season which alter evaporative emissions and derivability. According to altitude and temperatures, the volatility can be adjusted.

Fate of Oil

Once an oil spill occurs, the oil is transported and degraded according to numerous physical, chemical, and biological processes (**Figure 2**). The chemical properties of oil, as determined by its components, will determine its environmental fate. Moreover, other factors such as location of the spill (land, river, or ocean) and weather conditions (temperature, wind speed, humidity) will impact the distribution and fate of oil. If a spill occurs on land, it may be somewhat easier to control and remediate. However, when oil spills into a river or the open ocean, there are more factors to consider. Knowledge of the geographical location of the spill and an

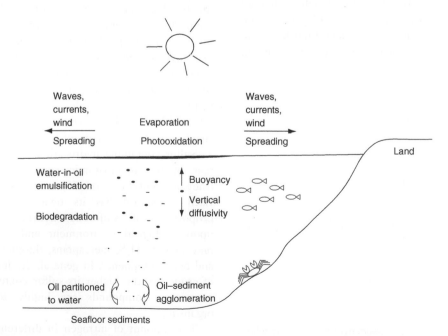

Figure 2 Transport and fate of oil in the ocean.

understanding of transport and fate could help ecologists determine the locations and species that might be affected. To illustrate the factors involved in the transport and the fate of oil, we will examine a more complicated situation where oil is spilled into the ocean.

Transport of Oil

Surface transport

If an oil spill occurs in calm water, it moves according to gravity and its resistance by inertia, surface tension, and viscosity. Movement is terminated when an oil-specific slick thickness (c. 1 mm) is attained. In reality, wind, waves, and currents diffuse and break up the slick, spreading it to cover a much wider area. Slick thickness (volume divided by area) is not distributed uniformly, as the shape and thickness are elongated in the direction of the prevailing wind. In general, 90% of an oil spill's volume is distributed over 10% of its area.

Vertical transport

The vertical transport of oil depends on both vertical diffusivity and buoyancy. Vertical diffusivity transports droplets deeper into the water, while buoyancy raises the oil to the surface. The depth of oil droplet penetration is determined by oil properties such as the concentration in the dissolved phase, concentration in the droplet phase, droplet diameter, and the vertical diffusion coefficient. Ocean conditions, such as wave height, length, and period, and water density will alter the depth of oil transport. The vertical distribution of oil droplets will determine greatly which types of marine species are affected. Sometimes, if oil is treated with surfactant-based dispersants, droplets become smaller and are able to penetrate the ocean more deeply. This has been a big concern for ecologists, since dispersants facilitate deeper transport of larger oil masses.

Horizontal subsurface transport

Under extremely calm weather conditions, vertically dispersed oil droplets can also rise to the surface, given sufficient time. However, in reality, it does not occur due to horizontal subsurface mixing which could ultimately dilute the oil. The process involved can be summarized in two steps: (1) scale-dependent diffusion and (2) shear dispersion. The latter mixing effect mainly results from the combination of velocity gradients with mixing in the direction of gradients.

Weathering of Oil

The physical–chemical property changes of oil in water are in sum total referred to as weathering, and include evaporation, photooxidation, water-in-oil emulsification, and biodegradation.

Evaporation

Evaporation (volatilization) is the most important and rapid step among the weathering processes. For a given amount of oil spilled into the ocean, evaporation can potentially remove almost half of crude oil and more than 75% of refined products, but less than 10% of residual fuel oils. Small molecular weight compounds evaporate immediately, while compounds with vapor pressures greater than n-C_{11} (b.p. < 204 °C) are generally lost within the first 20 days. The evaporation of low molecular weight compounds, including MAHs as well as two- and three-ring PAHs, can greatly reduce the toxicity of oil. The residual high molecular weight components become more viscous, precipitating to coat entrained water droplets into an emulsion. Speed of evaporation can be altered by temperature, wind speed, and sea conditions. When oil is mixed with water, forming a water-in-oil emulsion, or when a thick slick appears, the speed of evaporation decreases.

Photooxidation

Sunlight irradiation can change the physical–chemical properties of oil in the environment. Photooxidation can generate more polar compounds including aliphatic and aromatic ketones, carboxylic acids, aldehydes, fatty acids, esters, epoxides, sulfoxides, sulfones, quinine, phenols, and alcohols. Some of these products significantly contribute to the toxicity of oil to marine organisms. In general, aromatic compounds are most sensitive to photochemical oxidation, and alkyl substitution increases their reactivity. PAHs can be oxidized to more stable metabolites like quinones. Aliphatic sulfur compounds are oxidized more easily than the aromatic thiophenes; aliphatic sulfur is commonly oxidized to its corresponding sulfoxide, sulfonate, sulfone, and sulfate.

Sunlight wavelength greatly affects the rate of photooxidation; higher-energy light (wavelengths < 300 nm) increases the speed of photooxidation. In many cases, chemicals become more bioactive after photooxidation, which leads to greater toxicological effects in marine organisms. In addition to photooxidation, photosensitization is another important type of phototoxicity mechanism. While photooxidation modifies chemical structures to become more reactive, photosensitization transfers light energy to endogenous chemicals in tissues and can result in tissue damage. Both mechanisms are important pathways for cellular damage.

Water-in-oil emulsification

High molecular weight molecules with high boiling points (>350–400 °C), such as asphaltenes, resins, and waxes, are recognized as emulsification agents that can stabilize water-in-oil emulsification or 'mousse' formation. Those molecules orient within the oil phase at the oil–water interface and retard water droplets from forming separate water and oil phases. High viscosities of starting oil will stabilize the emulsions. The properties of starting oil, especially asphaltene, resin, and wax content, and the viscosity, greatly influence the stability of resultant emulsions. Unstable emulsions are those that separate to oil and water somewhat rapidly after the mixing energy is removed.

Biodegradation

The change in oil composition under certain conditions is often determined after analytical measurement. Besides physical–chemical factors that change oil composition, biodegradation also usually occurs. Several studies have utilized isotopically labelled substrates to study the biodegradation of specific components. The data provide insight regarding common biodegradation products and possible metabolic actions. However, it is difficult to predict the biodegradation of the remaining thousands of chemicals in most oils. Metabolic pathways commonly involved in oil degradation in the ocean include hydroxylation and oxidation by microorganisms (usually bacteria). For instance, *Alcanivorax* spp. are mainly responsible for alkane degradation, while *Cycloclasticus* spp. commonly contribute to MAH and PAH degradation.

Fate of Oil Droplets

Entrainment to water column

Some dissolved oil components may evaporate from water to air when droplets resurface by buoyant or mixing energy, while others enter in the water column. The amount and distribution of oil droplets entering the water column will determine the areas and organisms that are affected; entry rate is very sensitive to droplet size. Larger droplets resurface after a reduction in wind or wave turbulence, while smaller droplets (<60–80 μm) usually sink deeper into the water column and eventually dissolve. Soluble hydrocarbon content, viscosity, surface tension, and physical energy (i.e., wave and wind action) will affect droplet size and content. Evaporation and water-in-oil emulsification can increase the viscosity of a surface slick, preventing both wind and wave forces from forming oil droplets.

Partitioning, adherence, and sedimentation

Oil droplets may adhere to particulate matter in the water column according to equilibrium partitioning between the water and particles. Particulate matter bound with oil can penetrate deeper into the water column, carrying specific toxic components directly to sediments. The sources of both oil and the particulate materials affect the interaction, while the density of the combined particles ultimately determines the position of oil–sediment particles in the water column. Eventually, oil–sediment agglomerates and particles bound with oil settle out to bottom sediments. The oil–sediment agglomerate may be disturbed when large currents occur. Animals living in or near sediments may also mix them while burrowing, feeding, and passing water above the sediment layer.

Oil is usually spilled over an extended period of time and into a current of water. The transport and weathering mechanisms described above are usually in effect simultaneously. Physical–chemical forces determine oil distribution such as on a water surface, along a shoreline, within the water column, or in the sediment. Chemical and biological reactions further change the composition of spilled oil. Ultimately, both the distribution and composition of spilled oil determine the impacts observed for different species and their habitats.

Toxic Effects of Oils on Marine Organisms and Ecosystems

Once an oil spill occurs, primary responders need to immediately decide upon a course of action. Decisions to be made may involve ecological considerations and there is often a tradeoff between potentially lethal effects on different species and the potential impacts on natural resources. Ecological information such as the status of a population (endangered, threatened, or common species), prevalent life stage of the species, and the time of the year (spawning, nesting, or migration) will also affect decision making. An understanding of the ecological consequences and toxicological impacts between variable habitats and species will help agencies predict impacts on the ecosystem and make better response decisions. Although each oil spill is unique and it is difficult to extrapolate toxicity data from the laboratory to the field, a thorough toxicological assessment prior to a spill will facilitate more effective decision making. Ideally, toxicological tests (such as bioassays) should cover both acute and chronic effects and incorporate the most sensitive life stages of as many pertinent species as possible. The oil type, condition (fresh vs. weathered), and exposure regimen (continuous, pulsed, static renewal, etc.) should be taken into consideration. The most common toxicological endpoints include acute effects such as mortality, narcosis, and necrosis, and chronic effects such as impacts on development, behavior, and reproduction.

Bioavailability dictates how much of a toxicant is available for absorption by an organism – it greatly influences overall toxicity. Routes of exposure will determine the oil distribution as well as toxicity, and possible routes include the skin surface, gills, other exposed membranes, and gastrointestinal tract (via diet). Both dissolved hydrocarbon and suspended particulate material phases can be taken up by the organism and lead to potentially toxic effects. Dissolved hydrocarbons distributed in the body via blood circulation can interfere with physiological functions, while suspended particulate matter can physically impact an organism by coating body surfaces or gills, impairing respiratory gas exchange.

When considering oil toxicity, the adverse health impacts from both dissolved and particulate oil should be considered. In general, uptake of oil via diet is comparatively lower than from water. However, filter-feeding zooplankton and invertebrates ingest a large amount of oil by filtering the droplets. Species at higher levels of the aquatic food web (such as marine mammals or birds) are at much greater risk of exposure and toxic effects posed by biomagnification, since they feed on organisms which may accumulate the toxic constituents of oil.

General Toxic Mechanisms

Many oil components require phase I metabolism (via oxidation) to generate reactive metabolites; for instance, PAHs are metabolized by cytochrome P450 isozymes to form reactive epoxides which target important macromolecules such as nucleic acids (i.e., DNA, RNA) and proteins (**Figure 3**). While phase II conjugative enzymes can further metabolize reactive metabolites to more water-soluble forms that can be excreted, some conjugates can also react with macromolecules and lead to cellular damage. In general, invertebrates lower on the food web (such as mollusks) have lower metabolic activity when compared with vertebrates such as fishes.

The most basic form of acute oil toxicity is narcosis (i.e., the general solvent effect). Exposure to a variety of petroleum hydrocarbons, including alkanes, MAHs, and PAHs, can lead to this reversible state of nervous system sedation. Both polar and nonpolar hydrocarbons can lead to narcosis, while polar organics exhibit higher toxicity. In order to predict total oil toxicity via narcosis, some researchers have applied toxic unit models which assume that toxicity is additive (i.e., total oil toxicity can be summed by the

toxicities of the individual components). Once the sum reaches a threshold concentration which is species specific, mortality occurs. However, the narcosis model likely underestimates chronic oil toxicity, since narcosis is not the primary mode of action. Mortality is the most common way to access the acute impact of an oil spill on a given aquatic population. In laboratory studies, LC50 (the lethal concentration to 50% of a population) is also most commonly used to characterize the acute toxicological impacts on a species.

The long-term genotoxic impacts of oil can be assessed by studying the mortality of embryos from both invertebrates (such as sea urchins and mussels) and vertebrates (such as fish). Embryo mortality is linked to an increase in toxic oil components in a direct dose-dependent manner. In addition, micronucleus frequency, indicators of chromosomal damage, and DNA single-strand breaks are used as biomarkers for assessing genetic impacts. Some studies have reported significant genotoxic damage and impairment of development and/or reproduction in offspring of species living in the originally impacted area years after an oil spill. This demonstrates that residual levels of potentially genotoxic carcinogens from the spill may exist and continue to cause impacts, even though the extent of cytogenetic damage becomes less noticeable as recovery progresses.

Since petroleum and its products are complex mixtures of hydrocarbons, toxicologists have examined model components to try to predict the toxicity of these classes of compounds. Each group, such as the aliphatic hydrocarbons, PAHs, or metals, may potentially cause adverse health effects. Moreover, there are synergistic and antagonistic interactions of the various components. Therefore, it is very difficult to determine the cumulative toxicological effects among variable classes of chemicals. However, PAHs are believed to be the major contributors to oil toxicity.

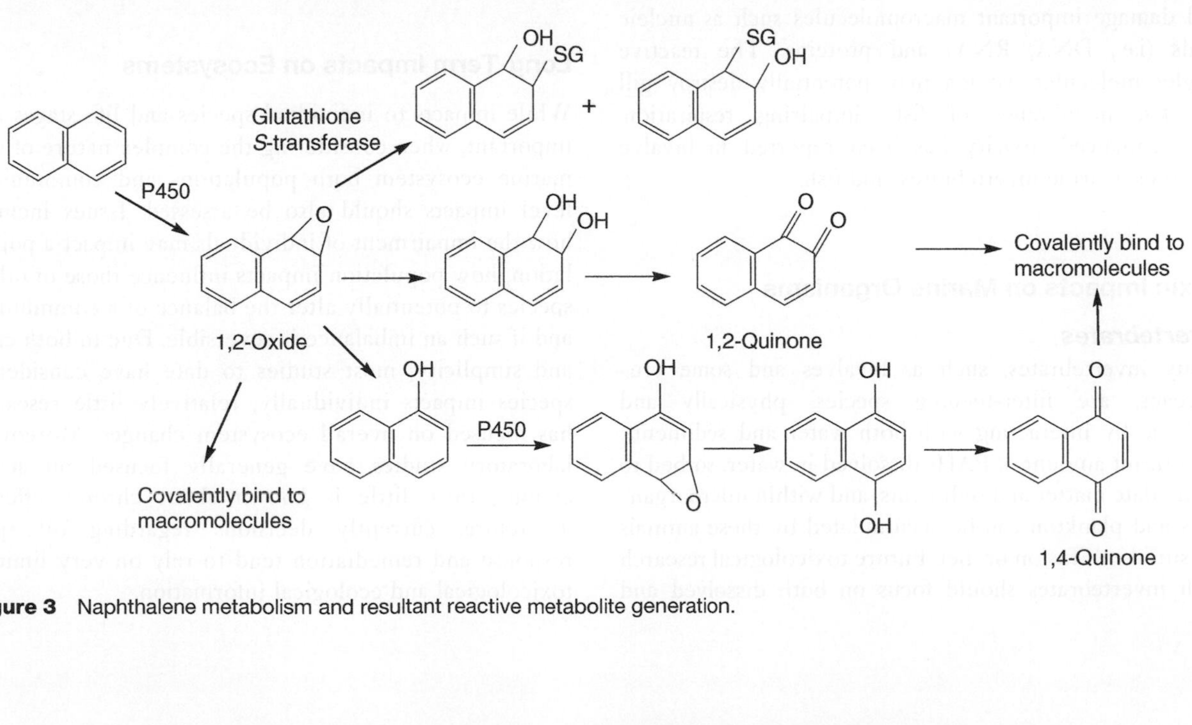

Figure 3 Naphthalene metabolism and resultant reactive metabolite generation.

PAH Toxicity

The PAHs represent a large family of compounds, ranging from the two-ringed naphthalenes to the ten-ringed derivatives of naphthalene. They can be modified by the presence of alkyl side chains or various functional groups which produce a large variety of physical–chemical properties and toxicities. While the most toxic components of oil, the MAHs, are relatively water soluble, they evaporate quickly after oil is spilled. At the other end, the nonvolatile high molecular weight PAHs cannot effectively dissolve in water. Therefore, only intermediate-sized PAHs (such as acenaphthene, phenanthrene, and fluoranthene) significantly influence the toxicity of oil to pelagic organisms in the water column. PAHs can lead to both carcinogenic and noncarcinogenic effects such as oxidative stress, suppression of the immune system, and impairment of endocrine regulation and development. They are metabolized by cytochrome P450 isozymes to reactive metabolites such as epoxides, which target biologically important macromolecules, such as DNA and proteins (**Figure 3**). The ability to generate reactive metabolites varies with species, mainly due to the varying types and amount of P450s within organisms. In general, invertebrates have lower enzymatic activity compared with vertebrates, but species variation is also significant. A few biological responses to PAHs have been suggested, such as an increased content of bile metabolites, induction of hepatic cytochrome P450s, increased DNA adducts in liver, and increased prevalence of liver cancer.

PAHs can degrade quickly via exposure to sunlight – in contrast, PAH half-lives in marine sediments range from months to years. In general, PAHs exposed to solar radiation can result in greater toxicity (phototoxicity), as free radicals which react with oxygen to form reactive oxygen species such as singlet molecular oxygen are generated. These reactive species can then target and damage important macromolecules such as nucleic acids (i.e., DNA, RNA) and proteins. The reactive singlet molecular oxygen may potentially destroy gill or skin membranes of fish, impairing respiration. Photoenhanced toxicity has been reported in bivalve embryos, marine invertebrates, and fish.

Toxic Impacts on Marine Organisms

Invertebrates

Many invertebrates, such as bivalves and some crustaceans, are filter-feeding species, physically and biologically interacting with both water and sediments. Significant amount of PAHs dissolved in water, sorbed to particulate matter and sediments, and within microorganisms and plankton can be accumulated by these animals via surface diffusion or diet. Future toxicological research with invertebrates should focus on both dissolved and sorbed PAHs to better characterize the dominant processes of accumulation.

Vertebrates – fishes

Since the metabolic capabilities of vertebrates are generally higher than those of invertebrates, potential toxicity from activated PAHs in vertebrates is higher than in that for invertebrates. PAH metabolites are commonly found in fish bile within 1 day of exposure. Common oil exposure biomarkers such as cytochrome P450 isozyme activity and formation of DNA adducts are generally induced after 1 week of initial exposure, but toxic manifestations in different species of fish are very different. Therefore, it is difficult to extrapolate toxicological data from one species to another based on these indicators. Numerous studies on pink salmon embryos following the Exxon Valdez oil spill suggest that toxicity was mainly from dissolved PAHs which partitioned from oiled substrates. However, since the salmon eggs were exposed to oiled gravel directly, PAHs were most likely directly absorbed by the eggs, leading to both high mortality and developmental abnormalities. Intergenerational toxicity in pink salmon was also suggested, as some studies reported an increased mortality of embryos exposed to oiled stream substrates years after the spill.

Marine mammals and birds

Marine mammals and birds routinely come in contact with the sea surface; therefore, there is a high risk to those species once an oil spill occurs. Oiled fur or feathers cause death due to hypothermia, drowning, smothering, and the digestion of toxic components through preening. Inhalation of toxic vapors and fumes can also lead to brain lesions and disorientation in marine mammals.

Long-Term Impacts on Ecosystems

While impacts to individual species and life stages are important, when considering the complex nature of the marine ecosystem both population- and community-level impacts should also be assessed. Issues include how the impairment of individuals may impact a population, how population impacts influence those of other species to potentially alter the balance of a community, and if such an imbalance is reversible. Due to both ease and simplicity, most studies to date have considered species impacts individually; relatively little research has focused on overall ecosystem changes. Moreover, laboratory studies have generally focused on acute actions; thus, little is known about chronic effects. Therefore, currently decisions regarding oil spill response and remediation tend to rely on very limited toxicological and ecological information.

The Exxon Valdez oil spill as an example

Some studies have addressed the ecosystem-level impacts of an oil spill. The most memorable and severe incident was the 1989 Exxon Valdez oil spill in northern Prince William Sound, Alaska. Forty-two million liters of North Slope Alaska crude oil were released into Prince William Sound, contaminating more than 1990 km of shoreline. It is estimated that more than 30% of the spilled oil evaporated into the atmosphere, leaving a viscous sticky fluid containing higher molecular weight hydrocarbons on the surface of the shoreline. It was estimated that up to 2800 sea otters, 300 harbor seals, and 250 000 seabirds succumbed to hypothermia, drowning, smothering, or digestion of oil in the days following the spill. Large numbers of plankton and benthic invertebrates were also killed from either smothering or the ingestion of toxic components. The acute mass mortality was severe, and the long-term ecosystem impacts continue to be studied.

Although debate continues as to whether or not the ecosystem has recovered, numerous studies suggest the persistence of oil, which continues to influence the region. Research has shown that the residual oil, at sublethal concentrations, has changed the size and structure of marine populations through compromised health, growth, and reproduction. Chronic effects are closely related to those organisms interacting with oil-contaminated sediments (such as clams and mussels), as they accumulate significant amounts of oil. Since they have low metabolic rates, they tend to accumulate PAHs in their parent forms. Predators such as sea otters thus continue to be exposed to large amounts of PAHs from both contact with the sediment and the consumption of invertebrates. As a result, the detoxifying enzyme hepatic cytochrome P4501A (inducible by PAHs) has been reported to be at elevated levels in sea otters in polluted areas compared with others residing in unpolluted areas. As much as 11 years after the Exxon Valdez spill, the sea otter population had still not recovered, remaining at only 50% of its pre-spill level, while the populations of otters in unpolluted areas have increased.

While many natural factors can potentially affect migrating seabird populations, several studies have reported the chronic effects of the Exxon Valdez oil spill. When feeding on intertidal benthic invertebrates, seabirds were found to suffer from exposure to residual oils. For instance, during the winter seasons of 1995 and 1997, harlequin ducks showed a higher mortality of adult females in heavily contaminated areas, as opposed to those in unpolluted areas. The increase was correlated with both an increase of hepatic cytochrome P4501A and a decrease in body mass, demonstrating the impact of residual oil on a sensitive population and potentially explaining their population decline. Some 10 years after the Exxon Valdez oil spill, adult pigeon guillemots foraging for benthic invertebrates also showed elevated levels of hepatic cytochrome P4501A1, while there was no evidence of residual oil

exposure to pigeon guillemot chicks (who consume fish). The population of Barrow's goldeneye also declined in oiled areas after the spill, and has still failed to fully recover.

The toxic effects observed in organisms that interact with sediments are also found in fish which lay eggs on gravel. Having embryos and larvae chronically exposed to residual oil during their developmental stages may explain the high mortality rates observed for pink salmon embryos years after the spill; weathered PAHs are much more toxic than their parent forms.

Other than mortality, effects on breeding, growth, and reproduction have been reported. Black oystercatchers, which prey on coastal mussel beds, showed increased mortality correlated with the degree of shoreline oil contamination in 1989 and 1990. In addition, a reduced breeding rate and abnormally small eggs were observed within a few months after the oil spill. Three years after the accident, delayed chick growth was also discovered. Both reduced growth rate and abnormal development were also reported for the pink salmon population; embryos and larvae exposed to oil displayed delayed growth, reproductive impairment, and higher mortality at later stages of life.

From the example of the Exxon Valdez oil spill, scientists realized that acute toxicity data alone are not sufficient for long-term risk assessment. In addition to mortality data, chronic effects such as impairment of growth and reproduction from sublethal exposures can impair a population. Moreover, since populations of different species are interdependent, both direct trophic effects and indirect interactions between species can change the balance of the ecosystem. Spill response agencies should consider the ecosystem, including delayed and chronic impacts, when decision making.

Biomarkers in Ecological Risk Assessment

There are two main components of risk assessment: toxicological identification and exposure assessment. Biomarkers responding to oil exposure can be applied to determine the amount of oil taken up by organisms for ecological risk assessment. Researchers can also evaluate the possibility of adverse effects on organisms or the ecosystem so that regulatory agencies can respond more effectively to a given situation. A good biomarker needs to be specific to oil exposure, be reproducible, sensitive, applicable to both the laboratory and field, not prohibitively expensive, and easy to use. The most common biomarkers for oil exposure include the hepatic cytochrome P450 isozymes, fluorescent biliary aromatic compounds, heat shock (stress) proteins, DNA and chromosomal damage, histopathology, and multiple xenobiotic metabolic enzymes. Each method is unlikely to satisfactorily address all of the criteria. There are also other concerns, such as biomarker quantification, timing of analysis, and whether or not responses are reversible. However, the above general biomarkers can provide a fast screen of exposure and indicate the need for further analysis when initial results produce measurable differences.

Currently, there exists little information on biomarkers of toxic responses for organisms exposed to oil. However, determining biological effects at sublethal concentrations will greatly assist risk assessment for chronic effects associated with development, reproduction, and survival. High-throughput analysis of genes, protein, or metabolites, followed by multivariate analysis, can potentially provide a powerful tool to accelerate the field of biomarker development. These high-throughput techniques can screen the changes of thousands of genes, proteins, or metabolites when organisms are exposed to oil at sublethal levels. The large mass of data will assess not only biomarker development, but mechanism studies. By studying the dynamic changes of biomarkers, researchers may better understand the biological impacts of oil in a more complete manner to predict the risk of different types of petroleum to different species.

Summary and Conclusions

Crude oil contains thousands of compounds with a wide range of physical–chemical properties and biological activities. Therefore, it is a challenge to predict the toxicity of such complex mixtures among the many varieties of marine organisms. Over the years, toxicologists have utilized both single- and multiple-compound oil exposures to examine toxic effects in numerous marine species. However, we need to recognize that species variation makes it very difficult to accurately predict toxicity. It is also very difficult to estimate the toxicity of a mixture by combining single-component information, due to chemical interactions within the mixtures.

Several oil spill impact models have been developed to estimate the fate of oil and its impacts on marine organisms and the ocean. The Spill Impact Model Application Package (SIMAP) developed by the US Natural Resource Damage Assessment for oil spills is one oil impact model attempting to incorporate as many factors as possible. Weather conditions, wind, wave and current actions, and the geological environment work in combination to determine the disposition of oil on the surface, in the water column, on the seashore, and in sediments. The model also attempts to evaluate biological exposure, considering both the movement of biota and oils, acute toxic effects (lethal and sublethal), indirect effects of acute exposure via food chain and habitats, non-density-dependent population level impacts from mortality and sublethal effects, and the biological effects of treatment (blooming and dispersant). Oil distribution is relatively easy to predict if accurate measurements of physical and geological factors are made. However, characterization of

toxicity traditionally focusing on acute LC50 (mortality) is problematic. In addition, impacts of development, reproduction, and animal behavior are difficult to estimate at sublethal levels of exposure. Ecological data, such as the distribution of wildlife in contaminated areas during an oil spill episode, are also very limited. Thus, when considering oil impacts on ecosystems, basic toxicological and ecological information is crucial for agencies to both make the best professional estimate of ecosystem impacts and further develop ecosystem-based oil spill management techniques.

Acknowledgments

Support was provided by the US NOAA/UNH Coastal Response Research Center, the California Department of Fish and Game, Office of Spill Prevention and Response, and California Department of Fish and Game's Oil Spill Response Trust Fund through the Oiled Wildlife Care Network at the Wildlife Health Center, School of Veterinary Medicine, UCD.

See also: Biogeochemical Approaches to Environmental Risk Assessment; Body Residues; Ecotoxicological Model of Populations, Ecosystems, and Landscapes.

Further Reading

Anderson JW and Lee RF (2006) Use of biomarkers in oil spill risk assessment in the marine environment. *Human and Ecological Risk Assessment* 12: 1192–1222.

Bolognesi C, Perrone E, Roggieri P, and Sciutto A (2006) Bioindicators in monitoring long term genotoxic impact of oil spill: Haven case study. *Marine Environmental Research* 62(supplement S): S287–S291.

Brannon EL, Collins KM, Brown JS, et al. (2006) Toxicity of weathered Exxon Valdez crude oil to pink salmon embryos. *Environmental Toxicology and Chemistry* 25: 962–972.

Fernandez N, Cesar A, Salamanca MJ, and DelValls TA (2006) Toxicological characterization of the aqueous soluble phase of the Prestige fuel-oil using the sea-urchin embryo bioassay. *Ecotoxicology* 15: 593–599.

French-McCay DP (2004) Oil spill impact modeling: Development and validation. *Environmental Toxicology and Chemistry* 23: 2441–2456.

Harayama S, Kasai Y, and Hara A (2004) Microbial communities in oil-contaminated seawater. *Current Opinion in Biotechnology* 15: 205–214.

Hylland K (2006) Polycyclic aromatic hydrocarbon (PAH) ecotoxicology in marine ecosystems. *Journal of Toxicology and Environmental Health – Part A: Current Issues* 69: 109–123.

Kingston PF (2002) Long-term environmental impact of oil spills. *Spill Science and Technology Bulletin* 7: 53–61.

Lee RF (2003) Photo-oxidation and photo-toxicity of crude and refined oils. *Spill Science and Technology Bulletin* 8: 157–162.

National Research Council of the National Academies (2005) *Oil Spill Dispersants: Efficacy and Effects.* Washington, DC: The National Academies Press.

Peterson CH, Rice SD, Short JW, et al. (2003) Long-term ecosystem response to the Exxon Valdez oil spill. *Science* 302: 2082–2086.

Simanzhenkov V and Idem R (2003) *Crude Oil Chemistry.* New York: Dekker.

Dioxin

R J Wenning and L B Martello, ENVIRON International Corporation, San Francisco, CA, USA

Introduction

This article provides a scientific overview of the current understanding of the toxicological effects typically observed in vertebrate and invertebrate wildlife exposed to polychlorinated dibenzo-*p*-dioxins (PCDDs) and dibenzofurans (PCDFs; collectively referred to as dioxins). This overview includes a brief consideration of plants, though data are limited, because of the importance of food chain transfer for bioaccumulative compounds such as the dioxins. The next section summarizes the physical and chemical properties of dioxins and their environmental fate, which are important to understanding how these compounds elicit toxic responses in birds, fish, mammals, plants, and other aquatic and terrestrial organisms. The subsequent section summarizes what is currently understood about the predominant acute and chronic ecotoxicological effects observed in different taxonomic groups. This article then gives an overview of toxic equivalency factors (TEFs) assigned to different dioxin and dioxin-like compounds and used to evaluate the significance of environmental exposures. The article concludes with a summary of current gaps in knowledge about dioxin exposure and ecotoxicological effects.

Environmental Fate and Exposure Considerations

Strictly defined, dioxins are a class of compounds consisting of 75 PCDD (dioxins) and 135 PCDF (furan) compounds (collectively referred to as dioxin congeners). The PCDD molecule consists of two phenyl rings joined by two oxygen bridges. The PCDF molecule comprises two phenyl rings joined by one oxygen bridge and one single bond (**Figure 1**). The individual dioxin congeners differ in their patterns of chlorine substitution. The degree of chlorination and the pattern of substitution on the two phenyl rings affect the stereochemistry of the congener, and are responsible for inter-congener differences in environmental behavior and toxicity. The 17 dioxin and furan compounds substituted only at the 2-, 3-, 7-, or 8-positions are widely considered the most toxic to humans and biota.

Increasingly, the term 'dioxin-like' is used to describe compounds that share structural similarities to PCDDs and PCDFs and share a common mechanism of toxic action. At present, the term is most often used to describe four non-*ortho*- and eight mono-*ortho*-substituted polychlorinated biphenyls (PCBs; see **Figure 1**). For simplicity, in this article, the 29 dioxin-like compounds (17 dioxin and furan congeners plus 12 PCB congeners) will be referred to as 'dioxins', unless a distinction is necessary to distinguish between individual compounds, homolog groups (i.e., mono- to octachlorinated congeners), or classes of congeners (i.e., PCDDs, PCDFs, and PCBs). **Table 1** provides information on the physical and chemical properties of dioxins and dioxin-like compounds.

Atmospheric Fate Considerations

Volatility (as measured by vapor pressure) affects environmental fate in two ways: by controlling the rate of partitioning between the vapor and the particle phases, and by controlling, together with water solubility (expressed as Henry's law constant), the rate of partitioning between the vapor phase in the atmosphere and the dissolved phase in water. Dioxins have a wide range of volatilities according to their degree of chlorination. In general, the higher-chlorinated dioxin congeners are less volatile than the lower-chlorinated congeners. PCDDs and PCDFs are included among compounds called semivolatile organic compounds (SVOCs); these substances have vapor pressures approximately between 10^{-4} and 10^{-11} atm (10^{1}–10^{-6} Pa) at ambient temperatures.

There are relatively few studies focusing on the vapor/particle partitioning of PCDD and PCDFs. In general, the hepta- and octachlorinated congeners are thought to be almost exclusively associated with atmospheric aerosols under ambient conditions. A measurable proportion of the tetra- and pentachlorinated congeners are present in the vapor phase. Partitioning between the

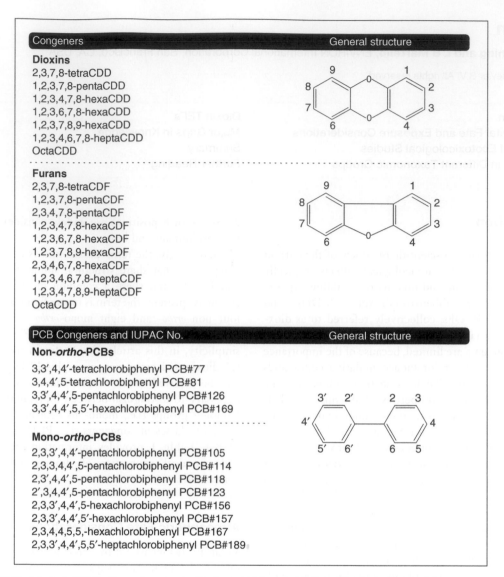

Figure 1 PCDDs (dioxins), PCDFs (furans), and polychlorinated biphenyls (PCBs) with 'dioxin-like' toxicity.

Table 1 Physical and chemical properties of dioxins and dioxin-like compounds

Property	PCDD[a]	PCDF[a]	PCB[b]
Octanol/water partition Coefficient (log Kow)	6–9	6–9	4.46–8.18
Water solubility (μg/L)	0.019	0.692	2.7–590
Vapor pressure (mmHg)	1.5×10^{-9} to 3.4×10^{-5}	9.21×10^{-7}	7.7×10^{-5} to 4.1×10^{-3}
Henry's law constant (atm m^{3}mol^{-1})	1.6×10^{-5} to 1.0×10^{-4}	1.48×10^{-5}	5.2×10^{-4} to 2.0×10^{-3}
BAF or BCF	130 000	61 000	60 000–270 000[c]

[a]USEPA (2002) *Dose-Response Assessment from Published Research of the Toxicity of 2,3,7,8-Tetrachlorodibenzo-p-dioxin and Related Compounds to Aquatic Wildlife – Laboratory Studies*. Cincinnati, OH: National Center for Environmental Assessment, Office of Research and Development (EPA/600/R-02/095).
[b]USEPA (2003) *Non-dioxin-like PCBs: Effects and consideration in ecological risk assessment*. Experimental Toxicology Division National Health and Environmental Effects Research Laboratory office of Research and Development. (NCEA-C-1340. ERASC-003. June). For selected Aroclors; water solubility, vapor pressure, and Henry's law measured at 25 °C.
[c]PCB BCF in fish for selected Aroclors.

vapor and particulate phases is related to temperature and atmospheric particle concentration. During the summer, when temperatures are higher, most of the less-chlorinated congeners tend to be in the vapor phase. In the winter, the less-chlorinated congeners tend to be split between the particulate and vapor phases. This can be important when considering human and wildlife exposure.

The most important pathway for removal of PCDDs and PCDFs from the atmosphere, and the main route through which dioxins enter aquatic and terrestrial environments, is by gravitational settling and washout in rain. Dioxins attached to particulate matter will tend to settle out under gravity, with larger, coarser particles deposited more rapidly and closer to emission sources than smaller particles. Dioxins bound to fine particulate and in gaseous form are more prone to long-range atmospheric transport. These particles will tend to be deposited by rain and snow, although they may have traveled far from the emissions source before they are eventually removed. With regard to sources of human and wildlife exposure, the inhalation pathway is generally not regarded to be a significant source of either human or ecological exposures relative to the levels likely encountered through the food chain.

Aquatic Fate Considerations

PCDDs and PCDFs are generally ubiquitous in the aquatic environment, particularly in sediments, and transported to and recycled within aquatic systems. PCDDs and PCDFs enter aquatic environments from wet and dry deposition, river inflows, groundwater flow, and direct and indirect discharges from industrial facilities. Dry and wet deposition may be the most important sources of PCDDs and PCDFs to water bodies with large surface areas (lakes and seas). Long-term or temporary sequestration of PCDDs and PCDFs in aquatic systems can occur when bound to particles that settle as sediment, or volatilized across the air–water interface, or by chemical and biological transformations. The latter two processes are possible but less significant than sedimentation.

Dioxins are highly lipophilic and hydrophobic compounds and, as such, are insoluble in water and generally have low Henry's law constant (H) values. The hydrophobicity of dioxin-like compounds can be ascertained using the octanol–water partition coefficient (K_{ow}), which provides insight on the relationship between bioconcentration and toxicity. Dioxin congeners are considered to be superhydrophobic because experimentally determined log K_{ow}s are typically greater than 6, and as high as 12.

PCDDs and PCDFs partition weakly between particulate and dissolved phases. In contrast, PCDDs and PCDFs associate strongly with organic matter; hence,

the degradation and mobility of organic carbon in sediment is a significant factor contributing to the mobility of PCDDs and PCDFs. A wide range of sorption partition coefficients have been reported for dioxins, with log K_{oc}s ranging from approximately 4 to 7.5. In surface waters, particulate-bound PCDDs and PCDFs may be incorporated in sediment and recycled at or near the sediment–water interface. In addition, PCDD- and PCDF-bound particles may become incorporated in deeper waters where resuspension and bottom currents are not strong enough to cause further transport, making sediments an important sink for dioxins. Agglomeration (a weak association, held together by surface tension and organic cohesion) or flocculation of particles held together by electrostatic forces also can be important when evaluating PCDD and PCDF fate in surface water. Particulate-associated PCDDs and PCDFs generally have short residence times in surface waters due to sedimentation or hydraulic flushing of particulates.

PCDD/Fs may be susceptible under certain environmental conditions to microbial degradation; however, research appears to be limited. It is suggested that certain bacteria have the ability to dechlorinate dioxins but the position and rate of dechlorination are dependent on a wide number of factors such as the availability of alternative nutrients, the species of bacteria, and the oxidative state of the sediment.

Terrestrial Fate Considerations

PCDDs and PCDFs are largely immobile once adsorbed to particles in the soil column. However, different PCDD and PCDF compounds have different solubilities, and factors such as the soil organic content, clay content, pH and moisture are important. Solute-transport models have shown that mechanisms such as wind and water erosion are likely to be more important than losses by movement within the soil. A soil dissipation rate of 0.069 3 yr^{-1} (which corresponds to a half-life of $c.$ 10 years) has been postulated under specific experimental conditions.

Rates of volatilization from soil are uncertain. In general, higher-chlorinated dioxins such as octachlorinated dibenzo-p-dioxins (OCDDs) do not vaporize under environmental conditions and have half-lives in soil measured in years. Some studies suggest that the half-lives of dioxins at the soil surface may be measurable in terms of weeks or months. When mixed in soils to depths generally below 5 mm, however, soil half-lives are likely measured in terms of years.

Rates of photolysis, involving removal of one or more chlorine atoms from the PCDD or PCDF molecule and resulting in higher-chlorinated compounds degrading to less-chlorinated compounds (referred to as dechlorination), are generally believed to be insignificant. Research has shown limited photodegradation of PCDDs and PCDFs from soil surfaces and sediments.

There is scarce information on microbial degradation of PCDDs and PCDFs. Limited data suggest that microbial degradation of PCDDs and PCDFs in soils is generally slow (measured in terms of years); microbial degradation rates generally decrease with increasing chlorination. Recent studies provide limited evidence that some fungi species, most notably white rot fungi, are able to mineralize some PCDD and PCDF congeners.

Bioaccumulation and Food Chain Transfers

PCDDs and PCDFs are known to bioaccumulate. Some of the highest concentrations occur in the top predator species of food chains where successive stages of bioaccumulation through the food chain can result in biomagnification. PCDDs and PCDFs can be mobilized from body fat into lactating females making this an important mode of transfer to offspring.

PCDDs and PCDFs are known to accumulate in fish and are concentrated in the fatty tissues. It is important to distinguish between two mechanisms of uptake in fish: bioaccumulation and bioconcentration. Bioconcentration involves direct uptake from water across the gill membrane and is distinct from bioaccumulation, which also includes dietary uptake. Bioconcentration factors (BCFs) developed to describe this process and are derived from the ratio of uptake and depuration rate constants. Published BCFs for PCDDs vary over 3 orders of magnitude from $c.$ 4 to as high as $c.$ 9000. Higher-chlorinated PCDDs and PCDFs tend to have lower BCFs than less-chlorinated compounds. This is likely due to factors such as differences in membrane transport, larger molecular sizes, lower solubilities, and the possibility for preferential metabolism of certain congeners.

Fish body burdens of PCDDs and PCDFs are likely due to dietary uptake rather than membrane transport across the gill surface. Environmental monitoring data generally indicate higher body burdens in benthic organisms and bottom-dwelling fish than in pelagic fish residing in the surface water. The differences are generally attributed to the close association of sediment-dwelling organisms with sediments and the generally low opportunity for exposure to PCDDs and PCDFs in the water column. Differences in BCF between species is often explained by different feeding strategies and, perhaps, by different rates of biotransformation or excretion. Metabolic transformations of certain dioxin compounds has been suggested as an important factor in explaining low bioconcentration and bioaccumulation factors in some fish species. For example, results from PCDD and PCDF measurements in fish and fish-eating birds from the Great Lakes, USA, demonstrated that 2,3,7,8-substituted congeners preferentially accumulate to a higher degree than non-2,3,7,8-substituted

congeners in the food chain. Further, there is some evidence in fish suggesting that lower-chlorinated 2,3,7,8-substituted congeners have longer half-lives (on the order of 50–100 days) than other PCDDs and PCDFs (on the order of several weeks or less).

Plant uptake of PCDDs and PCDFs from soil appears to be rare, or inefficient. Some differences in soil to plant uptake rates, however, have been observed. It is widely believed that PCDDs and PCDFs entering the plant root surface are effectively bound immediately for the life of the plant. Further, among the studies showing PCDDs and PCDFs bound to the root surface, few if any studies indicate migration into other plant compartments or translocation to aboveground plant tissues. For example, studies of 2,3,7,8-tetrachloro dibenzo-p-dioxin (2,3,7,8-TCDD) uptake by carrots grown in soil contaminated after the Seveso accident in Italy showed higher concentrations in the outer root surface than in other root and above ground tissues. Dry deposition and adsorption on the surface of foliage appears the most likely pathway of contact for vegetation, and the most plausible pathway for animal exposure.

Overview of Ecotoxicological Studies

Research findings on the molecular and cellular effects of dioxins suggests that the mode of action among different PCDD and PCDF compounds is broadly the same, at least among vertebrate animals. Most studies have been conducted on laboratory rodents and primates, which have traditionally been used as models to extrapolate the results to studies of potential human health effects. It has not been firmly established that the mode of action is the same in other vertebrate species or in invertebrates.

There are six main types of effects commonly ascribed to dioxins and dioxin-like compounds, all of which are exhibited in mammals, and most by other vertebrate groups. Numerous animal studies have shown that these six effects are not unique to dioxins. It has rarely been possible to demonstrate a clear cause and effect relationship between the biological responses observed in the laboratory and environmental exposures. These effects are summarized in **Table 2**.

Cytochrome P450-Mediated Effects

The mechanism of action for dioxin and dioxin-like compounds is generally accepted to function through binding to a specific protein in the cytoplasm of cells, the aryl hydrocarbon receptor (AhR). Once bound, the dioxin/AhR complex can bind to DNA in the cell nucleus, which results in the increased production of several proteins, particularly in the production of cytochrome P450 1A1. The binding of dioxins to the AhR is the essential step for the expression of dioxin-related effects in vertebrate species. Induction of this

Table 2 Six main types of effects commonly ascribed to dioxins and dioxin-like compounds

Possible effect	Current knowledge
Cytochrome P450 induction	It is believed that the root cause of many of the effects of dioxins lies in their ability to bind to a specific protein in the cytoplasm of body cells, the aryl hydrocarbon (Ah) receptor. This leads to the synthesis of P450-dependent enzymes, which, in turn, can affect the metabolism of useful substances like steroid hormones, leading to disturbances in critical biological functions
Immune system suppression	Dioxins are widely held to have effects on the immune systems of exposed animals. It is suspected that this type of effect contributed to the mass mortalities of seals and dolphins in European waters in the late 1980s and early 1990s. The mechanism for immune system effects is not well understood
Porphyria	Hepatic porphyria is a condition in which there is disruption of the process by which the liver produces a component of the blood pigment hemoglobin. Dioxins are known to disrupt the process leading to sensory disorders, paralysis, and psychological effects
Cancer promotion	An association between dioxins and cancer has been recognized for some time, TCDD causing skin and liver tumors in mice at lower concentrations than any other substance. It is felt by some authorities that dioxins are not mutagenic (i.e., do not initiate cancer development), but it is generally felt that they are strong promoters of tumor development
Disruption of vitamin A metabolism	Dioxins can inhibit the process by which vitamin A is stored in the liver. Decreased vitamin A storage, and increased levels in the blood, can result in fetal damage, growth disorders, and sterility
Sex hormone effects	Dioxins have been found to have significant effects on the sex hormones estrogen and testosterone. In rats, this has been found to result in decreased fertility and increased incidence of tumors in females and low testosterone levels in males

protein translates into an increased production of enzymes capable of oxidizing both foreign and endogenous substances. Although the function of this response may be to protect the cell from potential damage, cytochrome P450 induction can affect metabolism of endogenous substances like steroid hormones, leading to disturbances in critical biological functions. Taxonomic groups that possess the AhR such as fish, birds, and mammals show far greater susceptibility to dioxins than plants and most invertebrates, which generally do not possess this receptor.

Immune System Suppression

Dioxins have been shown to alter the immune system in a wide variety of animals. For example, dietary exposure to PCDDs and PCDFs and coplanar PCBs has been correlated with immunosuppression in field experiments with harbor seals. Dioxin and PCB-mediated immunosuppression is believed to have contributed to the mass mortalities of seals and dolphins in European waters in the late 1980s and early 1990s. PCDDs/Fs are known to cause hypertrophy of the thymus, which is the site of production of mature T-lymphocytes, which are indispensable for the development and maintenance of the T-cell-mediated portion of the immune system. Because dioxins affect maturation and specialization of T-cells, exposure is believed to reduce the organisms' ability to fight off harmful bacteria, viruses, and other substances.

Porphyria

Dioxins are known to cause the breakdown of the heme production process, resulting in the buildup of heme

precursors called porphyrins. Hepatic porphyria is a condition in which there is disruption of the process by which the liver produces heme, the active component of the blood pigment hemoglobin. The buildup of porphyrins in blood, tissues, and feces is used as a biomarker of dioxin exposure. In the Great Lakes, USA, porphyrin levels in Herring gull blood serum and tissue have been correlated with environmental exposure to PCDDs, PCDFs, and dichlorodiphenyldichloroethylene (DDE); it is not clear, at present, which compounds have been more responsible for declines in the population.

Cancer Promotion

The relationship between dioxins and carcinogenicity has been recognized for some time. The most toxic compounds among the dioxins, 2,3,7,8-TCDD, have been shown to promote the formation of skin lesions and liver tumors in rodents. While scientists continue to debate whether dioxins are mutagenic (i.e., do not initiate cancer development), it is generally acknowledged that dioxins are strong promoters of tumor development.

Disruption of Vitamin A Metabolism

Exposure to dioxins has been shown to disrupt the storage of vitamin A (retinol). Vitamin A is important to the immune system and to normal fetal development, growth, metabolism, and reproduction in vertebrates. Dioxins have been shown to inhibit the esterifying enzymes required for vitamin A storage in the liver. The disruption of hepatic storage and resulting increased levels of vitamin A in the blood has been associated with fetal damage, growth disorders, and sterility.

Hormone Effects

Dioxin, similar to several persistent organic pollutants, has been found to significantly affect estrogen- and testosterone-mediated processes. Some dioxin compounds have been shown to decrease the number of estrogen receptors in certain organs in female rats, possibly resulting in decreased fertility and increased incidence of tumors in these organs. In male rats, some compounds have been shown to reduce testosterone levels by preventing production of enzymes responsible for increasing testosterone synthesis when levels of the hormone get low. When administered prenatally to rats, lower sperm counts have been observed in male progeny and urogenital malformations have been observed in females.

Ecotoxicity in Different Taxonomic Groups

Plants

Toxicity data for plants are scarce and virtually no information is available on the relationship between dioxin exposure, accumulation in tissue, and biological response. Some species of terrestrial and aquatic plants have been observed to concentrate dioxins from their surroundings without any apparent toxic effects. For instance, some studies on food sources for dairy cows report that several species of grasses are able to incorporate dioxins in leafy tissue. Some species of aquatic plant such as the alga *Oedogonium cardiacum* and higher plants such as *Lemna trisulca* (duckweed) and *Potamogeton berchtoldii fieber* (pondweed) have been observed to sequester dioxins in leaf and stem tissues without any apparent toxic consequences.

Invertebrates

Relatively few experiments have investigated the toxicity of dioxins to invertebrates. Where work has been carried out the results generally indicate no susceptibility to dioxins. In one case, even where field evidence showed a correlation between exposure to dioxins and mortality in sediment-dwelling amphipods, laboratory experiments with spiked sediments resulted in no adverse effects and suggested that other factors contributed to mortality *in situ*. The few available studies have shown some evidence of toxic effects in invertebrates, including reduced reproductive success in sediment-dwelling worms and snails, and possible gene expression effects in clams. Soft-shelled clams were found to accumulate dioxins in gonad tissues after acute exposure in water, and showed possible alteration in gene expression associated with

increased cell cycling. Acute toxicity and cytochrome P450 induction have been demonstrated in crayfish.

Studies have shown that many invertebrate species do not possess a functional AhR, which may account for the apparent lack of susceptibility to dioxins reported in most invertebrate studies. Cytochrome P450 has been measured in the digestive organs of sea star and, in one study, CYP1A induction was found to be significantly related to environmental exposure to dioxin. However, before reaching any final conclusions about invertebrate sensitivity, additional long-term toxicity tests on a diverse set of aquatic invertebrates is needed.

Amphibians

Similar to invertebrates, few studies have examined the effects of dioxins on amphibians. In studies involving exposure of eggs, tadpoles, and frogs to dioxins either by direct injection to tissue or water, no significant mortalities or morphological abnormalities have been observed, except for an increased occurrence of lighter pigmentation in some frogs and tadpoles. Eggs and tadpoles eliminate dioxin relatively quickly, and appear to be about 100- to 1000-fold less sensitive to the deleterious effects of dioxin than early life stages of fish. Several explanations have been offered in the literature, including shorter birth cycle (3–6 days in most amphibians compared to 60 days or longer in fish) and reliance by fish on a yolk sac containing maternal lipids for 120 days after hatching, which greatly increases the likelihood for maternal transfer of dioxins.

Reptiles

With the exception of turtles and alligators, laboratory toxicity data are not available for reptiles. In snapping turtles (*Chelydra serpentina serpentine*) from the Great Lakes, USA, dioxins have been shown to affect hatching success and are believed to contribute to deformities including curled, bent, twisted, or absent tail; shortened or absent legs or digits; deformed eyes; recessed lower jaw; reduced body size; undeveloped carapace; presence of absence of scutes; unresorbed yolk sac; and missing claws. In American alligators, several studies suggest that dioxins adversely affect the endocrine and reproductive systems. Most notably, studies involving eggs treated with 2,3,7,8-TCDD show dose-dependent alterations in sex ratio, with a significant occurrence of female gonadal differentiation and increased numbers of females produced at birth despite ambient temperatures that should produce a higher percentage of male offspring.

Fish

Fish embryos and fry are known to be particularly sensitive to dioxins, particularly developmental processes. Symptoms of dioxin exposure in fish embryos and fry include edema of the yolk sac and pericardium, hemorrhaging in the head and tail regions, craniofacial deformities, and wasting syndrome. The cardiovascular system, in particular the vascular endothelium of the developing embryo, has been identified as a primary target of dioxin-induced toxicity. Several other tissues, including gill and digestive tract, are also known target tissues for AhR agonists.

The developing embryos of oviparous vertebrates, such as fish, are particularly prone to the adverse effects of dioxins due to their elevated exposure risk through water, sediment, and dietary sources of dioxin. Dioxin exposure has led to reduced and, in some cases, failed recruitment of young fish into breeding populations, which is believed to have occurred in lake trout in the Great Lakes, USA, during the 1960s and 1970s.

Dioxin toxicity in fish tends to be higher for congeners containing four, five, or six chlorine atoms. It appears that congeners with fewer chlorine atoms tend to be more rapidly metabolized and eliminated, while more highly chlorinated compounds have limited membrane permeability or bioavailability. As with higher mammal studies, it appears that 2,3,7,8-TCDD is the most toxic congener to juvenile and adult fish. Fish eggs appear to be highly sensitive, and likely represent the most important route of exposure for early life stages.

Birds

The most compelling case linking a specific suite of adverse biological effects such as embryonic and chick mortality, edema, growth retardation, and deformities (notably crossed bills) to residues of dioxin-like compounds can be made for piscivorous birds (e.g., herring gulls, Forster's terns, double-crested cormorants, and Caspian terns). In the Great Lakes, USA, deformities in aquatic birds were found to be similar to deformities induced in offspring of farm-raised chickens exposed to PCDDs and PCDFs in feed. However, it has been shown that most (>90%) of the TEQ found in the eggs of cormorants and terns in the Great Lakes is accounted for by planar PCBs, rather than PCDDs/Fs, which accounted for between 2% and 9% of the toxic equivalents.

Studies of European cormorants in Dutch field-collected eggs found that *in ovo* exposure to PCDDs and PCDFs may explain reduced reproductive success, but postnatal exposure and parental behavior may also be contributing factors. In fact, recent work has shown that *in ovo* exposure to dioxins can lead to grossly asymmetric development of avian brains – a phenomenon seen in field-collected great blue herons, cormorants, and eagles, as well as laboratory chickens.

Laboratory studies have shown other bird species to be susceptible to PCDD and PCDF exposure; however, LD50 values and biological responses vary widely among bird species. Studies involving chickens and pheasants suggest that, in general, birds display decreased egg production, embryotoxicity, and cardiovascular malformations. At least one study suggested that chickens may be more sensitive to dioxins and PCBs than wild birds, indicating that care should be taken when extrapolating ecotoxicity data between wild and domestic bird species.

Mammals

Generally, exposure to dioxins in mammals has a greater effect on growth, survival, and reproduction of animals than on tumor formation. Atrophy of the thymus appears to be the most consistent ecotoxicological effect in mammals. The developing mammalian fetus is especially sensitive to dioxins, and maternal exposure to high environmental levels has been shown to result in increased frequencies of stillbirths. At lower exposure levels, response in offspring generally involves teratogenic effects such as cleft palate and spinal deformities.

While laboratory findings using mice and rats appear to be applicable when extrapolating exposure to wild populations of mammals in field observations and experiments, there is some debate whether laboratory rodents are representative of all wild mammal species, or even the most sensitive species. It is generally recognized that mink is the most sensitive mammalian species to dioxin exposure, and may be about 1 order of magnitude more sensitive than laboratory rodents.

Dioxin TEFs

Dioxins occur in widely varying mixtures in the environment. This is because each source of dioxin generates the individual congeners in different proportions, which may change over time and with transport from one environmental compartment to another through differential degradation, metabolism, uptake, or elimination rates.

It is generally understood that different dioxin congeners are not equally toxic to biota, as defined by their ability to cause specific toxic effects in animals, and despite sharing a common mechanism of action. Therefore, to assess the likely toxicological effect of a particular mixture of dioxin congeners, scientists from the World Health Organization (WHO) and the US Environmental Protection Agency (US EPA) developed TEFs for individual 2,3,7,8-substituted dioxin congeners and dioxin-like compounds. The TEF represents the toxicity of a specific congener relative to 2,3,7,8-TCDD, recognized as the most potent dioxin from the earliest toxicological studies. Until recently, WHO and US EPA recognized different TEF values for some dioxin congeners due to different interpretations of the available toxicological studies. Further, WHO developed separate sets of TEFs for mammals, birds, and fish in response to recognition of differences in sensitivities and toxicity of certain congeners among different taxa. Dioxin TEFs are summarized in **Table 3**.

When interpreting the significance of environmental levels of dioxins, it is common practice to multiply the environmental concentration of individual congeners and their TEFs to derive a toxicity equivalent concentration (TEQ). The effective concentration of a mixture of dioxins is estimated by summing the TEQs of the individual congeners to derive a TEQ for the mixture (TEQ$_m$). That is,

$$\mathrm{TEQ}_m = \sum (\mathrm{TEF}_i \times c_i)$$

where c_i is the concentration of an individual compound and TEF_i is the corresponding factor.

The underlying premise for using a TEF scheme to evaluate dioxins is twofold: first, the mode of action of all 2,3,7,8-substituted dioxins and dioxin-like congeners is the same (i.e., AhR mediated); and, second, the combined effects of the individual dioxin congeners are dose additive. Additivity is an important prerequisite of the TEF concept. There is strong evidence supporting both of these assumptions for dioxins and dioxin-like compounds.

Table 3 World Health Organization (WHO) TEFs for mammals[a], birds[b], and fish[b]

Congeners	Mammals WHO (2005)	Birds WHO (1998)	Fish WHO (1998)
Dioxins			
2,3,7,8–TetraCDD	1	1	1
1,2,3,7,8–PentaCDD	1	1	1
1,2,3,4,7,8–HexaCDD	0.1	0.05	0.5
1,2,3,6,7,8–HexaCDD	0.1	0.01	0.01
1,2,3,7,8,9–HexaCDD	0.1	0.1	0.01
1,2,3,4,6,7,8–HeptaCDD	0.01	<0.001	0.001
OctaCDD	0.0003	0.0001	<0.0001
Furans			
2,3,7,8–TetraCDF	0.1	1	0.05
1,2,3,7,8–PentaCDF	0.03	0.1	0.05
2,3,4,7,8–PentaCDF	0.3	1	0.5
1,2,3,4,7,8–HexaCDF	0.1	0.1	0.1
1,2,3,6,7,8–HexaCDF	0.1	0.1	0.1
1,2,3,7,8,9–HexaCDF	0.1	0.1	0.1
2,3,4,6,7,8–HexaCDF	0.1	0.1	0.1
1,2,3,4,6,7,8–HeptaCDF	0.01	0.01	0.01
1,2,3,4,7,8,9–HeptaCDF	0.01	0.01	0.01
OctaCDF	0.0003	0.0001	<0.0001
PCB congeners IUPAC #			
Non-*ortho*-PCBs			
CB#77	0.0001	0.05	0.0001
PCB#81	0.0003	0.1	0.0005
PCB#126	0.1	0.1	0.005
PCB#169	0.03	0.001	0.00005
Mono-*ortho*-PCBs			
PCB#105	0.00003	0.0001	<0.000005
PCB#114	0.00003	0.0001	<0.000005
PCB#118	0.00003	0.00001	<0.000005
PCB#123	0.00003	0.00001	<0.000005
PCB#156	0.00003	0.0001	<0.000005
PCB#157	0.00003	0.0001	<0.000005
PCB#167	0.00003	0.00001	<0.000005
PCB#189	0.00003	0.00001	<0.000005

[a]Van den Berg M, Birnbaum L, Denison M, *et al.* (2006) The 2005 World Health Organization re-evaluation of human and mammalian toxic equivalency factors for dioxins and dioxin-like compounds. *Toxicological Sciences* 93(2): 223–241.
[b]Van Den Berg M, Birnbaum L, Bosveld A, *et al.* (1998) Toxic equivalency factors (TEFs) for PCBs, PCDDs, PCDFs for humans and wildlife. *Environmental Health Perspectives* 106: 550–557.

Some uncertainties in the TEF scheme are worth noting. The WHO concluded, for example, that there is considerable evidence that the relative toxicity of different congeners varies significantly among different taxonomic groups (e.g., birds, fish, and mammals). These differences may be even broader as more information becomes known regarding ecotoxicity to invertebrates, reptiles, and amphibians. In addition, the underlying toxicology studies used to derive TEF values rely on single large doses that are not typically encountered by animals; repeated exposure to lower doses is more likely, which may result in significant differences in tissue retention and biological response. It has been suggested, therefore, that additional studies are needed to define whether TEFs should be based on intake or tissue levels.

Finally, the assumption that the combined effects of the congeners are additive may not be true in all cases. There are studies providing evidence that non-dioxin-like AhR agonists and antagonists are able to increase or decrease the toxicity of 2,3,7,8-substituted compounds. In addition, there are natural nonchlorinated AhR agonists in the diets of many animals, and some studies have suggested that the potential effects of these may be significant. Several groups of compounds have been identified for possible future inclusion in the dioxin TEF scheme concept based on mechanistic considerations, including 3,4,4'-TCB (PCB 37) and certain polybrominated dibenzo-*p*-dioxins (PBDDs) and dibenzofurans (PBDFs), mixed halogenated dibenzo-*p*-dioxins (PXCDDs) and dibenzofurans (PXCDFs), polychlorinated and brominated naphthalenes (PCNs and PBNs), and polybrominated biphenyls (PBBs). However, for most if not all of these compounds, there is a distinct lack of human and wildlife exposure data.

Major Gaps in Knowledge

While scientific interest regarding the ecological effects of dioxins stems from interest in human heath, limited progress has been made in assessing dioxin risks to animals and plants. In fact, current information is insufficient to provide a thorough description of dioxin risks to wildlife. Consequently, the few available environmental quality guidelines for dioxins are based on limited ecotoxicity data. While the most sensitive and ecologically important endpoints for mammals and birds are associated with reproduction, there is a lack of reproduction bioassays and toxicokinetic information for most taxa and species to establish well-defined dose–response relationships.

Furthermore, it is generally acknowledged that assessing the probability of an individual organism experiencing harm (e.g., reproductive impairment or mortality) is not useful for wildlife assessment; risk is more appropriately assessed at the population, rather than the individual,

level. Population endpoints, however, tend to be difficult to assess, requiring the use of dynamic population models covering effects on survival, breeding success, and immigration. In general, well-validated population models do not yet exist for the majority of species, and it is difficult to estimate the extent of mortality or reproductive failure that could be incurred. Since population models are rarely available, it is more common for wildlife risk assessments to define the 'no observed adverse effect level' (NOAEL) for endpoints such as mortality or reproductive effects in individuals, and to assume that these may be used to set levels that will protect the whole population. Still, few NOAEL values for the dioxins are available, and those that are available are associated with high degrees of uncertainty.

Summary

Laboratory studies examining the toxicity of dioxins are available for only a limited number of species and very little data are available for wildlife. Laboratory studies show that dioxins are AhR agonists, which cause a wide spectrum of adverse effects in many vertebrate species, with embryos, fetus, and newborn being especially vulnerable to exposure during gestation and lactation. Dioxins are particularly potent developmental toxicants at low concentrations and can disrupt the development of the endocrine, reproductive, immune, and nervous system of the offspring of fish, birds, and mammals when exposed from conception through postnatal or post hatching stages. Outside the laboratory, it has not been possible to demonstrate a clear cause/effect relationship between biological response and the exposure to dioxins.

Mammals, birds, and fish vary among species in their sensitivity to dioxins. It has been suggested that, on average, humans are among the more dioxin-resistant species, but the human data are too limited to be conclusive. The most sensitive fish species are the salmonids, while the most sensitive avian species belong to the order of galliformes. The most sensitive mammalian species tested so far are the mink and the guinea pig. Aquatic algae and plants, invertebrates, and amphibians are much more tolerant of dioxins than fish, birds, and mammals. The observed lack of sensitivity of certain taxonomic groups is consistent with the view that the AhR is not present in these organisms.

The organisms at the top of the food chain, such as raptors and dolphins, generally accumulate relatively higher levels of dioxins as compared to lower-trophic-level organisms. However, localized dioxin hot spots cannot be ruled out as a source for dioxins in some individual animals with high levels, particularly those collected in urban/industrial environments. This is because dioxins are highly persistent in soil and sediment, and may still be present at high levels in some areas even though regulatory initiatives are in place to limit releases of dioxins to the environment.

Most existing contaminated sites are the result of historical rather than current emissions, arising from a time when there were fewer environmental controls.

Future work on dioxin ecotoxicology needs to take into account the effects of bioaccumulation of dioxins from the physical environment and from food. As with general toxicity, bioaccumulation appears to be low for congeners with fewer than four chlorine atoms or greater than six, because of rapid metabolism or elimination of sparingly chlorinated forms and poor bioavailability or cell membrane permeability of highly chlorinated forms. Similarly, only congeners with four to six chlorine atoms appear to biomagnify up the food chain.

Further Reading

Boening DW (1998) Toxicity of 2,3,7,8-tetrachlorodibenzo-*p*-dioxin to several ecological receptor groups: A short review. *Ecotoxicology and Environmental Safety* 39: 155–163.

Bosveld ATC and van den Berg M (1994) Effects of polychlorinated biphenyls, dibenzo-*p*-dioxins, and dibenzofurans on fish-eating birds. *Environmental Reviews* 2: 147–166.

Davies M (1999) *Compilation of EU Dioxin Exposure and Health Data Task 7 – Ecotoxicology*. European Commission DG Environment; UK Department of the Environment, Transport and the Regions (DETR) (97/322/3040/DEB/E1).

DeMarch BGE, DeWit CA, and Muir DCG (1998) Persistent organic pollutants. In: *AMAP Assessment Report: Arctic Pollution Issues*, pp. 183–371. Oslo: Arctic Monitoring and Assessment Programme (AMAP).

Gatehouse R (2004) Ecological risk assessment of dioxins in Australia. *National Dioxins Program Technical Report No. 11*. Canberra: Australian Government Department of the Environment and Heritage.

Giesy JP, Feyk LA, Jones PD, Kannan K, and Sanderson T (2003) Topic 4.8. Review of endocrine-disrupting chemicals in birds. *Pure and Applied Chemistry* 75(11–12): 2287–2303.

Giesy JP, Ludwig JP, and Tillitt DE (1994) Dioxins, dibenzofurans, PCBs and wildlife. In: Schecter A (ed.) *Dioxins and Health*, pp. 249–307. New York: Plenum.

Gray LE, Jr., Ostby J, Wolf C, Lambright C, and Kelce W (1998) The value of mechanistic studies in laboratory animals for the prediction of reproductive effects in wildlife: Endocrine effects on mammalian sexual differentiation. *Environmental Toxicology and Chemistry* 17: 109–118.

Grimwood MJ and Dobbs TJ (1995) A review of the aquatic ecotoxicology of polychlorinated dibenzo-*p*-dioxins and dibenzofurans. *Environmental Toxicology and Water Quality* 10: 57–75.

Henshel DS (1998) Developmental neurotoxic effects of dioxin and dioxin-like compounds on domestic and wild avian species. *Environmental Toxicology and Chemistry* 17: 88–98.

Loonen H, van de Guchte C, Parsons JR, de Voogt P, and Govers HAJ (1996) Ecological hazard assessment of dioxins: Hazards to organisms at different levels of aquatic food webs (fish-eating birds and mammals, fish and invertebrates). *The Science of the Total Environment* 182: 93–103.

Meyn O, Zeeman M, Wise MJ, and Keane SE (1997) Terrestrial wildlife risk assessment for TCDD in land-applied pulp and paper mill sludge. *Environmental Toxicology and Chemistry* 16: 1789–1801.

USEPA (1995) *Great Lakes Water Quality Initiative Criteria Documents for the Protection of Wildlife DDT; Mercury; 2,3,7,8-TCDD; PCBs*. Washington, DC: Office of Science and Technology for the Office of Water (EPA-820-B-95-008).

USEPA (2001) *Critical Review and Assessment of Published Research on Dioxins and Related Compounds in Avian Wildlife – Field Studies. External Review Draft*. Cincinnati, OH: National Center for Environmental Assessment, Office of Research and Development.

USEPA (2002) *Dose–Response Assessment from Published Research of the Toxicity of 2,3,7,8-Tetrachlorodibenzo-p-dioxin and Related Compounds to Aquatic Wildlife – Laboratory Studies*. Cincinnati, OH: National Center for Environmental Assessment, Office of Research and Development (EPA/600/R-02/095).

US EPA (2003) *Non-dioxin-like PCBs: Effects and consideration in ecological risk assessment. Experimental Toxicology Division National Health and Environmental Effects Research Laboratory office of Research and Development* (NCEA-C-1340. ERASC-003. June).

Van Den Berg M, Birnbaum L, Bosveld A, *et al.* (1998) Toxic equivalency factors (TEFs) for PCBs, PCDDs, PCDFs for humans and wildlife. *Environmental Health Perspectives* 106: 550–557.

Van den Berg M, Birnbaum L, Denison M, *et al.* (2006) The 2005 World Health Organization re-evaluation of human and mammalian toxic equivalency factors for dioxins and dioxin-like compounds. *Toxicological Sciences* 93(2): 223–241.

Endocrine Disruptors

S Matsui, Kyoto University, Kyoto, Japan

Introduction
Which Chemical Compounds Are Endocrine
    Disruptors?

Ecological Effects of Endocrine Disruptors
Further Reading

Introduction

Endocrine disruptors are chemical compounds that function as mimics of authentic agonist or antagonist ligands to endocrine receptors in animals. Under normal condition, any animal produces its authentic ligands of hormones and regulates homeostatic condition inside the body against external stress. Endocrine disruptors create a variety of functional troubles with animals including human beings. Among many types of hormones, reproductive hormones are important to wildlife, while human beings are concerned with hormone mimics of synthetic as well as natural chemicals in terms of endocrine, nerve, and immune system disruption. Irreversible endocrine disruption may occur at a sensitive development stage that is called a window period. Among development stages, fetus and perinatal stages are most sensitive to many disrupting chemicals. Those disrupting chemicals are effective in extremely low concentration ranges (picograms to nanograms per liter of blood) when they enter cells. There are optimum concentration ranges for different endocrine chemicals functioning inside cells, indicating lower concentration and higher concentration than the optimum range ineffective (called a reverse U-type concentration range). Endocrine chemicals function as information substances. After transferring information by coupling with hormone receptors, endocrine chemicals are easily oxidized and transformed into conjugates or broken amino acids, resulting in discharge out of cells and the body. Transferring information always induces necessary gene up- and downregulations, controlling gene functions for homeostatic condition. Endocrine systems evolved along evolution of animals. Authentic endocrine chemicals are produced from variety of external substances during metabolism. For example, many steroid hormones such as androgen, estrogen, corticoids, etc., are produced from cholesterol. Biogenicamine hormones such as serotonin, melatonin, etc., are produced from tryptophan. The fate of endocrine disrupting chemicals inside the cell is important to understand their toxic endpoints. During their metabolism, they may produce different toxic effects including DNA injury.

Which Chemical Compounds Are Endocrine Disruptors?

Table 1 gives an overview of the chemical compounds that are considered to be endocrine disruptors (see **Figure 1** also). The table indicates on which hormone and ligand the endocrine disruptors are acting.

Ecological Effects of Endocrine Disruptors

A number of ecological effects by endocrine disruptors have been observed, mainly as sexual abnormality in wildlife. **Table 2** gives an overview of the most important cases of ecological effects by endocrine disruptors.

Endocrine disruption may be associated with immune disruption with wildlife indicating weak resistance against virus infection. Further endocrine disruption may be associated with DNA injury. The injury can be explained by DNA adduct formation of endocrine disrupting

Table 1 Overview of the most important endocrine disruptors

Chemical compound	Acting on as ligands (either agonist or antagonist)
Coplanar PCB	Aryl hydrocarbon receptor – agonist
Hydroxide coplanar PCB	Thyroid hormone receptor – antagonist
Dioxin	Aryl hydrocarbon receptor – agonist
PAHs	Aryl hydrocarbon receptor – agonist
Nonylphenol	Estrogen receptor – agonist
Phthalates	Arachidonic acid hormone receptor – agonist or antagonist?
TBT	P450 aromatase inhibitor?
Insecticide (fenoxycarb, pyriproxyfen, etc.)	Juvenile hormone receptor-agonist

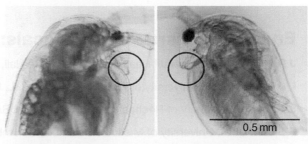

Figure 1 Male (left) and female (right) *Daphnia magna* (Crustacea, Cladocera) aged *c.* 24 h. Males can be distinguished from females by the length and morphology of the first antennae as shown in the circles. Juvenile hormones (ecdysterone) and their analogs endocrine disrupting chemicals (methylfarnesoate, methoprene, pyriproxyfen, and fenoxycarb) may easily change sex of *D. magna*. Adapted from Tatarazako N and Oda S (2007) The water flea *Daphnia magna* (Crustacea, Cladocera) as a test species for screening and evaluation of chemicals with endocrine disrupting effects on crustaceans. *Ecotoxicology* 16: 197–203.

Table 2 Overview of ecological effects of endocrine disruptors

Species	Description of effect
American alligator	Males show shortened penis
Roach	Males show quasi-female
Medaka fish (small herbivore fish)	Males show quasi-female
Mosquitofish (small carnivore fish)	Females show quasi-male
Daphnia Magna	Male offspring
Sea snails	Females show quasi-male
Flatfish	Males show quasi-female
Sea cormorant	Malformation of bill

chemicals themselves after oxidation, or formation of reactive oxygen species or reactive molecules during oxidation and reduction inside the cell. Those reactive molecules form a variety of DNA adducts. The DNA injury can be determined by liquid chromatography–tandem mass spectrometry (LC/MS/MS) analytical instruments.

See also: Effects of Endocrine Disruptors in Wildlife and Laboratory Animals.

Further Reading

Colborn T, Dumanoski D, and Myers JP (1996) *Our Stolen Future: Are We Threatening Our Fertility, Intelligence, and Survival? A Scientific Detective Story*, 306pp. New York: Dutton (ISBN 0452274141).

Endocrine Disruptors (2006) National Institute of Environmental Health Sciences, http://www.niehs.nih.gov/health/topics/agents/endocrine/(accessed on Jan 2008).

Executive Summary (2002) Global assessment of the state-of-the-science of endocrine disruptors. International Programme on Chemical Safety, World Health Organization, http://www.who.int/ipcs/publications/en/ch1.pdf (accessed January 2008).

Tatarazako N and Oda S (2007) The water flea *Daphnia magna* (Crustacea, Cladocera) as a test species for screening and evaluation of chemicals with endocrine disrupting effects on crustaceans. *Ecotoxicology* 16: 197–203.

Relevant Websites

http://ec.europa.eu – European Commission – Environment – Endocrine disrupters.

http://www.epa.gov/ – United States Environmental Protection Agency, Endocrine Disruptor Screening Program.

http://www.nrdc.org – Endocrine Disruptors FAQ, Natural Resources Defense Council.

Endocrine Disruptor Chemicals: An Overview

J P Myers, Environmental Health Sciences, White Hall, VA, USA

L J Guillette Jr., University of Florida, Gainesville, FL, USA

S H Swan, University of Rochester, Rochester, NY, USA

F S vom Saal, University of Missouri-Columbia, Columbia, MO, USA

Published by Elsevier B.V., 2010.

Introduction	The 'Low-Dose' Issue and Inverted-U Dose–Response
Definitions	Relationships for EDCs
Examples of Endocrine Disruption	Summary
Theoretical Concepts	Further Reading

Introduction

In a series of three articles we provide an overview of the impacts of endocrine disrupting chemicals (EDCs) on human health, laboratory animals, and wildlife. We begin with a brief definition of endocrine disruption and a consideration of several interrelated developments in the field, which necessitate important conceptual shifts in the theory and practice of toxicology. We then examine some of the data behind these conceptual shifts, first from studies of wildlife, then from studies of laboratory animals. Finally, we shift to a consideration of the considerable challenges this new science of endocrine disruption presents to epidemiology. We then briefly describe the progress being made toward a more 'environmentally sensitive epidemiology' and illustrate these concepts with data from a recent study that uses biomarkers of low environmental levels of EDCs to identify some of the impacts of these chemicals on human health.

Definitions

All living organisms depend upon a large and intricate array of chemical signaling systems to guide biological development and regulate cell and organ activity. Over the past two decades, scientific interest in the ability of many environmental contaminants to interfere with these sensitive systems has grown dramatically. A hybrid science, the study of endocrine disruption, has arisen from concerns about the effects of these phenomena on health and the environment. This science incorporates findings and methodologies from multiple disciplines including toxicology, endocrinology, developmental biology, molecular biology, ecology, behavioral biology, genetics, and epidemiology.

EDCs are chemicals that can disrupt any aspect of endocrine processes. Hormonal systems for which there is clear evidence of endocrine disruption include thyroid hormones, androgens, and estrogens. EDCs disrupt development by interfering with the hormonal signals that control normal development of the brain and other organ systems. EDCs can also affect adults by similar mechanisms, because these same hormones also play important regulatory roles in adults. EDCs can act at very low levels of exposure to produce profound effects on the course an organism follows from fertilized oocyte through to maturity, adulthood, and death. The effects of EDCs on developing organisms are of greatest concern, since the disruptive effects of developmental exposure, referred to as organizational effects are permanent and irreversible. EDC exposure also produces measurable, activational effects in adults that may be reversible (they often do not persist when the organism is no longer exposed). A related field of research, 'developmental origins of health and adult disease (DOHAD)', and the new DOHAD society, is converging with research on endocrine disruption. Work in this area shows that exposures during different stages of development, particularly during fetal life, contribute to adult chronic diseases, including obesity, heart disease, diabetes, decreased fertility, impaired immune function, and neurological deficits.

Examples of Endocrine Disruption

Data accumulated over the past two decades reveal substantial global contamination by EDCs. Contaminant dispersal is brought about by a combination of factors, including purposeful or accidental release into the environment, followed by long-range atmospheric transport. It also occurs because some EDCs have been incorporated both deliberately and inadvertently into consumer products. With regard to long-range transport, large masses of air have been tracked across the Pacific carrying a variety of pollutants from central Asia to the west coast of the US virtually undiluted, including ozone, heavy metals, and organochlorine compounds. In addition, so-called 'global distillation' processes – repeated sequences of volatilization and condensation – transport semivolatile compounds from sites of production, use, and disposal to colder regions, particularly at high latitude and altitude. The accumulation of vast amounts of plastic products in the oceans is yet another source of global pollution, due to leaching of endocrine disrupting chemicals from plastic.

Two of the many examples of inadvertent contamination of people due to the use of consumer products that contain EDCs include exposure to phthalates and bisphenol A. Phthalates are used as additives in cosmetics, intravenous medical tubing, and a wide variety of other products, including those made from polyvinyl chloride (PVC). Polyvinyl chloride products contain phthalates to soften the otherwise brittle PVC; so all PVC contains

some amount of a phthalate, the softer the product, the greater the amount of phthalate. Exposure to bisphenol A is also widespread. Bisphenol A is a monomer (not just an additive) used in the manufacture of resins that line the inner surface of food metal cans, and to manufacture polycarbonate plastic, which is hard and clear (although it can be colored), and is used to make food and beverage containers as well as a wide range of other products; bisphenol A is thus also used as an additive in many types of plastic, including PVC. Phthalates and bisphenol leach from these products and disrupt endocrine function. There are over 6 billion pounds of bisphenol A and in excess of 4 billion pounds of phthalates produced each year.

Coincident with emerging knowledge of the ability of EDCs to disrupt a range of developmental processes, increases have been reported in a wide range of human health diseases and abnormalities, though some remain controversial. These include increases in the frequency of (obesity and cognitive/behavioral dysfunctions, such as autism and attention deficit hyperactivity disorder (ADHD). There have also been increases in a cluster of male reproductive outcomes (cryptorchism, hypospadias, testicular cancer, and decreased sperm function) all of which are believed to originate *in utero*, and have been termed the testicular dysgenesis syndrome. . The strength of the epidemiological evidence demonstrating these epidemics varies. For example, there is little argument that there has been a widespread increase in rates of obesity and diabetes, but there is still significant debate about global decreases in reproductive function or increases in ADHD, due to limitations of historical data. While extensive study will be required to identify causes of these trends, their underlying biology suggests that alterations in inter- and intracellular signaling processes may be causally involved, and for each of the mentioned epidemics, data are available indicating one or more points of vulnerability to EDCs in the mechanisms of control. EDCs may also contribute importantly to geographic variability in these health endpoints.

Theoretical Concepts

New scientific findings on these issues are emerging at an exponential rate. Central to these findings is a reformulation of the traditional dichotomy between nature and nurture (the gene vs. environment argument) in the causation of disease (**Figure 1**).

That which is 'nature' is based on genes, while 'nurture' comes from the environment, *sensu latu*. Functional status and disease linked to genes were previously perceived as completely determined by heredity. Diseases traditionally viewed as nonhereditary ('environmental') can be caused by a wide array of exposures, stressors,

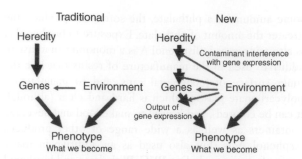

Figure 1 Contrasting traditional with new formulations of the interactions of genes and environment in the determination of phenotype. Traditionally, genetic diseases have been seen as determined by heredity. In the new formulation, genetic patterns of gene expression are vulnerable to disruption by environmental contaminants at multiple points in the sequence of steps that lead to gene expression, thereby rendering genetic diseases susceptible to modification by environmental factors.

experiences, nutrition, and other lifestyle factors. Concern about environment's interaction with the genetic determination of disease and functional differences has focused traditionally upon two pathways: (1) high-dose chemical exposures causing mutations and thus alterations in the base sequence of genes, and (2) genetic variation among individuals leading some to be more susceptible than others to certain contaminants.

The study of endocrine disruption today is dramatically altering this historical conceptualization. A property linked to a gene is, rather than simply being a factor determined by inheritance, one that is vulnerable to environmental disruption, particularly by EDCs. This is because EDCs at low levels can act during development to permanently interfere with gene expression and other cellular activities, resulting in abnormalities and disease that becomes apparent in adulthood . Thus, while some functional deficits and disease states are due to inherited mutations in genetic makeup, many diseases may also be associated with alterations in gene expression. These alterations relate to differential methylation of cytosine in the 5′ region of the promoter (where the presence of methyl-cytosine leads to gene silencing) as well as changes in acetylation/methylation of histone proteins that determine whether genes are able to be activated by transcription factors. These chemical modifications of chromatin and DNA are referred to as 'epigenetic' changes and are now recognized as a major factor in the process of cell differentiation as well as the ontogeny of cancer, in contrast to classical mutations, which involve base deletions and substitutions.

Initially, the majority of research on EDCs focused on interference with gene activation by the hormone 17β-estradiol (the most potent endogenous estrogen), and some initially assumed that endocrine disruptors were all environmental estrogens. In contrast to this assumption, it is now recognized that many EDCs can

stimulate genes and other cellular processes that have nothing to do with estrogen. Thus, over the last decade, EDCs have been shown to disrupt many other endogenous hormonal signaling molecules, including virtually all steroid hormones that have been carefully tested, as well as thyroid, retinoid, leptin, some transcription factors, growth factors, and other molecules not traditionally classified as hormones. One recent study even documents interference with chemical signaling between two symbiotic organisms, the bacterium *Rhizobium* and its leguminaceous host. The presumption now is that any chemically mediated signaling system is vulnerable, in principle, to disruption by chemicals to which wildlife and humans are exposed in their daily lives. However, there are, in fact, many EDCs that do act via cellular mechanisms that mediate the response to estradiol, whereas other EDCs antagonize estradiol or block the synthesis of estradiol. Whether estrogenic and antiestrogenic EDCs remain the largest group of chemicals that interfere with the endocrine system remains to be determined.

Given the enormous potential for EDCs to interfere with gene expression, how many of the 80 000+ chemicals registered for commercial use has endocrine disrupting activity? The vast majority of chemicals have not been tested in even the most basic way. Far fewer have been tested for endocrine disrupting effects, particularly during embryonic development, the most vulnerable time in life.

Altered gene expression during organismal development can induce dramatic and irreversible changes in developmental outcomes. Known effects of EDCs range from structural changes to functional deficits. For example, alterations in the production of hormone receptors in tissues through the alteration in the expression of genes for these receptors have been shown in experiments with laboratory animals, and these changes can then lead to altered responses to hormonal stimulation throughout the remainder of life. This can, in turn, lead to altered (increased or decreased) susceptibility to contamination with hormonal activity. .

Altered gene expression and cellular signaling subsequent to development can cause transient changes, termed activational responses, or particularly through carcinogenesis, permanent detrimental effects. For example, lifetime exposure to estrogen is the best predictor of breast cancer in women, and exposure to EDCs that are 'environmental estrogens' could plausibly increase breast cancer risk. Thus, the impact of EDCs vary depending upon a variety of factors, including when in the life cycle of the organism exposure occurs, as well as the duration and amount of exposure. Until recently, the great importance of life stage, the very great vulnerability of the embryo, and the fact that consequences of fetal exposure could be entirely different from those seen from adult exposure had not been appreciated.

Collectively, these new data from studies of EDCs are forcing a series of conceptual shifts that undermine long-held assumptions underlying toxicological studies and the applications of results from these studies to developing public health standards.

The 'Low-Dose' Issue and Inverted-U Dose–Response Relationships for EDCs

Foremost among these is a challenge to the operating assumption concerning doses appropriate for toxicological testing. Focusing on traditional toxicological endpoints, such as gene mutations, weight loss, and death, toxicologists customarily worked at what now are viewed as very high doses, typically in the range of parts per million (ppm) and parts per thousand levels. New data suggest that extremely low doses of EDCs (in the part per billion (ppb) and even part per trillion (ppt) range) can cause measurable and highly significant endocrine disruption. A growing array of studies reveals changes in gene expression, including both gene suppression and gene activation, as a result of low-level exposure to EDCs. For example, recent work on arsenic, long-established to be toxic at high doses, has revealed that at part per billion levels, arsenic can interfere with gluco-corticoid activation of genes involved in the control of metabolism, response to stress, immune function, and the suppression of tumor formation.

A second important conceptual shift arises from consistent findings that during the life cycle of an organism, developmental stages are typically far more vulnerable to signal disruption than adult stages. This is thought to occur for several reasons, including the absence of fully developed protective enzyme systems and higher metabolic rates. Most importantly, however, the events underway in development involve a series of organizational 'cross-roads' that are irreversible once the 'choice' in development is determined. In sharp contrast, in adults, the 'activational' processes at play can very often be reversed by removing the EDC, thus returning gene expression levels and organ function to normal. One recent example documenting extraordinary differential sensitivity of adult versus developing life stages was that adverse effects in tadpoles occurred at 1/30 000th of the lowest concentration of atrazine, a widely used herbicide, that was found to produce adverse effects in adults. Another related example from pharmacology is the synthetic estrogen diethylstilbestrol (DES), which caused irreversible reproductive changes and cancers in prenatally exposed females, but had little effect on the mothers who were exposed to massive doses throughout pregnancy.

One clear implication of this focus on low-level exposure during fetal and neonatal development is that levels of exposure that have been dismissed as 'background' and thus 'safe' may have deleterious effects. This was not recognized previously due to the absence of studies using these low doses, combined with the absence of studies of developmental effects at any dose level. Many laboratory studies now support the conclusion of high sensitivity of the embryo and neonate, as do some epidemiological data from human studies, which are discussed in more detail below. For example, a series of studies of children born to mothers exposed during pregnancy to polychlorinated biphenyls (PCBs) through consumption of fish and other food have shown that low parts-per-billion concentrations of these contaminants impair cognitive development, resulting in a decrease in IQ. One specific reported consequence of exposure to PCBs is a shift in the pattern of play behavior in boys toward patterns more typical of girls.

Toxicological experiments, particularly those used to develop regulatory standards of acceptable levels of exposure to environmental chemicals have been based upon the assumption that 'the dose make the poison,' which implies that high doses invariably cause more harm than lower doses. However, for hormonally active chemicals, as well as many well-recognized exposures such as vitamins, this assumption is not valid. In fact, endocrinologists and physicians have known for decades that very high doses of hormones and drugs can block rather than stimulate some responses, resulting in what is referred to as a non-monotonic dose–response relationship, that is a response in which effects initially increase and then decrease with increasing dose.

For example, recent work by Welshons and colleagues examined effects in response to estradiol exposure in human breast cancer cells (**Figure 2**). Estradiol levels between 0.1 and 100 ppt produced an increased growth response in breast cancer cells, because at these levels, an increase in exposure causes an increase in the number of estrogen receptors bound by estradiol, thus leading to increased gene activation. At exposure levels in the typical toxicological dose range (part per million range), further increases in the dose of estradiol began to produce cell death. This result is extremely important for regulatory toxicology, because the high-level exposures in these experiments are analogous to those used for the prediction of risk posed by low doses, but the actual effects of low doses predicted to be safe have, until very recently, never been examined experimentally. Changes in dose within this very high part per million dose range cannot reveal variations in receptor-mediated gene activation, since all receptors are already completely occupied (saturated) at much lower doses (a change in receptor occupancy is required to observe a change in response

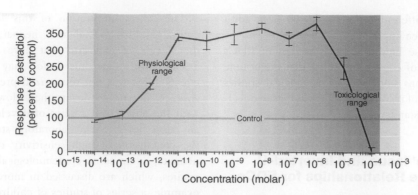

Figure 2 Proliferative response of MCF-7 cells to 17-β estradiol over 10 orders of magnitude. Responses in the 'physiological range' are mediated by binding with the estrogen receptor. Those in the 'toxicological range' reflect cell death. Adapted from Welshons WV, Thayer KA, Judy BM, et al. (2003) Large effects from small exposures. In: Mechanisms for endocrine-disrupting chemicals with estrogenic activity. *Environmental Health Perspectives* 111: 994–1006.

to a hormone or hormone-mimicking EDC). Hence, testing EDCs at only very high doses will miss receptor-mediated events that can be expected to occur at much lower levels of exposure.

These types of inverted-U dose–response curves have been reported for drugs, such as DES, which stimulated the growth of the fetal mouse prostate at a low oral dose fed to pregnant mice of 0.02 $\mu g\,kg^{-1}d^{-1}$, a maximum stimulatory response occurred at 0.2 $\mu g\,kg^{-1}d^{-1}$, and a complete inhibition of prostate development occurred at 200 $\mu g\,kg^{-1}d^{-1}$. There are also examples of inverted-U dose–response curves for environmental chemicals, such as the chemical used to make polycarbonate plastic, bisphenol A. For example, in rat pituitary tumor cells, bisphenol A significantly stimulated a rapid (within 30 s) influx of calcium followed by prolactin secretion at the lowest dose that was examined (0.23 ppt); the greatest response occurred at 230 ppt, while the magnitude of the response decreased at 2.3 ppb, forming an inverted-U dose–response curve. In another study, human prostate cancer (LnCAP) cell proliferation was stimulated by bisphenol A, with the amount of proliferation increasing between 23 ppt to a maximum effect at 230 ppt, after which the response decreased back to no proliferative effect at 23 ppb. The surfactant octylphenol was examined for effects on egg production in ramshorn snails (*Marisa cornuarietis*). Octylphenol produced a marked (eightfold) increase in egg production between 1 and 25 μg l^{-1}, but at 100 μg l^{-1} the stimulating effect markedly decreased relative to the 25 $\mu g\,l^{-1}$ dose. This stimulation of egg production during the sexual repose phase of the snails is detrimental to the affected females, because it causes a congestion of clutches in the pallial oviduct, leading to a rupture of the oviduct and ultimately to the female's death.

Identification of low-dose effects that are different from those seen at high doses, the importance of timing of exposure, recognition of the unique effects that can be disrupted during development, and genetic variation in genetically determined susceptibility, render the overly simplistic assumptions previously used in risk assessment invalid for many environmental chemicals.

Summary

In this article we have outlined evidence from diverse sources indicating that a variety of manmade compounds can interfere with reproductive system and brain development, resulting in reduced fertility, altered brain function, and behavior in wildlife, laboratory animals, and humans. Four summary points emerge based on the findings presented above and elsewhere in this encyclopedia:

- Contaminants at low levels can interfere with gene expression.
- Wildlife, laboratory animal, and human effects are strongly concordant.
- The available data are not consistent with several key assumptions traditionally used to guide regulatory science and regulations.
- Traditional epidemiology will have great difficulty establishing causation of effects of these chemicals in humans.

See also: Effects of Endocrine Disruptors in Wildlife and Laboratory Animals; Epidemiological Studies of Reproductive Effects in Humans.

Further Reading

Colborn T, vom Saal FS, and Soto AM (1993) Developmental effects of endocrine-disrupting chemicals in wildlife and humans. *Environmental Health Perspectives* 101: 378–384.

Colborn T, Dumanski D, and Myers JP (1996) *Our Stolen Future*. New York: Penguin.

Jacobson JL and Jacobson SW (1996) Intellectual impairment in children exposed to polychlorinated biphenyls *in utero*. *New England Journal of Medicine* 335: 783–789.

Michaels D (2005) Doubt is their product. *Scientific American* 292: 96–101.

Vreugdenhil HJ, Slijper FM, Mulder PG, and Weisglas-Kuperus N (2002) Effects of perinatal exposure to PCBs and dioxins on play behavior in Dutch children at school age. *Environmental Health Perspectives* 110: A593–A538.

Welshons WV, Thayer KA, Judy BM, *et al.* (2003) Large effects from small exposures. In: Mechanisms for endocrine-disrupting chemicals with estrogenic activity. *Environmental Health Perspectives* 111: 994–1006.

Relevant Website

http://www.environmentalhealthnews.org – Environmental Health News.

Halogenated Hydrocarbons

M A Q Khan, University of Illinois, Chicago, IL, USA

S F Khan, University of Denver, Denver, CO, USA

F Shattari, University of Boston, Boston, MA, USA

Introduction
Halogenated Aliphatic Hydrocarbons
Halogenated Aromatic Hydrocarbons
HHCs Used in Agriculture
HHCs in Polymers and By-Products
Naturally Produced HHCs

Environmental Fate of HHCs
Biotransformation and Biodegradation of HHCs
Toxicity and Mechanisms of Toxicity of HHCs
Ecotoxicology of HHCs
Further Reading

Introduction

The United States produces about $19 billion worth of organic (petroleum-based) chemicals annually. These organic chemicals are used as intermediates in synthesis of economical chemicals, such as plastics, nylon fibers, adhesives, agrochemicals, and pharmaceuticals. Many of these synthesis processes either use chlorine or halogens directly or as intermediates, such as chlorobenzenes in the manufacture of pesticides and dyes, and vinyl chloride (12 million tons yr^{-1} in USA) in the synthesis of polyvinyl chloride (PVC). The export of downstream products of these syntheses, such as ethylene dichloride, vinyl chloride monomer, and PVC polymer rose to 1229 million metric tons in 1995. This article identifies those economical HHCs, produced and used in the USA and abroad, which have caused and are causing ecological and human health (occupational, environmental) concerns.

Hydrocarbons in which the hydrogen (forming the carbon and hydrogen bond, R·C–H) is replaced with a halogen substituent (R·C–X or RX) become chemically more stable and hydrophobic. These properties can be affected by various other factors also, such as carbon chain length and its branching, functional groups, etc. This halogenation of aliphatic (acyclic and alicyclic) and aromatic (one and more benzene rings) hydrocarbons includes a diversity of economical chemicals, used globally in amounts of millions of tons per year (**Table 1**).

Halogenated Aliphatic Hydrocarbons

Large amounts of halogenated hydrocarbons (HHCs), including vinyl halides, are produced in massive amounts in the USA (**Table 2**). These HHCs can include alkyl halides, namely haloalkanes, haloalkenes, and haloalkynes. Some of the unsaturated aliphatic and alicyclic HHCs are used (**Table 2**) as industrial solvents, others as chemical intermediates (vinyl chloride, phosgene, hexachlorocyclopentadiene, chlorohydrins, etc.) in the synthesis of other economical compounds. Still others are used in the manufacture of various important HHCs, such as chlorinated paraffins (C_{10} to C_{30} chain length), chlorofluorocarbons (banned), hexachlorobutadiene, halothane anesthetics, insecticidal aliphatic and alicyclic cyclodienes, Toxaphene, mirex, kepone, hexachlorocyclohexanes, chlorinated pyrethroids, and Chlorpyriphos.

Various HHCs, especially saturated ones, are used as intermediates for nonhalogenated products (polyurethanes, polycarbonates), epichlorohydrin in resins, chlorohydrin for synthesis of propylene oxide, and for specialty chemicals, such as Grignard's reagent, Freidel–Crafts reaction. Phosgene (COCl) is extensively used on-site – 80% in producing isocyanates and di-isocyanates for synthesis of polyurethane (>2 million metric tons yr^{-1} globally), 10% in polycarbonate plastics (0.2 million metric tons yr^{-1} globally). The remaining phosgene is used in other chemicals, such as herbicides,

Table 1 Some abundantly used HHCs and their ecosystem health and concerns

Name	Uses	Effect(s)/system(s)	Human exposure
Aliphatic HHCs			
Acetyl chloride	In pharmaceutical and pesticide manufacture	HS, HW	
Bis(2-chloroisopropyl) ether	Solvents	SC, HS, HW, PTP	Occupational
Bis(chloromethyl) ether	In industry and in laboratory, an alkylating agent in the manufacture of polymers. As solvent of polymerization reactions, in the preparation of ion-exchange resin. Intermediate for organic synthesis	SC, HS, HW, PTP	Occupational
Bromodichloromethane		SC, PTP	
Bromoform	As a solvent, flame retardant, floating agent, rubber vulcanizing, mineral separation, chemical synthesis	C, HW, PTP	
Carbon tetrachloride	Specialty process solvent, in semiconductors, metal recovery, cables, as a catalyst, azeotropic drying agent for wet spark plugs, soap fragrance, extracting oils	Rat/mouse, NMC, causes tumors in liver	
Chlorinated paraffins	As plasticizers in plastics, in paints, as metal lubricant, in adhesives, as fire retardants		
2-Chloro-1,3-butadiene	Manufacture of artificial rubber		
Chloroform	In synthesis of fluorochemicals and fluoro polymers, in dry cleaning, as a rubber solvent	Rat/mouse liver, NMC, causes tumors, SC, HW, PTP	
Chloromethyl methyl ether	In the chemical industry for synthesis of organic chemicals	SHC, HW, PTP	
3-Chloro-2-methyl-1-propane (and allyl bromide, tetrachloroethylene)	Insecticidal fumigants		
2,3'-Dichloro-1,4-dioxane		AC	
1,1-Dichloroethane	Solvent, paints and varnish remover, in rubber cement, insecticide sprays, fire extinguishers, gasoline, high-vacuum rubber, ore-flotation, plastic and fabric spreading, produces vinyl chloride on thermal cracking		
Dichloroethyl ether		AC, HW, PTP	
Dimethyl carbamoyl chloride		APC, HW	
Ethylene dichloride	Paints, varnish and finishing remover	CP, HS, HW, PTP	
Hexachloro-1,3-butadiene	Solvent, intermediate in lubricants and rubber, fumigant	C, HW, PTP	
Hexachloroethane	Degreasing metals, in pyrotechnics, explosive and military	CP, HW, PTP, causes tumors in liver	
Methyl bromide	In degreasing wool, sterilizing food, extracting oils		
Methyl chloride	An intermediate in HHCs, in silicone industry, as a methylating agent, solvent and diluent for butyl rubber, foaming agent for plastics, in thermo-equipment fluids	SC, HW, PTP	
Methylene chloride	As a solvent, in paint stripping, extracting coffee and other chemicals from natural food, in manufacture of plastics, pharmaceuticals, in cleaning of metals, degreasing, in aerosols, adhesives	SC, HW, PTP	70 000 workers exposed
1,1,1-Trichloroethane	Restricted production and use, major industrial solvent, cleaning plastic mold, cold metal cleaning, coolant, lubricant for cutting oils, solvent for dyes, cleaning agent, in textile spotting fluid	C, HW, PTP	
1,1,2-Trichloroethane		ACP, HW, PTP, causes tumors in liver	
Trichloroethylene	In degreasing and dry cleaning, solvent, paint thinner, extracting caffeine, dental anesthetic, insecticide (fumigant)	CP, HS, HW, PTP, mouse liver, NMC, causes tumors in liver	

(Continued)

Table 1 Continued

Name	Uses	Effect(s)/system(s)	Human exposure
1',1,2'2-Tetrachloroethane		CP, HW, PTP	
Tetrachloroethylene	A solvent in cleaning cloths and metals, for degreasing	CP, HW, PTP, causes tumors in liver	
Vinyl cyclohexene dioxide		AC	
Vinylidene chloride		C, HS, HW, PTP	
Vinyl chloride	Synthesis of PVC	HC, HW, PTP	
Aromatic HHCs			
Benzoyl chloride	Textile and dye industries as a fastness compound		
Benzyl chloride	Intermediate in benzoil compounds (benzal chloride, benzyl alcohol, and benzaldehyde), manufacture of quaternary ammonium chloride, plastics, dyes, synthetic tannins, pharmaceuticals, perfumes, and resins	AC, HS, HW	
CCIU		C, HS, HW, PTP	
Chlorambem	Formulation or application of this pre-emergence herbicide	CP	Workers involved in the manufacture
Chlorambucil	Drug against cancer	HC, AC, HS	
Chlorobenzenes	In synthesis of chloroanilide herbicides, bacteriocides		
Chloronaphthalenes	Heat transformer media, solvents, lubricant, dielectric fluid, electric insulation		
Chloroprene	In the production of artificial rubber	C	2500 workers
Dichlorobenzene	Intermediate in polyphenylene insecticide for termites (fumigant), disinfectant		
Epichlorohydrin		APC, HS, HW	
Hexachlorobenzene	Fungicide, intermediate for dye and hexafluorobenzene, synthetic rubber, plasticizer in PVC, military pyrotechnic, electrodes	C	
Hexachlorophene	Topical anti-infective agent, detergent, antibacterial soaps, surgical scrubs, hospital equipment, fungicide		
Lindane		C(N,P), HS, HW, PTP, mouse liver, NMC	
Monochlorobenzene	In synthesis of nitrochlorobenzene as a dye intermediate, solvent		
O-Nitrochlorobenzene		APC	
P-Nitrochlorobenzene			
PCBs (polychlorinated biphenyls)		AC, HS, HW, PTP, rat/mouse liver, NMC	
P-Chlorobenzyl chloride		C, HS	
Pentachloronitrobenzene		APC, HW	
Picloram		CP	
TCDD		AC, HW, PTP, rat liver, lungs, NMC	
Trichlorobenzene	Heat transformer media, transformer fluid, and solvent		
Pesticides			
Aldrin	Pesticide manufacture, formulator and applicators	C, HS, HW, PTP	In air, water, and food
Allyl chloride	In the manufacture of epichlorophydrin and glycerol	HS	5000 workers
Chlordecone	Processing or formulating pesticides, control rootborers on bananas, wireworm control in tobacco and ants	C, HS, HW	600 workers
Chlorobenzilate		Rat liver, NMC	
p-Chlorobenzotrichloride	Pesticide manufacture	AC	
Chlorodane	Broad-spectrum insecticide for homes, gardens, and soil insects	CP, HS, HW	
Chloronaphazine	Treatment of leukemia and related cancer	HC, HS	
DDT		ASC, HS, HW, PTP	
DDVP		C, HS, HW, PTP	

(Continued)

Table 1 Continued

Name	Uses	Effect(s)/system(s)	Human exposure
1,2-Dibromo-3-chloropropane		Stomach, mammary gland	
1,2-Dibromoethane		Subcutaneous, stomach, blood vessels, lungs	
3,3'-Dichlorobenzidine and salts		AC, HW, PTP	
P, P'-Dichlorodiphenyl			
1,2-Dichloroethane		Subcutaneous, stomach, blood vessels, mammary gland	
Dichloroethylene		Mouse liver, NMC	
Dieldrin		Mouse liver, NMC	
HCB		C, HW, PTP	
HHC (hexachlorocyclohexane)		C, HW, PTP	
Heptachlore		C, HS, HW, PTP	
4,4'-Methylene-bis(2-chloroaniline)		C, HW	
Mirex		C, HS, HW	
Phenazopyridine hydrochloride		C	
2,4,5-T		C, HS, HW	
TDE		C, HW, PTP	
Toxaphene		C, HS, HW, PTP	
Trichlorophenyl		SC, HW, HS, PTP	

C, human carcinogen; AC, animal carcinogen; SC, suspected human carcinogen, PTP, USEPA's toxicity priority pollutant; HW, hazardous waste (USEPA).

pharmaceuticals, specialty chemicals, etc. Vinyl chloride (about 12 million tons used annually in USA) is used in PVC, the most abundantly used polymer. Other products include polyvinylidene chloride and chlorobutyl rubber. Saturated HHCs are used in metal-cleaning, fire-extinguishing compounds, rubber, plastic, paint and varnish, healthcare, textile, etc., as well as in agriculture (soil fungicides, insecticides).

Perfluoroacetic acids (PFAs), perfluorooctanoic acid, and perchlorooctone sulfonate have broad-spectrum industrial and consumer applications in surfactants, coatings (fabrics, upholstery, carpets, paper products), firefighting foam, and pesticides.

Halogenated Aromatic Hydrocarbons

Some of the commonly used aromatic HHCs include chlorinated benzenes, benzyl chloride, chlorinated phenols, chlorinated pyridines, chlorinated naphthalenes, chlorinated biphenyls, chlorinated dibenzodioxins, chlorinated dibenzofurans, chlorinated terphenyls, chlorinated benzoyl toluene, chlorinated phenoxyacetic acids, chlorobutyl rubber, Grignard's reagent, Friedel–Crafts's reagent, Diels–Alder Reaction, and alkanolamine.

Monochlorobenzene, dichlorobenzene (m- and p-dichlorobenzenes), trichlorobenzenes (1,2,3-, 1,3,5-, and 1,2,4-trichlorobenzenes), and their derivatives (1-chloro-3-nitrobenzene, 1-bromo-4-chlorobenzene) have been widely used as chemical intermediates and solvents. p-Dichlorobenzene is used as a fumigant (insecticide) and disinfectant. A mixture of trichlorobenzene isomers is used in termite control. 1,2,3- and 1,3,5-trichlorobenzenes have been used as heat-transfer media, transformer fluids, and solvents. Hexachlorobenzene (HCB) is used as a fungicide and as an intermediate for dyes and hexafluorobenzene, synthetic rubber, PVC, an additive for military's pyrotechnique compositors, and as a porosity-controlling agent in manufacture of electrodes.

Benzoyl chloride is an intermediate in the synthesis of benzyl compounds, quaternary ammonium chloride, dyes, tanning materials, pharmaceuticals, and perfume preparation. It is used in the textile and dyes industries as a fastness improver for dyed fiber and fabrics. Hexachlorophene is a topical anti-infective agent, used in detergents and as an antibacterial agent for soaps, surgical scrubs, hospital equipment, cosmetics, etc. It is used as a fungicide for vegetables and ornamental crops. Benzethonium chloride is used as a topical anti-infective agent in medicine and as a germicide for cleansing food and dairy utensils, and as a controlling agent for removing pool algae. It is an additive in deodorants

Table 2 HHC production in USA

Chemical	Current	Past
	Million metric tons yr^{-1a}	
Vinyl chloride	6–12 (10)	0.599 (694 in 1978)
Solvents		
Methyl chloride		0.198 (1971)
Methyl bromide		0.036 (1975)
Chloroethanes	2.1	
1,2-Dichloroethane	0.3	11 (1978)
1,2-Dibromoethane		0.230 (1978)
1,1,1-Trichloroethane	>1.0	0.644 (1978)
Hexachloroethane		0.228
Ethylene dichloride		0.016 (1980)
Ethylene dibromide		0.008 5 (1980)
Trichloroethylene	0.3	0.299 (1978)
Tetrachloroethylene	0.7	0.725 (1978)
Carbon tetrachloride	0.5	0.737 (1978)
Chloroform	0.3	0.349 (1978)
Other halogens		
Chlorobenzene		0.240 (1967)
p-Chloronitrobenzene		0.050 (1967)
o-Chloronitrobenzene		0.017 (1967)
Dichlorodifluoromethane		0.083 (1967)
Trichlorofluoromethane		0.120 (1967)
Chlorinated paraffins		0.028 (1967)
Hexachlorocyclopentadiene		0.050 (1967)
Total Pesticides		1.388 (1.0)
Herbicides		0.674
2,4-D		0.044 (1970)
2,4,5-T		0.012 (1970)
Propachlor		0.011
Insecticides		
Silvex		0.002 (1970)
Methoxychlor		0.005 5 (1975)
Chloropicrin		0.005 7 (1975)
Toxaphene		0.31
DDT		0.059 (1970)
Cyclodienes		0.89 (1975)
Aldrin/Dieldrin		0.009
Ethylene dibromide	0.28	0.088 (1975)
Chloral		0.026 (1967)
Alachlor		0.089 (1976)
Fungicides		0.143
Trifluralin		0.028
PCP		0.040 (1975)
HCB		
PCBs		
USA	0.013	900 (1978)
Rest of the world		1000 (1978)
Chlorofluorocarbons	0.12	

^{a}Data from Anonymous (1979) *Environmental Quality. The Annual Report of Council on Environmental Quality*, 816pp. Washington, DC: US Government Printing Office.

and hairdressing preparations. Tri-, tetra-, penta-, hexa-, and octachloronaphthalenes have been formerly used in heat-transfer media, solvents, lubricant additives, dielectric fluids, and electric insulating material.

In addition to the agrochemicals, a few other HHCs that have caused most serious ecotoxicological and health effects include polychlorinated biphenyls (PCBs), polybrominated biphenyls (PBBs), chlorinated naphthalenes, terphenyls, dioxins, dibenzofurans, and diphenyl ethers.

PCBs are mixtures of various congeners/isomers (**Table 3**). Mixtures of various congeners are marketed as Aroclors (Clophen outside USA) in which the last two digits indicate the degree of chlorination, such as Arochlor 1254 and 1260 (and Clophen A 60) (**Table 4**). These are chemically stable, unreactive, viscous liquids of low volatility. These have been used in closed, semiclosed, and 'open-end' systems, such as hydraulic fluids, coolants, electric transformers, capacitors, insulations (electric wire and cable), heat-transfer systems, fluorescent light ballasts, lubricating oils, plasticizers, paints, inks, paper, dielectric fluids, heat-transformer fluids, lubricants, vacuum pump oils, carbonless copying paper, wall coatings, surface treatment in textiles, wood, metal, and concrete; in caulking material, impregnated fruit wrappings, cutting oils, microscopic mounting and immersion oil, vapor suppressants; in insecticide and bactriocide formulations, etc., from agriculture to office buildings, automobiles, and homes. PCBs have caused widespread contamination. However, major environmental sources of PCBs include manufacturing wastes, careless disposition, and dumping. From 1929 to the mid-1970s, USA produced 1.4 billion tons of PCBs and rest of the world about the same amount. In USA about 300 000 t of PCBs have been disposed in dumps and landfills; and about 60 000 t in fresh water and coastal waters, while air contains less than 30 000 t of PCBs. The worst contamination with PCBs in USA occurred in Escamba Bay, FL; Lake Michigan, IL; the Ohio River, OH; Francisco Bay, CA; the Hudson River, NJ; and Puget Sound, WA (**Table 5**).

PBBs have been used mostly as flame retardants and have caused only local problems based on accidental contamination of cattle feed and waste disposition in Michigan. The brominated hydrocarbon flame retardant

Table 3 Numbers of possible congeners (isomers) of halogenated aromatics

Halogen substitution	Benzenes	Dioxin	Furans	Biphenyls
Mono	1	2	4	3
Di	3	10	16	12
Tri	4	14	28	24
Tetra	3	22	38	42
Penta	1	14	28	46
Hexa	1	10	16	42
Hepta		2	4	24
Octa		1	1	12
Nona		0	0	3
Deca		0	0	1
Total	13	75	135	209–265

Table 4 Percentage of various congeners in commercial PCB mixtures

Compound	Dichloro	Trichloro	Tetrachloro	Pentachloro	Hexachloro	Heptachloro	Octachloro
Arochlor-1242	16	49	29	8	1	0.1	ND
Ar-1254	0.5	1	21	48	23	6	ND
Ar-1260	ND	<0.5	1.5	12	38	41	8

Table 5 Sites with high levels of HHC

HHC	Site	Animal contaminants
PCB	Lake Michigan (Waukegan Harbor)	Fish, birds, human
	Great Lakes[a]	Fish, piscivorous birds
	St. Lawrence Seaway[a]	Fish, whales
	New York (Hudson R.)	Fish, birds, marine animals
	Wisconsin (Sheboygan R.)	Fish, birds
TCDD	Italy (Sveso)	Wildlife, livestock, humans
	Maryland (James R.)	Wildlife
Kepone	Maryland (Chesapeake Bay)	Invertebrates, fish, wildlife

[a]DDT, DDE, and dieldrin also present.

hexabromocyclododecane (HBCD) is blended with textile and polystyrene foam. It can leach out from these products and contaminate water and biota. Flame retardants bromodiphenyl ethers have become common environment and human contaminants.

Polychlorinated dioxins (PCDDs) and dibenzofurans (PCDFs) are present in halogenated aromatics as contaminants or are produced during their synthesis or incineration. Some of these are persistent and extremely toxic to wildlife, including higher animals and humans.

HHCs Used in Agriculture

HHC agrochemicals include insecticides, fungicides, and herbicides (**Table 2**). Dichlorodiphenyltrichloroethane (DDT) and its derivatives (dichlorodiphenyldichloroethylene (DDE), dichlorodiphenyldichloroethane (DDD), Kelthane, methoxychlor), cyclodienes (dieldrin, chlordanes, heptachlor, endrin, mirex, kepone), endosulfans, hexachlorocyclohexanes, ethylene dibromide, methyl chloride, Abate, and Chlorpyrifos are/have been most abundantly and indiscriminately used as residual insecticides and are/have been causing ecological and health problems. Cyclodienes were banned in the 1970s and Toxaphene and mirex in the 1980s. Only a few of these insecticides are allowed a restricted use because of their health and ecological effects. However, these (especially DDT) may still be used in developing countries (mosquito control in Africa) as well as in public

health and military for control of arthropod vectors and body lice.

Chlorinated phenols (especially PCPs) were more commonly used in lumbar treatment. Bleaching of pulp with chlorine can produce chlorophenols, polychlorinated dibenzodioxins (PCDDs), and polychlorinated dibenzofurans (PCDFs). Chlorophenols are acidic, water soluble, chemically reactive, but have limited persistence.

Chlorophenoxy herbicides are used to control broadleaf weeds (dicotyledons). These derivatives of phenoxyalkane carboxylic acids (2,4-D, 2,4,5-T, (4-chloro-2-methyl-phenoxy)acetic acid (MCPA), 2,4-DB) are water soluble as alkali salts and lipophilic as esters. These are readily biodegradable. These are of concern because of phytotoxicity, but mostly because of the serious ecological and health effects of their contaminant PCDDs, especially 2,3,7,8-tetrachlorodibenzo-p-dioxin (TCDD).

PCDDs are flat molecules with varying substituents of chlorine resulting in 75 possible congeners (**Table 3**). These are chemically stable and highly lipophilic with limited solubility in organic solvents. These are not commercial products and exist as by-products formed during synthesis of chlorinated phenols, as well as during the incineration of chlorinated aromatics and dispersion of chlorophenols. The most toxic PCDD is TCDD.

Chlorinated anilide herbicides, such as propanil, have been shown to be contaminated with extremely toxic tetrachlorazo- and tetrachloroazoxy-benzenes. The latter can also be produced on microbial degradation of these herbicides.

HHCs in Polymers and By-Products

PVC, one of the most versatile thermoplastics, accounts for 28% of the global annual chlorine consumption (19 million tons PVC). By-products, such as 1,1,2-trichloroethane, and gaseous and liquid wastes on incineration/oxidation are sources of concern. PVC accounts for $150\,000\ \mathrm{t\,yr^{-1}}$ of global production. Other end products using HHC include Neoprene or chlorobutyl rubber. Chlorohydrins are used in the production of propylene oxide to manufacture propylene oxide polymers. It is also used in synthesis of epichlorohydrin for resins and polymers, and specialty chemicals (alkanolamines). Other polymers include nylon-6, 6-hexamethylenediamine for carpeting, etc., and fluoropolymers (50 000 t yr^{-1}, globally), and Silicone (400 000 t yr^{-1}, globally).

Naturally Produced HHCs

About 1500 different HHCs, from simple alkanes to complex polyhalogenates, are produced naturally by a diversity of aquatic and terrestrial organisms in microgram per kilogram body mass levels. Methyl chloride, methyl iodide, and carbon tetrachloride, which are animal carcinogens, occur naturally in marine environment. Several of those produced in large amounts include chloromethanes, chlorophenols, and chlorinated humic/fulvic compounds. Largest amounts of these HHCs are produced in soil and sediments. Some of these HHCs include haloperoxidase enzymes, which yield volatile and phenolics. Reactions of chlorine with lignin pulp and humic/fulvic acids produce chloroliginin. Numerous other halogenated hydrocarbons, many structurally similar to known animal carcinogens and toxicants, occur in both terrestrial and marine organisms. Several halogenated hydrocarbons are essential for normal functioning of living organisms.

Environmental Fate of HHCs

Aliphatic HHCs are used mostly as solvents, and their production, use, and disposal can pose ecological and health (environmental and occupational) problems. Chlorination of drinking water produces chloroform and carbon tetrachloride, which, in USA alone, can continuously expose 29 million humans including infants and children. Tri- and tetrachloroethylenes used in dry cleaning are common environmental contaminants, especially at hazardous waste sites. These are present in groundwater and ambient and indoor air as well as in drinking water. Highly stable perfluoroalkyl acids, perfluorooctanoic acids, and perfluorooctane sulfone are present in environment, wildlife, and humans.

Aromatic HHCs have a wide range of physical properties, which determine their environmental fate and effects. These HHCs range widely in their distribution in different environmental compartments and phases, some of which is related with differences in their persistence. Low molecular weight HHCs are more volatile than higher molecular weight HHCs and can be present in occupational, urban, and industrial air. The major reactions during their transport in air involve hydroxyl radical and photolysis.

Estimates of sources of HHCs and of their concentrations can be more accurately predicted by using quantitative structure–activity relationship (QSAR) models. For example, the inputs of carbon tetrachloride in the

troposphere should be 1–2 million tons while anthropogenic sources amount to only one-tenth of this.

Environmetal sources of mono-, di-, and trichloroacetic acids are not fully known. Some of these can be biological and environmental products of perfluorooctanoic acids. These are present globally in ppb (μg l^{-1}) concentrations in rain and surface water, which are greater than any other low molecular weight HHC.

Trends in release of HHCs and by-products at manufacturing sites are more effectively controlled now and this has reduced their emissions and disposition. There is a dramatic reduction in PCDDs, PCDFs, and chlorophenol discharges from the paper and pulp industry due to the new delignification and bleaching technologies. Legislative restrictions and new technologies have also reduced emissions of PCDDs/PCDFs from waste incineration plants. HHC pesticide usage has declined tremendously. However, nonbiodegradable mirex is now distributed all over North America. Kepone had contaminated James River, VA. In the 1970s, the usage of these insecticides was minimized and most of these were banned by 1980.

Volatility and lipophilicity of polychlorinated aromatics (PCB, PCDD, PCDF, etc.) are associated with the degree of chlorination. Chlorination increases lipophilicity but also reduces volatility. Thus, low concentrations of HHCs, especially less chlorinated ones, are present in surface waters, while highly chlorinated aromatics remain in sediments and are picked up by invertebrates in high concentrations, as has been reported for 4-, 5-, and 6-chlorinated congeners of PCBs. In surface waters, photolytic *ortho*-dechlorination is common. The octanol:water partition coefficients for PCB 101, 128, 136, and 1 are 2.5, 5, 10, and 0.002, respectively. Bioconcentration factors for these chlorinated aromatics relate with their partition coefficients (**Table 6**).

Table 6 Biological concentration of HHC in biota

HHC	Bioaccumulation: × in water	
	Algae	Fish
PCP	1 240	350
2,4,6-TCP	800	230
HCB	24 000	400
2,4,2′,4′-Pentachlorobiphenyl	11 500	770
2,5,4-Trichlorobiphenyl	7 700	650
2,2′-Dichlorobiphenyl	3 050	830
Pentachloronitrobenzene	3 600	380
Hexachlorocyclopentadiene	1 140	308
2,4-D	6	2
p-Chlorobenzoic acid	63	1.1
2,4-DCBA	0.9	0.3

Biotransformation and Biodegradation of HHCs

Dichloromethane is metabolized to CO and HCHO (carcinogen). The metabolites of trihalomethanes start with loss of halide (via P4502E1), leading to CO production. Covalent bonding to macromolecules occurs via phosgene in case of chloroform and via dibromocarbonyl in the case of bromoform. Chloroform is metabolized to reactive phosgene, and carbon tetrachloride forms trichloromethyl, trichloromethylperoxy, and chlorine-free radicals. Several haloalkanes are first conjugated with GSH in liver and then metabolized to cysteine conjugates in kidney. The latter conjugates are converted, by β-lyase, to episulfonium ions.

Metabolism of haloethylenes (trichloroethylene, perchloroethylene, vinyl chloride, vinyl bromide, vinyl fluoride, vinylidene chloride, vinylidene fluoride) starts with epoxidation (by P4502E1). Resulting oxiranes are highly reactive and bind covalently to nucleic acids. Chlorinated ethylene epoxides breakdown as a function of halogenated substituents on the carbon. Compared to ethylene oxide, vinyl chloride and vinylidene chloride with increasing chlorination show an increase in bond strength of the chlorinated carbon to the oxygen and a decrease in the bond strength to nonchlorinated carbon. When each carbon is substituted with single chlorine, the C–O bonds become equally stable. When there are two chlorines on one carbon and one on the other, there is, again, a weakening of the less chlorinated C–O bond. The asymmetrical chloroethylene is more carcinogenic than symmetrical ones because of the bond breakage, which potentiates covalent bonding to DNA. There are limits to the effectiveness of highly reactive species, for example, the weak carcinogenic activity of vinyl chloride may be due to the instability of its putative metabolite (1,1-dichloroxirane), which is likely to be degraded rapidly. The less reactive oxiranes bind to DNA more readily and stably. Trichloro- and tetrachloroethylenes are metabolized to monochloroacetic acid and trichloroacetic acid, which are excreted as such or as their glucuronides and alcohols.

The environmental fate of most aromatic HHCs is determined by halogen substituents and the types of organisms exposed to them. The position of chlorination is more important than the degree of chlorination in biotransformation and degradation. Chlorination of biphenyls at *ortho*-position causes steric hindrance and decreases the degree of coplanarity at 2,2′,6,6′. Chlorinated dioxins and furans are held in coplanar orientation by the oxygen bridge. Aerobic bacterial degradation of PCBs involves dechlorination at *o*- and *m*-positions. In sediments, bacterial dechlorination is inversely proportional to their lipophilicity. In soils and sediments, biodegradation, hydrolysis, and photolysis determine their fate. The dechlorinated

products can be attacked by bacterial deoxygenases as well as by animal P450 oxygenases.

Animal metabolism of polychlorinated aromatics starts with attack by P450 monooxygenases causing hydroxylation (addition plus keto–enol rearrangement) via epoxidation, with or without displacement of the chlorine (NIH shift). Vinyl halogens favor epoxidation. PCBs with 4,4′ and/or 3,3′,5,5′-chlorination are refractory to biotransformation, while PCBs with four or more chlorines but with hydrogen in either 4 and/or in 4,5′-positions of one or both rings are readily metabolized. Polyhalogenated aromatics induce P450 monooxygenases; coplanar PCBs and TCDD induce P4501A1 while noncoplanars induce P4502B isozymic forms. P4501A1 can oxidize coplanar PCBs readily while P4502B preferentially oxidizes *o*-substituted congeners. Resulting hydroxylated metabolites can be conjugated with glucuronic acid and/or glutathione, which are excreted in bile; but glutathione conjugates are re-absorbed or metabolized to meracaptans, which are excreted as mercapturic acid or form corresponding sulfhydryl and methylated products. The methyl-thio-PCB can be further oxidized to corresponding sulfoxides and sulfones, which may be retained in tissues. The methylsulfonyl metabolites of PCBs, HCB, and DDT show high tissue-specific binding and toxic effects. Similar hydroxylations and dechlorinations and/or NIH shifts occur in dioxins and furans. The dihydroxy products following *p*-oxidation can also lead to ring opening and reduction of furans, which reduces their toxicity.

Toxicity and Mechanisms of Toxicity of HHCs

HHCs exert specific or broad-spectrum effects on living organisms. They can be acutely lethal or exert subchronic and chronic toxic effects, such as developmental (teratogenic), hormonal (endocrine disruptors), carcinogenic, neurotoxic, immunotoxic, and neurobehavioral effects.

Haloethylenes undergo epoxidation by P4502E1 and form oxiranes depending on halogen substituents, which are highly reactive and bind covalently to DNA. Compared to ethylene oxide, vinyl chloride and vinylidene chloride, with increasing chlorination, show an increase in bond strength of the chlorinated carbon to the oxygen and a decrease in the bond strength to nonchlorinated carbon. The asymmetric chloroethylenes are more carcinogenic than symmetrical ones because the bond breakage potentiates covalent bonding to DNA. There are limits to the effectiveness of highly reactive species, for example, weak carcinogenic activity of vinyl chloride may be due to instability of its putative metabolite (1,1-dichloroxirane), which is likely to be degraded rapidly. The less reactive oxiranes are more DNA binding.

Glutathione conjugates of several haloalkanes in liver are converted to cysteine conjugates in kidney, where their episulfonium ions bind covalently to DNA and proteins leading to nephrotoxicity and carcinogenicity. Vinyl chloride and bromide are genotoxic and carcinogenic. Trichloroethylene is not carcinogenic in liver while vinylidene chloride is. Dichloromethane is a central nervous system (CNS) depressant and may even suppress normal inspiration via this route. It is metabolized to CO, which forms COHb and HCHO (which can react with DNA). Chloroform affects liver and causes renal epithelial tumors. Its metabolite phosgene covalently binds to liver proteins and depletes hepatocyte GSH. Carbon tetrachloride forms trichloromethyl, trichloromethylperoxy, and chlorine-free radicals, which can attack unsaturated bonds of fatty acids in the endoplasmic reticulum leading to lipid peroxidation and membrane damage. It causes centrilobular necrosis and fatty liver.

In experimental mammals, HHCs that cause fatty liver without necrosis include chlorobromomethane, dichloromethane, cis-1,2-dichloroethylene, tetrachloroethylene, and 2-chloroethylene. Those HHCs which cause fatty liver and necrosis include carbon tetrachloride, carbon tetraiodide, carbon tetrabromide, bromotrichloromethane, chloroform, iodoform, bromoform, 1,1,2,2-tetrachloroethane, 1,2-dichloroethane, 1,2-dichloromethane, 1,1,1-trichloroethane, 1,1,2-trichloroethylene, 2-chloro-n-propane, and 1,2-dichloro-p-propane. Those HHCs which cause no liver hypertrophy but only necrosis are methylene chloride, methyl bromide, methyl iodide, dichlorodifluoromethane, trans-1,2-dichloroethylene, ethylene chloride, ethyl bromide, ethyl iodide, and n-butylchloride.

Aliphatic HHC insecticides can be neurotoxic (which inhibit GABA-A1 chloride ion channels in brain and ATP-ases in axons), respiratory poisons, endocrine disruptors, teratogens, carcinogens, etc. Some of these HHCs can cause liver enlargment and induction of cytochrome P-450 isozymic forms, mostly P4502B.

Aromatic halogens and polyaromatic HHCs also cause hepatic hypertrophy and induction of P450 isozymes, both P4501A and 2B. This induction of P4501A is mediated via transport of these HHCs by cytosolic Ah receptor (AHR, a member of the basic-helix-loop-helix/per, ARNT/AHR, SIM homology (bHLH/PAS) protein) inside the nucleus where heterodimer ARNT (Ah-receptor nuclear translocator) binds to specific nucleotide recognition sequence on regulatory region of the target gene DNA (dioxin response element, DRE). This leads to altered gene expression and results in hepatotoxicity, carcinogenicity, teratogenicity, etc. These and other responses are highly tissue, sex, age, and species specific. Differences in ARNT domain may be responsible for differential gene expression and altered sensitivities of strains and species, as well as to differences in genomic sequence at promoter and enhancer region. The gene expression of AHR and ARNT during embryonic development is stage and tissue specific. AHR mRNA appears in fetal mouse tissue between gestation days (GDs) 10 and 16. In DRE–lacZ mouse model, AHR is present in several developing tissues including genital tubercle, palate, and paws. In a transgenic mice model, the TCDD exposure *in utero* at GD 14.5 induced lacZ expression in the mesenchymal epithelium of developing paws and alteration in gene expression profile were observed after 24 h.

In liver cells of rats exposed subchronically for 13 weeks to AHR agonists TCDD (toxicity equivalent factor, TEF = 1), pentachlorobenzofuran (TEF = 0.5), 3,3′,4,4′,5-pentachlorobiphenyl (TEF = 0.1) several genes exhibited altered expression. Out of these genes, Serpia 7 (27-fold), Cyp3a13 (1000-fold), and Ces3 (fivefold) were downregulated. PCB126 and −153 had no effects but their mixture mimicked TCDD downregulation of 11 genes. Gender- and species-specific repression occurred within this subset of genes. In AHR knockout mice, seven of the 11 genes were downregulated. This early downregulation is followed by upregulation and liver carcinogenesis. *In utero* exposure of mice to TCDD upregulates amphiregulin (one of the epidermal growth factor signaling ligand) gene expression in fetal ureter. In mice, disruption of AHR signaling by TCDD decreases body weight and fecundity, and causes liver defects and disruption of cardiac and vascular development.

The α-, β/γ-, and δ-isoforms of the nuclear hormone receptor superfamily of ligand-activated transcription factors (peroxisome proliferator activated receptors) respond to perfluorooctanoic acid and perfluorooctane sulfonate (*in vivo* and *in utero*) and alter gene expression in various tissues during development and in adults leading to developmental toxicity and other adverse effects.

Tetrachloroazo- and tetrachloroazoxybenzenes inhibit adrenocorticotropic hormone release and affect thymus and other lymphatic systems. Toxic potency of congeners of polychlorinated aromatics is usually compared with that of TCDD and expressed as 'dioxin equivalent factor'. DEF depends on the coplanarity and chlorination in *para*- and *meta*-positions. The net toxicity of the PCB mixtures depends on these coplanar congeners, whose concentration in most commercial mixtures is very low. The DEF is reflective of the affinity of the congener for the cytosolic AHR causing its activation. The increased synthesis of the *CYP1A1* gene product P4501A1 increases the catalysis (hydroxylation) of coplanar HHCs. The interaction of HHCs with AHR initiates a constellation of effects, such as thymic atrophy, hypo- and/or hypertrophy of liver, chloracne, wasting, teratogenesis, etc. TCDD is 40–400 times more potent than other halogenated aromatics in binding with AHR. The non-AHR-type effects include hepatomegaly, induction of Phase II enzymes, and porphyria. However, the acute lethality of TCDD and other halogenated aromatics is dependent on species. In the case of PCBs, the confromational restrictions and hydroxylations are important in their estrogenic activity. Porpyhria-related diseases are species specific and frequently slow to develop. These polyhalogenated aromatics and their metabolites are, also, potent inducers of δ-aminolevolunic acid synthase, the rate-limiting enzyme of porphyrin synthesis.

Ecotoxicology of HHCs

Factors that affect toxicity of individual HHCs are increasing numbers of halogens in the molecule, increasing the size of the molecule, increasing ease of homolytic cleavage, electronegativity of halogens, and/or their chain length.

Differences in closely related compounds, such as positional isomers (**Table 8**) and functional groups, have pronounced effects on ecodisposition, bioconcentration, biodispersion, and biological effects. These major differences within subsets of generic properties are important in understanding the ecotoxicology of HHCs. The relation between structure, metabolism, and toxic action for one subgroup of HHCs may not be duplicatable in other groups. So caution must be taken in conclusively generalizing from experimental deductions.

Bioconcentration

Most of the serious ecological problems caused by HHCs are related with bioaccumulation and bioconcentration via food chain. For example, DDT, dieldrin, PCBs, etc., at their parts per trillion concentrations in water can be found in top carnivores in parts per million (ppm) concentrations, more than a millionfold ecological concentration. For example, terns in Lake Michigan can have 26 ppm in their fat and their eggs can have up to 100 ppm PCBs.

Effects on Aquatic Wildlife

The bodies of water close to industrial, agricultural, and urban centers show high levels of HHCs in water, sediments, and biota. Even in Arctic and Antarctic residues of PBBs, chlorobornanes, PBDEs, PCN, PCBs, and DDT can be detected in water and biota.

Embryos and young animals are generally more sensitive to HHCs and other toxicants. Bioconcentration of these stable compounds can lead to high concentrations in organisms higher up in the food chain. Each level of organisms may be affected by chronic exposures to these halogens. However, the effects on lower organisms are not easily available and recorded. HHCs differ widely in their toxicities to aquatic food chain organisms as exemplified by water flea *Daphnia magna* (**Table 7**).

Table 7 A list of commonly used halogenated hydrocarbons and their toxicities to *D. magna*[a]

HHC	LC_{50} ($\mu g\,l^{-1}$)
Alkanes	
Methyl chloride	
Methyl bromide, methyl iodide	
1,2-Dichloromethane	363
Ethyl iodide, ethyl bromide	
1,1-Dichloropropane	23
1,2-Dichloropropane	52
1,3-Dichloropropane	280
n-Butyl chloride	
Vinyl chloride	
Vinyl bromide	
2-Chloro-*n*-propane	
1,2-Dichloro-*p*-propane	
Chlorobromomethane	
Trichloromethane	198
1,1,2-Trichloroethane	70
1,1,1-Trichloroethane	530
Hexafluoropropane	
Tetrachloromethane	35
Tetraiodomethane	
Dichlorofluoromethane	
1,1,1,2-Tetrachloroethane	24
1,1,2,2-Tetrachloroethane	41
1,1,2,2,2-Pentachloroethane	12
Hexachloroethane	0.006
Alkenes	
1,3-Dichloropropene	6
Tetrachloroethene	14
Vinylidene chloride	
Hexachloropropene	
Alkynes	
1,1-Dichloroethylene	135
trans-1, 2-Dichloroethylene	220
1,1,2-Trichloroethylene	51
Methylene chloride	
Tetrachloroethylene	
cis-1,2-Dichloroethylene	
1,1,2-Trichloroethylene	
Monohaloethylenes	
Tetrafluoromethylene	
Chlorotrifluoromethylene	
1,1-Dichloro-2,2-difluoroethylene	
Chlorotrifluoroethylene	
2-Chloroethanol	
2,2,2-Trichloroethanol	
Halogenated aromatics	
Chlorobenzene	17
1,2-Dichlorobenzene	0.042
1,4-Dichlorobenzene	0.088
1,3-Dichlorobenzene	2.07
1,2,4-Trichlorobenzene	1.88
1,2,5-Tetrachlorobenzene	9.7
Pentachlorobenzene	5.3
Chlorinated phenols	
2-Chlororophenol	5.24
4-Chlorophenol	4.89
3-Chlorobiphenyl	5.30
2-Chlorobiphenyl	0.43
2,4-Dichlorophenol	0.26

(Continued)

Table 7 Continued

HHC	LC_{50} ($\mu g\,l^{-1}$)
3,4,5-Trichlorophenol	0.68
2,4,6-Trichlorophenol	3.03
2,4,5-Trichlorophenol	2.70
2,3,4,6-Tetrachlorophenol	0.29
Tetrachlorophenol	0.41
2,4,5,6-Tetrachlorophenol	0.56
Pentachlorophenol	0.73
Pentachlorophenol-sodium	0.40
Hexachloro-1, 3-butadiene	
Hexachlorocyclopentadiene	
Pesticides	
Dichlorovos	0.004
2,4-Dichlorophenoxyacetate	0.83
Picloram	59
Trichloropropane	0.117
Chloramp	212
Chloroxuron	2.95
Lindane	0.899
Dibenzofuran	0.005
DDT	0.003
Chloropheniphos	0.043
Endosulfan	0.283
Permethrin	0.002

[a]Data from Adams WJ, Champan GA, and Landis WG (eds.) (1988) *Aquatic Toxicology and Hazard Assessment*, Vol. 10. 579pp. Philadelphia: ASTM.

Fish, amphibians, and reptiles

Tumors in English sole (Puget Sound, WA), Medeka, and other species may be due to their exposure to HHCs along with polycyclic aryl hydrocarbons (PAHs). In addition to this, mortality of hatchlings (fry), adult survival, and reproductive effects (feminization) may occur. The decline in frog populations and their hind leg and other deformities have been assumed to be related with HHC pesticides. Alligators exposed to DDE and PCBs have been reported to show atrophy of their testes causing lowering of testosterone and feminization.

Water birds

Residues of HHCs in food and water of water birds has contaminated and affected fish-eating birds' reproduction (feminization of males, nest abandonment, ineffective incubation). This has been observed in Great Lakes Region in cormorants, terns, raptors, and other species, and in local and migratory waterfowl in Lake Michigan, Green Bay, Long Island Sound as well as in Adeline penguins and south polar skua. In Lake Michigan Basin, hatchling failure and malformations and sex reversal in male double-crested cormorants have been noticed. Foster tern eggs contain very high levels of PCBs leading to embryotoxicity, aberrant parental behavior, and hormonal changes. High concentration of DDT, dieldrin, and PCBs dangerously reduced the number of raptors in

North America and Britain and seriously affected their reproduction in the Great Lakes region.

Sea mammals

In St. Lawrence Seaway, whales/seals contain high concentrations of PCBs and PAHs, which are high in sediments: concentrations of PCB (180 ppm) in male belugas and in blubber of young ones (270 ppb) (young ones showing metastasized bladder carcinoma and males with other tumors). Juvenile females had about 600 ppm PCB in blubber. Mink whales from the Antarctic contained PCBs, HHCs, HCB, and DDT in blubber and food. Baltic seals are also contaminated with PCBs and DDT. Females with high concentrations of DDT and its metabolites showed pathological changes in their uteri. Those with high PCBs had implantation failure and adrenal cortex atrophy. California sea lions have also shown reproductive disorders (premature birth), immune suppression, gastric and intestinal ulceration, and liver, skin, and kidney lesions, which are also seen in adult beluga whales.

Land mammals

In Michigan, mink and ferret populations fed PBB/PCB-contaminated lake fish showed reproductive failure, induction of P450 in hepatocytes, degenerative renal changes, anorexia, gastric ulcers, loss of hair, and enlargement of liver, pancreas, and adrenals. The toxicities of various HHCs to experimental mammals (**Table 8**) indicate the severity of TCDD and its related coplanar chlorinated aryl hydrocarbons. Species-specific differences in sensitivity to these HHCs are very prominent (**Table 9**). In rats, dietary intake of PCBs for 130 days showed no symptoms at 0.001 $\mu g\,kg^{-1}\,d^{-1}$ (=0.13 $\mu g\,kg^{-1}$), but a 10 times higher dose reduced fertility in two to three generations and a 100 times higher dose affected neonatal survival. A 2-year dietary exposure (0.1 $\mu g\,kg^{-1}\,d^{-1}$) of rats caused several chronic effects, such as wasting, fatty

Table 8 Oral toxicity of TCDD in mammals

Animal	LD_{50} ($\mu g\,kg^{-1}$ body wt.) Average of male and female
Guinea pigs	0.6–2.1
Rat	22–45
Mice	283
Rabbit	115
Monkey	50–70
Hamster	100

LD_{50} values (mg kg^{-1}) for male rats for DDT, dieldrin, Toxaphene, dichlorovos, and lindane are 113, 46, 90, 80, and 200, respectively. Data from Kimbrough RD (1981) Chronic toxicity of halogenated biphenyls and related compounds in animals and health effects. In: Khan MAQ and Stanton RH (eds.) *Toxicology of Halogenated Hydrocarbons*, pp. 23–37. New York: Pergamon.

Table 9 Toxic effects of some commonly used HHCs in rodents

Effect	TCDD	PCB	PBB	Chloronaphthalenes
LD_{50}: $mg\,kg^{-1}$	0.004–0.01	4000–10 000	21 500	0.011–2000
Liver				
Porphyria	+	+	+	+
Necrosis	+	–	–	Atrophy
Vit. A deficiency	–	+	–	+
P450 induction	+	+	+	
Tumors	+ (+ Lung)	+	+	+ (+ Lung)
Immune suppression				
In mice		–	+	+
Hyperkeratosis				
In rabbit ear	+	+	+	+
Embryo toxicity	+ (+ Neonate, postnatals)	+	+	Neonates

Data from Kimbrough RD (1981) Chronic toxicity of halogenated biphenyls and related compounds in animals and health effects. In: Khan MAQ and Stanton RH (eds.) *Toxicology of Halogenated Hydrocarbons*, pp. 23–37. New York: Pergamon.

liver, lung lesions, hypothyroidism, chloracne, liver cancer, and mortality.

Humans

While adult humans can be exposed to HHCs through dermal absorption, inhalation, and dietary intake, the neonates can be exposed as fetuses *in utero* and via breast milk. Embryos, fetuses, infants, and young children are more sensitive than adults.

Clinical exposure of children to chloroform, dichloromethane, ethyl chloride, trichloroethane, and halothane is closely watched.

The effects of halogenated aromatics (PCBs, PCDDs, PCDFs) depend on the type of the compound. They cause acute irritation of eyes, mucus membranes, lungs, and gastrointestinal tract besides affecting nervous systems. These also cause acne (chloracne), liver dysfunction, porphyria, and malignancies (**Table 9**) and exert developmental effects. Chlorinated toluenes (benzyl and benzal chlorides, benzotrichloride) are group 2A carcinogens. Other HHCs with carcinogenic potential are listed in **Table 10**.

PCDD residues are widely distributed, especially in aquatic environment, fish, and fish-eating birds and mammals. Their residues in human food (**Table 11**) can have significant chronic effects, especially in the presence of other HHCs. TCDD causes several symptoms of toxicity, such as wasting, fatty liver, lung lesions, hypothyroidism, chloracne, liver cancer (**Tables 9 and 10**). TCDD-linked cancer has been reported in factory workers in the form of brain cancer and malignant melanomas. PCDF congeners are not as toxic as PCDDs. A list of effects in experimental mammals is summarized in **Tables 9 and 10**.

Table 10 A list of HHCs, which have tested positive for carcinogenicity

HHC	System	Nonmutagenic carcinogenicity[a]	Tumors[b]
Chlorobenzilate	Rat liver	+	
Chloroform	Rat/mouse liver	+	Kidney, thyroid, liver
Carbon tetrachloride	Rat/mouse	+	Liver
Dieldrin	Mouse liver	+	
P, P'-Dichlorodiphenyl dichloroethylene	Mouse liver	+	
Lindane	Mouse liver	+	
PCB	Rat/mouse liver	+	
TCDD	Rat liver, lung	+	
Trichloroethylene	Mouse liver	+	Liver
1,2-Dichloroethane			Subcutaneous, stomach, blood vessels, mammary gland
1,2-Dibromoethane			Subcutaneous, stomach, blood vessels, lung
1,2-Dibromo-3-chloropropane			Stomach, mammary gland
Hexachloroethane			Liver
Tetrachloroethylene			Liver
1,1,2-Trichloroethane			Liver

Table 11 PCDD and PCDF levels in food consumed in USA

Food	Conc. (pg/g fresh wt) Toxic equivalent
Beef	0.48
Pork	0.26
Chicken	0.19
Eggs	0.13
Dairy products	0.36
Milk	0.07
Fish	1.2

The daily dietary intake of TCDD in humans (about 118 pg (TEQ) in North America and the Netherlands) is mostly via meats, fish, eggs, milk, and other dairy products (**Table 11**). There is a federal warning about the consumption of Lake Michigan fish by pregnant women and those expecting pregnancy and of fish in Lake Michigan Basin by children and young adults. The high concentrations of these HHCs in human breast milk have advisories against breast-feeding of infants. *In utero* exposure to HHC has shown newborn and children and young adults exhibit neurobehavioral deficits, which become more severe with age.

The residues of stable/persistent and toxic pesticides used in past are still present in global ecosystems and humans. Their presence in human breast milk has created concerns about effects in neonates leading to neurobehavioral deficits in children and young adults. It will take several decades for their residues to completely disappear. Their residues are still biomagnified in top carnivores. These HHCs are nerve and liver toxins. The LD_{50} of DDT to mammalian species ranges between 113 and 450 mg/kg body wt. and that of cyclodienes between 40 and 60 mg/kg body wt.

Environmental products of cyclodiene insecticides (aldrin, dieldrin, heptachlor, chlordanes), namely their corresponding photoisomers, can become more toxic than the corresponding parent chemical and can be metabolized to even more toxic products. These toxic metabolites result from epoxidation of unsaturated bonds or oxidative dehydrochlorinations.

The metabolite of DDT, DDE, is more persistent in environment than DDT and is esterogenic. Methoxychlor, also, has estrogenic effects. DDD, a contaminant in DDT, causes adrenocorticoid atrophy.

PCBs have very low water solubility $(3-21\,\mu g\,l^{-1})$. Individual congeners with chlorine substitution in *ortho*-position on both rings vary in their persistence, stability, and toxicity. Coplanar PCBs have chlorine in *ortho*-position on both rings, such as 3,3',4,4'-tetrachlorobiphenyl. Planars have chlorine substitutions in 2-, 3-, or 4-*ortho*-positions, which moves rings out of phase because of interactions with adjacent chlorines in different rings. These have globular structures. PCBs are persistent and poorly metabolized in biota and thus are biomagnified in ecosystems harming top carnivores. Their half-life in humans varies from 30 days to 10 years. PCBs are transferred to offspring through the placenta and via milk. They cause neurobehavioral deficits in neonates due to their exposure *in utero* as fetuses. Birth defects and lowering of male sex hormones (testicular atrophy) are reported in several bird and mammalian species. PBBs are used as fire retardants (Fire master) and resemble PCBs in their effects. TCAB and TCAOBs are immunosuppressive agents and may even affect anterior pituitary secretion of ACTH.

See also: Reproductive Toxicity.

Further Reading

Adams WJ, Chapman GA, and Landis WG (eds.) (1988) *Aquatic Toxicology and Hazard Assessment,* vol. 10, 579pp. Philadelphia: ASTM.

Anonymous (1979) *Environmental Quality. The Annual Report of Council on Environmental Quality*, 816pp. Washington, DC: US Government Printing Office.

Anonymous (1985) The Effects of Great Lakes Contaminanats on Human Health. Report to Congress. USEPA 905-R-95-107, Sep. 1985. Chicago: Great Lakes National Program Office.

Anonymous (1988) The Great Lakes Clean-up effort. *Chemical and Engineering News Frb* 2: 22–27.

Avono S, Tanabe S, Fujise Y, Kato H, and Tatsukawa R (1998) Persistent organochlorines in mink whale (*Balaenoptera acutorostrata*) and their prey species from the Antarctic and the north Pacific. *Environmental Pollution* 98: 81–89.

Brown AWA (1982) *Ecology of Pesticides*, 340pp. New York: Plenum.

Brooks GT (1974) Chlorinated Hydrocarbons. Columbus: CRC Press.

Carey T, Cook P, Giesy J, *et al.* (1998) *Ecotoxicological Risk Assessment of the Chlorinated Organic Chemicals*, 375pp. Poncacola, FL: SETAC Press.

Colborn T, Dumanoski D, and Myers JP (1995) Our Stolen Future. New York: Dutton Press.

Court GS, Davis LS, Focardi S, *et al.* (1997) Chlorinated hydrocarbons in the tissues of south polar skuas (*Catharacta macocormicki*) and Adelie penguins (*Pygoscelis adeliea*) from Ross Sea, Antarctica. *Environmental Pollution* 97: 295–301.

Evenenset A, Christensen GN, and Kallenborn R (2005) Selected chlorobornanes, polychlorinated naphthalenes, and brominated flame retardants in Bjornoya (Bear Island) freshwater biota. *Environmental Pollution* 136: 419–430.

Hansen LG (1994) Halogenated aromatic hydrocarbons. In: Cockerham LG and Shane BS (eds.) *Basic Environmental Toxicology*, pp. 199–230. Ann Arbor, MI: CRC Press.

Home AS and Ho IK (1994) Toxicity of solvents. In: Cockerham LG and Shane BS (eds.) *Basic Environmental Toxicology*, pp. 157–184. Ann Arbor, MI: CRC Press.

Hutzinger O, Frei RW, Merian E, and Pocchiari F (1982) *Chlorinated Dioxins and Related Compounds*, 558pp. New York: Pergamon.

Khan MAQ (1976) *Pesticides in Environment*, 320pp. New York: Plenum.

Khan MAQ (1990) Biochemical effects of pesticides on mammals. In: Haug G and Hoffman H (eds.) *Chemistry of Plant Protection*, vol. 6, pp. 109–172. Berlin: Springer.

Khan MAQ and Stanton RH (1981) *Toxicology of Halogenated Hydrocarbons*. New York: Pergamon.

Kimbrough RD (1981) Chronic toxicity of halogenated biphenyls and related compounds in animals and health effects. In: Khan MAQ and Stanton RH (eds.) *Toxicology of Halogenated Hydrocarbons*, pp. 23–37. New York: Pergamon.

McEwen FL and Stephenson GR (1979) *The Use and Significance of Pesticides in the Environment*, 538pp. New York: Wiley Interscience.

Snyder R and Andrews LS (1995) Toxic effects of solvents and vapors. In: Klaassen CD (ed.) *Toxicology – The Basic Science of Poisons*, 5th edn., pp. 737–772. New York: McGraw.

Sun YV, Boverhof DR, Bourgon LD, Fielden MR, and Zacherowski TR (2004) Comparative analysis of dioxin response elements in human, mouse, and rat genomic sequence. *Nucleic Acid Research* 32: 4512–4522.

USEPA (1994) Health Assessment Document for 2,3,7,8-Tetrachlorodibenzo-p-dioxin (TCDD) and related compounds. Review draft. Vol. II, August 1994.

Walker CH, Hopkin SP, Silby RM, and Peakall DB (2006) *Principles of Ecotoxicology*, 3rd edn. New York: Taylor and Francis.

Weisburger HE (1981) Halogenated substances in environmental and industrial materials. In: Khan MAQ and Stanton RH (eds.) *Toxicology of Halogenated Hydrocarbons*, pp. 3–24. New York: McGraw.

Lead

A M Scheuhammer, Environment Canada, Ottawa, ON, Canada

W N Beyer, USGS Patuxent Wildlife Research Center, Beltsville, MD, USA

C J Schmitt, US Geological Survey, Columbia, MO, USA

Ecotoxicology of Lead

Further Reading

Ecotoxicology of Lead

Lead (Pb) is a naturally occurring metallic element; trace concentrations are found in all environmental media and in all living things. However, certain human activities, especially base metal mining and smelting; combustion of leaded gasoline; the use of Pb in hunting, target shooting, and recreational angling; the use of Pb-based paints; and the uncontrolled disposal of Pb-containing products such as old vehicle batteries and electronic devices have resulted in increased environmental levels of Pb, and have created risks for Pb exposure and toxicity in invertebrates, fish, and wildlife in some ecosystems.

Lead in the Aquatic Environment

The accurate measurement of Pb in water is extremely difficult due to pervasive environmental contamination with atmospheric Pb, which necessitates ultra-clean conditions for collection, storage, and analysis of samples. Analyses that have followed such precautions indicate that much of the Pb in natural waters is associated with particulate material. Dissolved Pb concentrations are typically <1 ng ml^{-1} in uncontaminated waters, but may reach or exceed 20 ng ml^{-1} in contaminated waters. Total Pb concentrations in uncontaminated waters are typically <20 ng ml^{-1}, but may be much greater in samples with high suspended sediment loads. In solution, Pb-containing salts dissociate and the free divalent ion, Pb^{2+}, can form stable complexes with carbonates, hydroxides, chlorides, other inorganic ions, and with naturally occurring organic substances. It is also readily adsorbed onto the surface of particulate material. In uncontaminated sediments, concentrations of total Pb in the $<63\,\mu M$ fractions are typically <4–$20\,\mu g\,g^{-1}$ dry weight, but may reach several thousand $\mu g\,g^{-1}$ in sediments containing mine tailings.

Aquatic geochemistry and toxicity

Toxicity of inorganic Pb to aquatic organisms is associated primarily with the concentration of the free dissolved ion. The toxicity of Pb^{2+} declines with increasing concentration of hardness ions (calcium, magnesium), and is also affected by other dissolved organic and inorganic substances that form stable complexes with Pb^{2+} (e.g., hydroxides, carbonates, chlorides, and dissolved organic compounds). Consequently, waterborne Pb is more bioavailable and toxic in soft freshwater than in either hard freshwater or saltwater. Acidic conditions such as those produced by acid rain and acid mine drainage can increase aqueous Pb^{2+} concentrations, but geochemical processes such as complexation and co-precipitation with iron and manganese oxides may remove Pb from solution.

The toxicity of waterborne Pb varies over several orders of magnitude depending on taxon, route and duration of exposure, water quality, and other factors. Daphnids and larval fishes are among the most sensitive freshwater organisms, and algae are the least sensitive. Amphibians are also not particularly sensitive to Pb except in combination with other metals, low pH, or both. However, differences among taxa are difficult to ascertain due to differing test protocols (duration of study, chemical form, route of exposure, endpoint, etc.) among studies. Diatoms and mysid shrimp are among the most sensitive marine organisms to inorganic Pb exposure, but organic Pb compounds (including those formerly used in leaded gasoline) are also highly toxic to the few marine organisms against which they have been tested.

Sediment chemistry and toxicity

Most of the Pb in sediments is associated with organic material, mineral sulfides, carbonates, and oxide coatings of silt and finer-sized particulates (<63 μM). Under anaerobic conditions, Pb concentrations in sediment pore waters are reduced by binding to insoluble acid-volatile sulfides. In sediments where Pb concentrations exceed the binding capacity of acid-volatile sulfides, Pb in sediment pore waters may reach concentrations that are toxic to benthic macroinvertebrates and other organisms. Particulate Pb in sediments may also be toxic to aquatic organisms through ingestion and incorporation into aquatic food chains.

Consensus-based guidelines for Pb concentrations in sediments at which effects on aquatic organisms are considered possible (threshold effect concentration = $35.8\,\mu g\,g^{-1}$ dry weight) or probable (probable effect concentration = $128\,\mu g\,g^{-1}$ dry weight) have been established. These values are commonly exceeded, especially in urban areas and in waters contaminated by industrial discharges and mining.

Lead in aquatic organisms

Some proportion of total Pb in water and sediment is bioavailable to aquatic plants and animals. In invertebrates and fish, the distribution and metabolism of Pb tends to follow that of calcium (Ca). Lead is concentrated in the liver or hepatopancreas, kidney, digestive tract, mucous coatings, and hard tissues (bone, scale, exoskeleton), whereas concentrations in muscle are comparatively low. Because Pb does not biomagnify in aquatic food chains, environmental Pb is generally considered a greater hazard to lower trophic level organisms (such as herbivorous waterfowl) than to top predators, including piscivorous wildlife. In aquatic ecosystems, Pb concentrations are typically greater in invertebrates and benthivorous fishes such as suckers (Catostomidae) and common carp (*Cyprinus carpio*) than in more predatory, fish-eating species such as black basses (*Micropterus* spp.). Background concentrations in whole fish are typically $<0.1–0.2\,\mu g\,g^{-1}$ wet weight, but may exceed $20\,\mu g\,g^{-1}$ in benthivorous fish inhabiting heavily contaminated waters.

Studies of Pb accumulation in aquatic organisms from mining districts illustrate the importance of dietary uptake to body burdens and toxicity. In areas where the Pb ores reside in an acidic host rock containing little carbonate buffering capacity, surface water Pb concentrations may be toxic to aquatic organisms. However, Pb deposits also occur in limestone–dolomite formations where surface waters are hard (typically $>180\,ng\,ml^{-1}$ as $CaCO_3$) and well buffered. In such areas, dissolved Pb concentrations rarely attain toxic levels, even in heavily contaminated streams. Nevertheless, Pb concentrations in aquatic biota from well-buffered, contaminated streams often reflect Pb levels in the sediment (**Figure 1**), and are elevated by several orders of magnitude relative to uncontaminated streams. Effects on the heme

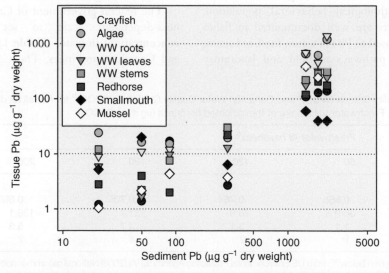

Figure 1 Concentrations of Pb in sediments and aquatic organisms from Missouri streams contaminated to differing degrees by historical and active Pb mining. Algae, *Cladophora* sp. (a filamentous chlorophyte); WW, water willow (*Justicia americana*) (an emergent macrophyte); mussel (*Lampsilis ventricosa*), a filter-feeding bivalve mollusk; crayfish (*Orconectes* spp.), an omnivore/detritivore; redhorse (*Moxostoma* spp.), a benthivorous fish; and smallmouth bass (*Micropterus dolomieu*), a predatory fish. From Schmitt CJ and Finger SE (1982) The transport, fate, and effects of trace metals in the Big and Black River watersheds, southeastern Missouri. US Fish and Wildlife Service, Columbia National Fisheries Research Laboratory, Columbia, MO. Project Completion Report to the St. Louis District, US Army Corps of Engineers, St. Louis, MO, 167pp.

biosynthetic pathway attributable to Pb in aquatic organisms inhabiting these systems are well documented.

Aquatic organisms exposed to elevated Pb concentrations may suffer a variety of toxic effects. Effects on enzymes involved in heme synthesis, especially δ-aminolevulinic acid dehydratase (ALA-D), have been reported in many species of fish and other aquatic and marine animals exposed to Pb in the laboratory and in the field. In fish, inhibition of ALA-D is associated with blood Pb concentrations of 0.1–0.5 $\mu g\,ml^{-1}$ and whole-fish concentrations of about 1.0 $\mu g\,g^{-1}$ (wet weight), which are commonly exceeded in Pb-contaminated waters. Behavioral changes and growth effects have been induced in several fish species by exposure to waterborne Pb in the laboratory, as have cytogenetic abnormalities and reduced hatchability and growth at various life stages. In field studies, reduced bone strength, which may impair swimming performance and ultimately lower the ability to escape predators, was reported in longear sunfish (*Lepomis megalotis*), in addition to effects on heme synthesis and elevated tissue Pb concentrations. Reduced body condition was associated with elevated Pb concentrations and effects on heme synthesis in two species of catfish (Pimelodidae) from a tailings-contaminated stream in Brazil in which the fish community was also depauperate. However, in general, effects at levels of biological organization higher than the individual are not well documented, or cannot confidently be ascribed solely to Pb exposure. Lead in ores, effluents, sediments, and surface waters is typically present in combination with other potentially toxic metals, especially cadmium, cobalt, copper, nickel, silver, and zinc. Although a variety of physiological, histopathological, behavioral, population, and community effects are well documented in fishes and invertebrates exposed to Pb via water, food, sediment, or a combination of pathways in field and laboratory studies, many reported effects undoubtedly reflect exposure to multiple stressors, such as combinations of metals as well as reduced ambient pH and degraded habitat. For example, populations of crayfish (*Orconectes* spp.) and sculpins (*Cottus* spp.) have been reduced or extirpated from streams contaminated by Pb mining, but it is difficult to ascribe this loss specifically to Pb because elevated levels of other metals, and the presence of other stressors, usually co-occur with Pb in such environments.

Environmental regulation of Pb in natural waters

In the US, the Environmental Protection Agency (USEPA) establishes national water quality criteria (WQC) under the Clean Water Act. The WQC are empirically derived from toxicity data for a wide variety of organisms, with the goal of protecting 95% of aquatic taxa. The 'criterion maximum concentration' (CMC) is the estimated highest surface water concentration to which aquatic communities can be exposed acutely without unacceptable negative effects; the 'criterion continuous concentration' (CCC) is that to which aquatic communities can be exposed indefinitely without harm. The CMC and CCC vary inversely with hardness, and are higher for saltwater than fresh due to proportionally lower concentrations of Pb^{2+}, and higher concentrations of stable $Pb-Cl^-$ complexes, in saltwater (**Table 1**). The values are also lower for filtered than unfiltered water due to the tendency of Pb to associate with suspended particulates, as indicated by the total:dissolved conversion factor (**Table 1**). The current WQC and the methodologies used to derive them are being reviewed by EPA, and a more mechanistically based replacement methodology is being considered for US waters.

The federal government of Canada also uses a hardness-dependent scale to set guidelines for Pb concentrations in water (**Table 1**), as do many US states and Canadian provinces. The Canadian Water Quality

Table 1 US Water Quality Criteria (WQC)[a] and Canadian Water Quality Guidelines (WQG)[b] for Pb (all as dissolved Pb, $ng\,ml^{-1}$) for protection of aquatic life. Freshwater values are at the indicated hardness ($\mu g\,ml^{-1}$ $CaCO_3$)

	Freshwater @ hardness[b]				Saltwater
	60	120	180	200	NA
US (WQC)[a]					
CF[c]	0.865	0.764	0.705	0.690	0.951
CMC[d]	36.9	78.9	121.7	136.1	210
CCC[e]	1.4	3.1	4.7	5.3	8.1
Canada (WQG)[b]	1	2	4	7	NV

[a]Computed from equations given below[d,e]; from US Environmental Protection Agency (EPA) (2006) National recommended water quality criteria. *Office of Water, Office of Science and Technology* (4304T). http://www.epa.gov/waterscience/criteria/wqcriteria.html#appendxa.
[b]For hardness categories <60, 60–120, 120–180, and >180; from Environment Canada/Canadian Council of Ministers of the Environment (CCME), (2005) *Canadian Environmental Quality Guidelines*. http://www.ccme.ca/assets/pdf/wqg_aql_summary_table.pdf.
[c]CF = 1.46203 − (ln hardness)(0.145712); from USEPA (2006).
[d]CMC ($ng\,ml^{-1}$, dissolved) = $e^{1.273(\ln\,hardness)\,-\,1.460}$ × CF; from USEPA (2006).
[e]CCC ($ng\,ml^{-1}$, dissolved) = $e^{1.273(\ln\,hardness)\,-\,4.705}$ × CF; from USEPA (2006).
CMC, criterion maximum concentration; CCC, criterion continuous concentration; CF, conversion factor for estimating dissolved concentrations or criteria from total recoverable concentrations or criteria; NA, not applicable; NV, no value specified.

Guidelines (WQG) are recommended 'not to exceed' concentrations, analogous to the CMC values, but the concentrations are similar to the CCC values (**Table 1**). The European Commission has identified Pb as a 'possible priority hazardous substance' for elimination or regulation under its Water Framework Directive, but has not yet drafted criteria for its regulation.

Lead in the Terrestrial Environment

Galena (lead sulfide; PbS) is the most common native (mineral) source of Pb in soil. Other Pb-based minerals commonly found in soils include cerussite (lead carbonate; $PbCO_3$) and anglesite (lead sulfate; $PbSO_4$). Lead in soils is usually present as Pb^{2+}, the most stable valence, and less frequently as Pb^{4+}. Similar to Pb in sediments, soil Pb is bound mainly to organic matter, or adsorbed onto clays, and ferric and manganese oxides. Generally, only a small proportion of Pb in soil is soluble in water, leaches through the soil profile, or is available in a chemically exchangeable form. The exchangeable fraction of Pb increases as soil pH declines; and may also increase in severely contaminated soils in which available binding sites have been saturated. Chemically active forms of Pb released into the environment gradually revert to more stable forms, such as $PbCO_3$, $PbSO_4$, and lead chlorophosphate (pyromorphite; $Pb_5(PO_4)_3Cl$). When contaminated sites are remediated, the mobility and availability of Pb can be decreased by amendments that form compounds such as $Pb_5(PO_4)_3Cl$, which is especially stable.

Metallic Pb (Pb^0) is relatively stable in the terrestrial environment, but is gradually weathered, dissolved, and transformed to various Pb salts over time. For example, metallic Pb is durable enough to have been used as a roofing material in the past; however, soils adjacent to such roofs have been found to be highly contaminated with Pb due to the slow oxidation of Pb^0 to Pb^{2+} and its transport onto soil in rain water runoff over many decades. Metallic Pb corrodes slowly, forming crusts primarily composed of Pb carbonates and sulfates. Complete oxidation of metallic Pb objects, such as Pb shotgun pellets and bullets, may take hundreds of years if buried deeply in alkaline, highly buffered soils that are not mechanically disturbed. Only decades may be required, however, if the soil is poorly buffered, acidic, and subject to frequent mechanical treatment, such as cultivation. The uppermost soil layer at shooting ranges can contain tens of thousands of $\mu g\,g^{-1}$ Pb from the corrosion of Pb shotgun pellets and Pb-containing bullets in some heavily shot-over locations.

The approximate average background concentration of Pb in crustal rock is $15\,\mu g\,g^{-1}$, and median concentrations of Pb in soils from rural, relatively uncontaminated, areas range from 10 to $60\,\mu g\,g^{-1}$. Median concentrations of Pb in urban soils range from background concentrations to several hundred $\mu g\,g^{-1}$, depending on their proximity to major roads or Pb-using industries. Because of extensive aerial deposition of Pb from various industrial activities and leaded gasoline combustion in motor vehicles, all soils have Pb concentrations that are at least somewhat elevated compared to concentrations that would have been present in pristine, pre-industrial-era soils. Soil Pb concentrations in old urban areas where buildings were painted with Pb-based paints can be very high, in the range of $100-1000\,\mu g\,g^{-1}$. Soil concentrations of Pb at base metal mining sites are variable, but may be in the tens of thousands of $\mu g\,g^{-1}$.

Lead in terrestrial plants and invertebrates

Soil Pb becomes concentrated around plant roots or in the root cortex, but only a small fraction of the Pb in soil is taken up internally by plants. Lead concentrations in plant stems tend to be lower than in roots, and lower still in seeds. High concentrations of Pb in soil are toxic to plants (generally $>1000\,\mu g\,g^{-1}$ Pb are required to produce toxic effects), and the potential for Pb poisoning of humans and animals consuming contaminated plants is of concern in ecosystems contaminated with Pb. When animals feed on plants growing in Pb-contaminated soil, the soil incidentally ingested while feeding usually contains much higher concentrations of Pb than does the plant tissue itself, especially when animals are feeding on tubers or roots. In the presence of substantial air pollution, more Pb tends to accumulate on the surface of the foliage than is taken up through the roots and transferred to the leaves. Foliage contaminated by Pb in this way can be a major pathway of dietary Pb exposure for herbivorous animals living near sources of airborne Pb release, such as base metal smelters.

The chemical forms of Pb normally found in soils are poorly absorbed by soil invertebrates. Consequently, concentrations of Pb in tissues of soil invertebrates, such as earthworms, are generally well below concentrations in soil; but some earthworms living in acidic soil and earthworms of the genus *Eisenoides* have been found to accumulate Pb to levels greater than the soils in which they are found. Earthworms consume soil; consequently, the toxicological risk to predators of earthworms is largely dependent on the amount of soil in the worm digestive tract and the Pb concentration of that soil.

Insects also tend to contain concentrations of Pb that are well below those in soil, and that decline with increasing trophic level. Most Pb in insects originates from dietary sources rather than from direct physical contact with environmental media, such as soil. For example, cicadas (*Magicicada septendecim*) living as larvae for 17 years in soil severely contaminated by smelter emissions nevertheless contained $<1\,\mu g\,g^{-1}$ Pb.

Lead concentrations in snails (*Cepaea nemoralis*) collected from contaminated areas were much lower than, and were only weakly correlated with, concentrations in soil. In feeding trials, snails (*Helix aspersa*) were found to accumulate Pb to concentrations below those contained in their diets, and when experimentally exposed to contaminated soil, they absorbed almost no Pb through the foot.

Although Pb does not generally accumulate to high concentrations in tissues of terrestrial invertebrates, high soil Pb concentrations can poison these organisms. For example, >1000 $\mu g\, g^{-1}$ Pb in soil can kill earthworms and springtails (Collembola), and sublethal effects occur at lower concentrations.

Importance of soil and sediment ingestion as a pathway of Pb exposure in wildlife

Excluding Pb shotgun pellets and other metallic Pb artifacts, the greatest exposure to Pb tends to be among those bird and mammal species that feed low in the food chain and, more especially, those that ingest relatively high amounts of soil or sediment in their diet. Moles, for example, have a greater exposure to Pb than most other species of small mammal because they ingest the soil present within the digestive tract, and outer surfaces, of earthworms. A similar risk exists for earthworm-consuming birds such as American woodcock (*Scolopax minor*) and American robin (*Turdus migratorius*). Grazers are exposed to Pb as they ingest soil adhering to blades or roots of grasses. Some animals intentionally ingest soil at salt licks. Sediments ingested from the Coeur d'Alene River Basin in Idaho have killed hundreds of tundra swans (*Cygnus cygnus*) and have caused widespread toxic effects in various other waterfowl species. Sediments in this area contained up to 3000–5000 $\mu g\, g^{-1}$ Pb from contamination associated with many decades of metal mining and smelting. Tundra swans were especially vulnerable to poisoning, as it was demonstrated that their diets comprised about 9% sediment, equivalent to about 400 $\mu g\, g^{-1}$ dietary Pb.

Lead toxicity in mammals

The toxic effects of Pb in mammals are well documented. Although generally less than 10% of ingested Pb is absorbed from the adult mammalian digestive tract, Pb resembles Ca in atomic structure, and the absorption of Pb can be greatly enhanced when dietary Ca is low. After absorption, Pb metabolism tends to follow Ca metabolism. Like Ca, Pb is stored in bone and other mineralized tissues, and interferes with Ca function in the body. Lead exposure can cause anemia, and it inhibits the enzyme ALA-D, whose measurement in blood is a sensitive indicator of the degree of Pb exposure. Additionally, Pb reacts with proteins in the kidney to form microscopic intranuclear inclusion bodies, and at higher concentrations may cause renal failure. Lead also interferes with the function of the central and peripheral nervous systems, causing lethargy, impaired movement, tremors, and ultimately death. Pathologists expect to see clinical signs of Pb poisoning

in mammals when Pb concentrations exceed 30 $\mu g\, g^{-1}$ in livers or 90 $\mu g\, g^{-1}$ in kidneys, on a dry weight basis.

Although Pb poisoning in domestic animals is relatively common, Pb toxicity in wild mammals has seldom been reported. Common sources of Pb exposure in domestic animals are paint and batteries. Lead poisoning has killed thousands of cattle from ingestion of contaminated crankcase oil and grease from farm vehicles, and environmental Pb from discarded batteries and Pb plumbing. Cases of horses being poisoned from ingestion of vegetation contaminated by Pb from base metal smelting have also been documented. Wild grazing mammals also risk exposure to Pb and other metals through ingestion of plant material contaminated with mining or smelting wastes.

In controlled dosing studies, Pb concentrations and toxic effects in tissues increase in a dose-dependent manner in small mammals. In the wild, however, mice feeding on seeds and insects generally do not accumulate toxicologically high concentrations of Pb, even at contaminated sites. Decreases in ALA-D activity and the presence of intranuclear inclusion bodies have been observed in wild populations of mice, such as *Peromyscus* sp., living near sites contaminated by smelters, mining wastes, and urban pollution; but, in general, only weak correlations have been reported between tissue-Pb and soil-Pb concentrations. Moles feeding on earthworms, and shrews feeding on a variety of soil invertebrates at contaminated sites, have been found to accumulate tissue concentrations of Pb that are much higher than those observed in mice and that are above toxicity threshold concentrations.

Although mainly an issue for birds, ingestion of Pb shotgun pellets has been documented in some mammals, particularly domestic cattle. Cattle fed chopped silage prepared from fields contaminated with Pb shotgun pellets from target shooting have experienced elevated Pb exposure and Pb poisoning.

Table 2 presents estimated tissue-Pb concentrations indicative of background exposure, moderate toxicity, and severe toxicity and death in wild mammals.

Lead toxicity in wild birds

Lead poisoning in wild birds is a much more common phenomenon than in wild mammals. Large numbers of wild birds of many different species have been exposed to lethal doses of Pb from several different sources, the most important of which include ingestion of soils or sediments contaminated with mining and smelting wastes; ingestion of paint chips near old buildings painted with Pb-based paints; and ingestion of small metallic Pb objects, such as shotgun pellets and fishing weights. Lead-contaminated sediments ingested from the Coeur d'Alene River Basin in Idaho have caused Pb poisoning of swans and various other waterfowl species. Birds that feed near major roadways may also ingest relatively high amounts of Pb from

Table 2 Approximate Pb concentrations (µg g⁻¹ dry weight; except blood, µg ml⁻¹) indicative of background exposure; elevated exposure for which toxic effects are likely; and exposure associated with severe Pb poisoning, for indicated tissues of wild birds and mammals. Values are estimates synthesized by the authors from various sources

	Background	Toxic effects probable	Severe poisoning (including death)
Birds			
Blood	<0.2	>0.5	>1.0
Liver	<8	>20	>50
Mammals			
Blood	<0.1	>0.4	>1.0
Liver	<1	>10	>30
Kidney	<2	>25	>90

Note: To confidently diagnose Pb poisoning mortality in wildlife, tissue Pb concentrations should be used in conjunction with other data (e.g., presence of Pb artifacts in the gastrointestinal system, or histopathology) whenever possible.

contamination of the local environment due to years of leaded gasoline combustion; however, overt toxicity and death from this source of exposure has not been documented. **Table 2** presents estimated tissue-Pb concentrations indicative of background exposure, moderate toxicity, and severe toxicity and death in wild birds.

Primary Pb shot poisoning

A major source of Pb exposure and toxicity for wild birds is the ingestion of Pb-based projectiles from ammunition, especially Pb shotgun pellets used in hunting and target shooting, and to a lesser extent Pb-containing bullets. Globally, tens of thousands of tons of metallic Pb projectiles are deposited into the environment every year from hunting and other shooting activities. Although many bird species actively ingest spent Pb shotgun pellets from soils and sediments, this phenomenon has been most extensively studied in waterfowls which frequently ingest these items from the bottoms of lakes, ponds, and marshes, and from fields, mistaking them for food items such as seeds, or for grit (small stones used to help birds grind up food). Once ingested, these pellets often become lodged in the muscular gizzard of waterfowl, where Pb dissolves over time as a result of the grinding action of the gizzard combined with the acidic environment of the upper digestive tract. Dissolved Pb is absorbed into the blood stream and is carried to various organs where it can exert toxic effects. Lead primarily targets the central and peripheral nervous systems, the muscles of the gizzard, the kidneys, and the blood cells, causing loss of coordination, difficulty ingesting food, anemia, emaciation, and ultimately death. Besides waterfowl, numerous other bird species have been documented to have ingested Pb shotgun pellets and bullet fragments from both aquatic (wetland) and terrestrial (upland) habitats.

Prior to North American regulations restricting the use of Pb-based ammunition for migratory bird hunting, it was estimated that 2–3% of the North American fall waterfowl migration died annually of Pb poisoning from shot ingestion. Since 1991 in the US, and 1996 in Canada, national regulations prohibiting the use of Pb for waterfowl hunting have been in effect, and have led to dramatic declines in the ingestion of Pb and the average concentrations of Pb in waterfowl tissues. A number of other nations (e.g., Denmark, Great Britain, Finland, the Netherlands, Norway, Sweden) have also established regulations prohibiting the use of Pb-based ammunition for certain activities, especially the hunting of waterfowl and other wetland shooting. A number of metals and alloys have been tested and approved as nontoxic when ingested as shotgun pellets at realistic rates by birds; approved materials include iron, tin, bismuth, bismuth/tin, tungsten, tungsten/iron, tungsten/nickel/copper, and several tungsten/plastic mixtures.

Secondary Pb shot poisoning

Birds of prey suffer Pb poisoning when they ingest Pb shotgun pellets or bullet fragments embedded in the flesh of dead or wounded animals shot with Pb-based ammunition. This source of exposure is responsible for virtually all cases of Pb poisoning in adult raptorial birds. Secondary Pb poisoning of bald eagles (*Haliaeetus leucocephalus*) that scavenge hunter-shot ducks was a major consideration leading to a national prohibition on the use of Pb shot for waterfowl hunting in the US. A similar phenomenon has also been documented in upland habitats where various eagle, hawk, and owl species that feed on terrestrial prey are documented to have died of Pb poisoning, probably from incidental ingestion of Pb shot lodged in carcasses of the upland game birds and mammals that they feed upon. Although the overall incidence of such poisonings is low for most raptor species, accounting for about 3–6% of total reported mortality, Pb poisoning from ingestion of Pb bullet fragments embedded in carcasses of hunter-killed animals is a major cause of mortality for California condors (*Gymnogyps californianus*) and has been an important factor limiting the successful reintroduction of this endangered species. Other large scavengers such as bald eagles and vultures that feed on large carrion killed using Pb ammunition are similarly at risk for this sort of Pb exposure and poisoning. Human communities, especially those that rely on subsistence hunting, are also at risk for increased dietary Pb exposure from consumption of game animals killed with Pb-based ammunition whose flesh consequently contains numerous small fragments of metallic Pb.

Lead fishing weights

Although ingestion of Pb shotgun pellets was for many years the main source of elevated Pb exposure and poisoning for waterfowl and most other birds, for some species (e.g., common loons, *Gavia immer*) ingestion of small Pb fishing weights (sinkers and jigs) is a more frequent cause of Pb poisoning. A 'sinker' is an object, usually made of Pb, fastened to a fishing line in order to sink the line; a 'jig' is a weighted hook used as part of a lure to catch fish. Globally, thousands of tons of Pb in the form of fishing weights are lost or discarded annually. Wildlife, primarily piscivorous birds and other water birds, sometimes ingest these items and suffer Pb poisoning as a result. Empirical evidence indicates that small (<30 g) Pb sinkers and jigs present the greatest potential for ingestion; however, almost all sinkers and jigs contain more than enough Pb to poison and kill water birds. In eastern North America, Pb poisoning from sinker or jig ingestion can be a major source of mortality for adult common loons during the breeding season in habitats that experience high recreational angling pressure, accounting for an average of about 20–30% of total reported mortality in these environments. In North America, in addition to loons, individuals of more than 20 species of wildlife (mainly water bird species) have been documented to ingest fishing sinkers and/or jigs. In Great Britain, extensive Pb poisoning of swans from sinker ingestion led to a regulation prohibiting the sale of small (≤28.35 g or 1 oz) Pb fishing weights. Since the ban on Pb sinkers, mute swan (*Cygnus olor*) numbers, which had been declining in Britain, have increased dramatically. A few other jurisdictions (Canada, Denmark, and some US states) have also established regulations prohibiting the use or sale of some Pb types and sizes of fishing weights.

See also: Acute and Chronic Toxicity; Bioavailability; Body Residues; Ecotoxicology: The History and Present Directions.

Further Reading

Ankley GT, DiToro DM, Hansen DJ, and Berry WJ (1996) Technical basis and proposal for deriving sediment quality criteria for metals. *Environmental Toxicology and Chemistry* 15: 2056–2066.

Clark AJ and Scheuhammer AM (2003) Lead poisoning in upland-foraging birds of prey in Canada. *Ecotoxicology* 12: 23–30.

DiToro DM, Allen HE, Bergman HL, *et al.* (2001) Biotic ligand model of the acute toxicity of metals. Part 1: Technical basis. *Environmental Toxicology and Chemistry* 20: 2383–2396.

Eisler R (1988) Lead hazards to fish, wildlife, and invertebrates: A synoptic review. US Fish and Wildlife Service. *Biological Report* 85(1.14): 134.

Eisler R (2000) *Handbook of Chemical Risk Assessment: Health Hazards to Humans, Plants, and Animals, Vol. 1, Metals*, 738pp. Boca Raton, FL: Lewis Publishers.

Environment Canada/Canadian Council of Ministers of the Environment (CCME) (2005) *Canadian Environmental Quality Guidelines*. http://www.cme.ca/assets/pdf/wqg-aql_summary_tablepdf (accessed September 2007).

Henny CJ (2003) Effects of mining lead on birds: A case history at Coeur d'Alene Basin, Idaho. In: Hoffman DJ, Rattner BA, Burton GA, Jr., and Cairns J, Jr. (eds.) *Handbook of Ecotoxicology,* 2nd edn., pp. 755–766. Boca Raton, FL: Lewis Publishing.

Kendall RJ, Lacher TE, Bunck C, *et al.* (1996) An ecological risk assessment of lead shot exposure in non-waterfowl avian species: Upland game birds and raptors. *Environmental Toxicology and Chemistry* 15: 4–20.

Ma W-C (1996) Lead in mammals. In: Beyer WN, Heinz GH, and Redmon-Norwood AW (eds.) *Environmental Contaminants in Wildlife: Interpreting Tissue Concentrations*, pp. 281–296. Boca Raton, FL: Lewis Publishers.

MacDonald DD, Ingersoll CG, and Berger TA (2000) Development and evaluation of consensus-based sediment quality guidelines for freshwater ecosystems. *Archives of Environmental Contamination and Toxicology* 39: 20–31.

Pattee OH and Pain DJ (2003) Lead in the Environment. In: Hoffman DJ, Rattner BA, Burton GA, Jr., and Cairns J, Jr. (eds.) *Handbook of Ecotoxicology* 2nd edn., pp. 373–408. Boca Raton, FL: Lewis Publishers.

Sanderson GC and Bellrose FC (1986) A review of the problem of lead poisoning in waterfowl. *Illinois Natural History Survey Special Publication* 4: 1–34.

Scheuhammer AM, Money SL, Kirk DA, and Donaldson G (2003) Lead fishing sinkers and jigs in Canada: Review of their use patterns and toxic impacts on wildlife. *Canadian Wildlife Service Occasional Paper no. 108*, 48pp. Ottawa: Environment Canada.

Schmitt CJ and Finger SE (1982) The transport, fate, and effects of trace metals in the Big and Black River watersheds, southeastern Missouri. US Fish and Wildlife Service. Columbia National Fisheries Research Laboratory Columbia, MO. Project Completion Report to the St. Louis District, US Army Corps of Engineers, St. Louis, MO, 167pp.

Schmitt CJ, Whyte JJ, Brumbaugh WG, and Tillitt DE (2005) Biochemical effects of lead, zinc, and cadmium from mining on fish in the Tri-States District of Northeastern Oklahoma. *Environmental Toxicology and Chemistry* 24: 1483–1495.

Stevenson AL, Scheuhammer AM, and Chan HM (2005) Effects of lead shot regulations on lead accumulation in ducks in Canada. *Archives of Environmental Contamination and Toxicology* 48: 405–413.

US Environmental Protection Agency (EPA) National recommended water quality criteria. *Office of Water, Office of Science and Technology* (4304T), http://www.epa.gov/waterscience/criteria/wqcriteria (accessed January 2008).

Wixson BG and Davies BE (eds.) (1993) *Lead in Soil: Recommended Guidelines*, 132pp. Northwood: Science Reviews.

Nitrogen

L van den Berg and M Ashmore, University of York, York, UK

Introduction

In contrast to toxicology, which deals with the effects of contaminants on individual organisms, ecotoxicology is concerned with the effects of contaminants on complex and interacting ecosystems. This implies that ecotoxicology not only considers direct effects on individual organisms but also pathways and processes through which contaminants affect other organisms and ecosystem functions. Since ecotoxicology considers ecosystem level effects, detrimental effects on individual organisms or species can be of less importance than effects on certain pathways or the development of an ecosystem. These can be very important ecotoxicological processes, even if they do not necessarily involve killing individual organisms. This is of specific interest when considering the ecotoxicological effects of nitrogen, since nitrogen is an essential nutrient and many ecosystems are nitrogen limited. Although increased nitrogen levels can increase the growth of some species, this can lead to adverse effects on other species, on characteristic species assemblages, and on ecosystem function.

For a clear understanding of the ecotoxicological effects of nitrogen in all its different forms, an understanding of nitrogen sources and cycling is essential. Nitrogen and nitrogen compounds cycle, like any other element, through the air, water, and soil. Although nitrogen in its inert, molecular, form (N_2) is the most common gas in the Earth's atmosphere, most organisms require nitrogen to be in a reactive form (bound to hydrogen, oxygen, or carbon), to be able to use it. The most common of these reactive forms are organic nitrogen compounds, nitrogen oxides (NO_y), nitrate (NO_3^-), and ammonium (NH_4^+) and ammonia (NH_3), together often referred to as NH_x. It is the oxidized and reduced forms of reactive nitrogen that are the most toxic forms to organisms.

Nitrogen Pollution in a Global Perspective

Nitrogen fixation (the conversion of inert nitrogen to ammonium, ammonia, nitrate, or nitrogen oxides) is a process which occurs naturally, but can also occur through anthropogenic processes. Natural nitrogen fixation is performed by a number of microorganisms, called diazothrops. Well-known examples of diazothrops are the *Rhizobium* bacteria, which can be found in the root nodules of higher plants (especially leguminous plants). A second form of natural nitrogen fixation is as a result of lightning, which converts inert nitrogen into nitrogen oxides. Human activities, however, now have by far the highest contribution to global nitrogen fixation. These include the synthesis of fertilizers, fuel combustion, and industrial nitrogen fixation.

As a result of such activities, the anthropogenic inputs of NO_y and NH_x to ecosystems have increased tenfold in 100 years, and at present account for the majority (65%) of the total global nitrogen budget. Estimates for the current global anthropogenic nitrogen fixation of $140\,Tg\,N\,yr^{-1}$ show that 15% originates from fuel combustion, 24% stems from the cultivation of rice and nitrogen fixating crops for food production, and 61% comes from the production of fertilizers via the chemical conversion of N_2 to NH_3 (the Haber–Bosch process). Of the different regions in the world, North America ($28.4\,Tg\,N\,yr^{-1}$), Europe and the former Soviet Union ($26.5\,Tg\,N\,yr^{-1}$), and Asia ($68.9\,Tg\,N\,yr^{-1}$) fix the majority of global reactive nitrogen.

Since the turnover of the reactive nitrogen to inert N_2 (via the process of denitrification) is lower than the production of reactive nitrogen, nitrogen may accumulate in different chemical forms and severely disrupt N cycles in the atmosphere, and in terrestrial and aquatic ecosystems. An overview of the global conversions and transfers of nitrogen between the atmosphere, and terrestrial and aquatic ecosystems, and its effects at different levels is given by the nitrogen cascade depicted in **Figure 1**.

Industrial nitrogen fixation and fuel combustion are together responsible for the vast majority of NO_y in anthropogenic nitrogen fixation. NO_y is almost completely emitted to the atmosphere, where it can increase concentrations of gases such as ozone via photochemical reactions. In the lower atmosphere, the troposphere, nitrogen oxides can form a variety of nitrate containing organic compounds (e.g., peroxyacytyl nitrate, PAN) and contribute to formation of excessive levels of ozone after reactions with volatile organic compounds (VOCs); both

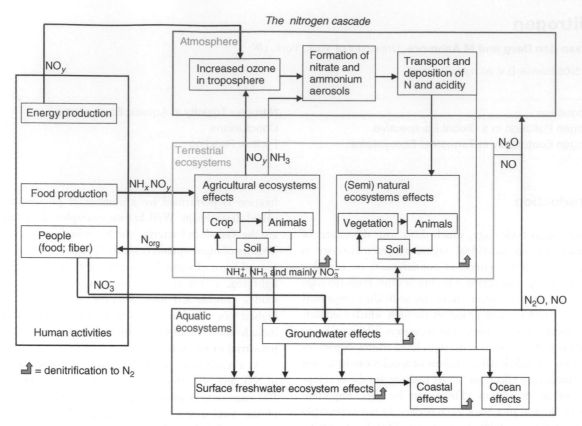

Figure 1 Diagram of the nitrogen cascade, illustrating the different sources and sinks of nitrogen forms in the atmosphere, terrestrial- and aquatic ecosystems. Adapted from Galloway *et al*. 2003. Some key processes are emitted from the original diagram because they are not relevant to this article.

PAN and ozone have ecotoxicological effects. In addition, nitrogen oxides can form the acidic HNO_2 and HNO_3 which can have ecotoxicological effects when deposited on aquatic and terrestrial ecosystems.

The production of food (which includes nitrogen fixation via fertilizer production and cultivation) and other agricultural practices account largely for the fixation of nitrogen as NH_3 and NH_4^+. Nitrogen in these forms can be released either directly into the soil and soil pore water (e.g., via application of fertilizers in agriculture), directly into aquatic bodies (e.g., via effluent from sewage works or from runoff from agricultural sources), or into the atmosphere (e.g., from industrial and agricultural sources). Subsequently, fixed nitrogen may be deposited from the atmosphere to terrestrial and aquatic ecosystems.

Although part of the anthropogenic reactive nitrogen is removed from ecosystems in the form of food and raw materials, this is only a small proportion of the amount which is fixed to produce the food and raw materials. Anthropogenic inputs to aquatic ecosystems from groundwater and surface runoff, from wastewater treatment runoff, and from uncontrolled sewage waters account for the majority of the remaining anthropogenic input of nitrogen into water systems.

Different forms of reactive nitrogen can cause either direct ecotoxic effects, or indirect effects by altering biogeochemical pathways or by interactions with secondary pollutants. In addition, it strongly depends on the source where, and in what nitrogen form, ecotoxic effects of nitrogen occur The ecotoxicity of reactive nitrogen is manifested differently within and between terrestrial and aquatic systems, and therefore these are now considered in turn.

Nitrogen Ecotoxicity in Terrestrial Ecosystems

Nitrogen applications to many managed systems are made to increase production and yield, and effects on biodiversity and ecosystem function in such systems are of lower concern. Atmospheric nitrogen deposition is the total of dry and wet deposition from the air and, in areas with little direct fertiliser nitrogen input to soils and waters, accounts for the majority of the annual nitrogen input. Dry deposition consists of gases, such as NO_2 and NH_3, and fine particles (including fine particle NH_4^+). Wet deposition (soluble forms in, for example, rain and snow) consists of NH_4^+, NO_3^-, and forms such as HNO_3^-, $(NH_4)_2SO_4$, and NH_4NO_3.

The deposition of reduced nitrogen (NH_x) is highest near its emission source, usually an agricultural source, whereas oxidized nitrogen compounds (NO_y), and especially NO_2, are transported over larger distances (so-called long-range trans-boundary pollutants). The deposition rate is also strongly influenced by the structure of the vegetation onto which it is deposited. For example, atmospheric nitrogen deposition onto vegetation with a rough canopy structure, for example, a pine forest, is considerably higher (>50%) than that onto smooth surfaces, for example, a calcareous grassland. When studying ecotoxicological effects of nitrogen, these differences must be considered.

The effects of increased nitrogen deposition on terrestrial ecosystems have been most studied in northern and Western Europe, partly because nitrogen input into terrestrial ecosystems from atmospheric deposition was extremely high in Europe in the 1980s and early 1990s, reaching $30-40\,kg\,ha^{-1}\,yr^{-1}$ in many areas of central Europe and exceeding $60\,kg\,ha^{-1}\,yr^{-1}$ in others. Although nitrogen deposition from the atmosphere declined after the 1990s as a result of the impact of new legislation, current deposition rates in, for example, the Netherlands are still very high (up to $42\,kg\,N\,ha^{-1}\,yr^{-1}$) and among the highest in the world. For this reason, ecotoxicological effects of nitrogen are most clearly understood in this region.

Direct Toxicity

Of all the reactive nitrogen forms, NH_3, NH_4^+, and NO_3^- are the most important in terrestrial ecosystems. Since ecotoxicity is very much a process depending on the dose and concentration of the contaminant, the deposition rate and the distance to the source determines to a large extent the potential for toxicity. Nitrogen in the form of nitrate (NO_3^-) is not considered toxic to plants, and in fact stimulates growth of many plant species. Since it does not bind strongly to soil particles, NO_3^- is easily leached to deeper layers and is therefore normally not present in high concentrations in the rooting zones of the soil. In contrast, NH_3 and NH_4^+ can be very toxic in terrestrial ecosystems. In the soil, NH_3 is readily converted into NH_4^+ which binds strongly to soil particles, resulting in accumulation of this form of nitrogen in high concentrations.

In general, three groups of plants can be distinguished, based on their nitrogen nutrition: those preferring NO_3^-, those preferring NH_4^+, and those using both NH_4^+ and NO_3^-. NH_4^+-tolerant species are usually found in acidic habitats and are likely to be adapted to NH_4^+ nutrition, since NH_4^+ is the dominant form of nitrogen in soils at low pH. These species are mainly slow-growing perennial species. In contrast, species from more buffered habitats usually prefer NO_3^- and are also usually faster-growing annual species. Despite these species preferences, NH_4^+ is readily taken up by both NH_4^+-tolerant and -intolerant species, as NH_4^+ uptake is energetically more favorable than NO_3^- uptake. Uptake of the different nitrogen forms can occur via the roots and via the leaves (foliar uptake). Following uptake, NH_4^+ is either assimilated into amino acids and other organic nitrogen compounds or accumulated as free NH_4^+. The assimilation of NH_4^+ takes place in the cells, mainly via the glutamine synthetase/glutamate synthase pathway and produces protons.

Accumulated NH_4^+ can cause severe toxicity symptoms in sensitive plant species, bryophytes, and lichens. The capacity to assimilate NH_4^+, and the rate at which this takes place, determines whether these species are sensitive to NH_4^+ or NH_3 nutrition. Direct NH_4^+ toxicity effects on vascular plants and mosses can be clearly shown in hydroponic experiments. Many characteristic species from species-rich grasslands are sensitive to NH_4^+ (e.g., *Antennaria dioica*, *Arnica montana*, and *Cirsum dissectum*), and NH_4^+ toxicity symptoms in these species are expressed as severe growth suppression and high mortality (**Figure 2**).

Antennaria dioica grown at pH 5.0 and 100 µmol l⁻¹ NH₄⁺

Antennaria dioica grown at pH 5.0 and 1000 µmol l⁻¹ NH₄⁺

Figure 2 Mountain Everlasting (*Antennaria dioica*) grown at two different NH_4^+ concentrations in a hydroponic experiment.

The mechanisms behind toxicity of NH_3 and NH_4^+ differ between species. In general, elevated NH_3 and NH_4^+ uptake decreases photosynthetic capacity and results in cation deficiency and chlorosis. At the same time, concentrations of the anions Cl^- and SO_4^{2-} can increase, resulting in a severe charge imbalance in the plant cells. Finally, NH_3 and NH_4^+ uptake results in disruption of cell membrane functioning. The decline of the brown moss *Scorpidium scorpioides*, for example, is directly related to serious disruption of cell pH regulation, reduction in nitrate reductase activity, and membrane dysfunction due to increased NH_4^+ nutrition. One of the key mechanisms of NH_4^+ toxicity in vascular plants is linked to NH_4^+ uptake mechanisms; with the uptake of NH_4^+, cations such as Ca^{2+}, Mg^{2+}, and K^+ (but mainly K^+) are actively excreted and uptake of cations (especially Mg^{2+}) is prevented, causing cation deficiencies and chlorosis in the leaves. Other recorded negative effects in higher plant species are a reduction in fine roots, a decline in mycorrhizal associations, and strongly reduced germination and seedling establishment.

Indirect Toxicity

In the air, nitrogen oxides dissolve in small water droplets and form the acidic molecules HNO_2 and HNO_3. As a result, nitrogen oxides cause a drop in pH of the rain. Natural rain has a pH in the range of 5.0–5.6, whereas acidified rain has a pH between 3.0 and 5.0. Sulfur dioxide (SO_2) also causes rain acidification, and was the dominant contributor in the late nineteenth century and the first half of the twentieth century, when acidification had devastating effects in many sensitive ecosystems in Europe and North America. As a result of strong reductions in sulfur emissions since the 1970s in these regions, much of the remaining acidification due to atmospheric deposition is caused by nitrogen-containing compounds. Nitrogen in the reduced (NH_x) form also causes acidification in the soil through both uptake by plants and increased nitrification (the microbial conversion of NH_4^+ into NO_3^-). Both processes produce protons and cause direct acidification of the soil.

Acidification seriously affects terrestrial ecosystems, especially slightly buffered systems. Following acidification, protons (H^+) are exchanged on the soil complex for macronutrients such as calcium (Ca^{2+}), magnesium (Mg^{2+}), and potassium (K^+), thereby buffering the soil against the acidifying processes. Subsequent leaching of Ca^{2+}, Mg^{2+}, and K^+ leads to loss of the soil's buffering capacity by base cations and to nutrient imbalances for plant growth. Buffering by cations and leaching will continue until all base cations are exchanged and continuing acidification will lead to a shift in the buffer range of the soil from cation buffering (pH 4.5–6.0) to aluminum (Al^{3+}) buffering (pH < 4.5). Consequently,

Al^{3+} and micronutrients such as manganese (Mn^{2+}) and iron (Fe^{2+}, Fe^{3+}) are exchanged for protons, resulting in increased concentrations of these free metal ions in the soil. Free metals are known to be highly toxic to most soil organisms. The decline of some characteristic plant species of species-rich acidic grasslands and heathlands in Western Europe is due to effects of aluminum toxicity and increased Al/Ca ratios in the soil, and to base cation depletion, as a result of increased acidification.

An increase in total nitrogen deposition drastically increases the availability of different forms (NH_x, NO_y) of nitrogen in the soil, either directly or indirectly through processes such as increased mineralization, litter turnover rates, and nitrification. Many natural and seminatural ecosystems (including forests, heathlands, and mat grass swards) are naturally nutrient poor (oligotrophic or mesotrophic) and nitrogen limited, which has resulted in a high plant diversity with characteristic species adapted to nitrogen-limited, nutrient-poor environments, and the low levels of competition typical of these conditions. An increase in nitrogen availability often results in a higher nitrogen uptake and higher plant productivity of competitive species for which elevated NH_4^+ is not directly toxic. For example, grasses such as the highly competitive species *Deschampsia flexuosa* do not respond negatively to elevated NH_4^+ and may even increase in biomass, indicating that the increased NH_4^+ input may eventually lead to an increase in competition for light and nutrients and hence to changes in species composition.

The decline of many characteristic species and the increase in highly competitive species (mainly grasses) is therefore often attributed to elevated nitrogen deposition. Such changes in species composition due to elevated nitrogen input have been observed in pine forests, heathlands, nutrient poor grasslands, wetlands, and bogs in Western Europe, where highly competitive grasses such as *Deschampsia flexuosa* and *Molinia caerulea* have increased at the expense of less competitive forb species. With increasing nitrogen loads, nitrogen becomes readily available and other nutrients become limiting. In oligotrophic and mesotrophic ecosystems, shifts from nitrogen limitation toward phosphate (P) limitation are commonly found. P availability can limit plant production and determine species composition in communities such as calcareous grasslands and calcareous dune grasslands. With increased nitrogen availability, the nitrogen content in these ecosystems increases relative to P content, affecting plants' susceptibility to herbivores and pathogens, as well as mycorrhizal infection rates, and sensitivity to frost and drought.

One example of these indirect mechanisms leading to ecotoxicological effects is a major shift in vegetation composition in Dutch heathlands that was linked to high atmospheric nitrogen input. Here a dominance of the shrub *Calluna vulgaris* was replaced by a dominance of the grass *Deschampsia flexuosa*. Although *D. flexuosa* is more competitive than the slow-growing *C. vulgaris*, in mature *Calluna* stands, invasion does not occur because of competition for light. Other key ecosystem effects were linked to increased nitrogen content of the vegetation, which increased the plants' susceptibility to herbivores like the heather beetle (*Lochmaea suturalis*). When *C. vulgaris* plants die as a result of such attacks, they leave large empty patches in which fast-growing grasses can outcompete the young *Calluna* plants if nitrogen availability increases. Grass species only started to dominate Dutch heathlands when the canopy of *C. vulgaris* was opened up by processes such as frost damage and herbivory.

Although direct toxicity effects might damage plants, they do not necessarily result in immediate mortality. For example, in Western European and North American coniferous forests (in particular pine and Douglas fir), nitrogen pollution has led to reduced growth rates, a higher percentage of brown and dead needles, and accelerated needle loss. In addition, acidic precipitation causes direct damage to the leaves by decreasing the wax layer, making the trees more vulnerable to disease. Nitrogen deposition also reduces the availability of base cations. In addition, geochemical conditions in the soil are affected as the concentrations of free metal ions (such as Al^{3+}) increase and pH levels decrease, causing serious deleterious effects to the roots. All these processes weaken the trees, making them vulnerable to drought, frost, diseases, and insects which are likely to ultimately kill them.

Nitrogen Toxicity in Aquatic Ecosystems

The effects of nitrogen in freshwater, estuarine, and coastal marine ecosystems are very different from those in terrestrial ecosystems. This is partly because the dominant nitrogen forms differ between the two types of ecosystems and also because these ecosystems differ in type of species found (e.g., angiosperms are not well represented in aquatic ecosystems whereas they dominate terrestrial ecosystems), and because they differ in terms of pollutant mixing, micro-niches/environments, etc. However, some general patterns in the effects of reactive nitrogen on organisms and ecosystem processes can be identified in terrestrial and aquatic ecosystems.

Direct Toxicity

Nitrogen enters aquatic ecosystems via surface and groundwater runoff, atmospheric deposition, nitrogen fixation by prokaryotes (cyanobacteria, blue-green algae), dissolution of nitrogen rich material, and decomposition. It occurs in water in four important, potentially toxic, forms: NH_4^+, NH_3, NO_2^-, and NO_3^-. NH_3 is the most toxic form to biota. However, since NH_4^+ and NH_3 are readily oxidized to NO_3^-, this is a more common form of nitrogen in water, although this strongly depends on factors such as sources and mixing, and abiotic parameters such as pH and temperature. A large amount of NO_3^- enters the aquatic environment directly via effluents from wastewater treatment. Background NO_3^- concentrations in natural waters typically range between 0 and $3\,mg\,l^{-1}$. However, due to anthropogenic sources, many waters exceed this concentration and concentrations above $100\,mg\,l^{-1}$ have been recorded. Concentrations of NO_3^- as low as $3\,mg\,l^{-1}$ can result in physical and behavioral abnormalities and egg mortality in amphibians in different life stages (e.g., eggs, tadpoles, and adults) and fish (e.g., rainbow trout). Elevated NO_3^- levels can also reduce the oxygen-carrying capacity of oxygen-carrying pigments (e.g., hemoglobin). High concentrations ($>10\,mg\,l^{-1}$) nearly always result in high mortalities even in the later life stages and subsequently a decline in amphibian and fish populations. An additional negative effect for amphibians is that adult fish are less susceptible to elevated NO_3^- concentrations. Increased NO_3^- concentrations will thus not only affect juvenile life stages directly through toxicity but also increase predation pressure on the tadpoles.

A more toxic component is the nitrite ion, NO_2^-. NO_2^- in water forms a chemical equilibrium with nitrous acid (HNO_2): $NO_2^- + H^+ \leftrightarrow HNO_2$. Both forms can be directly toxic to organisms, but the equilibrium strongly depends on pH and, in general, NO_2^- is a more common form than HNO_2. As with NO_3^-, NO_2^- strongly affects the oxygen-carrying capacity of organisms. In addition, NO_2^- is known to cause electrochemical imbalances in the cells, membrane malfunctions, and repression of the immune system. In high concentrations ($>0.25\,mg\,l^{-1}$), NO_2^- is toxic, and crustaceans (decapods, amphipods), insects (ephemeropterans), and fishes (salmonids) are among the most vulnerable groups. Since chloride (Cl^-) ions are taken up by fish via the same mechanisms in their gills, Cl^- uptake inhibits NO_2^- uptake and protects the fish against NO_2^- toxicity. Seawater therefore reduces the toxicity of NO_2^- considerably.

NH_3 mainly affects biota close to point sources such as agricultural and industrial effluents or sewage and wastewater treatment plants which lack nitrification steps. In these environments, plants are more tolerant of NH_3 than animals, and invertebrates are more tolerant than fish. Elevated NH_3 results in lower hatching and growth rates of fish. Normally, fish excrete NH_3 but with elevated concentrations in the water; excretion becomes difficult and high internal NH_3 concentrations can cause poor development and damage to gill, liver, and kidney tissues, repression of the immune system, and a reduction in oxygen-carrying capacity. Increased NH_4^+ concentrations in water occur mostly near agricultural point sources or in anoxic waters. Many freshwater plants (e.g., *Potamogeton* ssp., *Ranunculus* ssp.) and the salt water eelgrass (*Zostera marina*) all show poor growth rates, discoloration of the chloroplasts and a higher mortality at elevated NH_4^+ concentrations.

Indirect Toxicity

Aquatic ecosystems in Europe and North America, like terrestrial ecosystems, have suffered from acidification in sensitive poorly buffered areas due to nitrogen, as well as sulfur, deposition. A change in pH in streams and lakes can cause significant changes in plant and animal species composition. In general, juvenile life stages are more sensitive to pH changes than adult life stages, and a decrease in pH below 5.0, for example, has resulted in complete disappearance of snails in Norwegian lakes. Mollusks (snails, clams), arthropods (crustaceans, crayfish), and amphibians are particularly vulnerable to a pH drop to below 5. Direct negative effects of low pH on fish reproductive rate and survival rates occur, but reduced availability of their food source (crayfish and insects) also has an indirect negative effect on populations of numerous fish species. Increased acidification of aquatic environments results, as for terrestrial ecosystems, in increased solubility of metal ions such as Al^{3+} and trace metals (Cd, Cu, Pb, and Zn). This can cause direct metal toxicity to organisms both in the sediment and in the water. Increased Al^{3+} concentrations can also reduce phosphate availability and disrupt P cycling.

An increase in nitrogen availability in water can stimulate or enhance the growth and proliferation of primary producers (phytoplankton, benthic algae, and macrophytes). In nitrogen-limited freshwaters and coastal regions, this enhanced growth of highly competitive species reduces light penetration to the sediment and consequently, slow-growing and sensitive species may decline and disappear. The eutrophication of aquatic environments can result in huge expansive growth of primary producers (**Figure 3**). Some algal blooms (e.g., *Microcystis cyanobacteria* in freshwaters and *Alexandrium*

Figure 3 Algae bloom in Scandinavian lake.

dinoflagellates in coastal waters) are known to release toxic products which attack the nervous system, liver tissues, and cytoskeletons of many aquatic organisms. In addition, the decomposition of this algal organic matter when it dies and sinks to the bottom uses oxygen; with greatly increased rates of decomposition, the oxygen content of the water body is depleted. Because of the vast scales of such algal blooms, anoxic conditions can develop – a situation which is known as hypoxia. Hypoxia is known to be responsible for the death of a vast majority of the fish in these waters. Well-known cases of these hypoxic conditions as a direct result of nitrate occur in the Gulf of Mexico and the Baltic Sea. An additional negative effect of the hypoxic conditions is the formation of reduced compounds such as hydrogen sulfide (H_2S), which is responsible for acute lethal effects on fish and macrofauna, as it attacks the nervous system, and also causes root decay and plant mortality in wetlands.

Conclusions

Table 1a provides a summary of the direct effects of reactive nitrogen in aquatic and terrestrial ecosystems, which differ considerably. Reduced nitrogen is highly toxic in both ecosystems but, as its presence in aquatic ecosystems is restricted to areas close to sources or to anoxic conditions, its role in terrestrial ecosystems is more pronounced. In contrast to aquatic ecosystems, oxidized nitrogen is not considered toxic in terrestrial ecosystems. However, indirect effects of reactive nitrogen show considerable similarities between terrestrial and aquatic ecosystems for both oxidized and reduced nitrogen (**Table 1b**). Both types of ecosystem are indirectly affected by nitrogen oxides in atmospheric deposition causing acidification and an increase in free metal ions. In addition, increased nitrogen availability results in both systems in higher productivity, causing significant species composition shifts, in which highly competitive species start to outcompete less competitive species.

The impact of reactive nitrogen on sensitive aquatic and terrestrial ecosystems has led to the understanding that these ecosystems need protection from the anthropogenic input. In Europe in particular, there has been an emphasis on environmental management based on the long-term maintenance of the ecological status. In the case of deposition from the atmosphere, the convention on long-range trans-boundary air pollution has provided the framework within which ecotoxicological studies have been used to define critical loads of atmospheric nitrogen deposition below which ecosystems retain their biodiversity, functions, and characteristic species. As ecotoxicity strongly depends on specific characteristics of ecosystems, these critical loads vary between different aquatic and terrestrial ecosystems. Typical critical loads for nitrogen for sensitive ecosystems are listed in **Table 2**.

Table 1a Direct effects of different forms of nitrogen to terrestrial and aquatic ecosystems including mechanisms and examples of effects

Toxicity	Major direct effects	Mechanisms	Examples of sensitive species and species groups
Terrestrial ecosystems			
NH_3	Growth suppression, increased mortality, chlorosis	Base cation depletion, reduction of photosynthetic capacity, cell charge imbalances	Bryophytes, lichens
NH_4^+	Growth suppression, increased mortality, chlorosis	Base cation depletion, reduction of photosynthetic capacity, cell charge imbalances	*Antennaria dioca*, *Cirsium dissectum*
NO_3^-	Not toxic		
Aquatic ecosystems			
NH_3	Increased mortality, suppressed growth, poor egg development, damage to tissues	Charge imbalances in cells, direct toxicity due to accumulation	Fish, invertebrates
NH_4^+	Increased mortality, suppressed growth	Base cation depletion, reduced photosynthesis capacity, carbon limitation	Eelgrass, *Stratioites aloides*, *Potamogeton* species
NO_2^-, HNO_2	Physical and behavioral abnormalities, poor development, increased mortality	Reduces oxygen carrying capacity, electrochemical imbalances in cells, repression of the immune system	Crustaceans, insects, fish (salmonids) effects are more pronounced in freshwater
NO_3^-	Physical and behavioral abnormalities, increased mortality	Reduces oxygen carrying capacity	Amphibians, tadpoles, rainbow trout

Table 1b Indirect effects of different forms of nitrogen to terrestrial and aquatic ecosystems including mechanisms and examples of effects

	Major indirect effects	Mechanisms	Examples of sensitive species and species groups
Terrestrial ecosystems			
NH_3	Species composition changes, increased competition between species, growth suppression, mortality	Acidification of rhizosphere as a result of uptake by plants, depletion of base cation availability, decreased pH.	Decline of pine forests
NH_4^+	Species composition changes, increased competition between species, growth suppression, mortality	Acidification of rhizosphere as a result of uptake by plants and conversion via nitrification, depletion of base cation availability, decreased pH.	Changes in under story vegetation in pine forests and heathlands
NO_y (HNO_3, HNO_2, NO_2)	Species composition changes, increased competition between species, growth suppression, mortality	Acidification via acid rain, depletion of base cation availability, decreased pH.	Al^{3+} toxicity in *Arnica montana*
Total N	Species composition changes, increased competition between species, increased susceptibility to herbivores and pathogens	Increase of nitrogen availability	Grass encroachment in Dutch heathlands
Aquatic ecosystems			
NH_3	Inhibition of nitrification	Increases NH_4^+ concentrations	*Nitrosomonas* and *Nitrobacter* bacteria directly and all NH_4^+ sensitive species indirectly
NH_4^+	Growth suppression, higher mortality	Acidification via nitrification in the sediments	Mollusks, arthropods, amphibians

(Continued)

Table 1b Continued

	Major indirect effects	Mechanisms	Examples of sensitive species and species groups
NO$_y$ (HNO$_3$, HNO$_2$, NO$_2$)	Species composition changes, increased competition between species, growth suppression, mortality	Acidification via acid rain, heavy metal toxicity	Mollusks, arthropods, amphibians
Total N	Species composition changes, increased competition between species	Increase of nitrogen availability, hypoxia	Fish and invertebrate mortality in anoxic conditions

Table 2 Critical loads for different ecosystems based on empirical research and expert judgment as published by Swiss Agency for the Environment, Forests and Landscape (Environmental documentation no. 164, 2001)

Ecosystem	Critical load (kg N ha^{-1} yr^{-1})
Terrestrial ecosystems	
Temperate forests	10–20
Boreal forests	10–20
Tundra	5–10
Arctic, alpine, and subalpine scrub habitats	5–15
Northern wet heath	10–25
Dry heath	10–20
Semi-dry calcareous grassland	15–25
Dune grassland	10–20
Low altitude hay meadows	20–30
High altitude hay meadows	10–20
Wet oligotrophic grassland	10–25
Aquatic ecosystems	
Raised and blanket bogs	5–10
Poor fens	10–20
Rich fens	15–35
Mountain rich fens	15–25
Softwater lakes	5–10
Dune slack pools	10–20
Pioneer and low mid salt marsh	30–40

For aquatic systems, the new water framework directive of the European Union provides a framework within which environmental management needs to ensure the long-term ecological status of different water bodies. This leads to environmental standards for water quality (e.g., for ammonia, total inorganic nitrogen or dissolved oxygen) which vary according to the characteristics of the ecosystem and which specifically address the impact on key groups of plants and animals. In order to protect water ecology and water quality, aquatic ecosystems are required to have 'good ecological status' and 'good chemical status'. The ecological status is defined by biological parameters such as plant and animal species composition depending on the type of water body. Good chemical status is defined in terms of compliance with all the quality standards established for chemical substances at European level. The chemical status depends strongly on the type of water body, soil parameters, and other abiotic factors. However, plants and animals might respond to values much lower than these. Restrictions of the emissions and input of nitrogen to water bodies improve the nitrogen status of these waters.

Because reactive nitrogen is rapidly and widely dispersed through different ecosystems via the atmosphere and aquatic pathways and because the global anthropogenic input of nitrogen is expected to increase, initiatives such as those described above are needed worldwide to prevent and control ecosystem damage due to the effects of reactive nitrogen.

Further Reading

Bobbink R, Hornung M, and Roelofs JGM (1998) The effects of air-borne nitrogen pollutants on species diversity in natural and semi-natural European vegetation. *Journal of Ecology* 86: 717–738.

Britto DT and Kronzucker HJ (2002) NH$_4^+$ toxicity in higher plants: A critical review. *Journal of Plant Physiology* 159: 567–584.

Camatgo JA and Alonso Á (2006) Ecological and toxicological effects of inorganic nitrogen pollution in aquatic ecosystems: A global assessment. *Environment International* 32: 831–849.

Carpenter SR, Caraco NF, Correll DL, *et al.* (1998) Nonpoint pollution of surface waters with phosphorous and nitrogen. *Ecological Applications* 8(3): 559–568.

Fangmeier A, Hadwiger-Fangmeier A, Van der Eerden L, and Jäger HJ (1994) Effects of atmospheric ammonia on vegetation – A review. *Environmental Pollution* 86: 43–82.

Galloway JN, Aber JD, Erisman JW, *et al.* (2003) The nitrogen cascade. *BioScience* 53(4): 341–356.

Matson PA, McDowell WH, Townsend AR, and Vitousek PM (1999) The globalisation of N deposition: Ecosystem consequences in tropical environments. *Biogeochemistry* 46: 67–83.

Pearson J and Stewart GR (1993) Tansley review no. 56. The deposition of atmospheric ammonia and its effects on plants. *New Phytologist* 125: 283–305.

Relevant Website

http://ec.europa.eu – Water Framework Directive of the European Union.

Persistent Organic Chemicals

R Miniero and A L Iamiceli, Italian National Institute of Health, Rome, Italy

Chemical–Physical Characteristics of Persistent
  Organic Pollutants
Impacts of POPs on Organisms

Principal International Agreements on POPs
Further Reading

Glossary

bioaccumulation – Storage of a stable substance in living tissues, resulting in a much higher concentration than in the environment.

bioavailability – Ability of a substance to be taken up by living tissues.

biomagnification – Increase in the concentration of a substance through a food chain.

body burden – Amount of a substance present in a organism at a given time.

half-life ($t_{1/2}$) – Time required for the amount of a particular substance to be reduced to one half of its value when the rate of decay is exponential.

K-strategist – Species of organism that uses a survival and reproductive 'strategy' characterized by low fecundity, low mortality, longer life, and with populations approaching the carrying capacity of the environment, controlled by density-dependent factors.

lethal concentration (**dose**) – Concentration of a potentially toxic substance in an environmental medium that causes death following a certain period of exposure.

n-octanol–water partition coefficient (K_{ow}) – Ratio of the concentration of a chemical in *n*-octanol and that in water at equilibrium and a specified temperature.

population – A group of organisms, usually a group of sexual organisms that interbreed and share a gene pool.

r-strategist – Species of organism that uses a survival and reproductive 'strategy' characterized by high fecundity, high mortality, short longevity, and with populations controlled by density-independent factors.

TEQ – Toxic equivalent (TCDD equivalent).

Chemical–Physical Characteristics of Persistent Organic Pollutants

Persistent organic pollutants (POPs) are "chemical substances that persist in the environment, bioaccumulate through the food web, and pose a risk of causing adverse effects to human health and the environment" (United Nations Environment Programme, UNEP).

As with other chemicals, these compounds can be harmful, even if environmental contamination levels are low. Furthermore, prolonged exposure to the toxic substance increases the risk of damage.

POPs are, by definition, persistent, and this property is generally correlated to their chemical stability. In contrast to highly reactive substances which are relatively short-lived, persistent pollutants are highly resistant to biological, photolytic, and chemical degradation. This characteristic makes POPs persist in the environment for an extended period of time (i.e., with half-lives ($t_{1/2}$) greater than 6 months), thus representing a risk of a long-time exposure. The persistence of a chemical depends on the difficulty with which it is broken down and degraded into other less hazardous substances. This is connected with some structural characteristics of the molecule, such as the presence of aromatic systems and the substitution of one or more hydrogen atoms in the aromatic structure by one or more halogens. The carbon–halogen bond is in fact very stable and resistant to hydrolysis, and the greater the number of halogen atoms, the greater is the resistance to biological and photolytic degradation. From a technical point of view, the stability obtained by halogenation is a desirable characteristic. The fire resistance of polychlorinated biphenyls (PCBs), for instance, is increased by chlorination, while in the case of 1,1,1-trichloro-2,2-bis(4-chlorophenyl)ethane (DDT) it ensures longer-lasting effects on insect pests. On the other hand, the more stable the substances are the longer they resist degradation and the longer they remain in the environment. Furthermore, POPs can be found in every part of the world, even in areas where human activities are almost completely absent, for example, in the Antarctic and the Arctic. In fact, despite the fact that these molecules are present preferentially in soils, sediment, and living organisms, under normal conditions they can evaporate. In particular, pollutants with relatively low volatility do not remain long in the atmosphere, but are adsorbed onto the surface of airborne particles, which return them to the ground. However, more volatile POPs stay in the atmosphere for longer, even weeks, before settling back down to the Earth's surface. This means that winds can carry them thousands of kilometers away from their sources, to be found in similar concentration all over the world. In some cases, the concentrations of POPs in waters and sediments of Arctic and Antarctic can be higher than elsewhere. This phenomenon seems to be connected with the decrease of substance volatility with decreasing temperature, where pollutants that vaporize at warm latitudes are carried by winds to polar regions where they condense and concentrate. POPs are also characterized by their ability to bioaccumulate. The extent by which they bioaccumulate depends on the solubility of the substance in lipids. Highly lipophilic substances are substantially insoluble in water, as commonly shown by the high values of n-octanol–water partition coefficient (K_{ow}). Thus, in the aquatic enviroment, they display strong affinity for suspended particles, sediments, and living organism, where they reach concentrations many times higher than in the water itself. Halogenated organic compounds are less soluble in water and more soluble in lipids respect to their corresponding nonhalogenated compounds. This strengthens the tendency of these substances to be concentrated in the fatty tissue of living organism. PCB concentrations in fish are, for example, tens to hundreds of thousand times higher than in the water in which the fish live. Usually, the ability of halogenated organic compounds to bioaccumulate increases with the degree of halogenation, although in some cases, molecules with maximum degree of halogenation show a smaller tendency to bioaccumulate, due to their greater difficulty of passing through the cell membranes of living organisms.

As a result of bioaccumulation, some organic persistent substances are subject to biomagnification process, so these molecules are found at higher concentrations in animals at the highest levels of the food chain. This is connected with the fact that predatory animals eat hundreds of time their own weight in their prey and thus concentrations of persistent chemicals are far higher in the predators than in their prey.

In conclusion, the conventional POPs are highly chlorinated organic compounds with a molecular weight of 200–500 Da and a vapor pressure lower than 1000 Pa. **Table 1** lists the most significant chemical–physical properties of the 12 POPs considered to be of toxicological interest to the scientific community. It includes organochloride insecticides, such as dieldrin, DDT, toxaphene, and chlordane, and several industrial products or by-products such as PCBs, polychlorinated dibenzo-*p*-dioxins (PCDDs), and polychlorinated dibenzofurans (PCDFs).

Impacts of POPs on Organisms

The relation between the toxic action of chemicals on individual organisms and the performance of a population of these organisms is one of the central themes of ecotoxicology. To understand the environmental consequences of chemical pollution, the mechanisms that exist within a population to compensate or to magnify toxic effects on the members of that population must be known. How the populations respond depends on the priorities of the organism subject to toxicant stress. These priorities in a species-specific life history can be reproduction or growth and survival, which may be influenced in many different ways by toxicants exposure. In this sense, many studies in the field have reported effects of POPs directly related to survival of individuals and population level such as embryo lethality and developmental deformities.

The adverse effects of POPs on the organisms have been studied in two primary ways. First, by laboratory and controlled exposures of organisms to single congeners, mixtures, or matrice extracts; and second, by correlation of substance concentrations in the environment with abnormalities in birds, fish, and marine mammal populations, such as mortality during early development stages and modifications induced on the adrenal cortex in marine mammals. These two strategies have different application fields. Field research can integrate the knowledge of the impact of multiple environmental contamination on exposed organisms. Laboratory exposures, with both *in vivo* and *in vitro* studies, instead can identify the specific responses associated with exposure to a single toxicant, and determine the dose–response relationships for those responses. Anyway, both field and laboratory research are vital to understand the toxicity of chemicals.

Main Mechanisms of Toxic Action

Dioxin-like action chemicals

The mechanism of action of this group of chemicals, which includes the interaction between the 17 congeners of PCDD and PCDF, the 12 dioxin-like action PCBs with the citosolic receptor *Ah*, is well distributed among species. When the ligand substance has a planar structure, it binds the receptor with different affinities forming a complex that translocates in the cell nucleus, where it binds to specific DNA sequences, and induces changes in gene expression. Several DNA sequences of this type can be found in animals, therefore many gene expressions could be influenced. The most toxic substance among the mentioned compounds so far is TCDD or 2,3,7,8-tetrachlorodibenzo-*p*-dioxin. The relative toxic potency of the other compounds are expressed as proportions relative to TCDD and are referred to as toxicity equivalence factors (TEFs). **Table 2** shows the TEFs for mammals, birds, and fish. The

Table 1 Physicochemical properties of the most significant POPs

		Molecular weight (Da)	Water solubility	log K_{ow}	Vapor pressure (mmHg)	Half-life in soil (years)
Aldrin		365	$27\ \mu g\,l^{-1}$ (25 °C)	5.17–7.4	2.3×10^{-5} (20 °C)	
Chlordane		410	$56\ \mu g\,l^{-1}$ (25 °C)	4.58–5.57	0.98×10^{-5} (20 °C)	4
DDT		355	1.2–$5.5\ \mu g\,l^{-1}$ (25 °C)	6.19 (p,p'-DDT) 5.5 (p,p'-DDD) 5.7 (p,p'-DDE)	0.2×10^{-6} (20 °C)	15
Dieldrin		381	$140\ \mu g\,l^{-1}$ (20 °C)	3.69–6.2	1.78×10^{-7} (20 °C)	3–4
PCDDs		322–460[a]	0.074–$19.3\ ng\,l^{-1a}$ (25 °C)	6.80–8.20[a]	8.25×10^{-13} – 1.5×10^{-9a} (25 °C)	10–12[b]
PCDFs		306–444[c]	1.16–$419\ ng\,l^{-1c}$ (25 °C)	6.53–8.7[c]	3.75×10^{-13} – 1.5×10^{-8c} (25 °C)	

(Continued)

Table 1 Continued

	Molecular weight (Da)	Water solubility	log K_{ow}	Vapor pressure (mmHg)	Half-life in soil (years)
Endrin	381	220–260 $\mu g\,l^{-1}$ (25 °C)	3.21–5.34	2.7×10^{-7} (25 °C)	12
Hexachlorobenzene	285	50 $\mu g\,l^{-1}$ (20 °C)	3.93–6.42	1.09×10^{-5} (20 °C)	2.7–5.7
Heptachloro	373	180 $ng\,ml^{-1}$ (25 °C)	4.4–5.5	3×10^{-4} (20 °C)	0.75–2
Mirex	546	0.07 $\mu g\,l^{-1}$ (25 °C)	5.28	3×10^{-7} (25 °C)	10
PCBs	189–499	0.0001–0.01 $\mu g\,l^{-1}$ (25 °C)	4.3–8.26	0.003–1.6×10^{-6} (25 °C)	>6
Toxaphene	414	550 $\mu g\,ml^{-1}$ (20 °C)		0.2–0.4 (25 °C)	100 days to 12 years

[a]Data refer only to the seven toxic congeners.
[b]Data refer to TCDD.
[c]Data refer only to the ten toxic congeners.

Table 2 TEFs for PCDD, PCDF, and dioxin-like PCBs according to WHO risk-assessment approach (1997)[a]

Congener	Mammals	Birds[b]	Fish[b]
Polychlorodibenzodioxins			
2,3,7,8-T_4CDD	1	1	1
1,2,3,7,8-P_5CDD	1	1	1
1,2,3,4,7,8-H_6CDD	0.1	0.05	0.5
1,2,3,6,7,8-H_6CDD	0.1	0.01	0.01
1,2,3,7,8,9-H_6CDD	0.1	0.1	0.01
1,2,3,4,6,7,8-H_7CDD	0.01	≤0.001	≤0.001
O_8CDD	0.0001	≤0.0001	≤0.0001
Polychlorodibenzofurans			
2,3,7,8-T_4CDF	0.1	1	0.05
1,2,3,7,8-P_5CDF	0.05	0.1	0.05
2,3,4,7,8-P_5CDF	0.5	1	0.5
1,2,3,4,7,8-H_6CDF	0.1	0.1	0.1
1,2,3,6,7,8-H_6CDF	0.1	0.1	0.1
1,2,3,7,8,9-H_6CDF	0.1	0.1	0.1
2,3,4,6,7,8-H_6CDF	0.1	0.1	0.1
1,2,3,4,6,7,8-H_7CDF	0.01	0.01	0.01
1,2,3,4,7,8,9-H_7CDF	0.01	0.01	0.01
O_8CDF	0.0001	≤0.0001	≤0.0001
Non-ortho-polychlorobiphenyls			
[77] 3,3′,4,4′-T_4CB	0.0001	0.05	0.0001
[81] 3,4,4′,5-T_4CB	0.0001	0.1	0.0005
[126] 3,3′,4,4′,5-P_5CB	0.1	0.1	0.005
[169] 3,3′,4,4′,5,5′-H_6CB	0.01	0.001	0.00005
Mono-ortho-polychlorobiphenyls			
[105] 2,3,3′,4,4′-P_5CB	0.0001	0.0001	<0.000005
[114] 2,3,4,4′,5-P_5CB	0.0005	0.0001	<0.000005
[118] 2,3′,4,4′,5-P_5CB	0.0001	0.00001	<0.000005
[123] 2′,3,4,4′,5-P_5CB	0.0001	0.00001	<0.000005
[156] 2,3,3′,4,4′,5-H_6CB	0.0005	0.0001	<0.000005
[157] 2,3,3′,4,4′,5′-H_6CB	0.0005	0.0001	<0.000005
[167] 2,3′,4,4′,5,5′-H_6CB	0.00001	0.00001	<0.000005
[189] 2,3,3′,4,4′,5,5′-H_7CB	0.0001	0.00001	<0.000005

[a]Several TEF systems were defined during the last two decades. WHO proposed an update of its 1997 TEFs in 2005.
[b]Based on a limited data set.

biological matrices analyzed usually present complex patterns of the above-mentioned chemicals. The toxic potency of these chemicals can be calculated by multiplying the concentration of each compound by its TEF value and summing the products.

Endocrine disruptors

Many chemical substances of natural or anthropogenic origin are suspected or known to be endocrine disruptors. They can influence the endocrine system of organisms by a direct and/or an indirect interaction. Among the compounds which have been included in the 'old' POPs group are DDT and its metabolites, polychlorobiphenyl (PCBs), and PCDDs and PCDFs. Among the DDT isomers, the *o,p′*-DDT is the most active chemical species with estrogenic activity while an antiestrogenic effect has been reported for *p′,p′*-dichlorodiphenyldichloroethylene (DDE). In *in vitro* studies, chlorinated biphenyl mixtures (PCBs) show an affinity for the estrogenic receptors while in mammals, *in vivo*, they show an uterotrophic effect.

The antiestrogenic effects of TCDD have been investigated thoroughly in *in vivo* studies, where it has been determined that TCDD influences uterus weight and reduces the number of estrogen and progesterone receptors in organisms. *In vitro*, this compound also elicits an antiestrogenic effect, and finally a further effect which is called the 'downregulation' of estrogen receptors. In particular, TCDD is an exogenous agonist for the arylhydrocarbon (*Ah*)-receptor, which when activated increases the estrogen receptors degradation rate, induces the estradiol metabolizing enzymes, and inhibits gene expression controlled by estradiol or growth promoters.

Effects Observed Both in Laboratory and in the Field

Birds

Chick edema disease, which is characterized by ascites and by pericardial, subcutaneous, and pulmonary edema, is caused in offsprings of hens exposed to dioxin-like action chemicals in their feed. This pathology is of ectodermal origin in the same way as the developmental deformities, particularly abnormalities observed in the field in most of the embryos or chicks that die during early development. One abnormality that has been correlated with concentrations of the above-mentioned substances in bird eggs is the crossed-bill syndrome found in North American cormorants (*Phalacrocorax auritus*).

The mechanism of action eliciting the shell thinning by DDE appears to be associated with inhibition of the enzyme Ca^{2+}-ATPase in the eggshell gland. This action mode, studied in the laboratory, may be responsible for the effects observed in the field, such as the reduction of breeding success of several piscivorous and predator birds.

Fish

Field studies of wild fish populations provide evidence that dioxin-like action compounds may be impacting early life stage survival in the environment. An early development edematous syndrome prior to death has been described in *Salmo salar* sac fry where the suspected eliciting agents were PCDDs and PCDFs. In lake trout (*Salvelinus namaycush*) sac fry from Lake Ontario, an increased incidence of blue-sac disease, consisting of a yolk sac edema and hemorrhages, has been found, and retrospectively linked to PCDD and PCDF exposure. In the laboratory, many studies on fish developmental toxicity show that very low TCDD concentrations elicit similar effects in newly fertilized fish eggs and sac fry.

Effects Observed on Wildlife

The effects reported take into account a sequence of events beginning with the interaction of a molecule of a foreign substance with an endogenous molecule. This interaction begins its influence by assuming a form of disturbance at molecular level. When the organism's counteracting capacity is overloaded, the effect can be extended to cellular level. The extreme pressure at this level results in disturbance at tissue and organism levels. Furthermore, this effect can be propagated through the organism as a whole, and finally from the individual to the population.

Bird species

Many effects due to exposure to POPs have been observed in populations of wild bird species in North America and Northern Europe in the 1960s and 1970s and they are essentially of the correlational type. In the region of North American Great Lakes, the species involved were the herring gull (*Larus argentatus*), the Caspian (*Sterna caspia*) and the Forster's terns (*Sterna forsterii*), and double-crested cormorants, all piscivorous birds. In the gull, some colonies showed hatchabilities of less than 20% and a productivity of less then one fledged

young every ten nests was found. The authors believe that these effects were due to the presence of a few TCDD nanograms per tissue gram $(1-3\,ng\,g^{-1})$ as determined in the herring gull eggs. Similar effects were observed in the above-indicated species, although, in particular, the effects were more pronounced on Caspian terns. On this species, the effects included longer incubation times with smaller individuals due to a wasting syndrome, and growth deficiencies and deformities in the embryos. North European peregrine falcon (*Falco peregrinus*) and golden eagle (*Aquila chrysaëtos*) populations were deeply influenced in the 1960s following exposures to the pesticide dieldrin and *p,p'*-DDE, a persistent metabolite of *p,p'*-DDT. Dieldrin, in particular, was implicated in population reduction, and apart from acute toxicity the effects were to reduce the breeding success. Nesting peregrines lay one clutch of eggs per year, of three or four eggs. The mean clutch size is 3.5 eggs, with an average of 2.5 fledged young and only 4% of clutches with broken eggs. This percentage increased to 39% and appeared to be due to eggshell thinning. Similarly, the breeding success of golden eagle was greatly reduced in the 1960s when adults successfully reared offsprings in only 31% of the nests. Notwithstanding this phenomenon, the population did not decline, and a possible explanation for this trend was suggested some years ago as the population aged. The lifetime of this species spans as much as 30 years, so of this effect in the ecological perspective is that the golden eagle is a long-life *K*-strategist organism with a low reproductive rate. The effects on these organisms can be very dramatic, because when their population is reduced to low numbers it takes much more time to recover, and are thus subject to a greater extinction risk compared to an *r*-strategist. Anyway, the DDE effects were clearly demonstrated in the laboratory, confirming the historical observations made in the field, but with a strong species-specific effect pattern. Galliform species are very resistant to shell thinning whereas birds of prey are particularly affected. Until recently, as a result of the long transportation to Arctic and subarctic areas, hexachlorobenzene (HCB) and DDE levels were found along Norwegian coasts in *Larus marinus* eggs. These eggs showed a characteristic contamination pattern with highest levels found in samples collected in nearby Arctic areas. In particular, there was a significant positive relationship between blood concentrations of the cited substances in females, egg-laying dates, and probability of nest predation. In females with high levels of PCBs, there was also a decline in egg volume as egg laying progressed, compared to females with low OC levels.

Fish

Few field studies have been reported aimed at evaluating the effects on wild fish populations, but relevant data have been produced in laboratory studies dealing with effects at population level. In particular, PCDD, PCDF, and dioxin-like PCB congeners produce adverse effects that only occur days, weeks, or even months after exposure. Most meaningful comparisons can be made when toxic responses are based on the body burden of TCDD, and/or the other relevant PCDD and PCDF congeners, and when the fish are observed for many weeks or months after exposure. Fish are more sensitive to TCDD toxicity during very early development. For example, rainbow trout swim-up fry are approximately 10 times more sensitive to the lethal potency of TCDD than juvenile rainbow trout, but rainbow trout are even more sensitive when exposed as newly fertilized eggs, at levels of some hundred picograms of TCDD per gram of body weight, 20 times more sensitive than juveniles. Among the laboratory-tested species, the most sensitive, and most sensitive developmental stage to TCDD-induced lethality, are lake trout (*Salvelinus namaycush*) during an early development stage, such as newly fertilized eggs. A body burden of $0.04 \, \mathrm{ng \, g^{-1}}$ of fresh tissue showed a significant increase of sac fry mortality, with estimated TCDD lethal concentrations ranging from 50 to $60 \, \mathrm{ng \, g^{-1}}$ of eggs. Other effects potentially influencing the organism's energy requirements have been detected in fish, and include a decrease in feed consumption following exposure of juvenile fish to TCDD. Decreased feed consumption subsequently results in decreased body weight gain (wasting syndrome), which has been observed in yellow perch (*Perca fluviatilis*), rainbow trout (*Oncorhynchus mykiss*), and bluegill (*Lepomis macrochirus*), but the importance of these effects in the field still remains to be explained.

The ability of organisms to develop resistance to toxic chemicals has been recognized for decades, but the studies focused essentially on nonvertebrate groups. Among the few studies on the vertebrate group, researchers focused in particular on fish, and the chemicals most studied are PCDDs, PCDFs, and PCBs. This process, known as 'molecular drive', refers to the development of acquired resistance in geographically isolated wildlife populations, showing how pressure from environmental chemicals may permanently alter gene expression. Some indigenous nonmigratory fish populations from North America and Europe, exposed for generations, have shown resistance which is manifested by reduced mortality, reduced developmental abnormalities, and/or altered expression of toxicant-metabolizing enzymes. Resistance to chemicals like PCBs and TCDD most likely involves metabolic pathways shift, including altered sensitivity of target sites and/or balance of activation/detoxification pathways. The resistance developed may cause severe future environmental problems by reducing population fitness. A reduction of the reproductive output, a compromised immune function, and/or decreased growth rates have been documented. The low genetic heterogeneity found in resistant populations with a reduced fitness may generate resistant organisms very vulnerable to additional stress.

Mammals

Among mammals, effects on wildlife probably due to the exposure to POPs have been reported for marine – fish-eating – and flying mammals. In bats, a few studies have been produced that clearly demonstrate the role of lipids in the vulnerability of organisms. The principal effect determined on wild specimens was the mortality among migrating adults exposed to DDE. Starvation was induced by experimentation to stimulate the mobilization of lipid reserves for energy requirements, which is one of the conditions typical of migrating organisms. Before the experiment on field organisms, low levels of DDE residues in brain were identified. After induced exercise, the organisms died and the brain concentration increased considerably. Among marine mammals, in the 1960s and 1970s, similar to birds of prey, the ringed (*Phoca hispida*) and gray seal (*Halichoerus grypus*) populations of the Baltic Sea declined. The fertility of both species was reduced to *c.* 20% and the cause was *in uterus* abortion or fetus mortality. A further dramatic effect of the damaged uterus was the absence of further pregnancies. The damage was probably due to the effects of persistent pollutants on the adrenal cortex, a syndrome called hyperadrenocorticism. Among other symptoms, a large part could be attributed to the suppression of the immune system. In general, these effects coincided with changes in the POP concentrations in the areas of concern. However, among the considered chemicals, metabolites of DDT or PCBs seem to be primary responsible, due to their bioaccumulation in the adrenal cortex. In the 1990s, mature female Baltic Sea gray seals appeared to recover the original reproductive potentiality and this recovery appeared to be linked to the reduction of POP emissions in the Baltic Sea waters. The situation for ringed seals is very different, with occluded uterus still occurring (possibly due to the lifespan of the seals which started previously when POP levels where higher). Similarly, the population of American mink (*Mustela viso*), a fish-eating mammal, was severely impaired in the same period, and these effects have been confirmed in laboratory, where its reproductive capacity has been demonstrated to be disturbed by low concentrations of dioxin-like PCBs. Reproductive disturbances, probably due to toxic contaminants, have also been observed in common seals *Phoca vitulina* on the Dutch coast and in Californian sea lions *Zalophus californianus*. The population of common seal *Phoca vitulina* in the western part of the Wadden Sea, the Netherlands, declined sharply. Comparative studies on tissues of seals from the western and northern parts of the Wadden Sea showed that the PCBs were the most probable cause of reduced production of pups. PCB pollution predominantly is from the Rhine. Studies performed on laboratory mammals provide evidence of adverse effects of TCDD on male and female offsprings of pregnant dams. The effects associated with the lowest exposure concentrations span from a reduction of sperm count to a feminized sexual behavior in the male offsprings.

Factors Influencing the Exposure

The role of habitat selection

Habitats play a crucial role in mulding life histories, yet each organism's habitat, like each organism's life history, is unique. The role of a different habitat usage among populations of striped bass *Morone saxatilis* in relation to their PCB exposures has been studied in the area of the Hudson River Estuary and Long Island. The habitat usage history of individual fish was assessed by measuring the relative amount of strontium and calcium annual deposits within the otoliths. Since seawater is richer in strontium than freshwater, bony deposits, deposited when the fish is in marine waters, are more enriched in strontium than those deposited while in freshwater. Otoliths demonstrated that the striped bass population is comprised of individuals with widely varying migration patterns, showing both a 'typical' migration pattern and a residential behavior. This small group, constituted mainly of males, showed a highly significant positive correlation with PCB body burden, while confusing factors such as lipid content, length, age, sex, and weight were poorly correlated to the magnitude of contaminant body burdens. Striped bass, permanently residing in portions of the estuary adjacent to PCB sources, showed elevated total PCB levels with patterns mainly constituted of di-, tri-, and tetrachlorobiphenyls, whereas fish that spend the majority of their life in more saline waters of the estuary or in migrating contain lower PCB levels composed of more highly chlorinated congeners.

The role of food selection

Following restriction on PCB use, PCB levels in herring gull (*L. argentatus*) eggs declined until the early 1980s, when PCB decline rate slowed down except on Lake Erie where egg levels have continued to decline rapidly. The herring gull is a piscivorous species, but it also consumes a variety of other food types including garbage, small mammals, invertebrates, songbirds, amphibians, and vegetation. With the exclusion of Lake Erie colonies, in all the Great Lakes colonies the reduction of rates of PCB decline in eggs was *c.* 60%, while in the two Lake Erie colonies, the rate of decline increased by *c.* 20%. Feed values of the stable nitrogen and carbon isotopes in Great Lakes herring gull eggs changed through time in particular in eggs from Lake Erie, where the examination of stable isotope fingerprint suggested that the shifts in

herring gull egg values were probably the result of a shift in the herring gull diet from fish to terrestrial prey. In Northern Europe, a similar effect has been observed in Sweden on the peregrine falcon due to different feeding behaviors distributed along the country's north–south axis and in the migration pattern. The same may be said of the Mediterranean area for the gull species *Larus cachinnans* and *L. argentatus*. The first has a more variable diet including garbage and small mammals, whereas the second is essentially piscivorous. The TCDD-equivalent concentration levels in the eggs from *L. cachinnans* were 2 orders of magnitude below the concentration determined in the eggs from *L. argentatus*.

The role of lipids

Differing lipid contents of the animals play a relevant role in determining the differences in persistent pollutant levels which exist among them, and in particular the interindividual differences observed in the same species usually depend on this factor. In addition, the distribution of fat at intraindividual level determines the disposition of contaminants in any given individual. Cod liver, crustaceans' cuticles, and, in general, tissues rich in lipids, such as the blubber of marine mammals, may contain DDT, PCB, PCDD, and PCDF concentrations many orders of magnitude higher than the muscle flesh. Intraindividual variation is responsible for the seasonal dependence of such pollutant levels, which are higher during the spring spawning period than in autumn. It has been determined that a large proportion of herring (*Clupea harengus*) fat reserves are used up during spawning. This influences to a large extent the organism's energy reserves and indeed total DDT levels. Concentrations of PCBs and PCDFs in this species have been found to be seasonally dependent. Contaminant levels can depend on the age and size of the individual, as the concentrations of DDT, PCBs, and PCDDs are higher in older individuals than in younger ones. This is because the older animals have been exposed to pollutants for longer than the younger ones, but it may also be due to their migration patterns. It has, in any case, been proved that POP levels in females fall with increasing age and size, because the production of eggs/newborn and the milk production in mammals have to be considered clearance mechanisms whereas the size increase is primarily a dilution mechanism.

POPs and the Food Chain

Feeding relationships are more complicated than the simple definition 'food chain' suggests. Some species feed on organisms in two or more trophic levels and change their food sources during their life history. Studies in this area, that include both an assessment of the trophic position and the body burden of the contaminants of interest, are scarce. An extensive work has been done in recent years in the Baltic Sea. Pollutant concentrations in aquatic animals are determined by the levels both in the surrounding water and in their food. Most persistent organic pollutants are found in greater abundance in aquatic animals than in the planktonic algae, but their concentration levels are strongly influenced by their chemical–physical characteristics. According to the study carried out in the Baltic Sea, two of the more toxic PCB congeners occur in higher concentrations in consumer organisms such as herring and cod than in phytoplankton. The same tendency has been demonstrated for the three most toxic PCDDs, including TCDD, but other PCDD congeners exhibit a different behavior. Zooplankton and fish, for instance, show significantly lower levels of the fully chlorinated octachlorodibenzo-p-dioxin (OCDD) than phytoplankton. In general, the bioavailability of such a molecule appears to be limited by its comparatively large size and its extremely low water solubility, which can limit its passage from one link to another in the food chain; biological barriers are more difficult to cross for molecules with such properties. As a measure of positions in these food chains, concentrations of the nitrogen isotope ^{15}N, which becomes enriched compared with ^{14}N in conjunction with metabolism, have been determined. Fish are the main constituents of the diet of some mammals such as seals, otter (*Lutra lutra*), mink (*M. viso*) and in bird species, particularly birds of prey and water birds. It is among these animals at the top of aquatic food chains that the most serious

toxic pollutant problems have been found. A difference in biomagnification potentiality exists between aquatic and terrestrial food chains. In aquatic gill-breathing organisms, POPs do not undergo a marked biomagnification through the food chains because the species live in such close contact with the aquatic environment that the tendency for concentrations of pollutants to increase is counteracted by a constant tendency to achieve equilibrium with pollutant levels in the surrounding water (**Figure 1**). On the contrary, food is the principal source of persistent organochlorine insecticides for most trophic levels in terrestrial systems. Unlike gill-breathing animals, aquatic mammals do not take up persistent pollutants from the water, and they are equally unable to rely on the reverse process of their direct outward release from tissues into the water. In mammals and birds, concentrations of toxic substances are instead determined by the balance between their intake in food and detoxification mechanisms, which for persistent and bioaccumulating substances are slow. A further risk for nonaquatic animals is that many detoxification mechanisms include the formation of more polar metabolites to be excreted by water-based fluids, but the need for them to conserve water reduces this possibility. Among them, herbivores have more efficient systems for metabolizing foreign compounds than carnivores because of the nature of their diet. A critical window in the lifetime of these organisms is when they suffer starvation or a serious disease resulting in the use of their fat reserves, the tissue which have accumulated the highest levels of

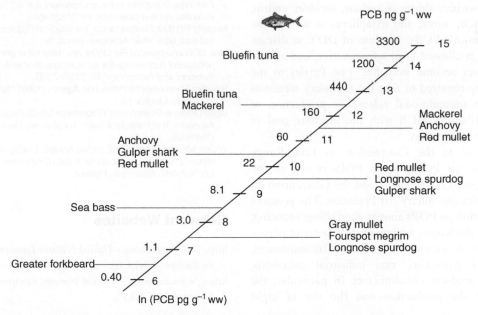

Figure 1 Indicative concentrations of total PCBs in muscular tissue or in the organism of Mediterranean ichthyofauna specimens. The concentrations increase as the trophic position, the organism size, and the species-specific lipid contents increase. In larger animals, the restoring of balance can take time. They may ingest more pollutants through the food web than they will ever release to the environment, and their POP body burden may be in an imbalanced state with those in the water. PCB total levels have been reported on a whole weight basis, thus encompassing the variability due to the species-specific content of lipids.

the contaminants, releasing a significant dose to reach the vital organs.

Principal International Agreements on POPs

Potential adverse effects on the environment and human health of exposure to POPs are of considerable concern among governments, nongovernmental organizations, and the scientific community. The persistence of such substances in the environment and their capacity to cover long distances from the point of release require that concerted international measures are adopted to efficiently control exposure. To this end, two international legally binding instruments have been negotiated and finalized: the global Stockholm Convention on POPs, opened for signatures in May 2001 and entered into force on 17 May 2004, and the Protocol to the Regional UNECE Convention on Long-Range Transboundary Air Pollution on POPs, opened for signatures in June 1998 and entered into force on 23 October 2003.

The Stockholm Convention on POPs provides an international framework, based on the precautionary principle, which seeks to guarantee the elimination or the reduction of production and use of POPs. Initially, the convention covers the 'dirty dozen' substances, but any signatory state of the convention may submit proposal to include new substances in the list covered by the Convention. The chemicals to be eliminated under the Stockholm Convention are those reported in **Table 1**, that is, the pesticides aldrin, chlordane, dieldrin, endrin, heptachlor, HCB, mirex, and toxaphene, as well as the industrial chemicals PCBs. The use of DDT as disease vector control is allowed until safe, effective, and affordable alternatives become available. The Parties to the Convention are required to take the necessary measures to reduce the unintentional release or production of PCDDs, PCDFs, and HCB with the ultimate goal of their complete elimination.

The protocol to the Convention on Long-Range Transboundary Air Pollution on POPs is one of the eight specific protocols which extend the Convention on Long-Range Transboundary Air Pollution. The protocol to the Convention on POPs aims at controlling, reducing, or eliminating discharges, emissions, and losses of persistent pollutants. It focuses on a list of 16 substances, comprising 11 pesticides, two industrial chemicals, and three by-products/contaminants. In particular, the protocol bans the production and the use of eight substances (aldrin, chlordane, chlordecone, dieldrin, endrin, hexabromobiphenyl, mirex, and toxaphene), with the elimination of others (DDT, heptachlor, HCB, PCBs) planned at a later stage. Finally, the protocol severely restricts the use of DDT, hexachlorocyclohexane (HCH, including lindane), and PCBs, and also obliges the states for which the convention is in force to reduce their emissions of PCDDs, PCDFs, polyaromatic hydrocarbons (PAHs), and HCB below their levels in 1990 (or an alternative year from 1985 to 1995).

The protocol to the Convention on Long-Range Transboundary Air Pollution and the Stockholm Convention are part of the numerous international treaties ratified on the environment in recent years, such as the Basel Convention on the Control of Transboundary Movements of Hazardous Wastes and Their Disposal (1992) and the Rotterdam Convention on the Prior Informed Consent Procedure for Certain Hazardous Chemicals and Pesticides (1998). Many new persistent organic pollutants as a result of actual industrial processes are entering the environment. However, their environmental effects are almost unknown.

See also: Dioxin.

Further Reading

Giesy JP, Ludwig JP, and Tillitt DE (1994) Dioxin, dibenzofurans, PCBs and colonial, fish-eating water birds. In: Schecter A (ed.) *Dioxins and Health*, pp. 254–307. New York: Plenum.

Lintelmann J, Katayama A, Kurihara N, Shore L, and Wenzel A (2003) Endocrine disrupters in the environment (IUPAC Technical Report). *Pure and Applied Chemistry* 75(5): 631–681.

Moriarty F (1999) *Ecotoxicology, the Study of Pollutants in Ecosystem*, 3rd edn. New York: Academic Press.

Muir DCG and Howard PH (2006) Are there other persistant organic pollutants? A challenge for environment chemists. *Environmental Sciences and Technology* 40: 7157–7166.

Swedish Environmental Protection Agency (1997) *Persistent Organic Pollutants*. Monitor 16.

United Nations Environment Programme (2003) Regionally Based Assessment of Persistent Toxic Substances. Geneva: UNEP Chemicals.

Walker MK and Peterson RE (1994) Aquatic toxicity of dioxins and related chemicals. In: Schecter A (ed.) *Dioxins and Health*, pp. 347–387. New York: Plenum.

Relevant Websites

http://www.unece.org – United Nations Economic Commission for Europe (UNECE).

http://www.unep.org – United Nations Environment Programme (UNEP).

Phenols

A J Stewart, Oak Ridge Associated Universities, Oak Ridge, TN, USA

R F Stewart, Bay Materials, LLC, Menlo Park, CA, USA

Introduction

This article summarizes information on phenols from two perspectives: (1) their significance as toxic man-made chemicals, and (2) their significance, ecologically and pharmacologically, as compounds produced naturally by plants.

Chemistry of Phenols

Man-made or natural phenols are planar organic compounds based on one or more six-carbon aromatic (benzene) rings to which one or more hydroxy groups (–O–H) are bonded (**Figure 1**).

The H atom of the –O–H group can dissociate from the O atom when the phenol molecule is dissolved in water, so phenols are weakly acidic. The term phenol now is used for any of a group of related acidic compounds that are hydroxyl derivatives of aromatic hydrocarbons. Included in this definition are a huge number of compounds of commercial importance such as cresols, catechols, quinols, xylenols, guaiacol, and resorcinol. These groups can be identified as follows:

- cresol – a phenol with one methyl group in the *ortho*- (*o*-), *meta*- (*m*-), or *para*- (*p*-) position;

- guaiacol – a phenol with one methoxy group (–O–CH_3) in the *o*-position;
- xylenol – phenol with two methyl groups (e.g., 2,4-dimethyl phenol);
- catechol – a phenol with a second hydroxy group in the *o*-position;
- resorcinol – a phenol with a second hydroxy group in the *m*-position;
- quinol (or hydroquinone) – a phenol with a second hydroxy group in the *p*-position;
- naphthol – two joined aromatic rings, with a hydroxy group on at least one ring.

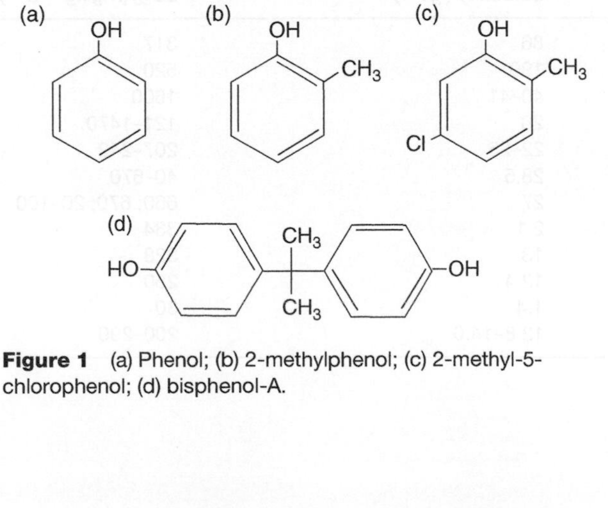

Functional groups, such as hydroxy, methyl (–CH_3), nitro (–NO_2), chloro (–Cl), fluoro (–F), or bromo (–Br) can substitute for one or more hydrogen atoms on a basic phenolic structure. A vast range of more complex substituents is also possible. These can include (but are not limited to) alkylated (H-saturated) carbon chains, branched carbon-based structures, or other simple or complex ring-based structures, with or without their own substituents. Two aromatic six-carbon rings, each with or without substituents, also can bond covalently to each other, creating various biphenols. Three-ring structures, such as flavones, are common naturally occurring phenolic compounds.

Substituted simple phenols are named according to the type of substituting groups and their positions relative to the initial hydroxy group. In this nomenclature, positions 2, 3, and 4 (the carbon to which the initial –OH is bonded is counted as the first position) are sometimes referred to as the *ortho*- (*o*-), *meta*- (*m*-), and *para*- (*p*-) positions, respectively. Substitutions can include sulfur (S) for the O of the defining –OH group, leading to thiophenols, which also can support substituents. This permits (for example) compounds such as 4-bromothiophenol, which contains an S substitution for O in the phenol-defining –OH group, and a bromine substitution at the 4 (or *p*-) position.

Figure 1 (a) Phenol; (b) 2-methylphenol; (c) 2-methyl-5-chlorophenol; (d) bisphenol-A.

Phenolics are sparingly to moderately soluble in water, but the acidity and aqueous solubility of the more highly substituted one-ring phenols are strongly affected by the types and positions of the substituent groups. The effects of substituents on acidity and solubility are especially large if the substituents are electron-withdrawing groups located at the o- or p-positions. The pK_a for phenol, for example, is 10.0, whereas the pK_a for p-nitrophenol is 7.2. Additional nitro groups radically lower the pK_a relative to the nonsubstituted parent compound (**Table 1**).

Analysis of Phenols

Many of the man-made phenols are best determined using gas chromatography (GC) followed by mass spectrometry (MS). As is true for quantification of other environmental pollutants, analysis of phenols in air, water, soils, or plant or animal tissues requires very careful attention to sample extraction and cleanup. Extraction and cleanup techniques vary depending upon the type of sample, the type of phenolic compound(s) under investigation, and the type of analysis method.

Flavonoids and other classes of naturally occurring plant phenolics increasingly are quantified using reversed-phase high-performance liquid chromatography (HPLC) with amperometric, ultraviolet (UV), photo-diode array, or electrochemical detection. Low concentrations of acetic acid, formic acid, or phosphate buffers in the mobile phase can markedly increase separation of flavonoids and other phenolics. UV detection is preferable when phenol concentrations are greater than about 1 ppm.

Man-Made Phenols and Drinking Water Quality

Phenols can impart unpleasant tastes and odors to water at concentrations lower than those considered to be of much toxicological significance, and phenols are not effectively removed from water by most conventional water treatment methods. Many simple phenols can be destroyed by strong oxidizing agents such as ozone. Phenolics can react with chlorine (such as that used to sanitize drinking water), thus creating more persistent chlorinated phenols. Trace concentrations of phenols generally can be removed from water by sorption to granulated activated carbon. The adsorbed phenols can be removed as phenates by rinsing the activated carbon with a solution of NaOH. Many natural (plant-derived) complex phenols undergo at least partial oxidation with exposure to UV wavelengths in sunlight. Some chlorinated phenolics also are photosensitive.

Toxicity of Phenols

The toxicity of even simple phenols varies considerably – about 50-fold for the few compounds given as examples in **Table 1**. Further, the toxicity values in **Table 1** are expressed as ranges in a few cases. This is done to firmly impress upon the reader the essential fact that there is no single toxicity value for a phenol, or indeed, for any other compound. The toxicity of a chemical is a function of the biological response to the chemical. Thus, toxicity depends upon the chemical, the duration of exposure to the chemical, the mode of exposure, the test conditions, the type of organism exposed to the chemical, the age, health, and gender of the test organism, and the property that is measured as an indicator of the biological response.

Table 1 pK_a values (i.e., ability of an ionizable group of an organic compound in aqueous media to donate a proton), water solubility, and toxicity (LD$_{50}$ values; acute, oral, rats) for selected phenols differing in the type and position of several substituents

Compound	pK_a	Solubility (g l^{-1})	LD$_{50}$ (mg kg^{-1} d^{-1})
Phenol	9.82–10.0	86	317
o-Methoxyphenol	10.0	190	520
p-Methoxyphenol	10.2	40–41	1600
o-Methylphenol	10.3	26	121–1470
p-Methylphenol	10.3	22–25	207–242
o-Chlorophenol	8.6	28.5	40–670
p-Chlorophenol	9.4	27	660; 670; 20–100
o-Nitrophenol	7.2	2.1	334
m-Nitrophenol	8.4	13	328
p-Nitrophenol	7.08–7.2	12.4	250
2,4-Dinitrophenol	3.94–4.1	1.4	30
2,4,6-Trinitrophenol	0.4	12.8–14.0	200–290

Data from various sources.

For these reasons, substantial variation is the rule rather than the exception. Toxicity values of course vary from species to species, for the same chemical – but they also vary from test to test, for the same species, exposed in the same manner, in tests of the same duration.

As a gross generalization, the toxicity of simple phenols decreases in the order:

phenol > *p*-cresol > *o*-cresol > *m*-cresol > catechol

These types of phenols do not bioaccumulate to any great extent, and when diluted tend to biodegrade to nontoxic products fairly rapidly (half-lives typically range from days to weeks).

In general, the toxicity of substituted phenols tends to increase with the bulk of the phenolic molecule (i.e., the extent of branching and the number of substitutions). Strongly electronegative atoms in the substituent positions increases a phenol's toxicity.

Halogenated phenols tend to be more toxic and less biodegradable than their nonhalogenated counterparts, and the toxicity of phenols and their resistance to biodegradation increase as the degree of halogenation increases. Pentachlorophenol (PCP), for example, is more toxic and resistant to biodegradation than phenol or *o*-chlorophenol.

The toxicity of man-made phenols to aquatic organisms is of particular interest because most phenols are at least moderately soluble in water (i.e., >1 ppm). Further, many phenols are used in very large quantities for diverse purposes. Thus, fugitive phenolics are inevitable and these can be moved about readily by water. In Canada, the level of hydroquinone and resorcinol deemed protective of freshwater systems is <2 and <12.5 $\mu g l^{-1}$, respectively. These two types of phenols can create unpleasant tastes in fish and shellfish, even at low concentrations. A level of $50 \mu g l^{-1}$ is considered to be protective of freshwater systems for most other types of phenols.

The more highly halogenated phenols pose special environmental problems: many are persistent and tend to bioaccumulate. Several types of man-made phenols also have endocrine-disrupting properties and are biologically active at low concentrations.

Quantitative structure–activity relationship (QSAR) modeling has been used to predict the toxicity of phenols, with reasonable accuracy.

Production and Use of Phenols by Man

The main method for commercial production of phenol is by peroxidation of cumene (isopropylbenzene) to cumene hydroperoxide, with subsequent cleavage to phenol and acetone. Feedstocks for this process include benzene and propylene. About 7.5 million metric tons of phenol were produced in 2004; worldwide phenol demand is expected to increase 4–6% annually over the next several years. Major uses of phenol include resins, polycarbonate and epoxy plastics, and some grades of nylon. Phenols are also produced as by-products in many industrial processes where organic chemicals are used. Butylated hydroxyl toluene (BHT) is an example of a phenolic antioxidant used as a food additive and preservative.

Chlorinated phenols are used as feedstocks to prepare many dyes, pigments, resins, pesticides, and herbicides. They are also used as flea repellents, fungicides, wood preservatives, mold inhibitors, antiseptics, disinfectants, and antigumming agents in gasoline. Pesticide products prepared from chlorinated phenols include 2,4-D (2,4-dichlorophenoxy acetic acid; a major component of Agent Orange), 2,4-dichlorophenol, 2,4,5-trichlorophenol, 2,3,4,6-tetrachlorophenol, and PCP.

Examples of Man-Made Phenols of Special Toxicological Interest

An alkylphenol is a phenol molecule to which a C_nH_{2n+1} chain is bonded (**Figure 2**).

About 500 million kg of alkylphenols are produced annually for use in surfactants, herbicides, gasoline additives, dyes, polymer additives, lubricant oil additives, and antioxidants. Nonylphenol (a 9-C alkyl chain) and related compounds have been much studied because they are widely used, relatively persistent, and have endocrine-disrupting (estrogenic) properties in fish, birds, and mammals.

Bisphenol A (BPA; **Figure 1**) is produced from phenol and acetone. Globally, very large quantities of BPA are used in polycarbonates (hard, clear plastics) and resins. More than 100 million kg of polycarbonates are produced annually; polycarbonates have many uses, including the electronic industry (CDs, for example) and the food industry (e.g., water bottles, can linings, etc.). BPA has estrogenic properties and has been investigated with respect to risks to mammals, fish, and amphibians. It can strongly affect reproduction in mollusks at concentrations

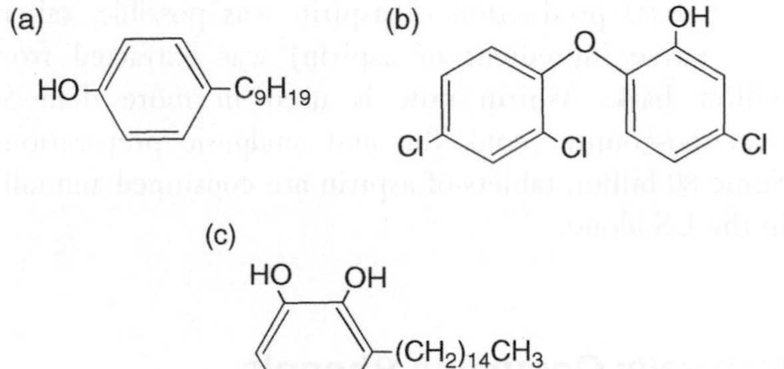

Figure 2 (a) 4-Nonylphenol, an example of an alkylphenol; (b) triclosan; (c) urushiol (one of the several active ingredients in poison ivy).

as low as $1\,\mu g\,l^{-1}$, and is under study because it appears to disrupt the function of estrogen in the developing brain, even at parts per trillion (ppt) concentrations.

PCP was widely used as a wood preservative (the purchase and use of PCP in the US was restricted in the 1980s). Commercial formulations of PCPs often contained other significant toxicants, such as dioxins, as production by-products. Based on freshwater field and laboratory studies, the probable no-effect concentrations for PCP have been shown to be on the order of $0.2-2\,\mu g\,l^{-1}$ for water and $12.4-124\,\mu g\,kg^{-1}$ for freshwater sediment. Some isolates of white rot fungi can biodegrade PCP efficiently via laccase, a phenol oxidase.

Brominated phenols and biphenols, and polybrominated diphenyl ethers, came into widespread use in the 1970s, especially as flame retardants (>200 million $kg\,yr^{-1}$). Some of these compounds are highly toxic and lipophilic; many are relatively slow to biodegrade, and some have endocrine-disrupting properties. Polybrominated biphenyls (PBBs), much like their polychlorinated counterparts (PCBs), are toxic and lipophilic: they are bioaccumulation and reproductive hazards.

Triclosan (**Figure 2**) has broad-spectrum antimicrobial properties and is used in many personal-care products, including deodorants, soaps, toothpaste, antiperspirants, lotions, and cosmetics. It is also used in the production of polymers, plastics, and textiles. Triclosan has been found in 58% of the natural waters in the US, and is troublesome in two ways: it is relatively persistent, and when in water, it can be converted to a dioxin by exposure to UV components of sunlight. Some dioxins are very toxic, others are less so: however, most dioxins readily bioaccumulate, and some of the less-toxic dioxins can convert to a more toxic form if they react with chlorine, which is often used as a disinfectant for drinking water.

Salicylic acid (2-hydroxybenzoic acid) is prepared by treating the sodium salt of phenol with CO_2 under pressure. This yields sodium salicylate, which is treated with sulfuric acid to form salicylic acid. The acetyl derivative of salicylic acid is aspirin. Aspirin is believed to act against fever, pain, and inflammation by interfering with the production of particular prostaglandins. Before the commercial production of aspirin was possible, salicin (the active ingredient of aspirin) was extracted from willow bark. Aspirin now is used in more than 50 over-the-counter cold, flu, and analgesic preparations. Some 80 billion tablets of aspirin are consumed annually in the US alone.

Naturally Occurring Phenols

Woody plants first appeared about 370 million years ago. The development of phenylpropanoid metabolic pathways allowed the synthesis of lignin, a major constituent of wood: this was a key evolutionary advance for land plants. The stiffer, stronger cell walls made possible by lignin allowed better support, larger size, and greater tissue differentiation (such as water-conducting elements). Phenylpropanoid metabolic pathways also allowed production of phenolics used extensively in pigmentation (leaves, fruits, and flowers), defense, and signaling. Detrital organic matter derived largely from plant phenolics is also functionally significant at the ecosystem level.

Production of Naturally Occurring Phenols

Phenolic compounds are produced almost universally by plants and can be formed by three known pathways: the shikimic acid pathway, the malonate–acetate pathway, and the isoprenoid pathway. The shikimic acid (or phenylpropanoid) pathway yields three aromatic amino acids that are important precursors of plant phenolics – tryptophan, tyrosine, and phenylalanine. Phenylalanine is modified progressively to hydroxycinnamic acids such as cinnamic acid, salicylic acid, p-hydroxybenzoic acid, p-coumaric acid, caffeic acid, sinapic acid, and ferulic acid. Coumarins (derivatives of p-coumaric acid) are synthesized primarily in leaves but can be found at high concentrations in fruits, stems, and roots.

The malonate–acetate (or polyketide) pathway is similar to fatty acid synthesis: it involves sequential additions of malonyl-CoA to create a polyketide chain, which then undergoes cyclization to form a phenolic ring structure. Malonyl-CoA, acetyl-CoA, and p-coumaric acid yield chalcone, which can undergo various modifications, allowing production of a diversity of flavanones, flavones, flaven-3-ols, anthocyanidins, and isoflavones.

Some phenolics are formed by combining products from both the shikimic and malonate–acetate pathways. About 8000 types of plant phenolics are known; about 3000 of these are flavonoids. Plant phenolics range from simple one-ring phenols to lignin, a huge polymeric structure composed of phenylpropanoid units cross-linked to each other through heterogeneous chemical bonds. Lignin, after cellulose, is the most abundant organic material on Earth; it decomposes slowly. The slow rate of lignin decomposition by fungi, actinomycetes, and bacteria is thought to be due to the complexity of its bonds and cross-linkages, and because it has a relatively low nitrogen content. Hydrolyzable and condensed tannins are produced from flavanones: the tannins contain free phenolic groups and generally are soluble in water. Tannins are present in both gymnosperms and angiosperms, but within angiosperms they are more prevalent in dicotyledons than in monocotyledons.

Even lower plants such as ferns, algae, and lichens (fungal–algal systems) are known to produce phenolics, but only one animal – a marine sponge – is thought to have this capability. Because larger, more-complex phenolic compounds are relatively recalcitrant to degradation by microbes, phenolics occur in association with fossil fuels such as coal, shale oil, and petroleum. In oil production fields, produced water (i.e., water pumped up along with petroleum) often contains elevated levels of phenolics. Produced water from the extraction of methane from coal-bed water also contains phenolics. The aromatic and phenolic compounds in produced water originating from oil and coal-bed methane extraction account for much of the water's acute toxicity.

Important classes of phenolics produced by plants are shown in **Table 2**. This list is generic only, as thousands of different types of phenolics have been reported.

Plant Phenolics as Pigments and Flavors

Condensed tannins (also referred to as proanthocyanidins) are flavonoid units linked by carbon–carbon bonds that are not susceptible to cleavage by hydrolysis. These compounds are responsible for the intense pink, red, purple, or blue colors of many flowers, fruits, and leaves. Flavonoids, mainly anthocyanins, are responsible for the bright autumn colors in many plant species. Other flavonoid pigments having antioxidant properties (quercetins) also increases dramatically in plants, especially in epidermal

cells, upon exposure to light in the UV-B range. These pigments are thought to have a photoprotective role.

Condensed tannins account for the astringent tastes of many fruits and wines. Many plants have fruits, leaves, or roots that are rich in flavorful phenolics. Consider, for example, the tastes of grapes, teas, cranberries, grapefruit, coffee, cinnamon, ginger, and vanilla.

Plant Phenolics as Natural Toxicants and Herbivore Deterrents

Plants and herbivores have a long history of coevolution. From a plant perspective, these interactions include both the beneficial activities of animals (e.g., pollination and seed dispersal), and their harmful activities, such as the consumption of plant tissues. Plants produce many chemicals that affect grazers. Alkaloids and phenolics are particularly important in this regard.

Tannins (condensed and hydrolyzable) can reduce the palatability, nutritional quality, and digestibility of forage to grazers, and, in some types of plants, tannins are inducible by grazing. The phenolic content of the leaves of northern birch, for example, increases rapidly (<48 h) in response to herbivory. But the phenolics contents of leaves can also be affected by light intensity and nutrient levels, so relationships between herbivores and phenolics

Table 2 Major classes of phenolic compounds in plants

No. of C atoms	Basic skeleton	Phenolic class	Examples
6	C_6	Simple phenols	Catechol, hydroxyquinone
		Benzoquinones	2,6-Dimethoxybenzoquinone
7	C_6-C_1	Phenolic acids	Gallic, salicylic
8	C_6-C_2	Acetophenones	3-Acetyl-6-methoxybenzaldehyde
		Tyrosine derivatives	Tyrosol
		Phenylacetic acids	p-Hydroxyphenylacetic
9	C_6-C_3	Hydroxycinnamic acids	Caffeic, ferulic
		Phenylpropenes	Myristicin, eugenol
		Coumarins	Umbelliferone, aesculetin
		Isocoumarins	Bergenon
		Chromones	Euenin
10	C_6-C_4	Naphthoquinones	Juglone, plumbagin
13	$C_6-C_1-C_6$	Xanthones	Mangiferin
14	$C_6-C_2-C_6$	Stilbenes	Resveratrol
		Anthraquinones	Emodin
15	$C_6-C_3-C_6$	Flavonoids	Quercetin, cyanidin
		Isoflavonoids	Genistein
18	$(C_6-C_3)_2$	Lignans	Pinoresinol
		Neolignans	Eusiderin
30	$(C_6-C_3-C_6)_2$	Biflavonoids	Amentoflavone
n	$(C_6-C_3)_n$	Lignins	
	$(C_6)_n$	Catechol melanins	
	$(C_6-C_3-C_6)_n$	Flavolans	

Adapted from Harborne JB (1980) Plant phenolics. In: Bell EA and Charlwood BV (eds.) *Encyclopedia of Plant Physiology, Vol. 8: Secondary Plant Products*, pp. 329–394. New York: Springer.

can vary with environmental factors, as well as with the type of plant and herbivore.

In poultry, tannins added to the diet at low concentrations (0.5–2%) reduce growth and egg production. Higher concentrations (3–7%) can cause death. Cattle also can die from consuming oak leaves (*Quercus incana* or *Q. havardii*).

Some animals have mechanisms allowing them to consume tannin-rich plant tissues as food. These mechanisms include an alkaline gut pH, a peritrophic membrane (this absorbs tannins, which can then be excreted with the feces), or the production of mucin (a tannin-binding, proline-rich protein in the saliva).

Sweet clover (*Melilotus* spp.) contains high concentrations of the phenolic, coumarin. If sweet clover is allowed to become moldy, the coumarin can be converted to dicoumarol by fungal enzymes. Dicoumarol is a potent anticoagulant. Cattle that are fed moldy sweet clover can become lethargic and anemic, and may suffer severe internal bleeding.

Many members of the Anacardiace (cashew family) contain phenolic compounds that can cause severe dermatitis. Urushiol, for example, is an alkylated catechol that is the main active ingredient of poison ivy, poison oak, and poison sumac, and the main constituent in Japanese lacquer (**Figure 2**). Urushiol also occurs in mangos, the seeds of ginkos, and the shells of cashew nuts. As little as 50 μg of urushiol is sufficient to cause a complicated delayed allergic reaction with the body's immune system in most humans. Mammals other than humans do not seem to be vulnerable to it.

Plant Phenolics as Antioxidants

Teas would not be much good without natural phenolics, and volumes have been written about the antioxidant properties of phenolics in teas, grapes (especially the skins), cranberries, and blueberries, and vegetables such as broccoli, onions, spinach, and kale. Dietary supplements of plant-derived phenolics are prevalent, and reported to function at least in part as antioxidants. In this capacity, they are presumed to react with, and deactivate, free radicals (a molecule containing an unpaired electron). Free radicals can cause cellular damage by oxidizing lipid, protein, and nucleic acids. Thus, the neutralization of free radicals by natural phenolics that have antioxidant properties is presumed to be the molecular basis for at least some of their health benefits.

Phenolics such as caffeic acid, phenethyl ester, and curcumin significantly lower the rate of tumor formation in intestinal tumor-susceptible mice.

Turmeric (ginger family) is widely used as a spice in curries and contains large quantities of the phenolic, curcumin. Curcumin has antitumor, antioxidant, antiamyloid, and anti-inflammatory properties. Curcumin and its derivatives also can correct at least some of the protein-folding defects associated with the most common form of cystic fibrosis.

Milk thistle (*Silybum marianum*) produces a flavonoid complex known as silymarin; silybin, the most abundant and bioactive component, has strong antioxidant properties. It has been reported to reduce liver damage caused by ingesting the death cap mushroom (*Amanita phalloides*), possibly by competitively binding to membrane receptors that otherwise would be available to the fungal toxins, amanititin and phalloidin.

Efforts are underway to genetically engineer plants to produce greater quantities of potentially beneficial phenolics.

Ecological Considerations

Plant Phenolics and Allelopathy

Allelopathy refers to a negative or positive effect on one type of plant, by a chemical produced by another type of plant. Various types of chemicals, including phenolics, hydroxamic acids, and short-chain fatty acids, have been identified as having allelopathic properties. Allelopathy is thought to be involved with plant species succession, and occurs both in agricultural and natural landscapes. Juglone, produced by walnuts, is an example of a well-studied phenolic having allelopathic properties (**Figure 3**).

Phenolics and Plant Signaling

In plants such as *Populus* spp., increases in defensive proteins and phenolics occur in response to wounding or herbivore attack. This response is triggered by jasmonic acid, a plant-produced signal of attack. Jasmonic acid and some of its close derivatives, such as methylester jasmonate or *cis*-jasmone, also attract aphid predators and parasitoids; jasmonic compounds also elicit the production of salicylate, a secondary phenolic (**Figure 3**). Salicylate is converted to methyl salicylate (oil of wintergreen), a volatile compound used by plants as an airborne signal of wounding. Many insect herbivores can detect and are repelled by methyl salicylate. Plants near the wounded plant emitting methyl

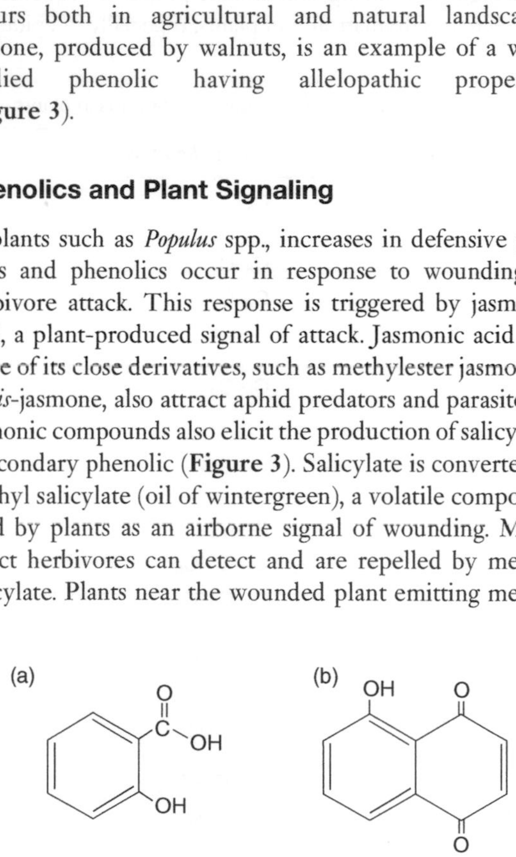

Figure 3 (a) Salicylic acid, a phenolic involved in plant signaling; (b) juglone, an allelopathic phenolic produced by walnut.

salicylate can detect the signal and are triggered to bolster their own chemical defenses.

Phenolics in Ecosystem Processes

Five characteristics of plant phenolics ensure that these compounds strongly influence system-level processes in soils and freshwater aquatic ecosystems: (1) the sheer abundance of phenolics produced by land plants; (2) the relative recalcitrance of detrital lignin and tannins to degradation by microbes; (3) the moderate solubility of phenolics in water; (4) the prevalence of carbon-to-carbon bonds and aromatic ring structures, which makes phenolics efficient at absorbing certain wavelengths of light; and (5) their chemical energy content, potentially available to decomposers. Characteristics (1) through (3) were described earlier in this article; characteristics (4) and (5) are summarized below.

The valence electrons of the carbon atoms of phenolic structures can delocalize over distances of several atoms. This tendency is referred to as resonance. Resonance is important because it explains how electrons shift about in the aromatic ring, when electron-withdrawing or electron-rich substituent groups are added to or removed from a phenolic structure. Essentially, the ability of the phenolic structure to move electrons about permits the structure to react with other dissolved materials, including cations (Zn^{2+}, Fe^{2+}, Ca^{2+}, etc.) and polar materials such as chlorinated pesticides.

Two classes of naturally occurring phenolic-based materials – humic and fulvic substances – are especially noteworthy with respect to ecosystem-level function. Humic and fulvic substances are chemically heterogeneous and arise from plant-derived partially degraded lignin, tannins, and other phenolics. Humics and fulvics are defined operationally by the procedure used to extract them from an environmental sample. Humic and fulvic acids are similar in that they dissolve in water under basic (pH 10) conditions. In water at pH 10, humic and fulvic materials are small enough to pass through a 0.5-μm-pore-size glass fiber filter: thus, they are said to be dissolved. Humic and fulvic acids differ in that humic acids (but not fulvic acids) precipitate from water at pH 2.

Humic substances are especially abundant in the humus-rich soils (humus, humic, human: a common etiology, stemming from "of the Earth"). Humic and fulvic acids have a net negative charge and readily complex metal ions. Humics become less soluble when complexed with common divalent cations that nullify the humic's net negative charge. The toxicities of metals such as Cu, Zn, and Cd are reduced dramatically by even low (<1 to $5\,mg\,l^{-1}$) concentrations of humic substances. Many highly chlorinated pesticides also become less biologically available to aquatic organisms when they bind to humic substances: the pesticide–humic complex is too large, apparently, to pass

through cellular membranes. Humic materials also bind to carbohydrates, amino acids, and proteins, making these compounds less available to bacteria and fungi.

Sphagnum bogs are good examples of humic-dominated aquatic systems. The brown-colored water of bogs is acidic (pH 4–5) and contains high concentrations of plant-derived organic acids. When exposed to light containing UV wavelengths, dissolved humic and fulvic materials can react with dissolved oxygen in an iron-mediated process involving singlet oxygen. Thus, the surface waters of bogs typically are undersaturated with respect to dissolved oxygen. Blackwater streams and rivers in tropical lowland forested areas carry water the color of well-steeped tea due to high concentrations of humic and fulvic materials.

See also: Bioaccumulation; Bioavailability; Biodegradability; Biodegradation; Endocrine Disruptor Chemicals: An Overview.

Further Reading

Dearden JC (2002) Prediction of environmental toxicity and fate using quantitative structure–activity relationships (QSARs). *Journal of the Brazilian Chemical Society* 13(6): 754–762.

Douglas CJ (1996) Phenylpropanoid metabolism and lignin biosynthesis: From weeds to trees. *Trends in Plant Science* 1(6): 171–178.

Egan ME, Pearson M, Weiner SA, et al. (2004) Curcumin, a major constituent of turmeric, corrects cystic fibrosis defects. *Science* 304: 600–602.

Galati G and O'Brien PJ (2004) Potential toxicity of flavonoids and other dietary phenolics: Significance for their chemopreventive and anticancer properties. *Free Radical Biology & Medicine* 37(3): 287–303.

Harborne JB (1980) Plant phenolics. In: Bell EA and Charlwood BV (eds.) *Encyclopedia of Plant Physiology, Vol. 8: Secondary Plant Products*, pp. 329–394. New York: Springer.

Niggeweg R, Michael AJ, and Martin C (2004) Engineering plants with increased levels of the antioxidant chlorgenic acid. *Nature Biochemistry* 22: 746–754.

Rappoport Z (ed.) (2003) *The Chemistry of Phenols (Chemistry of Functional Groups)*. New York: Wiley.

Roda AL and Baldwin IT (2003) Molecular technology reveals how the induced direct defenses of plants work. *Basic and Applied Ecology* 4: 15–26.

Servos MR (1999) Review of the aquatic toxicity, estrogenic responses and bioaccumulation of alkylphenols and alkylphenol polyethoxylates. *Water Quality Research Journal of Canada* 34(1): 123–177.

Thaler JS, Farag MA, Paré PW, and Dicke M (2002) Jasmonate-deficient plants have reduced direct and indirect defenses against herbivores. *Ecology Letters* 5: 764–774.

Tyman JHP (1996) *Synthetic and Natural Phenols*. New York: Elsevier.

Urquiaga I and Leighton F (2000) Plant polyphenol antioxidants and oxidative stress. *Biological Research* 33(2): 55–64.

Wetzel RG (2001) *Limnology: Lake and River Ecosystems*, 3rd edn. San Diego, CA: Academic Press.

Relevant Website

http://ctd.mdibl.org – The Comparative Toxicogenomics Database.

Pheromones

O Anderbrant, Lund University, Lund, Sweden

Further Reading

Pheromones, from Greek *pherein*, to transfer, and *hormon*, to excite, are intraspecific chemical signals, first defined in 1959 as "substances secreted to the outside by an individual and received by a second individual of the same species in which they release a specific reaction, for instance a defined behaviour (releaser pheromone) or developmental process (primer pheromone)." Since the first chemical characterization of a pheromone in 1959, pheromones have been identified, or at least proved to exist, in a couple of thousand species, terrestrial as well as aquatic. Pheromones can release a variety of behaviors in the receiving individual and often they are specified according to their function, for example, sex pheromone (mate attraction or other reproductive behavior – usually sex specific), aggregation pheromone (attraction of both sexes to a common resource), alarm pheromone (common in social insects), and host-marking or oviposition-deterring pheromones.

A pheromone may consist of a single substance (e.g., some alarm pheromones in social insects, transferring comparatively simple messages) or usually a blend of several compounds in a more or less specific ratio (most sex pheromones, where specificity is important). The chemical nature of the compounds varies depending on their origin and biosynthesis and covers several classes of chemicals. Several hundred different substances have been shown to elicit pheromone activity. Some substances are synthesized *de novo* by the emitting individual via primary biosynthetic pathways, whereas others may originate from compounds obtained through the food.

The size and structure of the molecules involved are important for the function of the pheromone. In terrestrial species, pheromones are often used for transferring messages over relatively large distances, typically 10–100 m, which means that the compounds constituting the pheromone have to be volatile. On the other hand, less-volatile compounds will remain for longer periods where they are released or deposited, which might be a desired character of, for example, host-marking pheromones or pheromones defining a territory border.

In moths (Lepidoptera), the most-studied group of insects from a pheromone perspective, the typical pheromone consists of two to four straight-chain fatty-acid derivatives, 10–18 carbon atoms long, with one or two double bonds and different functional groups, such as alcohol, ester, or aldehyde. Such long-chain molecules may also appear as ketones or epoxides. The biochemical basis of these moth pheromones is found in the fatty acid biosynthesis pathway. Many pheromone substances have a regularly branched carbon skeleton and several of these are terpenoid compounds, biosynthesized through the intermediate mevalonic acid. These terpenoid substances are built up by linking two or more isoprene (five-carbon) units. Yet other pheromone components are cyclic with carbons only in the ring or heterocyclic containing oxygen or nitrogen. Also polycyclic pheromone compounds exist. Many of the compounds identified from pheromones can exist in different structural forms, the so-called isomers. For instance, the double bonds in the carbon chain can be in different places (positional isomers) or the double bond is locking the molecule so that two different geometric isomers can exist. When a carbon atom binds to four different ligands (atoms or parts of molecule), we talk about optical isomers. The fate of the released pheromone molecules in the environment depends on their structure. Some are oxidized and others are sensitive to ultraviolet (UV) light and break down into more simple structures.

Pheromones usually show a high degree of species specificity, with few 'errors' made, although they might be exploited by natural enemies that have evolved capability to detect them. Furthermore, they are often produced and released in minute amounts, in insects typically 1–100 ng per day, and in correspondence to a sensitive and finely tuned detection apparatus in the receiving individual.

Several of the above-mentioned characteristics have made pheromones tractable for use in management of pest organisms, in particular insects. Pheromones are usually species specific, active at very low concentration, affecting critical steps in the life cycle such as mate finding or oviposition, and consist of naturally occurring, degradable, and nontoxic compounds. A disadvantage in connection to practical application is that each target species requires a new scientific effort in order to characterize the pheromone and sometimes also to synthesize it. Pheromones are used for various purposes in pest management.

For detection and monitoring. A trap is equipped with a dispenser (release device) loaded with the synthesized pheromone and attracts individuals of the target species. Since attraction is the key behavioral element for a trap to work, only sex and aggregation pheromones can be used for this purpose. The type of trap, trap position, dispenser, and release rate to be used vary among species, purpose, and habitat. Attractive traps may answer questions such as where, when, and how many. The first question may be relevant for mapping the distribution range or for detection of alien and potentially invasive species. Answers to the second question can be useful when timing treatments to control the target species and the third question is important when estimating population densities or establishing thresholds for future control. Pheromone-based attractive traps are commercially available for several hundred different insect species, notably moths and beetles, and are used on a large scale in forestry, agriculture, and for indoor pests.

For population control. Several methods have been developed to be able to manipulate a large-enough proportion of a population to affect its future damage potential. In essence, two strategies have been followed, one known as mass trapping, including attract and kill, and a second called mating disruption or confusion.

Mass trapping: The idea with this strategy is to catch as many individuals as possible in order to reduce the population below a level where it does not cause any harm. To date this has mainly been used against species using an aggregation pheromone resulting in catch of both sexes. Bark beetles infesting conifers in the northern temperate region have been subject to large-scale mass-trapping attempts. Mass trapping using sex-specific pheromones (e.g., female-produced sex pheromones in moths) is limited by the fact that most male moths can mate many times, and even with an efficient mass-trapping system the few non-trapped males may copulate with a large fraction of the females. 'Attract and kill' is a variant in which the pheromone is combined with a poison (e.g., an insecticide) which kills the attracted specimens. In this case no container or other device to catch the attracted individuals is needed.

Mating disruption: This method is used against species with sex pheromones, usually female-produced. The mate-finding behavior in males is disturbed or disrupted by the release of relatively large amounts of synthetic female sex pheromone (typically $100 \, g \, ha^{-1}$, effective over a period from weeks to a whole season). By sensing the smell of female "everywhere" the male is unable to find his way to the females, he is "confused." This leads to fewer matings within the treated area, fewer offspring, and less damage. Mating disruption is used on a large scale, that is, hundreds of thousands of hectares, against gypsy moths in North America and fruit and grape pests in many parts of the world.

Other pheromone-based methods include the push-pull strategy, in which an attractive pheromone is combined with a repulsive chemical signal in order to direct the target species to an area where it can be handled more easily.

From an ecotoxicological point of view, effects of pheromones have not received much attention so far and no ecotoxicological side effects have been reported. The reason for this is quite obvious considering the nontoxic chemical structures of most pheromone molecules, their degradability, and the small quantities in which they are used. A few reports exist in which a behavioral effect on the target species has been recorded after the pheromone dispensers have been removed. These reports are from mating-disruption treatments and presumably some pheromone has been adsorbed on plants or other material and released later on.

Two risk factors, though not ecotoxicological, are worth taking into consideration during repeated and long-lasting treatment with pheromones. First, there is a theoretical possibility of resistance development, that is, individuals with a pheromone deviating from that used in the treatment will be favored because they escape mass-trapping or mating disruption. However, this can be monitored and the synthetic pheromone composition altered when necessary. Second, nontarget species may be trapped and killed. This is most likely to hit natural enemies of the species in focus, but can sometimes be avoided by modifying, for instance, the trap design.

The use of pheromones in pest control, and possibly also in conservation biology, can be expected to increase worldwide in the future. An increased need for environmentally acceptable, albeit effective, pest control methods points toward use of various biologically based rather than traditional chemical methods. Pheromones and other semiochemicals (information-carrying chemicals, also including those transferring messages between species) provide one such strategy.

Further Reading

Agosta WC (1992) *Chemical Communication: The Language of Pheromones.* San Francisco: Scientific Library, W.H. Freeman.

Anon (2003) *Insect Pheromones: Mastering Communication to Control Pests.* National Academy of Sciences, USA. http://www.beyonddiscovery.org/content/view.article.asp?a=2702 (accessed January 2008).

Blomquist GJ and Vogt RG (eds.) (2003) *Insect Pheromone Biochemistry and Molecular Biology,* 745pp. Amsterdam: Elsevier.

Cardé RT and Minks AK (eds.) (1997) *Insect Pheromone Research, New Directions,* 684pp. New York: Chapman and Hall.

El-Sayed AM (2005) The Pherobase: Database of Insect Pheromones and Semiochemicals; http://www.pherobase.com/ (accessed January 2008).

Hardie J and Minks AK (eds.) (1999) *Pheromones of Non-Lepidopteran Insects Associated with Agricultural Plant,* 466pp. Wallingford: CABI Publishing.

Howse P, Stevens I, and Jones J (1998) *Insect Pheromones and Their Use in Pest Management,* 369pp. London: Chapman and Hall.

Witzgall P, Lindblom T, Bengtsson M, and Tóth M (2004) The Pherolist. http://www-pherolist.slu.se/pherolist.php (accessed January 2008).

Wyatt TD (2003) *Pheromones and Animal Behaviour, Communication by Smell and Taste,* 391pp. Cambridge: Cambridge University Press.

Phthalates

W J G M Peijnenburg, RIVM – Laboratory for Ecological Risk Assessment, Bilthoven, The Netherlands

Introduction
Emission of Phthalates in the Environment
Environmental Fate of Phthalates

Adverse Effects of Phthalates
Further Reading

Introduction

Phthalates are the esters of 1,2-dibenzene dicarboxylic acid; their general structure is given in **Figure 1**. They are produced by the addition of an excess of branched or normal alcohols to phthalic anhydride in the presence of a catalyst. Phthalates constitute a diverse family of industrial compounds that are by far the most widely produced plasticizers up to date to increase the flexibility and workability of high molecular weight polymers. Their low melting point and high boiling point make them also very useful as heat-transfer fluids and carriers. In some plastics, phthalates comprise up to 50% of the total weight. Both linear and branched phthalate esters are used in the manufacture of plastics; especially, linear esters provide superior flexibility at low temperatures and also have lower volatility. Phthalates with alkyl side chains lower than C6 are not often used as plasticizers because of volatility concerns. Phthalates can be found in ink, paint, adhesives, vinyl flooring, and even in some food products, cosmetics, and pharmaceuticals.

Exact numbers are lacking, but the estimated worldwide production is about $4\,300\,000\ t\,yr^{-1}$, of which $c.\ 90\%$ are used as plasticizers. The C8–C13 phthalate esters are the dominant vinyl plasticizers with di-2-ethyl hexyl phthalate (DEHP) and diisononyl phthalate (DINP) predominant. Currently, however, there is a clear trend of using technical mixtures of alcohols of similar boiling point as precursors, as a matter of course yielding a complex mixture of phthalate esters. Of the dozens of phthalates synthesized over the years, the most prolifically used in Western Europe are DEHP, diisodecyl phthalate (DIDP), DINP, and di-n-butyl phthalate

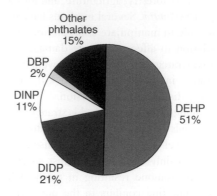

Figure 2 Approximation of the relative importance of the consumption of four of the main phthalates in the European Union in the 1990s.

(DBP), in this order (**Figure 2**). Most of the data generated on phthalates are derived from research on these four particular phthalates. It should be noted that DOP (dioctyl phthalate) is used as a synonym for DEHP.

Incorporating phthalate esters into a polymeric matrix reduces the glass transition temperature of the polymer. Phthalate esters are not chemically bound (covalent bonds) to the polymer and are therefore able to migrate to the surface of the polymer matrix. Here they may be lost by a variety of physical processes, albeit that various attractive forces hold the esters tightly within the vinyl matrix, so that migration occurs at a low rate. Nevertheless, concern has risen in the European Union on possible adverse effects on infants and consequently they are banned for use in toys.

Emission of Phthalates in the Environment

Phthalates can enter the environment through losses during manufacturing processes by air emissions, in water effluent, in solid waste, and by leaching from final products, because they are not chemically bonded to the polymeric matrix. Very little is released during manufacturing and processing as most of the phthalates released during production and processing are disposed of in waste

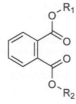

Figure 1 General chemical structure of phthalates. R1 and R2 = C_nH_{2n+1}; $n = 4 - 15$.

water and are either biodegraded or adsorbed to sludge. The major part of the phthalates found in the environment originates from the migration from plastic matter during use and disposal, also called weathering. The migration capacity varies in an inverse way to their molecular weight. As phthalates are not covalently bound in plastic matrices, they will migrate especially under conditions of high surface exposure and warm temperatures. This is especially so in case of exterior building materials, where phthalate esters can diffuse to air despite their rather low vapor pressure. Diffuse and pinpoint sources thus contribute to their dispersal, mainly from direct wet and dry atmospheric deposition, via surface water, and via indirect runoff and overland flow. As river inputs contribute to oceanic pollution, the Oslo and Paris Convention (OSPAR) for protection of the marine environment in the North-East Atlantic included DBP and DEHP on the list of priority chemicals of concern. In spite of bacterial degradation in surface water and degradation by photolysis, phthalates are at the present time detected in different compartments of the environment at concentrations ranging from 0.3 to 77 $ng m^{-3}$ in the atmosphere, from 0.3 to 98 $\mu g l^{-1}$ in surface water, from 0.2 to 8.4 mg per kg dry weight (DW) in sediment, and from 28 to 154 mg per kg DW in sewage sludge, albeit that at specific spots even higher levels may be found. The highest levels of dissolved DBP and DEHP are found in freshwater samples, whereas these compounds are usually below the limit of detection in marine water and sediment. Median levels are log–normal distributed, and usually the season in which the samples are taken is not a relevant determinant parameter. DEHP levels in pristine waters are generally found to be c. 0.01 $\mu g l^{-1}$ in surface water.

Environmental Fate of Phthalates

Physical–Chemical Properties Affecting Environmental Distribution

Physical–chemical properties dictate the behavior of a chemical in a given environment. The most important physical–chemical properties of several phthalate esters are given in **Table 1**. Phthalates differ greatly in chemical properties due to their varying chain lengths. Hence, the distribution of phthalates over the various environmental compartments also differs. Molecular weights range from about 194 to 550 $g mol^{-1}$, water solubility relates inversely to the molecular weight from about 4200 $mg l^{-1}$ to <0.001 $mg l^{-1}$ (i.e., a variance of over 6 orders of magnitude), but especially molar volume has been shown to be an excellent indicator of water solubility. As with most organic hydrophobic chemicals, phthalate esters are less soluble in saltwater than in freshwater. Experimental determination of the water solubility of phthalates meets difficulties as the esters are bipolar in nature. They may

Table 1 Physical–chemical properties of several phthalate esters

Name	Abbreviation	Molecular weight ($g mol^{-1}$)	Water solubility ($mg l^{-1}$)	log K_{OW}
Dimethyl phthalate	DMP	194.19	4200	1.61
Dibutyl phthalate	DBP	278.35	11.2	4.45
Diethylhexyl phthalate	DEHP	390.56	0.003	7.50
Diisononyl phthalate	DINP	425	<0.001	>8
Diisodecyl phthalate	DIDP	447	<0.001	>9

form micelles or other types of aggregates. Consequently, experimentally determined water solubilities for individual esters often span several orders of magnitude, and calculated water solubilities based on structure–activity relationships give generally lower estimates for water solubility than experimental results. The environmental distribution of phthalates is nevertheless driven by their partitioning to organic carbon. Octanol–water partition coefficients (K_{ow}) increase (i.e., hydrophobicity increases) with increasing chain length. The reason for this increase is that solubilities in water decrease more with increasing chain length than the solubilities in octanol. The high K_{ow} values of phthalates indicate that these substances are very hydrophobic and will adsorb strongly to organic matter and surfaces. In fact, a linear relationship would be expected between K_{ow} and K_{oc} (the organic carbon–water partition coefficient). However, a linear relationship fails for phthalates, with log K_{ow} exceeding 6. Instead, fairly constant log K_{oc} values of c. 6 are reported for most of the higher phthalates. This discrepancy may be caused by problems in measuring truly dissolved phthalate concentrations in water related to sorption to colloidal matter in the aqueous phase.

Octanol–air partition coefficients are appropriate to describe the partitioning between air and organic phases in soils, plants, and atmospheric aerosols. However, especially for phthalates with alkyl chains of more than six carbon atoms, limited vapor pressure data are available.

Degradation and Bioaccumulation of Phthalates

Monitoring data indicates that there exists a delicate balance between the quantities of phthalates entering the environment via various imission routes and the amounts of phthalates removed from the environment. This is despite the lack of monitoring data covering 'all' compartments on a global scale, and despite the enormous volumes of phthalates disposed of in wasted products and slowly weathering from their product matrix. To warrant that accumulation of chemicals as widely used as phthalates in either the environment, or in biota and humans, does not take place, it is essential that such high-production-volume chemicals are easily degraded. Degradation cannot occur if the chemical is unavailable to the active chemical agents or microorganisms, and will be limited if the concentration truly in solution is less than the total concentration on the other hand. Factors reducing the effective concentration of a chemical with regard to degradation are also reducing bioaccumulation and hence toxicity. The 'bioavailability' of a chemical is most notably impacted by solubility, dissolution rate, and sorption to dissolved or particulate organic matter. Especially for higher molecular weight phthalates, degradation as well as bioaccumulation may be reduced by low bioavailability.

Atmosphere

Photodegradation is the dominant degradation pathway in the atmosphere with predicted half-lives of c. 0.2–4.0 days. Direct photolysis rates are in general too low to significantly impact the overall phthalate fate. Indirect photolysis in air due to hydroxyl radical attack in both vapor phase and particle-sorbed phthalates may have a significant role in the removal of phthalates.

Water

Chemical hydrolysis of the ester bond of phthalate esters is known to be a slow process with half-lives of many years. On the other hand, there are numerous reports that phthalates undergo rapid ultimate biodegradation in various screening and simulation tests. Most phthalate esters meet all the criteria for ready biodegradability in tests strictly adhering to regulatory guidelines, but it cannot be ruled out that these laboratory test systems are not predictive of environmental degradation rates. In general, biodegradation rates in aerobic aquatic environments vary from 0.4 to 36 days. The little data available on degradation in anaerobic environments as well as in nutrient-poor or cold environments indicate that under these conditions too degradation is possible, albeit at significantly lower rates.

The metabolic pathway for phthalate ester biodegradation is via a stepwise hydrolysis of the two ester bonds, first giving the monoester plus the free alcohol, followed by hydrolysis of the second ester bond to give phthalic acid and alcohol (**Figure 3**). Various bacteria are capable of metabolizing phthalic acid and alcohols. As expected on the basis of this general metabolic pathway, rates of primary biodegradation of a series of *n*-alkyl phthalate esters were found to be similar.

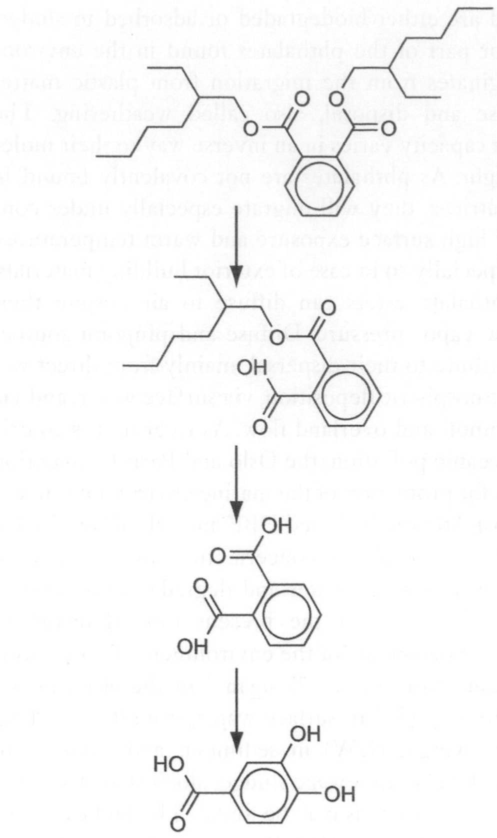

Figure 3 Metabolic pathway for degradation of DEHP.

Soil

Mobility studies in soil have provided evidence that especially low molecular phthalate esters biodegrade quickly in soil, with half-lives in soils from <1 week to several months. Bioavailability differences as well as variance in biomass and temperature are important factors affecting rates of biodegradation in solid phases. The transformation pathways for degradation of phthalates in soil are similar to those in water, and abiotic processes are of limited importance. Roughly speaking, half-lives in soil vary as much as 1–400 days.

Bioaccumulation of Phthalates

Bioaccumulation quantification describes how it can be that concentrations in biota sometimes are higher than those in the surrounding environment or in the prey or food of organisms. The general model for bioaccumulation of hydrophobic organic chemicals stipulates that organic chemicals bioconcentrate in lipid tissues mainly by passive exchange processes from and to water. The magnitude of bioconcentration largely depends on the hydrophobicity, expressed via the octanol–water partition coefficient, and the lipid content of the organisms.

Laboratory and field studies indicate that phthalate esters do not biomagnify in aquatic food webs:

1. High molecular weight phthalates show evidence of trophic dilution in aquatic food webs. Laboratory and modeling studies indicate that metabolic transformation is a key mitigating factor in this respect.
2. Phthalate esters with intermediate molecular weights (like DBP, diisobutyl phthalate, and butyl benzyl phthalate) have bioaccumulation patterns that are consistent with the general lipid–water partitioning model.
3. Low molecular weight phthalates display bioaccumulation factors that are greater than predicted from a lipid–water partitioning model, and a considerable variation in field-derived bioaccumulation factors for low molecular weight phthalate esters across aquatic species suggests that species-specific differences in metabolic transformation capacity have a significant effect on observed bioaccumulation.

With some exceptions, bioaccumulation factors of phthalates are below 5000. The low bioavailability of the high molecular weight phthalate esters, as a consequence of their low water solubility and strong adsorption to (dissolved) organic carbon, is the main reason why these compounds too do not bioaccumulate to a considerable extent. It should be noted that although theoretical studies show that as much as 60% of the exposure of high molecular weight phthalates in predators could be derived from the diet, only limited data on dietary bioaccumulation have been reported.

Adverse Effects of Phthalates

Human Health Effects

Humans are exposed to phthalates through ingestion, inhalation, and dermal exposure during their whole lifetime. Preventive limit values, such as reference dose (RfD) of the US Environmental Protection Agency (US EPA) and tolerable daily intake (TDI) of the European Union, are 20 μg per kg of body weight per day and 37 μg per kg of body weight per day, respectively. It has been shown that the general population can be exposed to DEHP to a much higher extent than previously believed, and exposure of children at levels twice as high as the exposure of adults with respect to their body weight has been observed.

Phthalates are rapidly metabolized in humans to their respective monoesters, which, depending on the phthalate, can be further metabolized to oxidative products of their lipophilic aliphatic side chain. Monoesters and the oxidative metabolites of phthalates may be conjugated as the glucuronide (phase II biotransformation), and both free and conjugated metabolites can be excreted in urine and feces.

In animal studies, a wide variety of adverse effects were linked to phthalate exposure; the level of exposure in these experiments was, however, several times higher as in actual day-to-day exposures.

Phthalates are carcinogenic to animal and can cause fetal death, malformations, testicular injury, liver injury, anti-androgenic activity, teratogenicity, peroxisome proliferation, and especially reproductive toxicity in laboratory animals. Toxicity profiles and potency vary by specific phthalate. The extent of these toxicities and their applicability to humans remains incompletely characterized and controversial.

Aquatic Toxicity

Acute toxicity data for lower molecular weight phthalate esters for freshwater organisms are abundant, and the various taxonomic groups (protozoa, algae, crustaceans, insects, and fish) are comparable in their sensitivity. Less data are available for marine species, but the available acute data point toward similar sensitivity as for freshwater organisms. Chronic toxicity data for freshwater organisms are available for algae, crustaceans, fish, and amphibians, with differences between taxonomic groups being less than a factor of 30. The limited chronic data for saltwater species again point toward similar sensitivity as for freshwater organisms. In Table 2, an overview is given of the chronic and acute toxicity data of DBP to aquatic organisms. The data presented in Table 2 were used as input for the derivation of maximum permissible concentrations (MPCs) for DBP.

For the higher phthalate esters, no observed effect concentrations (NOECs) exceeding the water solubility are often reported (these values are usually discarded for risk-assessment purposes). Micelle formation and direct physical effects like surface entrapment of the test organisms in layers of the phthalate esters on top of the water column often induce adverse effects at short-term exposure. Such physical effects clearly are not indicative of inherent toxicity. In general, it is found that higher molecular weight phthalates do not exhibit short-term and chronic toxicity to aquatic organisms at concentrations below the aqueous solubility, for the organisms and exposure times tested.

Although elevated levels of phthalates may be measured in sediments, there is no evidence as to the potential or real damage caused to species in the sediment.

Soil Toxicity

Phthalate esters, especially DBP, have been identified as being toxic to some plants in areas of restricted ventilation. Other studies indicate that effect levels exceed $100 \, \text{mg} \, \text{kg}^{-1}$. In general, however, toxicity data are scarce or are to be considered unlikely. An example of an

Table 2 Overview of toxicity data used as input for the derivation of maximum permissible concentrations for DBP in the Netherlands

Taxonomic group	Species	NOEC (mg l^{-1})
Chronic toxicity of DBP to aquatic organisms		
Algae	*Chlorella emersoni*	2.8
Algae	*Pseudokirchnerellia subspicata*	0.77
Algae	*Scenedesmus subspicatus*	6.1
Algae (saltwater)	*Dunaliella parva*	0.28
Crustacea	*Daphnia magna*	0.88
Pisces	*Oncorhynchus mykiss*	0.1
Pisces	*Pimephalis promelas*	0.56

Taxonomic group	Species	L(E)C50 (mg l^{-1})
Acute toxicity of DBP to aquatic organisms		
Bacteria (saltwater)	*Vibrio fisheri*	11–23
Protozoa	*Tetrahymena pyriformis*	7
Algae	*Pseudokirchnerellia subspicata*	0.4
Algae	*Scenedesmus subspicatus*	4.2
Algae (saltwater)	*Gymnodinium breve*	0.05
Crustacea	*Daphnia magna*	3.9
Crustacea	*Gammarus pseudolimnaeus*	2.1
Crustacea (saltwater)	*Artemia salina*	8
Crustacea (saltwater)	*Mysidopsis bahia*	0.5
Crustacea (saltwater)	*Nitroca sinipes*	1.7
Insecta	*Chironimus plumosus*	2.5
Insecta	*Paratanytarsus parthenogenica*	6.3
Pisces	*Brachydanio rerio*	2.2
Pisces	*Lepomis macrochirus*	1.5
Pisces	*Oncorhynchus mykiss*	2.3
Pisces	*Perca flavescens*	0.35
Pisces	*Pimephalis promelas*	1.2
Pisces (saltwater)	*Ictalurus punctatus*	1.2

NOEC = no observed effect concentration; L(E)C50 is the concentration at which an effect is observed for 50% of the test species (EC50), or the concentration at which 50% of the show mortality (LC50).

unlikely result is a study in which an influence of DEHP on respiration in soil is reported at a concentration of 49 000 mg per kg DW. Assuming respiration inhibition to be induced via the porewater, assuming the highest value of K_{oc} reported in literature (510 000 l kg^{-1}), and assuming a high concentration of organic carbon in the soil of 10%, this value would be equal to a porewater concentration of at least 1 mg l^{-1}. This is well above the water solubility of DEHP.

Risk Assessment

Comparing the MPCs for DBP and DEHP in water of 10 and 0.19 µg l^{-1}, respectively, with measured concentrations, shows that the aquatic ecosystem is seldom at risk due to the presence of DBP. However, reported concentrations of DEHP are often higher than the calculated MPC; in the Netherlands, for instance, the MPC of DEHP in surface water in 1997 exceeded that in the main rivers by a factor of 3–10. Measured phthalate concentrations exceed the MPC of sediment to a large extent, although it should be noted that due to a lack of toxicity data for sediment, gross assumptions on the distribution and the corresponding distribution coefficients of phthalates had to be made.

Endocrine-Disruptive Effects of Phthalates

General

There are gender similarities and differences comparing male and female reproductive responses to phthalates. In rats, DEHP works by similar mechanisms and it is a reproductive toxicant in both genders. Conversely, DBP and monobutyl phthalate, its active metabolite, produce developmental effects in males and reproductive tract effects in females. When DBP is administered to pregnant females, it induces a syndrome resembling testicular dysgenesis syndrome in the rat male offspring. This occurs because phthalates can suppress endogenous testosterone production by the fetal testis, thus interfering with sexual differentiation.

The developmental and reproductive effects of DEHP, whose toxicity seems to occur at an older age in females than in males, are especially under scrutiny. DEHP and its metabolite MEHP appear to target analogous sites within the testis and ovary. Hormonal disturbance of the fetal testis development by DEHP has been reported and is dependent on the stage of development at exposure.

Prolonged DEHP exposure increases serum concentrations of both gonadotropin luteinizing hormone and sex hormones (testosterone and 17β-estradiol (E2)) in rats, which suggests the possibility of multiple cross talks between androgen, estrogen, and *steroid hormone receptors. On the other hand, the presence of estrogen receptors in other tissues, for example, cardiovascular system and bones, implies that the increases in serum E2 levels have implications beyond reproduction, including systemic physiology.

Data available from *in vitro* experiments

In vitro assays to screen for (anti-)estrogenic/androgenic potency can roughly be classified in one of the following groups:

1. *Receptor binding affinity tests.* The binding affinity to a receptor is measured with competitive ligand-binding techniques. This assay type does not discern agonists from antagonists.

2. *Cellular proliferation assays.* These measure proliferation in cell lines that are dependent on hormones.
3. *Gene expression tests.* Measurement of the gene expression after exposure to contaminants by determining amounts of mRNA, or gene products (e.g., an enzyme), is made. Sometimes, the cell lines have been constructed with the help of recombinant techniques.

In test types 2 and 3, receptor binding of the ligand as well as further events such as interaction with the responsive element on the DNA, DNA transcription, and the resulting production of proteins are measured, whereas in test type 1 only ligand binding is measured. *In vitro* test systems cannot replace *in vivo* tests as the sole base for screening chemicals for potential (anti-)estrogenic/androgenic action. The major objections to using solely *in vitro* systems are the differences in metabolism, bioavailability, and toxicokinetics between *in vitro* and *in vivo* test systems. In addition, intercellular interaction and mechanisms related to endocrine homeostasis are absent in *in vitro* systems. However, *in vitro* test systems can give a prediction about the possibility that compounds are able to exert endocrine-disruptive effects. This means that if they are not positive in any of the *in vitro* tests, it is not likely that the tested substances will be endocrine disruptors *in vivo*. *In vitro* tests are therefore used in so-called 'high-throughput screening' of high numbers of chemicals.

For several of the phthalate esters tested, no effects were found at the highest tested concentration in any of the tests. For dibutyl, diethyl, dihexyl, butylbenzyl, di-2-ethylhexyl, diisobutyl, and methyl ethyl phthalate, endocrine-disruptive effects are reported. The relative potency related to the potency of the natural estrogen, 17β-estradiol, is always low (10^{-4}–10^{-8}). This relative potency appears the highest for butyl benzyl phthalate (10^{-4}–10^{-6}); for DEHP, the reported relative potencies are 10^{-5}.

The three major groups in which the *in vitro* tests are classified do not always show the same picture. For DBP, endocrine-disruptive effects are obtained in all types of *in vitro* systems, while for DEHP effects are observed in receptor binding assays but not in the proliferation or gene expression assays. Apparently, DEHP is able to bind to the estradiol receptor, but further events such as binding of the ligand–receptor complex to the DNA and resulting transcription and protein production do not take place.

Despite some inconsistencies in the data, the results reported do give rise to the idea that some phthalate compounds are able to act as xeno-estrogens, and thereby are able to act as endocrine disruptors.

Data available from in vivo experiments

Existing mammalian test methods (in general rat and mice) are in general useful for screening of potential (anti-)estrogenic/androgenic action. It is not yet clear however if these methods are predictive for other classes of vertebrate wildlife. The most commonly used *in vivo* assays in mammals for estrogenic effects are the uterotrophic assay in which the uterine weight is determined, and the vaginal cornification assay. For fish and birds, full and partial life-cycle tests seem suitable assays.

Most *in vivo* data on endocrine-disruptive effects are available for mouse and rat. Endocrine-disruptive effects are found in *in vivo* tests for DBP, DEHP, butyl benzyl, diethyl, and dihexyl phthalate, but not for other phthalates tested. Especially the two-generation reproduction studies appear sensitive in detecting endocrine disruptive effects. None of the tests with uterine weight showed positive results. The distinction between phthalates which can or cannot act as endocrine disruptors appears to be the same from *in vitro* and *in vivo* tests.

Significance for environmental risk assessment

Although *in vivo* testing is restricted to mammals, it is possible to calculate worst-case concentrations in the environment that are protective against endocrine-disrupting effects. However, several assumptions are needed for this purpose, such as

1. similar sensitivity of mammals tested *in vivo* and mammalian wildlife,
2. worst-case values of the biomagnification factor (BMF),
3. worst-case values of the biota-to-soil or -to-sediment accumulation factor (BSAF), and
4. complete lack of biotransformation.

From these assumptions, it can be concluded that MPCs as derived from 'regular' toxicity testing (i.e., not focusing on endocrine-disrupting activity) are sufficiently protective against endocrine-disruptive effects. As a matter of course, this conclusion can be invalidated upon generation of new data for mammals or aquatic species not yet tested.

See also: Biodegradability; Biodegradation; Ecological Risk Assessment; Reproductive Toxicity.

Further Reading

Ankley G, Mihaich E, Stahl R, *et al.* (1998) Overview of a workshop on screening methods for detecting potential (anti-)estrogenic/androgenic chemicals in wildlife. *Environmental Toxicology and Chemistry* 17: 68–87.
European Union (2002) Risk assessment report on bis(2-ethylhexyl) phthalate (DEHP). EU Regulation 793/93. Brussels: European Union.
Fürthmann K (1993) Phthalate in der aquatischen Umwelt. Düsseldorf, Germany: Landesamt für Wasser und Abfall Nordrhein-Westfalen.
Latini G (2005) Monitoring phthalate exposure in humans. *Clinica Chimica Acta* 361: 20–29.
Mackay D (1991) *Multimedia Environmental Models. The Fugacity Approach.* Chelsea, MI: Lewis.
OECD (1984, 1992) *Guidelines for Testing of Chemicals.* Paris: OECD.
Staples CA (2003) *Phthalate Esters.* Berlin: Springer.
van Leeuwen CJ and Vermeire T (2007) *Risk Assessment of Chemical.* Berlin: Springer.

Plutonium

R M Harper, Western Washington University, Bellingham, WA, USA

R M Tinnacher, Colorado School of Mines, Golden, CO, USA

Plutonium Sources
Plutonium Properties
Environmental Transport and Pathways of Exposure
Mechanisms of Plutonium Toxicity

Plutonium Ecotoxicity
Environmental Risk – Humans
Further Reading

Plutonium Sources

Plutonium is commonly considered a man-made element, as the vast majority of plutonium found today has been artificially generated in nuclear reactors (**Table 1**). Small quantities of naturally occurring plutonium isotopes, however, result from the remains of supernova explosions (Pu-244 and Pu-239) and the continuous spontaneous fission of U-238 in uranium-rich ores (Pu-239).

Based on the history of plutonium contamination in the environment since the 1940s and an evaluation of future potential releases, three important sources of plutonium contamination are: (1) the introduction into the atmosphere through nuclear weapons testing; (2) the direct accidental discharge into soils and waters from various civil and military facilities; and (3) the potential accidental losses from the nuclear fuel cycle and from underground nuclear waste depositories. The relative contribution of plutonium release has changed over time with atmospheric nuclear weapons testing predominating in the past, but constituting a smaller portion of the releases as several countries shifted to underground nuclear weapons testing in 1963. Since then, plutonium air concentrations have followed an approximate exponential decrease, with an apparent

half-life of around 4 years decreasing to levels below $0.1\,\mu Bq\,m^{-3}$ ($10^{-7}\,Bq = 2.7 \times 10^{-18}\,Ci$) in 1993.

Atmospheric nuclear weapons testing is the primary source of plutonium on a global scale. This has resulted in worldwide deposition of plutonium from atmospheric fallout (fallout plutonium) and in localized high soil contamination levels at the test sites. It has been estimated that since 1945 atmospheric nuclear weapons testing caused the release of approximately $1.33 \times 10^{16}\,Bq$ ($3.59 \times 10^{5}\,Ci$) or 6 t of plutonium into the environment. The volatilization of both unfissioned plutonium and plutonium created by neutron irradiation of U-238 contaminates the atmosphere with several isotopes, including Pu-238, Pu-239, Pu-240, and Pu-241. Pu-239 and Pu-240 are usually reported together as 'plutonium' since they are the most abundant isotopes and cannot be easily distinguished by alpha spectrometry.

Based on the relative surface areas of land and sea, approximately two-thirds of plutonium fallout is deposited into the oceans. The measured Pu-239/Pu-240 concentrations in seawater are significantly lower than in soils, which is in part due to dilution effects. In addition, the strong affinity of plutonium for mineral surfaces results in deposition and incorporation of significant amounts of plutonium into sea bottom sediments, making this a predominant sink for plutonium (**Table 2**).

Table 1 Plutonium isotopes in the environment

Isotope	Main sources	Half-life (years)	Principal mode of decay	Decay energy (MeV)	Remaining after 10 000 years[a] (%)
Pu-238	WF[b]	87.7	Alpha to U-234	5.593	~0
Pu-239	WF, NI[c], N[d]	24 110	Alpha to U-235	5.245	~75
Pu-240	WF, NI	6 564	Alpha to U-236	5.256	~35
Pu-241	WF	14.35	Beta to Am-241	0.021	~0
Pu-244	N	80 000 000	Alpha to U-240	4.666	~100

[a]The US Environmental Protection Agency proposed a time-frame of 10 000 years or longer as a possible compliance period for risk assessment of the planned Yucca Mountain, NV nuclear waste depository (http://www.epa.gov/).
[b]WF = weapons fallout.
[c]NI = nuclear industry releases.
[d]N = natural sources.
Eisenbud M (1987) *Environmental Radioactivity*, 3rd edn. San Diego, CA: Academic Press. http://atom.kaeri.re.kr.

On a local scale, several cases of accidental plutonium release include those with the construction of plutonium-producing reactors and the extraction of plutonium from irradiated uranium at nuclear research and production centers (e.g., Hanford (Washington) and Los Alamos (New Mexico) in the United States). Later examples of plutonium releases are the nuclear reactor accidents at Windscale, UK (now Sellafield), in 1957 and at Chernobyl, Ukraine, in 1986, the accidental re-entry of a radioisotope-powered navigational satellite (SNAP 9A) over the Indian Ocean in 1964, the crash landing of a B52 bomber in Greenland in 1968, and the gradual release of plutonium from the processing plant at Rocky Flats, USA.

Due to the cessation of atmospheric testing and the potentially increasing utilization of nuclear power, the life cycle of nuclear fuel generation, storage, and use, and the containment of nuclear waste will become the primary (potential) source of plutonium contamination. The nuclear power industry discharged approximately 220 000 t of spent fuel containing about 1400 t of plutonium into temporary storage facilities through the year 2000. Future production is estimated to generate approximately 11 000 t of spent fuel per year. The largest contributor to the radiotoxic inventory (a measure of radiotoxic effects of spent fuel, calculated as the weighted sum of all toxic ingredients) will be Pu-239 and Pu-240, formed from U-238 by the absorption of neutrons in nuclear reactors. These isotopes will account for 80% of the inventory after 300 years, and for 90% after 500 years.

Plutonium Properties

Several chemical characteristics affect plutonium solubility, its fate and transport in the environment, the route of exposure, biological uptake efficiency, biological half-life, and retention in target organs. First, plutonium shows strong affinities for mineral, biological, and other surfaces. This leads to its association with aerosols and particles in the atmosphere, plant surfaces and soil particles, sediments in rivers and oceans, as well as mineral colloids and bacterial membranes in groundwater systems. Second, the redox chemistry and chemical speciation of plutonium in aqueous systems, such as natural waters, biological media, or liquid-particle aerosols, are complex. For example, although under most oxidizing conditions the dominant oxidation states are IV and V, plutonium can potentially coexist at up to four oxidation states (III, IV, V, VI) in the same solution. Furthermore, organic and inorganic ligands, such as humic substances, carbonates, sulfates, and phosphates form complexes with plutonium, which may change its sorption and bioavailability characteristics. Plutonium hydrolysis reactions also play an important role due to the low solubility of the Pu(IV)-hydrolysis products. Ultimately, all these processes depend on solution conditions such as pH, redox conditions, ionic strength, and temperature. For atmospheric aerosols, however, photochemical reactions and interactions with free radicals may also be of importance.

Table 2 Environmental plutonium concentrations

	Pu-238		Pu-239/Pu-240	
Environmental media	(Bq kg^{-1})	(mol kg^{-1})	(Bq kg^{-1})	(mol kg^{-1})
Plutonium in soils				
Natural Pu-239	n/a	n/a	1.0E–04[a]	1.0E–16[a]
Man-made Pu: average	~0.26[b]	~1.7E–15[b]	1.4–7.1[b]	2.5E–12 to 1.3E–11[b]
Man-made Pu: point sources	n/a	n/a	0.2–466[c,d]	4E–13 to 8.5E–10[c,d]
Plutonium in sediments				
Man-made	n/a	n/a	0–1750[d]	0 to 3E–09[d]
	(Bq m^{-3})	(mol m^{-3})	(Bq m^{-3})	(mol m^{-3})
Plutonium in oceans				
Man-made	n/a	n/a	2E–03 to 17[d]	4E–15 to 3E–11[d]
Plutonium in rivers				
Man-made	n/a	n/a	8E–02 to 0.5[d,e]	1.5E–13 to 9E–13[d,e]

[a]Taylor DM (2001) Environmental plutonium – Creation of the universe to twenty-first century mankind. In: Kudo A (ed.) *Plutonium in the Environment: Edited Proceedings of the Second Invited International Symposium*, pp. 1–14. Amsterdam: Elsevier Science.
[b]Hardy EP, Krey PW, and Volchok HL (1973) Global inventory and distribution of fallout plutonium. *Nature* 241: 444–445.
[c]Liator MI (1999) Plutonium contamination in soils in open space and residential areas near Rocky Flats, Colorado. *Health Physics* 76: 172–179.
[d]Watters RL, Edington DN, Hakonson TE, *et al.* (1980) Synthesis of research literature. In: Hanson WC (ed.) *Transuranic Elements in the Environment*, DOE/TIC-22800, pp. 1–44. Oak Ridge, TN: US Department of Energy.
[e]Trapeznikov AV, Pozolotina VN, Chebotina M Ya, *et al.* (1993) Radioactive contamination of the Techa River in the Urals. *Health Physics* 65: 481–488.

Environmental Transport and Pathways of Exposure

Humans, believed to be far more sensitive to radiation than plants and lower animals, were historically the focus of plutonium exposure scenarios and regulations have been designed to minimize human risk as a result of plutonium exposures (**Table 3**). More recently, however, the distribution of plutonium in other biological systems has been considered (**Table 4**).

Atmospheric sources of Pu (i.e., from fallout plutonium) can lead to human and animal exposure in the form of direct radiation, inhalation of respirable plutonium-containing dust particles and aerosols, and deposition on the skin (**Figure 1**). In the lung, the size distribution of plutonium-containing particles is directly related to the radiological dose received, the retention rates, and the distribution to target organs. Furthermore, plutonium solubility in lung tissue after inhalation affects the exposure time, with less-soluble compounds (e.g., Pu-239 dioxide) being retained longer than the more soluble forms (e.g., Pu-239 nitrate). Soluble compounds, however, are more available for absorption into the cytoplasm of a cell. This decreases the distance between the ionizing radiation and sensitive cellular structures such as DNA, which results in an increased chance of mutations. With dermal contact, the penetration of plutonium through the tissue is generally slight, which limits the exposure to cells near the site of contact and reduces systemic effects.

Deposition of fallout plutonium onto plant surfaces, soils, and surface waters can cause increased human

Table 3 Examples of plutonium concentration limits used by regulatory agencies

US Environmental Protection Agency	
Air: maximum dose of radionuclides to humans	10 mrem[a] (0.1 mSv)
Drinking water: max. contam. level for gross alpha emitters	15 pCi l^{-1a}
Example for soil action levels[b]	
Rocky Flats, USA	35–600 pCi g^{-1c}
Enewetak Atoll, Pacific	40–60 pCi g^{-1c}
European Union – Council Directive 98/83/EC	
Drinking water: total indicative dose	0.1 mSv yr^{-1d}
World Health Organization – Guidelines for Drinking Water Quality[e]	
Recommended RDL for drinking water[f]	0.1 mSv yr^{-1}
Guidance level for individual Pu-238, Pu-239, Pu-240 conc.	1 Bq l^{-1}

[a]http://www.epa.gov/
[b]Soil action levels are determined individually for contaminated sites based on site-specific risks.
[c]http://www.pacificislandtravel.com
[d]http://www.lenntech.com
[e]http://www.who.int/en/
[f]RDL = Reference dose level from a year's consumption of drinking water.

Table 4 Plutonium bioconcentration factors

Sample description	Bioconcentration factor
Flesh of freshwater fish	350[a]
Fish in water of low mineral content	50[b]
Fish in water of high mineral content	10[b]
Piscivores fish	5–250[c,d]
Planktivores fish	25[d]
Bottom-feeding fish	250[d]
Plants	2E−03[e]

[a]National Research Council of Canada (NRCC) (1982) Data sheets on selected toxic elements, no. 19252. Ottawa: National Research Council.
[b]Canadian Standards Association (CSA) (1987) *Guidelines for Calculating Derived Release Limits for Radioactive Material in Airborne and Liquid Effluents for Normal Operation of Nuclear Facilities*. National Standard of Canada, CAM/CSA-N288.1-M8. Rexdale, Toronto: Canadian Standards Association.
[c]Myers DK (1989) *The general principles and consequences of environmental radiation exposure in relation to Canada's nuclear fuel waste management concept*. AECL 9917. Chalk River, ON: Atomic Energy of Canada Limited, Chalk River Nuclear Laboratories.
[d]Poston TM and Klopfer DC (1985) A literature review of the concentration factors of selected radionuclides in freshwater and marine fish. PNL-5484. Richland, WA: Pacific Northwest Laboratory.
[e]Garten CT, Jr., Bondietti JR, Trabalka RL, *et al.* (1987) Field studies of terrestrial behavior of actinide elements in East Tennessee. In: Pinder JE, Alberts JJ, McLeod KW, and Schreckhise RG (eds.) *Environmental Research on Actinide Elements*, pp. 109–119. Washington, DC: US Department of Energy.

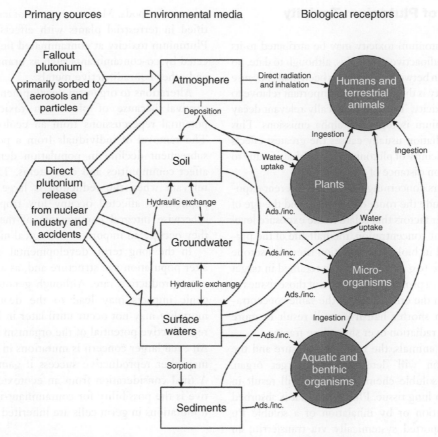

Primary sources | Environmental media | Biological receptors

Figure 1 Plutonium in the environment: sources of release, potential distribution, and pathways of exposure to biological receptors. 'Ad./inc.' indicates 'adsorption or incorporation'.

uptake through ingestion of contaminated food and water. Whereas the total plutonium intake by ingestion is significantly higher than by inhalation, the fraction of ingested plutonium that is absorbed into the blood and distributed to target organs is relatively low (0.000 5). Inhaled material thus contributes to over 90% of fallout plutonium retained in the human body.

In a terrestrial environment, the fate and transport of plutonium are related to the chemical speciation and the partitioning of plutonium between soil and porewater. This determines the potential for removal from soil by processes such as plant uptake, water percolation through soil, or groundwater flow. Plant uptake of plutonium in terrestrial systems is relatively limited, due to the formation of insoluble Pu-hydrolysis products in soil solutions, and the fact that most plants can discriminate against and exclude plutonium at the root membrane level. However, if the plant absorbs the contaminant, its distribution in the plant tissue is minimized due to Pu stabilization by natural ligands and metabolites, causing most plutonium to be retained in the roots. As a consequence, the exposure to and accumulation by organisms feeding on the above-ground foliage is minimal, resulting in low plutonium concentrations in terrestrial food webs.

Up to 98% of the Pu in contaminated soils is immobile and is concentrated in the top few meters or centimeters. Recent field studies show, however, that plutonium can penetrate at least 2 m into the ground, and be transported with mineral or organic colloids in groundwater flow. Once plutonium reaches aquatic systems, by either direct release or mobilization from contaminated soils, it is distributed in the ecosystem over various pathways, as depicted in **Figure 1**. In freshwater and marine systems, plutonium has been shown to adsorb to and/or bioaccumulate and bioconcentrate in plants (e.g., seaweed), protists (e.g., algae), invertebrates (e.g., zooplankton, crayfish, snails, clams, mussels), and vertebrates (e.g., fish, fish eggs, and sharks), with benthic organisms and submerged plants accumulating the greatest concentrations. Uptake of plutonium is typically greatest from sediment and food sources, although assimilation efficiency is typically low. To a lesser extent, water can also provide a bioavailable source of plutonium depending upon environmental conditions. Typically, biomagnification does not occur. The majority of accumulated plutonium is deposited and sequestered in shell and bone, as opposed to the soft tissue of vertebrates and invertebrates.

Mechanisms of Plutonium Toxicity

Mechanisms of plutonium toxicity may be attributed to its chemical and/or radioactive properties, although to date, no thorough distinction between the two has been made because the chemical toxicity is thought to be unimportant relative to the radiological toxicity. The toxicologically relevant decay products of plutonium isotopes are alpha emissions. This highly ionizing radiation usually causes the greatest effects in the immediate vicinity of plutonium contamination due to the short penetration distance of alpha particles.

Important factors concerning toxicity in different exposure scenarios include the route of exposure and the age of the organism. Other factors that relate to the exposure level are the contaminant concentration, the half-life of the particular isotope, and its biological retention time. An isotope with a long half-life (e.g., Pu-239) that is retained in target organs will result in a prolonged exposure of those tissues to alpha radiation. On the other hand, at the same concentration, isotopes with shorter half-lives will result in more emissions of alpha radiation over short time frames.

In humans and animals, the route of exposure and the chemical speciation will determine the target organ. Inhalation of less-soluble chemical species will result in localized effects on lung tissue. Plutonium that is absorbed (whether by ingestion or by inhalation of a soluble Pu species) and transported systemically via transferrin, an iron carrier protein, can be stored in the bone. Other target organs include the liver and the lymph nodes.

Depending on the exposure scenario, the resulting effects of the ionizing radiation include cytotoxicity and tissue necrosis (at higher doses) or genotoxicity (at lower doses), with cancer being the best documented in animal studies. In inhalation studies conducted with beagles, in order of frequency, bone, lung, and liver tumors were identified and determined to be the cause of death in many cases, with liver tumors being the least likely to result in death. The mechanisms of action for tumor development (and other non-tumor-forming mutations) are either direct radiation damage of DNA strands or indirect toxicity with strand damage after free radical generation. Other documented effects of plutonium exposure in (nonhuman) mammals include altered immune system structure and function, and reproductive effects.

Plutonium Ecotoxicity

The understanding of plutonium ecotoxicity is limited by the number and types of studies conducted with the radionuclide. In aquatic systems, increased mortality and developmental effects in carp and fathead minnows occur at activity concentrations as low as $5.97 \times 10^5\ \mu\mathrm{Bq\ ml}^{-1}$. In terrestrial systems, decreases in population density have been observed in earthworms, mites, and microarthropods. Moreover, genotoxicity has been identified in terrestrial plants with effects to somatic cells. Plutonium toxicity at contaminated field sites is complicated by co-contaminants such as uranium-235 and -238 and other nonradioactive metals.

Alterations to organism development and decreases in survival because of plutonium toxicity have several potential repercussions from an ecological perspective. The removal of individuals from a population and the subsequent decline in population density can directly affect communities and ecosystems. This is particularly relevant when the reduction is large enough and the populations affected occupy low trophic levels or are otherwise integral components of the food web, or if they occupy an important ecological niche.

In the long term, developmental effects potentially alter population age structure and, as a consequence, the net reproductive rate. Although genotoxic and carcinogenic impacts may lead to the death of individuals, mortality may not occur until later in life, when the full reproductive potential of the organism has been realized. An even larger concern is mutations in germ cells, which may alter reproductive success if gametes are affected. A final consideration from an ecotoxicological perspective is the possibility for contaminant-induced evolution if mutations in germ cells are inherited by offspring.

Environmental Risk – Humans

It has been determined that humans have always contained a baseline level of naturally occurring plutonium (Pu-239 and Pu-244) and fallout plutonium has contributed some plutonium to these body burdens. Total levels, however, are typically low and range from 0 to 120 mBq/kg fresh weight of tissues (measured as Pu-239/Pu-240), depending upon the population investigated and the specific tissue analyzed. At measured levels, incidence of bone tumors is expected to be extremely low and is estimated to increase by 2 in 100 000 cases per 50 years if tissue levels were to increase by 200 times the current Pu-239 levels in humans. Furthermore, from an epidemiological point of view, links between human exposure to plutonium and human cancers at the levels measured are often confounded by factors such as cigarette smoking. Investigations into the potential for increased toxicity caused by eating contaminated marine organisms have resulted in low risk estimates because of the low level of plutonium present in the organism's tissue and the low bioavailability of plutonium when ingested.

See also: Bioaccumulation; Bioavailability; Ecological Risk Assessment; Ecotoxicology: The Focal Topics; Ecotoxicology: The History and Present Directions;

Exposure and Exposure Assessment; Mutagenesis; Radioactivity; Risk Management Safety Factor; Uranium.

Further Reading

Canadian Standards Association (CSA) (1987) Guidelines for calculating derived release limits for radioactive material in airborne and liquid effluents for normal operation of nuclear facilities. National Standard of Canada, CAM/CSA-N288.1-M8. Rexdale, Toronto: Canadian Standards Association.

Clarke RH, Dunster J, Nenot J, Smith H, and Voeltz G (1996) The environmental safety and health implications of plutonium. *Journal of Radiological Protection* 16: 91–105.

Driver CJ (1994) *Ecotoxicity Literature Review of Selected Hanford Site Contaminants*. US Department of Energy, PNL-9394, UC-600.

Eisenbud M (1987) *Environmental Radioactivity*, 3rd edn. San Diego, CA: Academic Press.

Frontasyeva MV, Perelygin VP, and Vater P (eds.) (2001) *Radionuclides and Heavy Metals in the Environment*. Dordrecht, The Netherlands: Kluwer Academic Publishers.

Hardy EP, Krey PW, and Volchok HL (1973) Global inventory and distribution of fallout plutonium. *Nature* 241: 444–445.

Kudo A (ed.) (2001) *Plutonium in the Environment. Edited Proceedings of the Second Invited International Symposium*. Amsterdam: Elsevier Science.

National Research Council of Canada (NRCC) (1982) *Data sheets on selected toxic elements, no. 19252*. Ottawa: National Research Council.

Noshkin VE (1972) Ecological aspects of plutonium dissemination in aquatic environments. *Health Physics* 22: 537–549.

Pillay KKS and Kim KC (eds.) (2000) *Plutonium Futures – The Science: Topical Conference on Plutonium and Actinides, Santa Fe, New Mexico*. Melville, NY: American Institute of Physics.

Poston TM and Klopfer DC (1985) *A literature review of the concentration factors of selected radionuclides in freshwater and marine fish. PNL-5484*. Richland, WA: Pacific Northwest Laboratory.

Romney EM and Davis JJ (1972) Ecological aspects of plutonium dissemination in terrestrial environments. *Health Physics* 22: 551–557.

Swift DJ and Pentreath RJ (1988) The accumulation of plutonium by the edible winkle (*Littorina littorea* L.). *Journal of Environmental Radioactivity* 7: 29–48.

Taylor DM (1995) Environmental plutonium in humans. *Applied Radiation and Isotopes* 46: 1245–1252.

US Environmental Protection Agency (1972) Estimates of ionizing radiation doses in the United States 1960–2000. *Report of Special Studies Group*. Washington, DC: Division of Criteria and Standards, Office of Radiation Programs.

Relevant Websites

http://www.atsdr.cdc.gov – Agency for Toxic Substances and Disease Registry.

http://www.osti.gov – Information Bridge (DOE Scientific and Technical Information), US Department of Energy Scientific and Technical Information.

http://www.iaea.org – International Atomic Energy Agency.

http://atom.kaeri.re.kr – Table of Nuclides, Nuclear Data Evaluation Lab.

http://www.epa.gov – US Environmental Protection Agency.

http://www.who.int – World Health Organization.

Polychlorinated Biphenyls

G O Thomas, Lancaster University, Lancaster, UK

Properties, Production, and Environmental Sources of PCBs
Exposure of Biota to PCBs

Behavior of PCBs in Biota
Effects of PCBs on Populations
Further Reading

Properties, Production, and Environmental Sources of PCBs

Structure and Properties

Polychlorinated biphenyls (PCBs) are a group of industrial chemicals that have not been found to occur naturally. The generic structure of PCBs is shown in **Figure 1**, and consists of biphenyl chlorinated in one or more positions. The positions which can be chlorinated are numbered separately on each phenyl ring, using a prime (′) to differentiate between the rings. For the purposes of assessing reactive sites the relative positions on the rings are of most importance and are most easily represented as *ortho*, *meta*, and *para* to the bond between the phenyl rings. There are 209 different PCBs

(called congeners), due to the structural isomers possible at each chlorination level (homolog). There are also a number of PCBs which exist as enantiomers, the chiral center being the bond between the phenyl rings, for which free rotation is precluded by steric hindrance

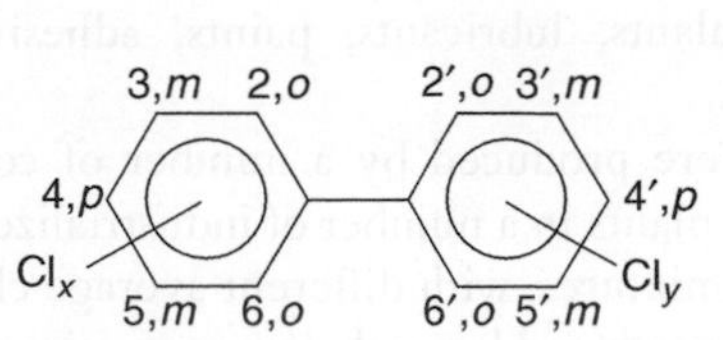

Figure 1 Chemical structure of polychlorinated biphenyls, showing numbered and named ring positions, where x is 1–5 and y is 0–5. Named ring positions: $o =$ ortho; $m =$ meta; $p =$ para.

when there are three or four chlorines present at *ortho* positions. PCBs were produced industrially by the chlorination of biphenyl, leading to complex (but racemic) mixtures of congeners. Individual PCB congeners are generally named in two ways, a IUPAC structural name (e.g., 2,2',5-trichlorobiphenyl) and a numbering system originally devised by Ballschmitter and Zell, and adopted by IUPAC (e.g., PCB 18, or #18), in which PCB 1 is 2-monochlorobiphenyl and PCB 209 is decachlorobiphenyl.

PCBs are chemically stable (e.g., they are stable to both strong acids and strong bases, are not easily oxidized or reduced, are nonflammable and thermally stable), are soluble in a wide range of organic solvents, and have low electrical conductivity. Pure PCBs are either liquids or noncrystalline solids at room temperature, and most commercial mixtures are viscous liquids (although the highest chlorine content mixtures are solids). The vapor pressures of PCBs range from 1.2 Pa for monochlorinated congeners to 7×10^{-6} Pa for decachlorobiphenyl at $25\,^\circ$C, and the log octanol–water partition coefficients (log K_{OW}) of PCBs are in the range 4.2 (monochlorobiphenyl) to 8.3 (decachlorobiphenyl) at $25\,^\circ$C. PCBs are poorly soluble in water, with a solubility of 2×10^{-6} mol m^{-3} for decachlorobiphenyl to 29 mol m^{-3} for monochlorobiphenyl, at $25\,^\circ$C.

Production, Use and Legislation

Since industrial production of PCBs began in 1929 it is estimated that approximately 1.2–1.5 million metric tonnes have been produced. Production peaked in the late 1960s, and ended in most countries in the 1970s and 1980s. However, it is not certain whether production has completely ceased in all countries, even in 2006. It is estimated that approximately one-third of all PCBs produced have been released into the environment. This, combined with their properties (notably their stability, vapor pressures and K_{OW}), has led to them becoming ubiquitous in the environment, having been found in biota and other environmental media from all regions of the globe and in all habitats.

Their dielectric properties and fire resistance led to PCBs being used widely in electrical equipment, such as transformers and capacitors, predominantly in enclosed systems. However, PCBs have also had a very broad range of other, often unenclosed, uses, including in heat transfer liquids, sealants, lubricants, paints, adhesives, and as plasticizers.

PCBs were produced by a number of companies, at production plants in a number of industrialized countries. A range of mixtures, with different average chlorine contents, were produced by each company – in each mixture the range of individual PCB congeners present was different, and there are likely to have been differences in congener composition between different batches of the same mixture. Each company used their own names for the mixtures they marketed, so there are a great number of trade names for PCB-containing commercial products, the most well known being Aroclor, Clophen, Kanechlor, and Phenoclor. The use of PCBs was overwhelmingly concentrated in industrial countries, meaning that use was concentrated in the temperate regions of the Northern Hemisphere where most industrial regions are situated.

In the 1960s and 1970s, following the discovery of the widespread environmental occurrence of PCBs, and industrial accidents which led to acute human and environmental exposure to PCBs, many countries introduced legislation to control the manufacture, use, and disposal of PCBs. This was most notable after 1973, after recommendations from the OECD. PCB production and use was banned in Japan in 1972, and the leading manufacturer in the USA ceased production in 1977. Most other countries banned PCBs and introduced special disposal requirements (typically highly controlled incineration) in the late 1970s and 1980s, although previously established enclosed use continues in many countries, and production and new uses may still occur in others.

In addition to national controls, international efforts have been made to control PCBs, and the United Nations Environment Programme's Stockholm Convention on 'persistent organic pollutants' (POPs) is a global agreement to strictly control, with the aim of eliminating, PCBs and other persistent organic chemicals (including some organochlorine pesticides and 'dioxins' – polychlorinated dibenzo-*p*-dioxins and dibenzofurans), to which only a handful of countries are not either party or signatory.

Sources to the Environment

The major primary sources of PCBs to the environment include vaporization of PCBs from unenclosed uses, inappropriate disposal practices, volatilization and runoff from landfills containing PCB waste, accidental release of PCBs from facilities where they are used, and incineration of wastes containing PCBs. Since the implementation of legislative and control measures, and the ending of PCB production, in most countries primary sources of PCBs to the environment have declined, and secondary sources (e.g., the remobilization of PCBs from contaminated land and water bodies) have become increasingly important.

Exposure of Biota to PCBs

Environmental Transport and Abiotic Degradation

The properties which made PCBs useful industrial chemicals (particularly their chemical inertness) make them persistent within the environment. PCBs are volatile enough and soluble enough to be transported effectively in the air and in water bodies, but because they partition strongly to organic material they are readily concentrated into biota and other organic material (particularly the organic matter present in soils and sediments). PCBs are found in all regions of the globe, including the Poles, which implies that air transport is the most important mode of global movement, but at the local and regional scale transport in water bodies is likely to be important, and indeed transport in migratory animals may have an impact in certain (particularly relatively uninhabited, nonindustrial) regions.

PCBs partition between the vapor and particulate phases in air, with generally increasing partitioning to particles with increasing level of chlorination (the octanol–air partition coefficient, K_{OA}, has been shown to be a good descriptor of the partitioning of PCBs between the vapor and particulate phases in air). At most environmentally relevant temperatures and particulate loadings PCBs are found predominantly in the vapor phase. However, at high atmospheric particulate loadings and low temperatures the particulate bound fraction will be appreciable. Transport of PCBs from source regions in the air, followed by partitioning to vegetation and soils and deposition with precipitation, is likely to be the primary source of PCBs to the Poles and other cold environments (including high mountains) – the low temperatures in these environments strongly favoring partitioning to the solid (especially organic) phase. This process has been proposed for a range of organic chemicals with similar properties, and has been called 'cold-condensation'.

The relatively high concentrations of suspended organic matter in water bodies, coupled with the K_{OW} of PCBs, generally leads to higher concentrations of PCBs in sediments and suspended matter than in the dissolved phase. It has been estimated that most of the PCBs that have been released into the environment are now held in aquatic sediments. These sediments act predominantly as sinks, but can release PCBs to organisms or back to the water column.

PCBs partition strongly to soils, which act as long-term, high-capacity reservoirs. Leaching from soils may be a source of PCBs to groundwater. Volatilization from soils can also be a source of PCBs to the atmosphere, as can the atmospheric entrainment of soil particles by the wind. Much higher concentrations of PCBs have been measured in soils from the temperate Northern Hemisphere, where they were predominantly used, than in other regions, implying that the bulk of PCBs have not, on the whole, been transported very far from their primary sources.

Differences in the K_{OA} and K_{OW} of individual PCBs cause changes in the patterns of PCBs seen in the environment after the 'weathering' processes which control

environmental transport have taken place. Since both the water solubility and vapor pressure of PCBs become lower with increasing chlorination level (and vary to an extent with other structural differences), in general, the more-chlorinated PCBs are less mobile in the environment than the less-chlorinated PCBs. Thus, a PCB mixture that has been in the environment for a time will become deficient in the less-chlorinated PCBs compared to the original mixture. Equally, PCBs transported to nonsource regions are likely to be enriched in the less-chlorinated PCBs compared to the original mixtures released into the environment.

In general, the persistency of PCBs increases with the level of chlorination. The vapor phase reaction with hydroxyl radicals in the atmosphere is probably the major abiotic degradation pathway. Direct photolysis of PCBs is likely to be a less important mode of degradation in the atmosphere, but is likely to be the major mode in water. Degradation rates are dependent on the structure of individual PCB congeners, and therefore degradation causes changes in the patterns seen in PCB mixtures over time.

Concentrations of PCBs in many environmental media in many regions (notably in human foodstuffs and air in industrial countries) have fallen dramatically since the introduction of controls and legislation. However, in some regions temporal trends cannot be established because few measurements have been made, and in others (including some equatorial areas) PCB concentrations do not yet show significant reductions.

Exposure Routes

PCBs partition strongly to organic media from water and air, and bioconcentrate and bioaccumulate efficiently in biota.

Most plants are not able to significantly translocate PCBs, so transfer is generally from the air (or water for aquatic plants) to external surfaces, with some transfer to internal leaf tissues, and from the soil (or sediment) to the external surfaces of roots and tubers.

In aquatic systems exposure at low trophic levels is predominantly directly from the water, with efficient bioconcentration taking place. However, strong bioconcentration in prey species leads to bioaccumulation from food becoming more important at higher trophic levels. Air-breathing marine animals are exposed almost entirely through the diet, as air concentrations and air intake are relatively low. Because of the strong tendency to biomagnify in aquatic systems, and the high lipid content of most marine animals (especially mammals), marine top predators tend to have much higher PCB concentrations than equivalent terrestrial animals. In higher organisms in terrestrial systems exposure is predominantly through the diet. However, many soil invertebrates may absorb PCBs directly from contact with the soil. PCBs bioaccumulate strongly in terrestrial foodchains.

In animals, PCBs are effectively translocated by the blood, and concentrate in lipid-rich tissues. Eggs and milk are rich in lipids, and so the young of many species are often exposed to relatively high doses of PCBs from their mothers. This is particularly true for many marine mammals, which produce extremely lipid-rich milk for their young.

Behavior of PCBs in Biota

Absorption and Excretion

PCBs are readily absorbed through many tissue membranes, including by microorganisms, the gills of fish, the skin of many animals, and the gastrointestinal tract. In plants PCBs may enter the cellular structure, but are more likely to be contained within the cuticular layers. PCBs accumulate in lipid-rich tissues, and in animals it is usual to express concentrations on a lipid weight basis. Lipid-rich tissues are the main storage sites in animals, and the liver (being the first organ chemicals absorbed into the blood from the gastrointestinal tract encounter, highly blood-perfused, and the main metabolic site) is generally relatively enriched in PCBs on a lipid-weight basis. The release of PCBs occurs when fat is redistributed due to food shortage, poor health, hibernation, lactation, incubation, and migration. However, PCBs are not necessarily released *pro rata*, and adipose PCB concentrations tend to increase at these times. PCB concentrations found in tissues of a dairy cow in the UK in 1996 are shown in **Table 1**. It can clearly be seen that in the subcutaneous fat and internal organs PCBs have been distributed relatively evenly, on a lipid-weight basis, and that concentrations in liver are much higher than the other tissues, again on a lipid-weight basis.

Excretion of PCBs is generally poor, and highly dependent on the structure of individual PCB congeners. Plants can lose PCBs by volatilization to air (or dissolution to water) (due to a change in a parameter controlling equilibrium – e.g., PCBs may partition back to the air after an increase in temperature, or after a change to low PCB concentrations in air relative to the plant), by shedding of surface cuticular fragments and waxes, and by senescence of leaves. PCBs are generally not accessible to the sites of metabolism in plants. Animals do not excrete PCBs effectively in feces (except those unabsorbed from the diet) unless partitioning to feces across the gastrointestinal tract is favored (e.g., if the diet has very low PCB concentrations compared to those accumulated in the body). Excretion of metabolized PCBs is the dominant elimination route for animals, but there is very wide variation in the ability of organisms to metabolize PCBs, and wide variation in the efficiency of metabolism of different PCB congeners. The mechanisms of metabolism of PCBs by animals is dealt with in more detail below. Female animals can excrete PCBs to their offspring in the lipid-rich tissues of eggs, and female mammals can, additionally, pass PCBs to their offspring during gestation and lactation.

Metabolism

Some microorganisms have been shown to degrade many of the less-chlorinated PCBs (with one to four chlorines) at varying rates, but are not able to degrade more-chlorinated PCBs effectively. The chlorine-substituted positions on the biphenyl structure are very important in determining which congeners are degraded, although different microorganisms may attack different positions preferentially. Mostly, chlorine atoms at the *para* positions are preferentially degraded. In anaerobic conditions dechlorination of PCBs has been shown, and thus the more-chlorinated PCBs can be dechlorinated to less-chlorinated PCBs and subsequently be degraded.

In animals, the main metabolic pathway is ring hydroxylation, mediated by the cytochrome P450 enzyme system (which can also be induced by PCBs, increasing its activity, and the rate of metabolism), to form hydroxy-PCBs (PCBs in which one ring is phenyl, the other phenol). The dominant oxidation sites are where there are adjacent unchlorinated positions – that is, *ortho-meta* or *meta-para* unsubstituted positions, the *para* position being particularly important. In general, cytochrome P450 1A1 (CYP1) can metabolize only coplanar PCBs (those with no, or only one, *ortho*-substituted chlorine atom) whereas cytochrome P450 1A2 (CYP1A2) can metabolize coplanar and noncoplanar PCBs. Hydroxylated PCBs are not very soluble in water, but are more soluble than PCBs, and can be excreted from animals slowly in the feces and urine (or through the gills in aquatic species). Hydroxylated PCBs, and some reactive intermediates formed during hydroxylation, can be conjugated with water-soluble derivatives (e.g., glucuronide) and are excreted in the urine and to some extent in the feces (via bile). Intermediate products in the hydroxylation of PCBs can be sulfonated, which reduces rather than increases their water solubility, and these can be stored within adipose

Table 1 Concentrations of PCBs in tissues from a dairy cow

Tissue	[PCB] (pg g^{-1} fresh)	[PCB] (pg g^{-1} lipid)
Liver	210	14 000
Kidney fat		4 600
Kidney	65	4 200
Diaphragm	140	4 100
Subcutaneous fat		5 100

Adapted from Thomas GO, Sweetman AJ, and Jones KC (1999) Metabolism and body-burden of PCBs in lactating dairy cows. *Chemosphere* 39: 1533–44.

tissues – for example, methyl-sulfonyl PCBs are nonpolar and persistent.

Since metabolism is easiest at adjacent unsubstituted positions, the less-chlorinated PCBs are most easily metabolized – as the level of chlorination increases fewer and fewer PCBs have the unsubstituted positions needed for efficient metabolism. Thus there are specific pattern changes that can generally be seen through foodchains, with the less-chlorinated PCBs becoming relatively reduced in concentration, and the least easily metabolized PCB congeners becoming dominant. In general the higher animals have more capacity to metabolize PCBs than the lower animals, and some specific species have unusually high or low metabolic capacity (e.g., cetaceans metabolize PCBs relatively slowly). It should be noted that within a species there are likely to be P450 polymorphisms (inherited genetic differences in the efficiency of certain P450 enzymes), which may lead some groups of individuals to have unusually high or low capacities for the metabolism of PCBs. It is also important to note that there can be wide variation in the relative efficiencies of specific metabolic pathways between species.

In **Figure 2** a comparison of PCB patterns in the diet (fish), feces, and blubber of a gray seal from the North Sea (UK coast) in 2001 is shown. The relative loss of the more easily metabolized PCB congeners can easily be seen by comparing the fish PCB congener pattern with the

blubber pattern. This is most notable for most of the less-chlorinated (lower numbered) PCBs (below hexa-chlorobiphenyl), but also affects some highly chlorinated PCBs that have chlorine-substitution patterns that allow effective metabolism. It is also evident that some less-chlorinated PCBs (e.g., PCB 99) are relatively resistant to metabolism. The PCBs that are relatively resistant to metabolism in the gray seal are marked 'R' on the blubber PCB pattern.

Biomagnification

The physical and chemical properties of PCBs has made them useful in the development of models for the environmental behavior and transport of chemicals, including models for bioconcentration, bioaccumulation, and risk assessment.

In aquatic organisms PCBs bioconcentrate strongly, as they are within the molecular size range and K_{OW} range suited to efficient absorption through membranes. At higher trophic levels and in the terrestrial foodchain the diet is the most important source of PCBs, and PCBs strongly bioaccumulate. Absorption into an organism is controlled by the relationship between the amount of PCB present in the absorbing part(s) of the organism and the PCB concentrations encountered, which can be explained using the 'fugacity' concept. For example, an

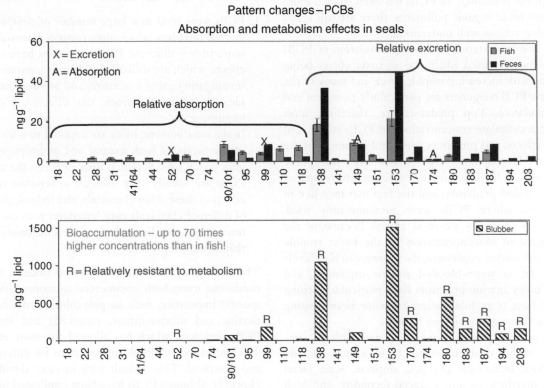

Figure 2 PCB concentrations in fish, seal feces, and seal blubber from a gray seal from the North Sea (captured on the east coast of the UK).

organism with large fat reserves will accumulate more PCBs in total from its environment or diet than an organism with smaller fat reserves because the fat reserves in the two organisms will reach equilibrium at the same concentration (or achieve steady state at similar concentrations). Efficient metabolism of particular PCB congeners has a strong effect on absorption (a positive effect) and biomagnification (a negative effect) because strongly metabolized PCBs are kept at lower concentrations within the organism. This is illustrated in **Figure 2**, where it can be seen that seal faecal concentrations are higher than the diet (fish) for the unmetabolized PCBs, because they are at relatively higher concentrations in the seals blood – they are absorbed less strongly than the metabolized PCBs, which are at relatively lower concentrations in the seals blood. Because of the role of K_{OW} in the absorption of PCBs from either water or the diet, the more-chlorinated PCBs generally biomagnify more efficiently than the less-chlorinated PCBs.

Toxicology

PCBs have a relatively low acute toxicity to vertebrates, and highly variable toxicity to other organisms. PCB congeners show different modes of toxic action, and have been shown to have a wide range of chronic toxic effects, including carcinogenicity, and effects on the immune, reproductive, endocrine, and nervous systems. Although the toxicology of PCBs has been studied more than many other organic pollutants, there are still many aspects that are not well understood.

The bioconcentration and bioaccumulation of PCBs leads to the increased likelihood of toxic effects being expressed with increasing trophic level, and many of the most toxic PCB congeners are particularly persistent and bioaccumulative. Top predators are, therefore, most likely to accumulate concentrations of PCBs which lead to toxic effects, and predatory birds and mammals in the Northern Hemisphere are particularly susceptible because of their high food intake requirements (compared to cold-blooded predators) and the fact that they live in the regions where PCBs were predominantly used. Marine top predators are most at risk because of the high degree of bioconcentration in the lower trophic levels of the marine ecosystem, the presence of high levels of body fat in warm-blooded aquatic organisms, and because many marine predators have particularly strong seasonal fasts (e.g., hibernation in polar bears, fasting during lactation and moulting in seals).

After acute exposure of mammals and birds to commercial mixtures, documented toxic effects include weight loss, behavioral changes, alopecia, acne, facial oedema, diarrhea, anemia, reduced fecundity, and high mortality of adults and offspring. However, it should be noted that some of the toxic effects seen may be due to the action of impurities in the PCB commercial mixtures, especially of 'dioxins' (polychlorinated dibenzo-p-dioxins and dibenzofurans) which are often produced during the production of PCBs, and can also be produced during certain uses in which high temperatures are encountered.

The effects of chronic PCB exposure which have been documented in vertebrates include

- reproductive effects, including reduced conception and live birth rates, reduced birth weights, and reduced sperm counts;
- neurological effects, including reduced neurological development (including on eyesight and memory);
- effects on the endocrine system, including decreased thyroid hormone levels;
- immune system effects, including decreased size of the thymus, reduced immune system response, and reduced resistance to viral and other infectious agents;
- cancer, particularly of the liver.

Many of the toxic effects of PCBs do not have a 'no observable effect dose', meaning that there is no safe minimum dose below which the effect does not cause an increase in the likelihood of the effect occurring.

Some of the toxic effects of PCBs are likely to be primarily, or partly, caused by the hydroxymetabolites of PCBs, and possibly the short-lived reactive intermediates produced during metabolism.

Determining the toxic effects of PCBs is complicated by two very important factors:

1. PCBs were used as a large number of complex commercial mixtures, which often contain potentially toxic impurities – different PCB congeners have different effects, which are difficult to identify separately when investigating using a mixture, and are also difficult to identify separately from the effects of different impurities
2. In the environment, biota are exposed to a wide range of chemicals, of both natural and anthropogenic origins, which have potential toxic effects – the effects of PCBs are usually very difficult to separate from the effects of these other chemicals, and indeed, the effects of different chemicals may 'interfere' with each other; interactions may be synergistic, or antagonistic, rather than being simply additive.

The toxicity of PCBs has been tested under a range of conditions using both commercial mixtures (often with specific impurities, such as polychlorinated dibenzo-p-dioxins and dibenzofurans, removed) and individual PCB congeners, which has allowed the toxic effects of different classes of PCB congeners to be differentiated and described. This is dealt with in more detail below. However, although PCBs have been implicated in certain toxic effects in the field, the actual impact of PCBs on field observations of toxic effects has not been satisfactorily quantified as there is insufficient data and knowledge about the effects of the chemical mixtures present for firm epidemiological conclusions to be drawn. For example, there is evidence that PCBs are immune suppressants in mammals (from controlled studies, e.g., in one study seals were fed fish containing high concentrations of PCBs), but a number of field studies of seals have been unable to separate potential immune-suppressant effects of PCBs from other persistent pollutants, such as organochlorine pesticides, which are often also present at relatively high concentrations, and were controlled and regulated at similar times leading to similar temporal changes in concentrations.

Toxic modes of action for coplanar PCBs

PCBs that do not possess chlorine atoms substituted at the *ortho* positions are called 'coplanar' congeners, as they can take a planar molecular shape. These congeners can bind to the cytosolic aryl hydrocarbon receptor (Ah receptor), and exhibit toxic effects similar to 2,3,7,8-tetrachlorodibenzo-*p*-dioxin (the most toxic of the 'dioxins' – polychlorinated dibenzo-*p*-dioxins and dibenzofurans). This form of toxicity is often manifest as hepatotoxicity, embryo mortality, and induction of CYP1A1 and CYP1A2. Induction of the CYP enzymes can lead to synergistic effects between PCBs and other chemicals, as the activation of some mutagens and carcinogens may become enhanced, leading to increased DNA adduct formation. Some hydroxymetabolites of coplanar PCBs are structurally similar to thyroxine (T_4), and can compete with T_4 for binding sites on transthyretin, with associated physiological effects.

The toxicity of 2,3,7,8-tetrachlorodibenzo-*p*-dioxin is also expressed to different degrees for a number of different polychlorinated dibenzo-*p*-dioxins and dibenzofurans, and coplanar PCBs (i.e., each has a different level of toxicity relative to dose). However, the total 'dioxin-like' toxicity of a mixture has been found to be simply (or predominantly) additive, and is directly related to the sum of the dose of each of the chemicals with dioxin-like toxicity multiplied by a factor to correct for the relative toxicity of each component. The correction factor for each chemical component in this calculation is derived from *in vitro* and *in vivo* tests, and is called a toxic equivalence factor (TEF). The TEF multiplied by a dose or concentration gives the 'toxic equivalent' (TEQ) – the dioxin-like toxicity from that chemical, expressed as an equivalent dose or concentration of 2,3,7,8-tetrachlorodibenzo-*p*-dioxin, in the same units. It has been found that, in environmental samples, the total dioxin-like toxicity from PCBs can actually exceed that of PCDD/Fs themselves. The current TEF values for 'dioxin-like' PCBs in fish, birds, and mammals (including humans), assigned by the World Health Organization in 1997, and reassigned for mammals in 2005, are shown in **Table 2**. It can be seen that there are marked differences in the relative toxicities of PCBs to fish, birds, and mammals. It should be noted that TEF values are intended for estimating dioxin-like toxicity from oral ingestion only.

Toxic modes of action for noncoplanar PCBs

Although the dioxin-like toxicity of coplanar PCBs has received a great deal of attention, it has become apparent that the noncoplanar PCBs also have a range of toxic effects. It is notable that, in commercial mixtures and the environment, the coplanar PCBs are present at much lower concentrations than many of the noncoplanar

Table 2 TEF values for PCBs, assigned by the WHO in 1997 and 2005

PCB name	PCB number	WHO 1997 TEF fish[a]	WHO 1997 TEF birds[a]	WHO 2005 TEF humans and mammals[b]
3,3',4,4'-tetrachlorobiphenyl[c]	PCB 77	0.0001	0.05	0.0001
3,4,4',5-tetrachlorobiphenyl[c]	PCB 81	0.0005	0.1	0.0003
3,3',4,4',5-pentachlorobiphenyl[c]	PCB 126	0.005	0.1	0.1
3,3',4,4',5,5'-hexachlorobiphenyl[c]	PCB 169	0.00005	0.001	0.03
2,3,3',4,4'-pentachlorobiphenyl[d]	PCB 105	<0.000005	0.0001	0.00003
2,3,4,4',5-pentachlorobiphenyl[d]	PCB 114	<0.000005	0.0001	0.00003
2,3',4,4',5-pentachlorobiphenyl[d]	PCB 118	<0.000005	0.00001	0.00003
2,3',4,4',5'-pentachlorobiphenyl[d]	PCB 123	<0.000005	0.00001	0.00003
2,3,3',4,4',5-hexachlorobiphenyl[d]	PCB 156	<0.000005	0.0001	0.00003
2,3,3',4,4',5'-hexachlorobiphenyl[d]	PCB 157	<0.000005	0.0001	0.00003
2,3',4,4',5,5'-hexachlorobiphenyl[d]	PCB 167	<0.000005	0.00001	0.00003
2,3,3',4,4',5,5'-heptachlorobiphenyl[d]	PCB 189	<0.000005	0.00001	0.00003

[a]Adapted from van den Berg M, Birnbaum L, Bosveld ATC, *et al.* (1998) Toxic equivalency factors (TEFs) for PCBs, PCDDs, PCDFs for humans and wildlife. *Environmental Health Perspectives* 106: 775–792.
[b]Adapted from van den Berg M, Birnbaum LS, Denison M, *et al.* (2006) The 2005 World Health Organization reevaluation of human and mammalian toxic equivalency factors for dioxins and dioxin-like compounds. *Toxicological Sciences* 93: 223–241.
[c]Non-*ortho* substituted.
[d]Mono-*ortho* substituted.

PCBs, so even noncoplanar PCBs with relatively low toxicity may have appreciable effects at the doses received compared to coplanar PCBs. Some of the toxic effects of noncoplanar PCBs are the same as, or similar to, the non-dioxin-like toxic effects of coplanar PCBs.

Toxic effects of *ortho*-substituted PCBs that have been documented include neurotoxicity (e.g., decreased catecholamine levels in the brain and behavioral changes), effects on insulin production, and effects on the endocrine system (e.g., oestrogen-like activity of PCBs and their metabolites). The mechanisms leading to these effects are not well understood, but are likely to be related (at least partly) to hydroxy-metabolites, and include interference in signal transduction pathways and intracellular Ca^{2+} homeostasis (for neurological problems), and the rigidity of noncoplanar PCBs (and metabolites) allowing action on steroid and hormone receptors.

It is important to note that the toxic effects of noncoplanar PCBs are not dioxin-like, and are not encompassed at all in the use of TEFs described above. Similar effects may be seen from exposure to a range of biogenic and anthropogenic chemicals present in the environment, and, as such, the mixture toxicity effects of additivity, synergy (enhancement), and antagonism (reduction) may be seen. Because of this, and the relatively low concentrations of PCBs generally present in the environment, it is also extremely difficult to quantify the effects of PCBs separately from other chemicals, and even large-scale (extremely expensive) epidemiological studies in the field may not elucidate the impact of each component separately.

Effects of PCBs on Populations

Researchers have suggested PCBs as potential causative or contributing agents in a number of population declines. However, as discussed above, positively attributing effects on either individuals or population to PCBs is very difficult because of the impurities contained in the commercial mixtures which are the source of environmental PCBs, the co-occurrence of a number of other potential toxic agents (some of which came under controls and reduction measures at a similar time to PCBs), and the generally low incidence (even when enhanced, and potentially problematic) of the effects. Although a very substantial amount of work has been performed on the toxicity of PCBs in controlled (predominantly) laboratory studies, on a range of species, it is especially difficult to extrapolate effects seen in these studies to effects in wild populations, even when the population is studied in great depth and over an extended period.

In areas affected by acute exposure (i.e., in highly contaminated environments, such as waste outlets at PCB production or use facilities) the microbial population can be selected for PCB tolerance, and the ability to use PCBs as a source of energy (e.g., some anaerobic microbes may derive energy by dechlorinating PCBs). However, at more generally encountered environmental concentrations no effects on microbe population have been proven.

As noted above, it is difficult to identify and quantify the effects of PCBs in wild populations, and the few cases where PCBs have been accepted as potential causes of population effects have been in birds and mammals, and assessing the 'weight of evidence' from a range of sources has been very important in identifying PCBs as likely causative factors. Although few specific population declines have been attributed specifically to PCBs, it appears to be generally accepted that PCBs may cause long-term, relatively subtle effects within populations. These may affect many types of biota within an ecosystem, but any measurable evidence for adverse effects is most likely to be found in top predators, particularly birds and mammals, due to the effects of bioaccumulation. However, since PCBs are now under strict control in most countries, and effective disposal measures are in place, only populations that are exposed to relatively high concentrations of PCBs at particularly contaminated sites, or due to the effects of environmental redistribution and biomagnification, are likely to be affected in the future.

Some of the effects on bird and mammal populations for which PCBs may be contributory agents are

1. In the 1980s a range of predominantly reproductive effects (including reproductive impairment, high embryo and chick mortality, deformities including club feet and crossed bills and decreased body weight) were seen in herring gulls, Forster's terns, cormorants, and Caspian terns from the North American Great Lakes, causing population decline. The effects were linked to exposure to coplanar PCBs and dioxins. **Figure 3** shows the dramatic decline in PCB concentrations in birds around Lake Erie (North America) since the 1970s.

2. In the 1970s the populations of wild mink and otters in the North American Great Lakes region, and of otters in Europe, declined dramatically. Farmed mink fed on Great Lakes fish (containing high concentrations of organochlorine chemicals) suffered reproductive failure. Laboratory studies of the effects of PCBs on mink caused embryo death and reduced survivability and growth. The population declines in wild mink and otters were accepted as being related to the effects of PCBs and dioxins.

3. Baltic ringed and gray seals suffered population declines in the 1960s due to a high incidence of

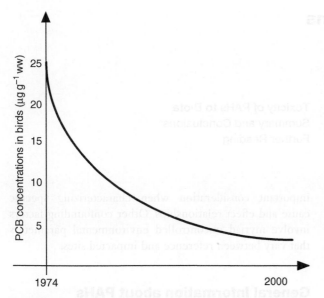

Figure 3 Changes in PCB concentrations in birds from around Lake Erie (N. America) over time.

abortion and sterility. There were also bone lesions and claw deformities. It was found that the seals had high concentrations of p,p'-DDE and PCBs. After controlled studies in seals and mice the effects were linked to the methylsulphonyl metabolite products of p,p'-DDE and PCBs.

4. In 1988 and 2002 common (harbor) seals in the North Sea suffered dramatic epizootics of phocine distemper virus (in 1988 around 50% of the Western Atlantic population died). The seals were found to contain high concentrations of PCBs and other organochlorine pollutants, and there is a suspicion that these chemicals may have suppressed the seal immune systems, making them more susceptible to the phocine distemper virus. No definite link between PCB exposure and the epizootics has been proven.

5. Harbor porpoises from the UK caught between 1989 and 2002 found to have died of infectious disease were compared to healthy porpoises that had died of physical trauma (e.g., by being caught in commercial fishing nets). A threshold for health effects in marine mammals of $17 \, \mathrm{mg \, kg^{-1}}$ lipid was proposed, and some porpoises in both groups had PCB concentrations above that. In the porpoises with PCB concentrations higher than the proposed threshold, higher PCB concentrations were found in porpoises that had died from infection than trauma, whereas for the porpoises below the proposed threshold PCB concentration there was no statistically significant difference in PCB concentrations between the two groups. It has been suggested that this is consistent with an immunotoxic effect of PCBs reducing survivability of porpoises.

6. In the 1990s, a lack of older females and high juvenile mortality were noticed in the polar bear population of Svalbard (Spitzbergen). It has not been possible to determine the cause of these effects, but there is a suspicion that they may be linked to the high concentrations of PCBs and other pollutants found in polar bears in that area. The high pollutant concentrations may be causing reproductive impairment, lowered survival rates of cubs, increased mortality of females, and immunotoxic effects.

See also: Bioaccumulation.

Further Reading

Bernhoft A, Skaare JU, Wiig O, Derocher AE, and Larsen HJS (2000) Possible immunotoxic effects of organochlorines in polar bears (*Ursus maritimus*) at Svalbard. *Journal of Toxicology and Environmental Health – Part A* 59: 561–574.

Brunström B and Halldin K (2000) Ecotoxicological risk assessment of environmental pollutants in the Arctic. *Toxicology Letters* 112–113: 111–118.

Derocher AE, Wolkers H, Colborn T, et al. (2003) Contaminants in Svalbard polar bear samples archived since 1967 and possible population level effects. *Science of the Total Environment* 301: 163–174.

Fischer LJ, Seegal RF, Ganey PE, Pessah IN, and Kodavanti PRS (1998) Symposium overview: Toxicity of non-coplanar PCBs. *Toxicological Sciences* 41: 49–61.

Jepson PD, Bennett PM, Deaville R, et al. (2005) Relationships between polychlorinated biphenyls and health status in harbor porpoises (*Phocoena phocoena*) stranded in the United Kingdom. *Environmental Toxicology and Chemistry* 24: 238–248.

Safe S (1993) Toxicology, structure–function relationship, and human and environmental-health impacts of polychlorinated-biphenyls – progress and problems. *Environmental Health Perspectives* 100: 259–268.

Skaare JU, Bernhoft A, Derocher A, et al. (2000) Organochlorines in top predators at Svalbard – occurrence, levels and effects. *Toxicology Letters* 112: 103–109.

Thomas GO, Sweetman AJ, and Jones KC (1999) Metabolism and body-burden of PCBs in lactating dairy cows. *Chemosphere* 39: 1533–44. van den Berg M, Birnbaum L, Bosveld ATC, et al. (1998) Toxic equivalency factors (TEFs) for PCBs, PCDDs, PCDFs for humans and wildlife. *Environmental Health Perspectives* 106: 775–792.

van den Berg M, Birnbaum LS, Denison M, et al. (2006) The 2005 World Health Organization re-evaluation of human and mammalian toxic equivalency factors for dioxins and dioxin-like compounds. *Toxicological Sciences* 93: 223–241.

World Health Organization (1993) *Environmental Health Criteria 140 – Polychlorinated Biphenyls and Terphenyls*, 2nd edn. Geneva: WHO.

Relevant Websites

http://www.chem.unep.ch – Persistent Organic Pollutants, UNEP Chemicals.

http://www.pops.int – Stockholm Convention on Persistent Organic Pollutants POPs.

http://www.epa.gov – US Environment Protection Agency.

Polycyclic Aromatic Hydrocarbons

J P Meador, NOAA Fisheries, Seattle, WA, USA

Introduction

The following is a general review describing the current state of knowledge on the toxic responses observed for a wide variety of species exposed to polycyclic aromatic hydrocarbons (PAHs). The term 'ecotoxicology' mainly refers to the effects that toxicants have on ecosystems; however, this is a very broad concept. Even though humans are generally considered a component of ecosystems, such evaluations are usually not a part of ecotoxicological investigations. This article focuses on the adverse effects from exposure to PAHs that occur in most species except *Homo sapiens*. Ecotoxicology as a field is broad and covers the direct and indirect effects on ecosystems, in addition to laboratory studies that focus on specific contaminants and their effects on whole organisms. As ecotoxicologists, we are generally interested in any adverse responses that affect population integrity; therefore, we generally focus on effects relating to survival, growth, and reproduction. Over the last several years, however, other relevant responses have been elevated in importance because of their potential effect on population dynamics. For example, alterations to the immune system or organismal behavior can certainly lead to an increase in mortality for individuals. For most of these 'endpoints', ecotoxicologists study not only the higher-level responses, such as reduced biomass, immobilization, or fewer offspring, but many of the physiological and biochemical mechanisms that lead to the higher-level responses. In most cases, the decisions regarding toxic concentrations are made on statistically significant changes for the higher-level responses and some physiological alterations. Unlike toxicity profiling for humans, evidence of biochemical change is usually considered not sufficient for regulatory action to protect an ecosystem.

Most of the data we have for ecotoxicological responses come from controlled laboratory studies, which will be the focus of this article. Field studies demonstrating alterations to species or ecosystems are inherently difficult to perform and interpret. Such studies usually suffer from confounding factors that preclude definitive conclusions. In most cases, multiple contaminants occur at a site and these are often correlated with each other, which is an important consideration when characterizing specific cause and effect relationships. Other confounding factors involve myriad uncontrolled environmental parameters that vary between reference and impacted sites.

General Information about PAHs

What Are PAHs?

PAHs are compounds with two or more fused benzene rings and often contain alkyl side groups. PAHs are divided into two groups, low molecular weight PAHs (LPAHs) and the high molecular weight PAHs (HPAHs). LPAHs contain two or three aromatic rings and HPAHs are those with four or more rings. The distinction of low and high molecular weight is subtle because the actual difference in molecular weight between 2- and 6-ring compounds is not dramatic ($\approx$120–300 Da). Even though these designations seem unimportant, they are relatively important for toxicity considerations because the LPAHs and HPAHs generally exhibit very different physicochemical properties, bioavailability, bioaccumulation, and toxic potential. Another important consideration is the designation of parent PAH versus alkylated homologs. Parent PAHs are those compounds consisting of the basic ring structure and no side groups. Many of the parent compounds exist as alkyl homologs. For example, the parent compound phenanthrene (Phen) may be alkylated producing several unique combinations such as methylphen, dimethylphen, ethylmethylphen, and others. There can be dozens of alkylated homologs for many of the parent compounds. One confusing aspect of all this is that many alkylated LPAHs have molecular weights that are higher than many of the so-called HPAH parent compounds and exhibit physiochemical properties that are more similar to parent HPAHs. In many cases, chemists will group the alkylated PAHs based on the total number of carbons in the alkyl moiety, for example, C1, C2, C3, or C4 Phen. A C2 Phen can be ethylPhen or dimethylPhen.

No one knows exactly how many PAH compounds exist; however, hundreds have been identified. In general, ecotoxicologists focus on a small number of PAHs (up to $\sim$40) because of their relatively high frequency of occurrence. More recently, with the advancements made for gas chromatogram/mass spectrometers and standard reference materials that contain known quantities, an expanded list of PAHs can be identified and quantified for ecotoxicology assessments.

Alkylated PAHs

As mentioned above, many parent PAHs contain alkyl side groups including naphthalenes, phenanthrenes, fluorenes, chrysenes, dibenzothiophenes, and others, which increase the hydrophobicity and bioaccumulation potential with increasing alkylation. Consequently, when only a limited set of parent PAH compounds are considered, the bioaccumulation and toxicity potentials can be severely underestimated.

Only a few studies have highlighted the degree to which organisms are exposed to alkylated PAHs. Additionally, very little information on the metabolism of alkylated PAHs is available, which would be useful for understanding the accumulation of these compounds in organisms and those at higher trophic levels. Many studies, including large monitoring programs, have found that alkylated PAHs often comprise a very large percentage of the total PAH concentration in stomach contents or tissues.

Very few studies have examined the toxicity of alkylated PAHs, even though they are expected to be as toxic (or more toxic) than their respective parent compound. As an example, one study found that a single oral dose of Prudhoe Bay Crude oil produced adverse effects in herring gull nestlings. Upon further analysis, these authors found that the fraction containing the methyl homologs of several HPAHs was the most toxic. Another fraction containing alkylated LPAHs did not appear to have any effect on growth, indicating that the alkyl homologs of the four- to six-ring PAHs were more toxic than the alkyl homologs containing two or three rings.

Physicochemical Properties Important for Toxicity Assessment

One of the more important chemical differences between PAHs is their solubility in water, which can be expressed in terms of hydrophobicity. In general, PAHs are more hydrophobic as molecular weight and alkylation increase. The most common way to express hydrophobicity is with the octanol–water partition coefficient (K_{ow}), which can be obtained from many sources. The K_{ow} is the dominant physical parameter that explains a substantial amount of the partitioning behavior exhibited by PAHs in the environment, which is a crucial feature for understanding toxicity from exposure to PAHs found in water or sediment. The K_{ow} values for all PAHs vary by approximately 4 orders of magnitude ($\approx 10\,000$-fold).

Another useful partition coefficient is the ratio between PAH concentrations in aquatic sediment or soil organic carbon and water (K_{oc}), which is determined by the concentration of PAH per gram of organic carbon in sediment or soil divided by the concentration of PAH in either overlying or interstitial water extracted from the sediment or soil. For example, this coefficient is useful for predicting the amount of waterborne PAH for a given sediment concentration under equilibrium conditions. The K_{oc} for a PAH may be predicted with the K_{ow} because organic carbon behaves similarly to octanol as a partition phase. This partitioning behavior is also similar for the lipid found in organisms. Organic carbon and lipid are hydrophobic phases that PAHs prefer and will accumulate according to these physical–chemical properties (K_{ow}, water solubility, and other factors). Based on these partitioning properties, predictions for the distribution of PAHs between water, sediment, and tissue can be generated, which is very useful for predicting bioaccumulation factors and toxicity values.

Sources of PAHs

PAHs are ubiquitous in the environment and can often occur at high concentrations. They are spread around the environment by stormwater runoff, atmospheric transport, petroleum operations (including drilling, refining, transport, combustion, spills, and natural seeps), municipal and industrial wastewater, and combustion of most organic material. In most cases, PAHs are often concentrated in urban environments and their concentrations are increasing primarily due to increases in urbanization and automobile use. One exception is the often high concentrations found in areas of recent forest fires.

Why Are PAHs Important for Ecotoxicology?

PAHs are a major focus for ecotoxicologists because they are common in the environment and are a complex mixture of many compounds. Additionally, toxicity assessment is very complicated, which is mainly due to a wide range in effects, toxicity at low doses, and variable metabolic capability among species. The following information is provided below to help the reader understand their toxic potential, general toxicity features, important differences among species, and why we should be concerned about these compounds.

Factors That Affect Bioaccumulation and Toxicity

Biotransformation

Many species, especially vertebrates, are capable of substantial metabolism of PAHs leading to high variability in effect concentrations (e.g., the LC_{50}), precluding the determination of a toxic tissue concentration, which can be useful for assessing toxicity. Interestingly, PAHs can cause adverse effects in some species even though measured tissue concentrations are extremely low. One of the most important considerations is the amount of accumulated PAH that is metabolized, especially for species that have strong biotransformation capabilities for PAHs. These species may accumulate large amounts of these compounds, show effects, but not contain any measurable concentration of the parent compounds. In those species that are able to metabolize these compounds, mutagenic metabolites are often formed, which can be more toxic than the parent compounds.

Biotransformation of PAHs by various species has been well studied. The metabolites produced from PAH biotransformation are changed chemically and are rapidly excreted. For example, many studies have shown that a high percentage (>75%) of the total PAH dose given to different species was found in the bile after a short time due to biotransformation. Once the PAHs have been transformed to more water soluble metabolites, they are accumulated in bile and excreted. Because of the increase in polarity for these hydrophobic compounds, the rate of excretion is expected to increase greatly leading to the elimination of metabolites. This is supported by studies showing a large increase in clearance of PAH metabolites by the kidney.

Invertebrates, which are a very diverse group, exhibit a great deal of variability in their ability to metabolize PAHs. Among invertebrates, species of crustacea generally possess some of the highest rates of biotransformation for PAHs, whereas mollusks frequently have weak to nonexistent activation of detoxifying enzymes. Even within taxa, high variability exists. A few studies have reported that annelid and crustacean species display highly differential rates of biotransformation for PAHs. For example, one study of four annelid species found a large range in the percentage of benzo[*a*]pyrene (BaP) converted to metabolites, with one species at 7%, two species at around 40%, and one species able to convert 96% of the total BaP. Additionally, this study found high variability for two crustacean species, with metabolites accounting for 20% and 60% of the BaP accumulated after 7 days.

One of the most pressing problems for PAH toxicity assessment is the determination of exposure. Because most species can effectively metabolize PAHs, determining if individuals have been exposed is often very difficult. In some species (mostly invertebrates) that do not extensively metabolize PAHs, a tissue concentration–response relationship can be established. Future work linking biological responses to biomarkers of exposure or specific metabolites found in bile may be valuable in defining dose–response relationships for vertebrates exposed to PAHs. For example, the determination of fluorescent aromatic compounds (FACs) in bile as a biomarker for

PAH exposure may be a useful way to link exposure and response metrics. Other biomarkers of exposure include DNA adducts in various tissues and activation of certain enzymes, such as cytochrome P450–1A. At this point, there are no generalized correlations between the actual dose ($\mu g\,g^{-1}$ or $\mu g\,ml^{-1}$ in water, food, or sediment/soil) and biomarkers of exposure. For most of these biomarkers, such correlations have rarely been explored. For compounds that are metabolized extensively, such as PAHs, the administered dose or throughput of a compound (e.g., microgram of toxicant/gram organism/day) may be a more important factor than the actual whole-body tissue concentration or the amount present at the site of toxic action and should be considered as a viable dose metric.

Route of Exposure

The route of exposure is an important consideration in the determination of toxicity for PAHs, especially for aquatic species. In the aquatic environment, species exhibit a variety of feeding modes, including sediment ingestion (selective and nonselective), detritus feeding, predation, suspension feeding, and filter feeding. Each of these modes may have a dramatic impact on the degree that the organism is exposed to contaminants and final bioaccumulation values. There are several studies that demonstrate the importance of the mode of feeding by invertebrates in relation to the degree of bioaccumulation. An additional consideration is the physical–chemical properties for PAHs. Toxicants, such as PAHs, usually exist in all phases (water, sediment, air, tissue) in various proportions, depending on their fugacity (f), which is a property that determines the degree to which a specific compound will occur in each phase. Several simulation models exist that use chemical, environmental, and organism-specific properties (e.g., f, K_{ow}, and rate of elimination) to predict the occurrence of individual contaminants in these various compartments. These models and experimental work have found that many contaminants with low hydrophobicity will be taken up predominately by aqueous exposure. This generally occurs when water is passed over the gills (ventilation). Other compounds that are more hydrophobic will be mostly accumulated from dietary sources. This is generally true because the concentration of the compound is orders of magnitude higher in food and sediment that is ingested compared to the amount found in the water phase. Knowing the route of exposure is often important when assessing the likelihood of environmental contaminants to accumulate and cause adverse effects.

Lipid

Lipid normalization (expressing the toxicant concentration as a lipid value; e.g., μg PAH in tissue/g lipid) may be useful for comparing responses among species on a body-residue basis for hydrophobic toxicants; however, if the compound is not lipid soluble (e.g., $\log_{10} K_{ow} < 2$), then normalization may have little utility. The lipid content is important for many hydrophobic compounds because it often determines the rate of elimination, which is important for determining the bioaccumulation factor and time to a steady-state tissue concentration. Also, the more lipid an organism contains, the more a contaminant will bioaccumulate. Essentially all PAHs exhibit a log $K_{ow} > 2$, therefore this is an important parameter, especially for those species that exhibit weak or nonexistent biotransformation for these compounds. It is not known how important lipid content is in those species that metabolize PAHs. Because these species do not accumulate PAHs, the bioaccumulation factor is not a useful parameter and the amount accumulated has little to do with the whole-body lipid content. It is possible however that lipid is important for these species because as PAHs are taken up, there is likely some partitioning occurring within the body before they are metabolized. In general, lipid normalization may improve comparisons between species because storage lipid will likely reduce the amount of chemical that is circulating in plasma, which is related to the amount that will reach the site of toxic action (e.g., a specific enzyme or cell membrane).

Trophic Transfer

Food web transfer of PAH metabolites is another area that has received little attention. Because many species can metabolize PAHs, it is generally assumed that trophic transfer and biomagnification (an increase in tissue concentration over two or more trophic levels) of these compounds is not important for food webs. Trophic transfer and biomagnification of PAHs is not expected for food webs involving vertebrates; however, species from the lower trophic levels that are not able to effectively metabolize these compounds may exhibit food web transfer if vertebrates are not involved. Even though parent PAHs may not be biomagnified, prey species may contain high levels of metabolites that could be accumulated by predators. A few studies have found that fish accumulate PAH metabolites after ingesting invertebrates that had metabolized these compounds. Because these metabolites can be very toxic, additional research examining bioaccumulation of PAH metabolites from ingested prey is very important for a complete understanding of trophic transfer and the potential for toxic effects.

Photoactivation Toxicity

A number of studies over the last 20 years have demonstrated increased toxicity of PAHs when they are exposed to ultraviolet (UV) light. Some PAHs are known to absorb UV light, which alters the reactivity of the PAH and can render it more toxic. One of the main determinants of photoactivation is the HOMO–LUMO gap (highest occupied and lowest unoccupied molecular orbitals). PAHs that possess a gap in the range of 6.8–7.6 eV are known to be activated by UV light. Anthracene, benz[a]anthracene, fluoranthene, and benzo[b]fluoranthene are susceptible to UV activation, whereas this has not been observed for chrysene or dimethylphenanthrene. One critical feature of PAH phototoxicity is that these compounds must be bioaccumulated and retained in the tissue for photoactivation toxicity to occur. For many species, this does not occur, implying that photoactivation is only a minor concern. Few detailed studies have been conducted on those species that are known to bioaccumulate PAHs and are exposed to relatively high levels of UV radiation. The results from these and future studies become increasingly important as the protective ozone layer breaks down, allowing higher levels of UV radiation to reach the biosphere.

The environmental relevance of phototoxicity has been addressed recently. Some authors acknowledge the laboratory studies and mechanistic explanation for PAH phototoxicity; however, they conclude that several factors in the environment will protect species from the expected increase in tissue damage. They also point out that very few, if any, field studies have demonstrated adverse effects. Because UV light attenuates rapidly in water and soil, the most logical place to look for adverse effects would be in species that live aboveground or in shallow pools.

Environmental Stressors

Environmental factors such as pH, temperature, dissolved oxygen, salinity, and additional toxicants are always important considerations when characterizing bioaccumulation and toxicity. Most of these environmental variables will affect the rates of uptake and elimination (toxicokinetics) of a toxicant, which can greatly affect the amount accumulated and the resulting toxic response. Additionally, even when two individuals contain equal concentrations of a toxicant in their tissues, many of these environmental factors can affect the potency (toxicodynamics) of the compound by one of several actions, such as altering biochemical rates or changing membrane permeability. Additionally, these parameters can affect the organism's sensitivity. For example, if a species is expending additional energy to osmoregulate in an environment that is outside its normal range of salinity, the additional stress invoked by the toxicant may affect energy-generating pathways that will certainly exacerbate the organism's ability to maintain homeostasis.

Toxicity of PAHs to Biota

Known Mechanisms of Toxicity

Mechanisms of toxicity generally refer to the actual biochemical event, such as binding to a specific enzyme receptor or alteration of gene transcription. The mode of toxic action is a more general descriptor. For example, uncoupling of oxidative phosphorylation is considered a mode of toxic action that affects energy production by organisms. The particular mode of action can result from several different biochemical mechanisms. This distinction is important for determining when toxicants are additive. Toxic responses are a higher level of classification. For example, growth inhibition is a toxic response that may be caused by several different modes of action.

Many environmental toxicologists consider the non-specific mode of action (also known as baseline toxicity or narcosis) the primary toxic action for PAHs. Acute, or short-term, baseline toxicity occurs when whole-body tissue concentrations reach approximately $2-8 \,\mu mol \, g^{-1}$ ($\approx 400-1600 \,\mu g \, g^{-1}$ for PAHs), which is nonspecific and reversible. At these levels, the concentrations are high enough to affect cell membranes by disrupting their integrity, which usually leads to death if the source is not removed. Because of the nonspecific nature of this mode of toxic action, all PAHs are essentially equipotent, when considered as lipid-normalized tissue concentrations. Death from PAH exposure may occur by other modes of action such as impairment of the immune system that may lead to lethal infections. This is an indirect response that is likely caused by a specific mechanism of action, which requires a longer period of time to develop than the acute, baseline toxic response.

Sublethal responses generally occur by more specific mechanisms of toxic action; however, very few have been described. The best example is mutagenicity due to metabolites of some PAHs that make a covalent bond with DNA. This can occur at very low tissue concentrations. In many cases, the mechanisms of toxic response that are responsible for the observed effects are not known or are only partially understood. For example, we know that PAHs interact with the aryl hydrocarbon receptor (AhR), which appears to be important for toxicity. If the receptor is blocked, the response is often ameliorated. Many toxicants are known to be AhR agonists, including dioxins, furans, polychlorinated biphenyls (PCBs), and PAHs. A recent review provides a list of PAH potency factors for fish, which are indexed to the toxicity response for 2,3,7,8-tetrachlorodibenzo-p-dioxin (TCDD).

Knowing that PAHs interact with the AhR is just part of the story. Demonstrating activation of the AhR does not necessarily lead to adverse biological responses. Sometimes the fact that certain PAHs can affect a biochemical pathway is useful for understanding the potential response. For example, PAHs are known to affect the biosynthesis of several hormones related to reproduction. With this information, studies on reproductive effects due to PAH exposure can be designed and the results supported with plausible, mechanistic interactions. When considering the acute lethal response, all PAH congeners may be considered additive due to the nonspecific mode of action. For sublethal responses, such as growth impairment, immunotoxicity, and mutagenicity, only certain PAHs are likely to cause these effects. When specific mechanisms of toxic action cause adverse effects, only those PAHs that are known to elicit a given response would be considered additive. If toxicity is to be characterized for an adverse response caused by a specific mechanism of toxic action, the differential potency among the various PAH congeners in a mixture must be considered.

Common Biological Responses

Biological effects of exposure to PAHs are diverse. These compounds are known to affect the immune system, impair growth and reproduction, cause tumor formation, and of course lead to death if concentrations are high. Some of the known responses are discussed below. Toxicity values are generally highly variable among species and compounds. This is primarily a result of the inherent toxicity of compounds, species susceptibility, the length of time for exposure, and the uptake and elimination kinetics (toxicokinetics) found among species. A brief general description of the observed toxicity values are presented here, with additional detail for the most commonly reported responses. Most studies on PAHs examine responses to water concentrations, dietary input, or sediment/soil exposure. It appears that the majority of toxicity information on PAHs has been generated for aquatic invertebrates and fish; therefore, most of the specific information and examples are from these studies. Where appropriate, information from studies on terrestrial invertebrates, amphibians, reptiles, birds, and mammals are included, in addition to a general section, for some of these groups, that is presented below.

As mentioned above, hydrophobicity for PAHs ranges about 10 000 fold, which is also generally true for the LC_{50} toxicity values when water concentrations for a given species are considered. In general, uptake clearance (k_1; ml water $g^{-1} h^{-1}$) for water exposure of PAHs increases with increasing K_{ow} while the rate of elimination (k_2; h^{-1}) decreases with increasing K_{ow}. The bioconcentration factor (BCF) at steady state (a condition where uptake and elimination are balanced) can be determined by the equation k_1/k_2. In combination, these opposing trends will lead to large differences in the BCF and toxicity values over K_{ow}. Assessment of bioaccumulation and toxicity based on dietary uptake (sediment or prey) is more complicated and may not show the same range over K_{ow} as was shown for water exposure. The route of uptake generally has no effect on the rate of elimination (k_2), which is mostly a function of K_{ow} and the rate of biotransformation.

The rate of elimination (k_2) is a parameter that can provide important information about bioaccumulation and toxicity. The half-life for a toxicant in tissue and the time it takes the tissue concentration to reach steady state are both determined solely by k_2. The equation for half-life for any toxicant in tissue is determined with the equation $0.693/k_2$, which also equals the time it takes to reach 50% of steady state. The time to 'complete' steady state (actually $\approx 95\%$) is defined as $2.99/k_2$, which is characterized by a constant and maximum tissue concentration and constant rate of toxicity (e.g., unchanging LC_{50}).

The lethal response

In general, when aquatic organisms are exposed to waterborne PAHs, acute (short-term, e.g., $\leq 96\,h$) lethal values (e.g., the LC_{50}) occur in the $100-5000\,ng\,ml^{-1}$ range for many fish and invertebrates. The LC_{50} is a measure of a compound's toxicity that is based on the water or sediment/soil concentration at which half of the individuals die. Ideally, the LC_{50}, or any such measure of toxicity, should be determined at steady state (i.e., no change over time); however, this is infrequently done necessitating the designation of time (e.g., $96\,h\,LC_{50}$). As mentioned above, all PAHs exhibit the same lethal toxicity when expressed as a tissue concentration. Most organic compounds that occur at this tissue concentration cause lethality. Because of this commonality in tissue concentrations causing lethality, the observed variability in PAH toxicity based on water, sediment/soil, or prey concentrations as the exposure metric is solely due to differences in bioaccumulation potential and the organism's ability to metabolize these compounds.

The response concentration for tissue described above is designated as the lethal residue (tissue concentration) that causes mortality in 50% of the organisms (LR_{50}) and is expressed as $\mu g\,g^{-1}$ or $\mu mol\,g^{-1}$. The LR_{50} is considered the 'acquired' dose or the amount found in tissue that is correlated to the response. This is distinguished from the administered dose which is generally the amount of a toxicant given at one or several time periods. This is termed lethal dose (LD_{50}) and is expressed as $\mu g\,g^{-1}d^{-1}$ or just $\mu g\,g^{-1}$ (μg or μmol administered/g body weight (bw)). This is an important distinction because the LR_{50} is the final concentration associated with the response and the LD_{50} is the amount given, which can be very different compared to the actual tissue concentration causing the response. Once a compound is administered to an organism it may be metabolized, excreted, or depurated through passive diffusion. Expressing toxicity as the

LR_{50} is often more informative for interspecies comparisons and toxicity evaluation of organisms in the field because of the similarity in toxic potential and concentration among compounds. The LR_{50} is generally more common for invertebrates and much less so for vertebrates.

Most of the lethal toxicity data for vertebrates are based on administered dose (LD_{50}). In general, LD_{50} values are determined for mammals and wildlife such as birds, reptiles, and amphibians by a variety of methods such as dietary uptake, injection, and gavage (introduction by tube to the stomach). Only a few studies have established LD_{50} values for laboratory animals based on oral exposure to PAHs. Values for the LD_{50} range from $50\,\mu g/g\,bw$ for BaP to $2000\,\mu g/g\,bw$ for fluoranthene. Even fewer LR_{50} values are available because it is difficult to achieve a lethal tissue concentration in vertebrates from dietary uptake and only a few researchers measure tissue concentrations in experimental animals. For vertebrates in the field, mortality from PAH exposure is not likely to occur because of the high levels needed to achieve the lethal tissue concentration; however, mortality from secondary effects such as oiled fur in marine mammals, suffocation, tumor formation, or lung damage from inhalation is possible. Acute mortality is, however, very possible for invertebrates and those species with inefficient biotransformation of PAHs.

Sublethal responses

The lowest observed effect concentrations (LOECs; lowest treatment concentration that is statistically different from the control exposure) for PAHs frequently occur in the low part per billion (ppb; $ng\,ml^{-1}$) range for water concentrations and the low part per million (ppm; $\mu g\,g^{-1}$) or high ppb range for sediment/soil and dietary concentrations. LOECs are generally applied to sublethal toxicity tests; however, they can be used to express a low percentile for the lethality response. Several recent review articles provide summary data showing biochemical, histopathological, immunological, genetic, reproductive, and developmental effects due to PAH exposure in a variety of species exposed to individual PAHs and mixtures by several methods (e.g., water, sediment, dietary, and injection).

A number of review articles report a wide range in the administered dose ($0.04-3300\,\mu g/g\,bw$) causing adverse effects for lab animals and wildlife that are species and compound specific. Consequently, this variability precludes any generalizations that can be made about sublethal toxic responses to PAHs. Because PAHs are readily metabolized by most species, there are very few established tissue concentrations associated with sublethal responses that could be used to determine a threshold concentration for environmental protection.

Immunotoxicity

The potential of PAHs to cause impairment to the immune system has been demonstrated for many species. These studies generally examine specific (e.g., antibody response) and nonspecific (e.g., T-cells, phagocytic activity, secondary plaque-forming cells, and B-cell-mediated immunity) components of the immune system. Very few studies examine whole organism responses to immunological impairment. This would be accomplished in a controlled experiment by dosing a species with a range of PAH concentrations, then exposing them to a known pathogen to test for the *in vivo* response.

There are only a handful of studies examining immunocompromise in fish exposed to PAHs and even fewer for invertebrates. Some studies have documented immunological responses at very low concentrations of PAHs. For example, one study found alterations to biochemical and molecular processes related to the immune system in fish and noted an LOEC of $17 \, \text{ng ml}^{-1}$ for rainbow trout exposed to creosote in water. When only the PAH fraction was considered, which is likely the toxic component, the effect concentration was 0.61 ng of total PAH per ml. This LOEC is a concentration that is often observed in urban waterways.

Vertebrates in general appear to exhibit immunological responses at very low PAH exposure concentrations. One article reports an effective dose for a 50% response (ED_{50}) based on immunosuppression in deer mice in the range of 0.026–0.14 µg/g organism/day for three different PAH compounds (methylcholanthrene, dibenz[*a*]anthracene, and dimethylbenz[*a*]anthracene).

Only a handful of studies have examined immunotoxicity for invertebrates exposed to PAHs. One notable study found severe immunological alterations in oysters exposed to oil from a spill. The authors also suggested that environmental stress (e.g., altered temperature and salinity) may exacerbate the immunological response and should always be considered in such an assessment.

Growth and development

PAHs have been shown to inhibit growth in a number of species. Some studies on fish have reported growth inhibition or changes to physiological parameters associated with energy allocation at low exposure concentrations. Sublethal responses for fish have been found at doses occurring as low as 0.1–1.0 µg total PAH/g fish/day. Interestingly, recent research reports alterations to several growth-related functions in fish (e.g., lipid content and metabolic enzymes), indicating that PAHs can cause a starvation syndrome even though the rate of food ingestion is not affected.

A few studies have examined growth effects for invertebrates. Studies with an earthworm (*Eisenia veneta*) exposed to eight different PAHs reported that most of the EC_{10} values were around 30 µg/g dry weight of soil.

For comparison to other studies, such as those above for fish, the ingestion rate and uptake efficiency of the PAHs from the ingested soil would have to be known. Another study reported dramatic reductions in the response concentration (several thousand-fold) when growth was assessed in juvenile bivalves (*Mulinia lateralis*) exposed to individual PAHs in combination with UV radiation (as compared to the response to PAHs without UV). In that study, only minor effects were observed when exposed to various petroleum products, indicating that these complex mixtures do not contain sufficient levels of the most phototoxic PAHs.

The effects on development due to PAH exposure have been studied in several species (mostly vertebrates), often reporting adverse responses at very low exposure concentrations. For example, one study reported several adverse effects (skeletal abnormalities, yolk sac edema, pericardial edema, genetic damage, and behavioral effects) in Pacific herring (*Clupea pallasi*) when eggs were exposed to crude oil for 16 days. These responses were observed at concentrations as low as $0.4 \, \text{ng ml}^{-1}$ for 'weathered' crude oil, which is a natural process that results in the more volatile PAHs being lost, increasing the percentage of HPAHs and alkylated homologs. Recent work with zebrafish (*Danio rerio*) demonstrated that only some PAHs were able to induce a suite of developmental abnormalities that were triggered by a direct effect on cardiac function. The authors noted these effects for the PAHs dibenzothiophene and phenanthrene, but not for pyrene, which induced a different suite of adverse effects similar to those acting through the AhR.

Endocrine disruption and reproductive effects

Several recent reviews provide mounting evidence that PAHs can act as endocrine disruptors, alter reproductive hormones, and cause adverse effects in various reproductive functions. These compounds are likely to act by several different toxic mechanisms that would result in altered reproduction. One of the most important modes is their role as endocrine disruptors. As noted in one of the review articles, there are several studies demonstrating endocrine and reproductive effects in small mammals and humans from PAH exposure (mostly for BaP). The responses include adverse effects on thymic glucocorticoid receptors, placental function, and oocyte and follicle integrity. Also, most rapidly proliferating cells are likely to be susceptible to PAHs. It is likely that these responses to PAH exposure are dependent on the life stage of the species that is exposed. Based on the known toxicological effects for PAHs, the early stages of life (development) are more susceptible to PAH exposure than other stages, such as juveniles and adults.

There are many reports of adverse effects on reproduction in a number of soil invertebrates that were exposed to several PAH compounds. For example, several PAHs elicited effects at concentrations ranging from 5 to 100 μg/g soil, with variability among species, compounds, endpoints, soil characteristics, and exposure time. A number of field studies have shown reproductive effects in flatfish due to PAH exposure. Even though field studies do not often result in definitive conclusions about specific contaminants, a few provide convincing cause and effect relationships that support PAHs as the causative agents.

Mutagenicity and neoplasia

There is overwhelming evidence that certain PAHs are well-known mutagens that can cause genetic damage, toxicopathic lesions, and tumor formation. Most of the studies describe the mutagenic properties of PAHs in mammals, which has been directed at human health. Several review articles describe many of the relevant studies for ecotoxicologists concerning PAH-induced abnormalities.

Potential mutagens are those PAHs containing four to six rings (HPAHs). Additionally, the alkyl moiety often increases the mutagenic potential. Based on the diverse sources of PAHs in the environment and urban areas, there is ample opportunity for both humans and other species to be exposed. Much of the mammalian literature is directly applicable to wildlife, including most vertebrates. Because metabolic activation of PAHs is necessary for the expression of mutagenic properties, well-developed biotransformation capabilities are needed. For that reason, many invertebrate species do not exhibit alterations to DNA from exposure to PAHs.

Several field studies have observed histopathological changes and tumor formation in fish associated with sites contaminated with PAHs. Also, one recent study examined cancer in beluga whales and concluded that PAHs were the likely cause. As noted by one author, several factors relating to PAH exposure and toxicity, such as specificity of response, dose–response relationships, experimental evidence, and others, strongly support the conclusion of PAH-induced neoplasia for fish from contaminated sites. This is one of the few examples where a specific (or group of) contaminant(s) can be associated with adverse effects in organisms collected in the field, which is likely applicable for other species.

General Toxicity Information

Most of the information above on toxic responses are described with examples using invertebrates, fish, and mammals, which reflects the volume of toxicity information for PAHs. This section provides additional toxicity information for those groups (plants, invertebrates, birds, reptiles, and amphibians) that were briefly covered above.

Plants

The literature concerning toxicity of PAHs indicates that these compounds are toxic to plants only at very high concentrations and there are several examples of plants accumulating high concentrations. Photoactivation of PAHs is likely to be an important process for plants due to their high exposure to UV radiation and high tissue concentrations. One review cites a number of studies that demonstrate a substantial increase in toxicity to plants when photoactivation is considered. Several bioassays have been developed that are based on physiological and biochemical processes, including chlorophyll (Chl *a*) fluorescence, which appears to be a very useful bioindicator of PAH toxicity because it has been linked to altered growth. As an example, one study reported aqueous EC_{50} values for several plant functions in the low μg ml^{-1} range when aquatic macrophytes where exposed to creosote, a PAH-rich mixture. It should be noted that these values are relatively high compared to the EC_{50} values observed for many faunal species.

A substantial body of literature concerns the remediation of soils and aquatic sediments by plants. Several plant species appear to be effective for removing these contaminants from areas where concentrations are likely toxic for some faunal species but not so for plants. One facet of this research that has received little attention is the increase in availability of soil or sediment PAHs to animals. When certain plant species are introduced to an area for PAH remediation, an increase in animal abundance, especially wildlife, may follow. By accumulating and concentrating PAHs in their tissues, especially the very toxic photoactivated PAHs, animals that graze on these plants may be acquiring a higher dose of these compounds than they normally would in areas not densely populated by these plants.

Invertebrates
Terrestrial invertebrates

Several studies show high variability in toxic response among species and PAH compounds. For example, the soil LC_{50} for many of the LPAHs were in the range of 40–167 μg/g soil for the springtail *Folsomia fimetaria*; however, for some of the more hydrophobic HPAHs (e.g., BaP and benz[*a*]anthracene), the 21 day LC_{50} was greater than 800–1000 μg g^{-1}. At these concentrations, the HPAHs should be toxic, which implies a reduced bioaccumulation for these compounds. As is the case for many species, sublethal effect concentrations occur at levels usually 10–100 times lower than those causing mortality. Sublethal effects, such as reproductive and growth abnormalities, exhibit effect concentrations in the low ppm range (5–100 μg PAH/g of soil).

Aquatic invertebrates

As mentioned above, when exposed to aqueous solutions of individual PAHs, the range in lethal toxicity can be up to 4 orders of magnitude for a given species over all congeners. This is best shown for aquatic invertebrates (and fish) because of the large amount of mortality data that has been generated. While the concentrations for sublethal responses will be lower, it is not known if this same pattern of variability will hold. For many sublethal responses, specific mechanisms of toxicity are responsible for the effect which may result in differential toxicity among the various PAHs.

A large effort has been applied to determining the concentrations of contaminants in bedded sediment that are toxic to aquatic invertebrates. These are known as sediment quality guidelines, which have been developed for a large number of compounds including individual and total PAHs. A number of methods and approaches have been used and most of these indicate adverse effects to invertebrates at concentrations ranging from 2 to 4 μg total PAH/g dry sediment. These values are intended to be protective of most adverse effects.

Vertebrates

Birds

Toxic effects in adult birds appear to be quite variable, which may be a function of dosing scheme, PAH profile, biotransformation rate, lipid content, receptor affinity, or potency differences. For example, one study dosed adult mallards with a diet containing 0.4% PAHs (4000 μg g^{-1}) for 7 months and found only minor effects. Even though the mixture of PAHs in this diet was composed mostly of LPAHs, these concentrations are extremely high, which may be a testament to the biotransformation ability for adults of this species. Another study with blackbirds and sparrows found LD$_{50}$ values in the 100–200 μg g^{-1} range for various PAH compounds. It is unclear if the method of administration for this study (gavage) would produce a different result compared to that obtained for the mallard that was exposed via dietary ingestion.

PAHs appear to be very toxic to bird embryos, and the mallard has been shown to be one of the more sensitive species. Severe adverse effects have been observed when PAHs were added externally to a small area on mallard eggs. Increased mortality, decreased embryonic body weight, and increased abnormal survivors have been observed at several doses of chrysene, BaP, and 1,2-dimethylbenz[*a*]anthracene. Measured concentrations of PAHs in eggs indicate that effects were occurring at concentrations as low as 0.005–0.04 μg g^{-1}. Of course, measured concentrations should be adjusted for metabolism, especially if the tissue is sampled several days after the exposure. Bird eggs have been shown to be very efficient at metabolizing PAHs; one study showed

that chicken eggs could metabolize more than 90% of the accumulated PAHs after 2 weeks. Once researchers determine the concentrations in eggs known to cause effects, comparison with the large database of measured values in field-collected animals will be very informative for determining risk for adverse effects.

Reptiles

Very few studies have been conducted on the responses of reptiles to PAH exposure, which is confirmed by several review articles. It appears that many of the studies that have been conducted examined the responses of marine reptiles (snakes and turtles) to petroleum exposure. These studies generally describe the direct effects of oil on these species; however, one study did expose juvenile loggerhead turtles to weathered crude oil (which is enriched with HPAHs and alkylated homologs). These authors noted several abnormalities including a fourfold increase in white blood cell count, a reduction in red blood cells, abnormal glucose levels, and numerous inflammatory and histologic responses in various tissues. Biotransformation rates of PAHs for reptiles (and amphibians) are generally lower than what is measured in mammals. Consequently, many of these species are less likely to form tumors because of the reduced potential for conversion to reactive metabolites. Conversely, because reptiles (and amphibians) do not biotransform PAHs as rapidly as other species, they are susceptible to other adverse effects, because tissue concentrations can accumulate to much higher levels than that expected for other vertebrates.

Amphibians

There are very few studies on the reactions of amphibians to PAHs. One study concluded that amphibians were more resistant to PAH-induced carcinogenesis compared to mammals. Phototoxicity may be an important effect for amphibians, especially larval forms, because they can be exposed to high concentrations of PAHs and live in shallow pools. One research study exposed the larvae of three species (two frogs and one salamander) to fluoranthene and then subjected them to UV light. This study found a greater sensitivity to fluoranthene in the presence of natural UV light (outdoors). For two of the species, the time to death was reduced by approximately 20-fold (*Ambystoma maculatum*) and 70-fold (*Xenopus laevis*) over that observed for the same PAH exposure without UV radiation. These are striking differences that are supported by other work demonstrating a dose–response relationship between the median time to death and exposure (=UVA intensity times tissue concentration) in larval *Rana pipiens*.

Summary and Conclusions

Assessing the toxic potential for PAHs is a complicated process. Species exhibit wide variation in susceptibility and response. Currently, the most appropriate method for protecting against adverse effects from toxicant exposure is to test many species and construct species-sensitivity distributions (SSDs), which are basically empirical cumulative frequency distributions. By testing a number of species from a variety of taxa, the distribution of responses can be examined and an appropriate level of protection (e.g., 5th percentile) can be set.

PAH contamination is a problem worldwide, occurring almost universally in urban areas and occasionally in non-urban areas due to oil spills, forest fires, and atmospheric fallout. Surprisingly, for such a common and abundant class of contaminants, there is a paucity of toxicity information for species exposed to PAHs. Considering that very low concentrations can lead to serious adverse effects, many populations are simply not protected. Surprisingly, there are few environmental statutes that contain regulatory values for protecting ecosystems from PAH exposure. Within the United States, ambient water quality criteria are promulgated by the US EPA under the Clean Water Act to protect aquatic life. Under this statute, several criteria values have been developed for various toxicants; however, none are currently in place for any of the PAHs. Canada has issued interim aquatic life guidelines for several PAH compounds in freshwater and many have been set at very low concentrations (0.012–0.04 ppb). The Netherlands also has guidelines for several PAH compounds in freshwater that are very low (most between 0.001 and 0.05 ppb).

In the US, the EPA is actively working to develop sediment criteria for PAHs, and at least one state, Washington, has codified sediment guidelines. Canada has interim sediment guidelines for several individual PAHs, many set at relatively low values (6.2–111 ng/g dry sediment). In most cases, if ecotoxicologists want to assess the toxic potential for PAH contamination in a given system, they must turn to the sparse literature that is available. Of course, most sites contain multiple toxicants that can interact and cause effects at lower than expected concentrations. Only after significant progress is made in understanding how individual PAHs act on myriad biological systems can we begin to tackle the problem of complex mixtures containing PAHs and other toxicants.

See also: Bioaccumulation; Body Residues; Crude Oil, Oil, Gasoline and Petrol; Dose–Response; Ecological Risk

Assessment; Ecotoxicology Nomenclature; Ecotoxicology: The Focal Topics; Ecotoxicology: The History and Present Directions; Reproductive Toxicity.

Further Reading

Agency for Toxic Substances and Disease Registry (1995) *Toxicological profile for polycyclic aromatic hydrocarbons*. Atlanta, GA: US Department of Health and Humans Services, Public Health Service.

Di Toro D M, McGrath JA, and Hansen DJ (2000) Technical basis for narcotic chemicals and polycyclic aromatic hydrocarbon criteria. Part I: Water and tissue. *Environmental Toxicology and Chemistry* 19: 1951–1970.

Douben PET (ed.) (2003) *PAHs: An Ecotoxicological Perspective*. London: Wiley.

Eisler R (2000) Polycyclic aromatic hydrocarbons. *Handbook of Chemical Risk Assessment: Health Hazards to Humans, Plants, and Animals, Vol. 2: Organics*, pp. 1343–1411. Boca Raton, FL: Lewis Publishers.

Fairbrother A, Smits J, and Grasman KA (2004) Avian immunotoxicology. *Journal of Toxicology and Environmental Health* 7B: 105–137.

Galloway TS and Depledge MH (2001) Immunotoxicity in invertebrates: Measurement and ecotoxicological relevance. *Ecotoxicology* 10: 5–23.

Geschwind SA, Eiseman E, Spektor D, and Hudson A (1999) *The Impact of Endocrine Disrupting Chemicals on Wildlife: A Review of the Literature 1985–1998*. Washington, DC: Science and Technology Policy Institute, http://www.rand.org/pubs/monograph_reports/2005/MR1050.0.pdf (accessed Oct. 2007).

Irwin RJ, Van Mouwerik M, Stevens L, Seese MD, and Basham W (1997) *PAHs. Environmental Contaminants Encyclopedia*. Fort Collins, CO: National Park Service, Water Resources Division, http://www.nature.nps.gov/hazardssafety/toxic/index.cfm (accessed Oct. 2007).

Mahler BJ, van Metre PC, Bashara TJ, Wilson JT, and Johns DA (2005) Parking lot sealcoat: An unrecognized source of urban polycyclic aromatic hydrocarbons. *Environmental Science and Technology* 39: 5560–5566.

Matthiessen P (2003) Historical perspective on endocrine disruption in wildlife. *Pure and Applied Chemistry* 75: 2197–2206.

McDonald BG and Chapman PM (2002) PAH phototoxicity – An ecologically irrelevant phenomenon? *Marine Pollution Bulletin* 44: 1321–1326.

Meador JP, Stein JE, Reichert WL, and Varanasi U (1995) A review of bioaccumulation of polycyclic aromatic hydrocarbons by marine organisms. *Reviews of Environmental Contamination and Toxicology* 143: 79–165.

Myers MS, Johnson LL, and Collier TK (2003) Establishing the causal relationship between polycyclic aromatic hydrocarbon (PAH) exposure and hepatic neoplasms and neoplasia-related liver lesions in English sole (*Pleuronectes vetulus*). *Human and Ecological Risk Assessment* 9: 67–94.

Pauli BD, Perrault JA, and Money SL (2000) RATL: A database of reptile and amphibian toxicology literature. *Technical Report 357, National Wildlife Research Center, Canadian Wildlife Service*. Hull, QC: Environment Canada.

van Metre PC, Mahler BJ, and Furlong ET (2000) Urban sprawl leaves its PAH signature. *Environmental Science and Technology* 34: 4064–4070.

Varanasi U (ed.) (1989) *Metabolism of Polycyclic Aromatic Hydrocarbons in the Aquatic Environment*. Boca Raton, FL: CRC Press.

Radioactivity

D B Chambers, H Phillips, S Fernandes, and A Garva, SENES Consultants Limited, Richmond Hill, ON, Canada

Introduction

Ionizing radiation is ubiquitous and all living things are, and always have been, exposed to naturally occurring radiation and radioactivity. However, people's activities have added to the levels of radiation and radioactivity globally through fallout from aboveground nuclear (weapons) testing and locally through activities such as mining, production of phosphate fertilizers, oil and gas off-shore platforms, and nuclear fuel cycle activities among others.

Until quite recently, the prevailing view has been that, if humans were adequately protected, then "other living things are also likely to be sufficiently protected" (International Commission on Radiological Protection, 1977) or "other species are not put at risk" (International Commission on Radiological Protection, 1991). Consistent with this view, a great deal of effort has been made over the past 50 years or so to study the behavior of radioactivity in the environment, particularly as it relates to potential pathways of exposure to humans. However, by the 1990s, attempts were made to look in general at the effects of radiation on plants and animals at levels implied by the radiation protection standards for humans. Events such as the United Nations Conference on Environment and Development in Rio de Janeiro, Brazil, in 1992 and the Convention on Biodiversity gave impetus to environmental protection in general and stimulated international and national agencies to become more active in radioecotoxicology.

The 1996 report of the United Nations Scientific Committee on the Effects of Atomic Radiation (UNSCEAR) was of seminal importance as it was the first time that the UNSCEAR Committee had examined the effects of ionizing radiation on nonhuman biota. Until that time, UNSCEAR, along with other international and national organizations, had considered living organisms primarily as part of the human food chain.

Since 1996, there has been increasing attention given to the potential effects of ionizing radiation on the environment; this increasing interest since 1996 is well illustrated by the numerous activities of national and international organizations such as the Canadian Nuclear Safety Commission, the European Community, the US Department of Energy, the United Kingdom Environmental Agency, and the International Atomic Energy Agency, which have undertaken initiatives to investigate the effects of ionizing radiation on the environment.

The following sections build on the dialog that has taken place over the last 11 years and provides an overview of the current generic approach to assessing risks to nonhuman biota from ionizing radiation and radioactivity in the environment, including discussion of selected issues unique to ionizing radiation.

General Approach to ERA for Radiation and Radioactivity

In general terms, the current approach to assessing potential risks to nonhuman biota from radiation and radioactivity in the environment involves:

- estimation of the levels of radioactivity in the biota arising from levels of radioactivity in the environment;
- estimation of the consequent dose or dose rate to biota arising from both the internally deposited radioactivity and the external radiation from the environment in which it lives; and
- comparison of the calculated dose (dose rate) to a reference dose (dose rate).

To facilitate the implementation of this approach, five key elements are defined and are summarized in **Table 1**. Further discussion on each of these aspects is provided in the following sections.

Table 1 Simplified framework for assessing risks to plants and animals

Conceptual site model	Define pathways of exposure
	Identify possible exposed biota and characterize reference biota or indicator species
	Consider individual biota when species are rare or endangered
Exposure	Estimate levels of radiation and radioactivity in the environment by use of measured data and/or models
	Determine spatial and temporal patterns of radiation and radioactivity
	Determine uptake by organism
Dosimetry	Estimate absorbed dose (by whole body or by tissue/organ)
	Perform geometry corrections for size of organism
	Utilize radiation weighting factors for internally deposited radionuclides
Reference dose rate	Define appropriate population-level effects such as mortality or reproductive capacity; and corresponding reference doses
	Establish corresponding reference dose rates below which effects on populations of biota are unlikely, in many cases a value of $10\,mGy\,d^{-1}$ is appropriate
Assessing effects	Compare estimated dose rate to reference dose rate and develop a screening index value which takes into account:
	spatial and temporal aspects of exposure
	natural population variability
	background radiation levels
	Determine the possibility for effects in biota

Conceptual Site Model

Figure 1 provides a schematic illustrating the various ways in which biota can be exposed to radiation or radioactivity in the environment. There is transfer between abiotic environmental media; in the simplified representation provided in **Figure 1**, this is shown by radioactivity present in the air and in rain being transferred into the aquatic environment. The aquatic environment has different components: in the case of the freshwater environment this includes streams, rivers, lakes, and sediments; and in the marine environment includes tidal zones, coastal waters, and marine sediments. Radioactivity can then be transferred into biota such as plants, phytoplankton, zooplankton, macro-invertebrates, sessile aquatic plants, fish, water-based amphibians, crustaceans, mammals, and birds which obtain dietary components from the aquatic (freshwater/marine) environment.

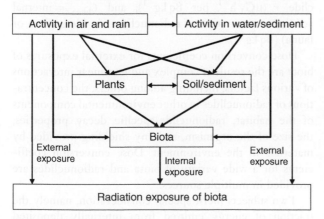

Figure 1 Simplified schematic illustrating pathways of exposure for nonhuman biota.

Radioactivity can also be transferred into the terrestrial environment which includes soils which can then be taken up into terrestrial biota such as plants, invertebrates, and vertebrates (mammals, birds, reptiles, and land-based amphibians).

It is important to note that the terrestrial and aquatic environments are not totally separate since some terrestrial biota obtain food or drinking water from the aquatic environment. For example, bears eat fish and moose and waterfowl feed on aquatic plants. Therefore, care is needed in defining the ways in which such animals may be exposed within the conceptual model.

As discussed above, there is a vast array of biota that can be considered in the development of the conceptual model. However, it should be recognized that it is not possible to develop information on all biota; therefore, the concept of a set of reference organisms that are representative of common ecosystems has been developed to focus the effort on methods and data needed to assess these reference organisms. The results of such dose assessments for predefined reference organisms facilitate a basic assessment of possible biological effects. This approach provides a strategy that allows assessment efforts to be focused and thereby reduced to a more manageable size.

The International Commission on Radiation Protection (ICRP) defines "a Reference Animal or Plant (RAP) as a hypothetical entity, with the assumed basic characteristics of a specific type of animal or plant, as described to the generality of the taxonomic level of family, with precisely defined anatomical, physiological, and life-history properties that can be used for the purposes of relating exposure to dose, and dose to effects, for that type of living organism" and describes the reference animals and plants in groups (family or taxonomic level). The reference organisms selected by the ICRP cover a range of global ecosystems and taxonomic families.

Therefore, the first step in the environmental risk assessment (ERA) process involves developing a conceptual model of the study area, an understanding of the sources and routes of exposure, and, based on this understanding, the selection of representative organisms for assessment purposes. It is important to understand that biota considered as reference organisms in a particular ERA need to be representative of the particular location and, therefore, the reference organisms used will vary from assessment to assessment. In selecting reference organisms for the assessment, consideration should be given to the value of the biota to the local ecosystem. For example, special consideration is generally given to rare or endangered species as these need to be protected.

Exposure

Biota may be exposed to ionizing radiation from the environment from both external and internal exposure from radionuclides taken into the organism. For example, terrestrial organisms are exposed externally to radiation from the nearby soil, and aquatic organisms are exposed externally from the radioactivity in the water. Bottom dwellers or benthic organisms are exposed externally by the radioactivity in sediment and both terrestrial and aquatic organisms are exposed internally by radionuclides taken up by the organism. Therefore, in evaluating exposure it is necessary to obtain or estimate the levels of radioactivity in the abiotic and biotic components of the ecosystem. In some cases, there may be measured data available; however, often models are necessary to provide an estimate of radioactivity in various compartments.

As indicated earlier, a great deal of work has been carried out to estimate the concentrations of radionuclides in biota that are part of the human food chain. In the absence of actual measurements of radionuclide levels in biota, models developed to estimate the levels of radioactivity in biota for use in assessing dose to humans are also useful in assessing exposure to plants and animals. The work of the International Atomic Energy Agency (IAEA) has been particularly helpful in summarizing methods and data from around the world. Although dynamic models are needed to evaluate the effects of transient concentrations in the environment, equilibrium models, where the radionuclides are assumed to reach equilibrium with each of the environmental compartments, are widely used for assessing risks to nonhuman biota.

Another aspect in determining exposure is that radiation levels can vary with distance (spatially) and with time (temporally). These two considerations are important in the determination of exposure.

Dosimetry

Dosimetry models used to assess the potential adverse effect of environmental radioactivity on biota and to demonstrate compliance with biota radiation dose standards are based on radioactivity concentrations in biota and in environmental media (air, water, soil, sediment) in direct contact with biota. Models for biota are not as well developed as compared to those for humans; this is in part due to the almost infinite diversity of species of nonhuman biota (anatomical, morphological, and physiological aspects). In practice, simplifying assumptions have to be made in the models to approximate the ecological system.

Models are needed to convert exposure from both external and internal radiation to absorbed dose. Factors important for external dosimetry include geometrical relations between the source of the radiation and the biota, the size of the biota, and characteristics of the radionuclides (e.g., radioactive decay characteristics). Factors important to estimating dose from internal radionuclides include chemical species and the possibility of a nonuniform distribution of radionuclides within the organism, the fraction of energy emitted from internally deposited radionuclides that is absorbed in the biota (a function of the size of the biota and the energy of the radiation(s)), and the relative biological effectiveness (RBE) of different kinds of radiation.

The radiation dose received by the biota (or some organ or tissue of the biota) is the sum of both external and internal exposures. In general terms, the dose rate D (μGy h^{-1}) can be calculated according to

$$D = \sum_r \left[\mathrm{DCC}_{\mathrm{ext},r} \cdot C_{\mathrm{soil,sediment,water},r} + \mathrm{DCC}_{\mathrm{int},r} \cdot C_{\mathrm{biota},r} \right] \quad [1]$$

where $\mathrm{DCC}_{\mathrm{ext},r} =$ dose conversion coefficient for external exposure to radionuclide r (μGy h^{-1} per Bq kg^{-1}); $C_{\mathrm{soil,sediment,water},r} =$ activity concentration of radionuclide r in soil, sediment, or water (Bq kg^{-1}); $\mathrm{DCC}_{\mathrm{int},r} =$ dose conversion coefficient for internal exposure to radionuclide r (μGy h^{-1} per Bq kg^{-1}); and $C_{\mathrm{biota},r} =$ internal activity concentration of radionuclide r in biota (flora or fauna) (Bq kg^{-1}).

Dose conversion coefficients for external exposures of biota are the result of complex and nonlinear interactions of various factors, including among others, the concentrations of radionuclides in other environmental components of the habitat, radionuclide-specific decay properties, the size of the organism, and any shielding provided by materials in the environment. Dose conversion coefficients for a wide variety of biota and radionuclides are provided in multiple sources.

Two subjects merit special consideration, namely the fraction of energy emitted from internally deposited radionuclides that is retained in the organism (absorbed fraction) and the weighting factor that should be applied to account for RBE of the various kinds of internally deposited radionuclides.

Absorbed Fraction

For small organisms, some of the energy emitted from internally deposited radionuclides will escape from the organism and thus the absorbed fraction will in general be less than unity. For alpha radiation, the absorbed fraction is normally assumed to be unity unless the dimensions of the organism are very small, for example, phytoplankton, zooplankton, or fish eggs.

The National Council on Radiation Protection and Measurements (NCRP) reports absorbed fractions for beta radiation and gamma radiation for several reference biota species calculated using the point source dose distribution method. The absorbed fractions for beta radiation are very close to unity except for small organisms such as insects and larvae and for beta (max) energies greater than 1 MeV. However, for gamma radiation, the absorbed fractions are less than 0.1 for the species modeled at gamma energies above 0.1 MeV.

A simple approach that is sometimes used for screening purposes is to assume that all the energies emitted by the radionuclide are absorbed within the organism's tissue or organ under consideration, that is, assume the absorbed fractions for all radiation (alpha, electrons, gamma) to be unity. While this approach is reasonable for a conservative screen for beta radiation, it leads to a significant overestimate for gamma radiation. For radionuclides that emit high-energy gamma such as ^{60}Co, the internal dose calculated using this approach can be overestimated by an order of magnitude or more if this assumption is used.

Weighting Factor for Internally Deposited Radionuclides

For internally deposited radionuclides, potential effects depend not only on the absorbed dose, but also on the type or 'quality' of the radiation. This issue is addressed through consideration of the RBE of the various radiations relative to a reference radiation. The reference radiation is generally taken to be 250 kVp X-rays or gamma rays from ^{137}Cs or ^{60}Co. However, it is important to understand that the choice of reference radiation will affect the RBE. For example, it has been shown that at low doses, X-rays are about twice as effective as gamma rays in producing damage, and hence an RBE established by comparison to gamma rays will be about twice than if X-rays are used as the reference radiation.

The concept of RBE can be defined as

$$RBE = \frac{\text{dose of the reference radiation needed to produce the same effect}}{\text{dose of the given radiation needed to produce a given biological effect}}$$

A number of authors have reported evaluations of published data on RBE for internally deposited alpha emitters and nominal values are reported in **Table 2**. In considering these values, it is important to note that the experimental RBEs are specific to the endpoint studied, the biological environment, exposure conditions (e.g., reference radiation, dose rate, dose, etc.), and other factors; therefore, there is some uncertainty associated with the selection of an appropriate radiation weighting factor for use in an ERA.

A nominal alpha radiation weighting factor of 5 for population-relevant deterministic and stochastic endpoints has been suggested based on literature reviews for use in ERAs; however, to reflect the limitations in the experimental data, uncertainty ranges of 1–10 and 1–20 for population-relevant deterministic and stochastic endpoints, respectively, can be applied.

The European Community Framework for Assessment of Environmental Impact (FASSET) program suggests to modify the absorbed dose by a radiation weighting factor of 10 for alpha radiation, 3 for low energy beta radiation ($E < 10$ keV), and 1 for both beta radiation with energies > 10 keV and gamma radiation in order to illustrate the effect of radiation quality of emissions from internally deposited radionuclides.

Although somewhat contentious, for present purposes, an RBE value of 10 is suggested for internally deposited alpha emitters. For gamma emitters and internally deposited beta emitters, noting that deterministic effects (e.g., reproductive capacity) are thought to be of greatest significance, a value of 1 is suggested as a nominal RBE, although a factor of 3 may be appropriate for tritium.

Reference Dose Rates

The ultimate goal of ecological protection is to ensure that communities and populations of organisms can thrive and all the component parts will be self-sustaining. This simple principle requires a shift of focus, typical of most chemical criteria, from individual organism protection to community protection. Focus on individuals, usually the most sensitive individual, is a natural product of controlled laboratory testing. While laboratory data have been essential to developing environmental criteria based on the sensitivity of species to single or multiple stressors, they do not ensure ecological protection. On the other hand, the ICRP indicates that effects on ecosystems are usually observed at the population or higher levels of organization but information on dose responses to radiation is usually obtained experimentally at the level of individual organisms.

The effects of ionizing radiation on biota have been reviewed several times by national and international authorities. Critical radiation effects for natural biota are those that directly affect reproductive success, via significant

Table 2 Radiation weighting factors for internal alpha radiation for deterministic effects in nonhuman biota (relative to low-LET radiation)

Nominal value	Comment
1[a]	Built-in conservatism in dose model
20[b]	Keep same as for humans
2–10[c]	Nonstochastic effect of neutrons and heavy ions
5[d]	Average for deterministic effects
10[e]	Deterministic population-relevant endpoints
20[f]	Likely to be conservative for deterministic effects
40[g]	Includes studies with high RBEs
5–20 (10)[h]	5–10 Deterministic effects (cell-killing, reproductive)
	10–20 Cancer, chromosome abnormalities
	10 Nominal central value
5–50 (10)[i]	10 To illustrate effect of alpha RBE

[a]United States National Council on Radiation Protection and Measurements (1991) Effectys of ionizing radiation on aquatic organisms. *NCRP Report No. 109.*

[b]International Atomic Energy Agency (1992) Effects of ionizing radiation on plants and animals at levels implied by current radiation protection standards. *Technical Reports Series No. 332.* Vienna: IAEA.

[c]Barendsen GW (1992) RBE for non-stochastic effects. *Advances in Space Research* 12(2–3): 385–392.

[d]United Nations Scientific Committee on the Effects of Atomic Radiation (UNSCEAR) (1996) *Effects of Radiation on the Environment.* Annex to Sources and Effects of Ionizing Radiation (Report to the General Assembly, with one Annex), Scientific Committee on the Effects of Atomic Radiation. New York: United Nations.

[e]Trivedi A and Gentner NE (2002) Ecodosimetry weighting factor for non-human biota. In: IRPA-10. *Proceedings of the International Radiation Protection Association,* Japan, 14–19 May 2000 (available on CD-ROM).

[f]Copplestone D, Bielby S, Jones SR, *et al.* (2001) *R&D Publication 128: Impact Assessment of Ionising Radiation on Wildlife.* Bristol: United Kingdom Environment Agency.

[g]Environment Canada (2000) Priority substances list assessment report (PSL2). *Releases of Radionuclides from Nuclear Facilities (Impact on Non-Human Biota).* Ottawa: Environment Canada and Health Canada.

[h]Advisory Committee on Radiological Protection (2002) *Protection of Non-Human Biota from Ionizing Radiation.* Canadian Nuclear Safety Commission (CNSC), INFO-0703, March.

[i]Framework for Assessment of Environment Impact (FASSET) (2003) Deliverable 3: Dosimetric models and data for assessing radiation exposures to biota (Pröhl G, ed.).

impairment of gametogenesis or embryonic development and survival, for example. Measurable responses to radiation exposure (e.g., biochemical changes, histological changes in kidney tubules) can occur at exposure levels well below those that actually impair reproduction or survival at any life stage. Such changes are usually regarded as 'biomarkers' of exposure, and in general are considered as poor endpoints for ecological risk assessments.

UNSCEAR has noted that the sensitivity of an organism to radiation depends on the life stage at exposure and that embryos and juvenile forms are more sensitive than adults. Overall, UNSCEAR concluded that "the available data indicate that the production of viable offspring through gametogenesis and reproduction is a more radiosensitive population attribute than the induction of individual mortality."

Based on a number of literature studies, UNSCEAR concluded that:

- Exposure to chronic dose rates less than $400 \, \mu Gy \, h^{-1}$ ($10 \, mGy \, d^{-1}$) would have "effects, although slight, in sensitive plants but would be unlikely to have significant deleterious effects in the wider range of plants present in natural plant community".
- Chronic dose rates below $400 \, \mu Gy \, h^{-1}$ are unlikely to cause adverse effects in most sensitive animal species.

- Maximum dose rates of $400 \, \mu Gy \, h^{-1}$ to a small proportion of the individuals were unlikely to "have any detrimental effects at the population level" in aquatic organisms.

With the exception of observations from the Chernobyl accident, few new data on effects of exposure to ionizing radiation are available since 1996. Much of the information on radiation levels and effects on biota observed in the region around the Chernobyl nuclear reactor has been reported by the Chernobyl forum. The main observations from the followup studies at Chernobyl can be summarized as follows:

- Irradiation from radionuclides released from the Chernobyl accident caused numerous acute adverse effects in the biota up to distances of tens of kilometers from the release point.
- The radio-ecotoxicological environmental response to the Chernobyl accident involved a complex interaction among radiation dose, dose rate, and its temporal and spatial variations, as well as the radiosensitivities of the different taxons. Both individual and population effects caused by radiation-induced cell death were observed in plants and animals as follows:

1. increased mortality of coniferous plants, soil invertebrates, and mammals;

2. reproductive losses in plants and animals;
3. chronic radiation sickness of animals (mammals, birds, etc.).

• No adverse radiation-induced effect was reported in plants and animals exposed to a cumulative dose of less than 0.3 Gy during the first month after the accident.

• Following the natural reduction of exposure levels due to radionuclide decay and migration, populations have been recovering as a result of the combined effects of reproduction and immigration.

In general terms, these observations from the Chernobyl accident appear to support the 1996 conclusions of UNSCEAR and suggest that dose rates below about $400\,\mu Gy\,h^{-1}$ (i.e., $0.3\,Gy/30\,days/24\,h$) are unlikely to effect results in populations of biota.

Assessing Effects

For assessment purposes, ERAs commonly use the concept of a screening index (SI) which is simply the ratio of the estimated dose rate (to an individual biota) to the reference dose rate, viz.

$$SI = \frac{\text{Estimated dose rate}}{\text{Reference dose rate}} \quad [2]$$

This comparison assumes that the numerator and the denominator of the SI are based on a common assessment of dose relevant to the endpoint of interest (e.g., mortality, reproductive capacity).

When the estimated SI is below 1, it is considered that an effect to a (population of) biota is unlikely. When an SI is estimated to be greater than 1, an effect may be possible and further detailed evaluations are carried out to investigate whether the SI being greater than 1 is an artifact of undue conservative assumptions and the nature of the calculation, or whether an actual effect might be possible. At this time, the reference dose rates described in the previous section provide an appropriate basis for assessment.

The interpretation of SIs also needs to consider aspects such as the spatial and temporal distribution of radiation, the natural cycles of populations, and background radiation levels.

Arctic caribou provide one example of a situation where natural background can result in quite high doses to biota. Caribou and reindeer subsist on lichens during the winter months but must graze over a wide area to obtain enough nourishment. Lichens do not have a root system but are highly efficient at collecting and retaining nutrient material deposited on their surfaces. Furthermore, lichens do not turn over annually but integrate all deposited materials, including radionuclides, over several decades. Hence this food chain integrates radioactivity not only over time, but also over spatial area. Estimated radiation doses to various tissues of caribou taken from areas with natural

radiation in Nunavat, Canada, are over the reference dose rate. Therefore, it can be important to consider natural background in ERA calculations.

It is evident that there is a wide range of complex factors to account for in the extrapolation of dose estimation from an individual level to the population level. When applying the SI approach to nonhuman biota at the individual level, great caution is therefore necessary about the interpretation of the predicted outcomes (as in most cases they would be overly conservative at the population level). A reliable interpretation also requires knowledge on the extrapolation from the individual level to the population level of both nonhuman biota dose rates and the related potential effects.

Final Comments

As previously noted, ionizing radiation is ubiquitous and all living things are, and have always been, exposed to naturally occurring radiation and radioactivity. Moreover, natural background levels of radiation vary widely from place to place; levels of radioactivity in soils, sediments, or pathways of exposure may result in elevated intake of natural radionuclides by biota. In this respect, it is important to understand that there may be practical difficulties in establishing protection criteria for radiation which is ubiquitous in the natural environment in comparison to many chemicals that are not naturally occurring.

In addition, nonhuman biota are exposed to external radiation from sources outside of the organism and from radiation arising from radionuclides taken into the organism. As noted earlier, many simplifying assumptions are needed to allow a practical estimation of doses from external and internal radiation, one example being the common assumption that radionuclides taken into the organism are uniformly distributed throughout the organism. Such assumptions introduce uncertainty into the assessment of risks to nonhuman biota.

Further uncertainty arises from the selection of a reference dose rate below which population-level effects are unlikely. As discussed in an earlier section, such data are not available for all biota and endpoints and professional judgments must be made to select the appropriate reference dose rate. At this time, the reference dose rates established by UNSCEAR are suggested for use in ecological risk assessments.

Recently, the European Community Sixth Framework Programme project ERICA (Environmental Risks from Ionizing Contaminants Assessment and Management), the successor to the FASSET project, adopted an ERA tiered methodology that requires risk assessment screening dose rate values for the risk characterization within tiers 1 and 2. ERICA adopted a $10\,\mu Gy\,h^{-1}$ screening dose rate based on the analysis of the chronic exposure data and incorporating safety factors.

The concept of an SI is widely used as the basis for risk assessment. Such an evaluation can be done using either very conservative assumptions about exposure or with more realistic assumptions about exposure conditions in a stepwise manner depending on the type of risk assessment. The key to the stepwise (tiered or graded) approach is to progress from a very conservative screen and, if necessary, move progressively to more realistic and less conservative assessments. The caveat in the process is that as one moves to more realistic assessment, there is an increasing demand for data and evaluations that involve detailed site-specific models.

ERAs are expected to continue evolving in light of recent developmental work (such as the ERICA project), the forthcoming update of the UNSCEAR report on effects of ionizing radiation on nonhuman biota, and further developments by ICRP in this area of emerging expertise. Moreover, the IAEA coordination group on the radiological protection of the environment provides a relevant platform for discussions about the international acceptance and harmonization of these developments as they arise. Although still under development, such approaches may find increasing utility in the future.

See also: Dose–Response.

Further Reading

Advisory Committee on Radiological Protection (2002) *Protection of Non-Human Biota from Ionizing Radiation*. Canadian Nuclear Safety Commission (CNSC), INFO-0703, March.

Barendsen GW (1992) RBE for non-stochastic effects. *Advances in Space Research* 12(2–3): 385–392.
Copplestone D, Bielby S, Jones SR, *et al.* (2001) *R&D Publication 128: Impact Assessment of Ionising Radiation on Wildlife*. Bristol: United Kingdom Environment Agency.
Environment Canada (2000) Priority substances list assessment report (PSL2). *Releases of Radionuclides from Nuclear Facilities (Impact on Non-Human Biota)*. Ottawa: Environment Canada and Health Canada.
Expert Group on Environment (EGE) (2005) Environmental consequences of the Chernobyl accident and their remediation: Twenty years of experience. *Report of the Chernobyl Forum Expert Group 'Environment'*. Vienna: IAEA (STI/PUB/1239).
Framework for Assessment of Environmental Impact (FASSET) (2003) Deliverable 3: Dosimetric models and data for assessing radiation exposures to biota (Pröhl G, ed.).
International Atomic Energy Agency (1992) Effects of ionizing radiation on plants and animals at levels implied by current radiation protection standards. *Technical Reports Series No. 332*. Vienna: IAEA.
International Commission on Radiological Protection (1977) ICRP Publication 26: Recommendations of the International Commission on Radiological Protection. *Annals of the ICRP* 1(3).
International Commission on Radiological Protection (1991) ICRP Publication 60: 1990 recommendations of the International Commission on Radiological Protection. *Annals of the ICRP* 21(1–3).
Trivedi A and Gentner NE (2002) Ecodosimetry weighting factor for non-human biota. In: IRPA-10. *Proceedings of the International Radiation Protection Association*, Japan, 14–19 May 2000 (available on CD-ROM).
United Nations Conference on Environment and Development (1992) *Convention on Biological Diversity*. New York: United Nations.
United Nations Conference on Environment and Development (1992) *Rio Declaration on Environment and Development*. New York: United Nations.
United Nations Scientific Committee on the Effects of Atomic Radiation (UNSCEAR) (1996) *Effects of Radiation on the Environment*. Annex to Sources and Effects of Ionizing Radiation (Report to the General Assembly, with one Annex), Scientific Committee on the Effects of Atomic Radiation. New York: United Nations.
United States National Council on Radiation Protection and Measurements (1991) Effects of ionizing radiation on aquatic organisms. *NCRP Report No. 109*.

Synthetic Polymers

K H Reinert[*], AMEC Earth & Environmental, Inc., Plymouth Meeting, PA, USA

J P Carbone, Rohm and Haas Company, Spring House, PA, USA

Introduction
Bulk Solid Polymers/Plastics
Emulsion/Dispersion Polymers
Water-Soluble/Swellable Polymers

Biodegradable Water-Soluble Polymers
Charged/Uncharged Polymers
Fluoropolymers
Further Reading

Introduction

Although the combinations of various monomers is almost limitless, synthetic polymers, often due to their size, regulatory exemptions, and seemingly low environmental toxicity,

[*] This manuscript does not necessarily represent the view or opinions of the employers of the authors.

do not have a substantial environmental toxicity database for many categories or classes. For this article, we have chosen to concentrate on the following categories because data are available for them. There may be additional categories that become subject to testing and become available in the future due to regulatory changes, persistency issues, and/or novel chemistries and uses for these important materials. The following categories are included in this article:

- polycarboxylates;
- polyesters;
- styrene–acrylics;
- charged polymers: anionic, cationic, amphoteric;
- polyamines;
- biodegradable polymers; and
- fluoropolymers.

The list is not exhaustive and includes a mixture of polymer categories identified by physical property, chemical composition, chemical characteristics, and end use. As far as intrinsic characteristics are concerned, polymers can be categorized in several ways.

Addition versus condensation polymers. Addition polymers are formed by a reaction in which monomer units simply add to one another. The monomers usually contain carbon–carbon double bonds. Addition polymers include polystyrene, polyethylene, polyacrylates, and methacrylates. Condensation polymers are formed by the reaction of bi- or polyfunctional molecules, with the elimination of some small molecule (such as water) as a by-product. Examples include polyester, polyamide, polyurethane, and polysiloxane.

Based on their physical properties. Materials where noncovalent bonds and intermolecular attractions are broken during heating are thermoplastic (see **Table 1**). Examples include polyethylene and nylon. Materials that melt initially but upon further heating become permanently hardened (cross-linked) are known as thermoset plastics. They cannot be softened and remolded without destruction of the polymer because covalent bonds are broken. Polymers without functionality that are incapable of dissociation are nonionic. If ionizable functionality is present, the polymer can be further characterized as cationic, anionic, or amphoteric (exhibit both cationic and anionic aspects).

Based on environmental toxicology. Polymers could be grouped based on their water extractability (thus potential bioavailability), stability to chemical or microbial degradation, or ability to adsorb/desorb. Charged polymers could be found/used in any physical form, but we expect that polymer charge would have the highest impact in water-soluble polymer systems. Therefore, a discussion of cationic/anionic/nonionic/amphoteric polymers could be done under that physical form category. Applications that had an especially high ecotoxicity

Table 1 Polymer categorization based on their physical form

Physical form	Chemistry	Application
Bulk solid polymer/plastics	Polyolefins	Synthetic rubbers
	Polyethylene	Plastic articles/parts
	Polypropylene	
	Polyisobutylene	
	Acrylonitrile copolymers	
	ABS (acrylonitrile–butadiene–styrene)	
	SAN (styrene–acrylonitrile)	
	Polyvinylchloride	
	Polystyrene	
	Styrene butadiene (SBR rubbers)	
	Polyesters/polycarbonates	
	Polymethylmethacrylate	
	Biodegradable plastics (e.g., polylactic acid)	
	Polysiloxanes	
Emulsion/dispersion	(Meth)acrylic copolymers	Paints
	SBR emulsions	Caulks
	Vinyl acetate–acrylic copolymers	Adhesives
	Styrene–acrylic copolymers	Paper coating
	Vinyl acetate homopolymer	Carpets
	Vinyl acetate–ethylene copolymers	Textiles
		Other
Water soluble/swellable	Polyacrylic acid	Superabsorbants
	Polymaleic acid and copolymers	Water treatment
	Biodegradable water-soluble polymers	Dispersants
	Charged/Uncharged Polymers	
	Cationic – include quaternary aminoesters	
	Anionic polymers	
	Nonionic polymers	
	Polyethylene oxide (PEO)	
	Polypropylene oxide (PPO)	
	PEO–PPO copolymers	
	Amphoteric polymers	

impact for any physical form/chemistry classification are highlighted in our discussion.

Bulk Solid Polymers/Plastics

Just about all solid polymers, plastics, and other high molecular weight polymers are insoluble in water, have low chemical reactivity (unless functionalized), exhibit considerable mechanical strength, and are usually resistant to degradation (biological, hydrolytic, and photolytic). They also typically have low direct environmental toxicity due to the above properties. Indirect hazards due to persistence, entanglement of marine organisms (birds and mammals; and, to a lesser extent, aquatic birds and mammals), and visual deterioration of the environment due to debris occurs. Environmentally responsible disposal practices and the development of economically viable biodegradable plastics and solid polymers will minimize these risks to the environment in the future.

As a special case of high molecular weight polymers, polyacrylate superabsorbents are acrylic polymers able to absorb and retain high levels of water or aqueous solutions. These superabsorbents, often used in various disposable products (e.g., diapers), find their way to landfills ($\sim$60%), and are virtually insoluble in water with the exception of non-cross-linked polymer (usually <10%) and traces of residual monomer.

These water-soluble components may occur in landfill leachate and wastewater treatment facilities; however, they are typically considered relatively nontoxic to aquatic and terrestrial receptors.

Emulsion/Dispersion Polymers

Many polymers are generally classified as water-based emulsion polymers. These dispersion polymers are often used as key ingredients in water-based paints, adhesives, floor coatings, caulks, and related products. The resulting polymer typically contains $\sim$40–50% polymer solids, 40–50% water, $\sim$1% dilute surfactants, varying levels of residual monomers, and typically less than 0.1% of additives such as salts and biocides. As emulsions, the environmental toxicity of both anionic and nonionic polymers is comparable to cationic polymer toxicity; however, cationics typically are more toxic to environmental receptors (**Table 2**).

Polymer backbones containing (1) monomers with nonreactive functional groups or (2) monomers with reactive functional groups other than cationic groups are of low concern for acute aquatic toxicity (class I). Polymer backbones considered of low concern include carboxylic acids, acrylates, methacrylates, acrylamide, acrylonitrile, styrene, butadiene, and vinyl acetate monomers. Emulsion polymer backbones containing monomers with cationic nitrogen-containing functional groups may be of higher concern for acute aquatic toxicity (class II). Finally, emulsion polymers associated with long-chain ethoxylated amines used as cationic counterions may be of a higher acute aquatic toxicity (class III) concern than those using monoalcohol amines.

Acute Aquatic Toxicity

Polymers with significant dispersibility tend to be correlated with higher environmental concern than insoluble polymers like plastics. Often, more physical effects like

Table 2   Aquatic toxicity of emulsion/dispersion polymers

Polymer	Class	D. magna (48 h EC50)	P. promelas (96 h LC50)	L. macrochirus (96 h LC50)	O. mykiss (96 h LC50)	Algae (96 h EC50)
Poly(ethyl acrylate)	I	22	5.3			
Poly(butyl acrylate/vinyl acetate)	I	71	>1000			
Poly(ethyl acrylate/methacrylic acid/methylmethacrylate)	I	>1000	>1000			
Poly(ethyl acrylate/quaternary ammonium salt of alkylamino-substituted methacrylate)	II	6.5	4.7			
Poly(ethyl acrylate/dimethylamino propyl methacrylate/MMA)	II	220		36		3.9
Poly(acrylonitrile/ethylhexyl acrylate/ methyl methacrylate/styrene) with ethoxylated amine counterion	III	81			31	
Poly(methacrylic acid/methyl methacrylate/methyl mercaptopropionate/styrene) with monoethanolamine counterion	III	>1000	>1000			>1000

All units are mg l^{-1}.

gill clogging at high concentrations (e.g., >1000 mg l^{-1}) are observed, unless functional groups on the polymer cause primary toxicity. Assimilation through biological membranes can occur for polymers with molecular weights ≤500 Da and decrease rapidly above that molecular weight. In general, dispersible and insoluble polymers are a low concern from an aquatic risk assessment perspective unless low molecular weight components (residual monomer, dimers, trimers) are toxic or the polymer has functionalized groups present. For cationic polymers, mitigation using dissolved organic carbon (DOC) like humic and fulvic acids has been shown to reduce toxicity. For anionic and amphoteric polymers, essential divalent sequestering may be reduced by adding Ca^+ and/or Mg^+ to the test medium. With respect to both mitigation approaches, these effects are minimized when testing under more environmentally relevant conditions.

These findings, conducted under laboratory conditions, are consistent with the hypothesis that reactive monomers, reactive functionalities, and reactive counterions may increase the aquatic toxicity of dispersion or emulsion polymers. Often these reactive polymers strongly sorb to organic materials found in natural environments, reducing exposure and risk to aquatic receptors. Although the cationic and cationic counterionic polymers may be of higher aquatic toxicity concern, destabilization and adsorption in natural aquatic systems should mitigate this potential. In addition to the potential for direct toxicity, some of these emulsion polymers can produce physical, indirect, or secondary toxicity, for example, particles clogging feeding mechanisms in *Daphnia magna*, coating gills, or even chelating nutrients essential for algal growth (i.e., sequestering essential divalent cations, Ca^+ and Mg^+).

In addition, six nonionic emulsion polymers were systematically tested for acute aquatic toxicity and biodegradability. The polymers were:

- poly(ethyl acrylate/methyl methacrylate/acrylamide);
- poly(butyl methacrylate/methyl methacrylate/styrene);
- poly(butyl acrylate/methacrylic acid);
- poly(butyl acrylate/methyl methacrylate);
- poly(butyl methacrylate/ethylhexyl acrylate/methacrylic acid/methyl methacrylate/styrene); and
- poly(ethyl acrylate/methyl methacrylate).

All polymers were practically nontoxic (EC or LC50 > 100 mg l^{-1}) to *Selenastrum capricornutum* (72 h algae), *D. magna* (48 h), and *Onchorhynchus mykiss* (96 h) as well as *Photobacterium phosphoreum* (15 min; Microtox®). In addition, an OECD 209, activated sludge respiration inhibition study, produced practically nontoxic results of EC50 > 100 mg l^{-1} for all polymers tested. Finally, using OECD 302, modified Zahn–Wellens inherent biodegradation testing, the polymers were found not to be inhibitory to activated sludge inoculum and ranged from not readily bioeliminable to readily bioeliminable via sorption to the test system sludge. All were considered low concern for aquatic toxicity.

Approximately 200–300 mg l^{-1} of acrylic dispersion used in floor coatings was required to produce an EC50 in microbial glucose uptake (measures reduction in microbial aerobic metabolism) with a no observed adverse effect concentration (NOAEC) of 60–100 mg l^{-1}. These results are consistent with other studies where Microtox EC50 results were >100 mg l^{-1}.

Other components, such as surfactants, may cause enhanced aquatic toxicity. However, when nine acrylic dispersion polymers were tested, the type of dispersion surfactant did not influence the aquatic toxicity (EC or LC50 > 100 mg l^{-1}).

Chronic Aquatic Toxicity

Most acute aquatic toxicity data point to levels of >100 mg l^{-1}. Chronic toxicity to algae, aquatic invertebrates, and fish are also considered low and acute to chronic ratios (ACRs) of ∼1 are typically observed. *Ceriodaphnia dubia* exhibited a 7 day no observable effect concentration (NOEC) of 1 mg l^{-1} and *S. capricornutum* exhibited a 96 h NOEC of 29 mg l^{-1} to an insoluble, anionic styrene–acrylic dispersion polymer with a molecular weight of 50–60 kDa used in floor finishes. The *C. dubia* value was caused primarily by physical effects. Additionally, *C. dubia* is often more sensitive than other aquatic invertebrates to many polymers, possibly due to a relatively high body surface to volume ratio. As for the algal NOEC, the NOEC increased to 500 mg l^{-1} after a 4× increase in hardness to 140 mg l^{-1} as $CaCO_3$. The ACR for these two organisms ranged from 26 to 910, probably reflecting the cited physical toxicity issues that would typically be ameliorated in the natural environment.

Another anionic, insoluble (at pH ≤ 8) anionic styrene–acrylic dispersion polymer with a molecular weight of 4500–9000 Da used in graphic arts, produced a 7 day NOEC in *C. dubia* of 26 mg l^{-1} (ACR 12), 7 day NOEC in *Pimephales promelas* of 190 mg l^{-1} (ACR 2), and 96 h NOEC of 310 mg l^{-1} in *S. capricornutum* (ACR 2). In this case, increasing the hardness by ∼4× did not increase the EC50 or the NOEC.

In *O. mykiss*, cationic polymers primarily affect the gill. This reversible effect disrupts gill function (e.g., ion regulation) and does not produce systemic toxicity.

A dispersible polymer, the sodium salt of poly(acrylic acid/maleic acid), with a molecular weight of 70 kDa is often used as a detergent additive, a builder to aid in soil removal, to prevent soil redeposition, and to prevent precipitation of inorganic salts on fabrics. This polymer was not toxic to activated sludge bacteria.

Sediment and Terrestrial Organism Toxicity

Retention of dispersion polymers by adsorption to sediment, soil, or landfill organic compounds is not expected to cause adverse effects on plants, soil organisms, or benthos. Acute and chronic studies with a styrene–acrylic polymer used in floor finishes and *Chironomus riparius* (midge larvae) produced an acute NOAEC of $690 \, mg \, kg^{-1}$ and a chronic NOEC of $>718 \, mg \, kg^{-1}$. The same polymer produced an earthworm (*Eisenia foetida*) acute LC50 of $520 \, mg \, kg^{-1}$ and a chronic NOEC of $250 \, mg \, kg^{-1}$. In addition, terrestrial plant species (corn, ryegrass, cucumber, lettuce, and radish) were not affected by this polymer at levels up to $72 \, mg \, kg^{-1}$ activated sludge in soil (from a 60 day semi-continuous activated sludge (SCAS) study using $3{-}6 \, mg \, g^{-1}$ polymer).

A similar lower molecular weight styrene–acrylic polymer used in the graphics industry exhibited the following toxicity results:

- *C. riparius* LC50 of $>911 \, mg \, kg^{-1}$; chronic NOEC of $264 \, mg \, kg^{-1}$ (ACR 3);
- *E. foetida* LC50 of $>1100 \, mg \, kg^{-1}$; chronic NOEC of $>1100 \, mg \, kg^{-1}$ (ACR 1);
- terrestrial plants (corn, ryegrass, cucumber, lettuce, and radish) NOEC of $64 \, mg \, kg^{-1}$ activated sludge in soil (from a 60 day SCAS study using $3{-}6 \, mg \, g^{-1}$ polymer).

Water-Soluble/Swellable Polymers

The following sections focus on the water-soluble and swellable polymers. Water-soluble polymers have a variety of functions from detergent builders and soil deposition inhibitors to wastewater treatment polymers used as flocculants or coagulants. Swellable polymers are the superabsorbent polymers used most frequently in wipe and diaper applications. These materials are generally comprised of homopolymers of acrylic acid (pAA) or copolymers of acrylic acid and maleic acid (pAA–MA).

Polyacrylic Acid

Polycarboxylates have supplanted phosphorous-bearing polymers as builders in laundry detergents because of environmental concerns regarding phosphate-induced eutrophication of water bodies. The polycarboxylates used in detergents are comprised of homopolymers of acrylic acid (pAA) or copolymers of acrylic acid and maleic acid (pAA–MA) among other materials. Polycarboxylates are produced in industrial processes by free radical polymerization of monomer acrylic acid or the monomers acrylic acid and maleic anhydride. The polycarboxylate materials, used in low-phosphate and phosphate-free detergents, have been formulated into the detergent to inhibit the deposition of inorganic precipitates into the fabric and/or as soil/dirt dispersants. Polyacrylates are also used in the manufacture of super-absorbent materials, for example, disposable diapers. The methylated form of acrylic acid, methacrylic acid, is produced in the millions-of-kilogram range and is used primarily as a precursor for the formation of methyl-methacrylate and other methacrylate esters.

Because of the nature of the functional groups along the polymer chain, polycarboxylates are polyelectrolytes. In particular, polycarboxylates comprised of polyacrylic acid or polyacrylic acid maleic anhydride are anionic due to the high density of anionic carboxylate groups.

Aquatic acute ecotoxicity

Wastewater treatment will be an important removal process for p(AA) materials. Therefore an analysis of the toxicity of these polycarboxylates to wastewater treatment facility (WWTF) biomass is of critical importance. An IC50 of $>100 \, mg \, l^{-1}$ p(AA) 4500 has been reported and thus the likelihood of significant biomass toxicity as a result of entry into a WWTF is minimal.

A number of acute fish and aquatic invertebrate toxicity studies are available for polyacrylates. Ninety-six hour LC50 data for rainbow trout (*Salmo gairdneri*, *O. mykiss*) and bluegill sunfish (*Lepomis macrochirus*) were reported as 315 and $>450 \, mg \, l^{-1}$ p(AA) 4500. Generally these materials are characterized according to United States Environmental Protection Agency (USEPA) Toxic Substances Control Act (TSCA) criteria as of low toxicity to aquatic organisms. **Table 3** illustrates additional acute fish and invertebrate (*D. magna*) toxicity data for a variety of molecular weight polyacrylates.

A detailed survey of the toxicity of p(AA) polymers and additional wastewater treatment polymers chemistries such as methacryloyloxy trimethyl ammonium chloride (METAC), acryloyloxy trimethyl ammonium chloride (AETAC), epichlorohydrin/dimethylamine (EPI/DMA), diallyldimethyl ammonium chloride (DADMAC), melamine formaldehyde and mannich (secondary amines) in *Daphnia pulex*, and the fathead minnow (*P. promelas*) has been conducted. LC50 values for the *D. pulex* ranged from 0.06 to $2.0 \, mg \, l^{-1}$ for the majority of the polymers tested. *Daphnia pulex* was much less sensitive to melamine formaldehyde and mannich polymers (LC50 values: 12.1–$70.1 \, mg \, l^{-1}$). Some of the *D. pulex* toxicity was apparently the result of physical entrapment of the organism by the polymer. The fathead minnow was generally less sensitive than the invertebrate. Polymer toxicity to the fathead minnow appeared to be related to the charge of the polymer. Cationic polymers were found to be clearly more toxic to *P. promelas* than anionic polymers with polymer toxicity generally increasing with increasing positive charge density.

Table 3 Acute aquatic toxicity of polyacrylates in fish and *D. magna*

Polymer tested (molecular weight, Da)	Zebra fish (Brachydanio rerio) 96 h LC50 (mg l^{-1})	Golden orfe (Leuciscus idus melanotus) 96 h LC50 (mg l^{-1})	Bluegill (Lepomis macrochirus) 96 h LC50 (mg l^{-1})	Trout (Salmo gairdneri, Oncorhynchus mykiss) 96 h LC50 (mg l^{-1})	Daphnia magna 48 h EC50 (mobility) (mg l^{-1})
p(AA) 1000	>200		>1000	>1000	>200
					≥1000
p(AA) 2000	>200				>200
p(AA) 4500	>200		>1000	700	>200
			>1000		>1000
					>450
p(AA) 9400	>1000				
p(AA) 10 000	>1000				
p(AA) 23 000	>1000		>1000		
p(AA) 78 000	>1000	1590			>750
p(AA) 111 000	>1000				
p(AA) 152 000	>1000				
p(AA) 215 000	>1000				
p(AA) 12 000	>200				>200
p(AA) 70 000	>100	>200			>200
	>1000				>100
					>200
					>908

Data from tables 6 and 7 of ECETOC (1993) *Joint Assessment of Commodity Chemicals, No. 23: Polycarboxylate Polymers as used in Detergents.* Brussels ECETOC (ISSN-0773-6339-23).

Algal toxicity

The toxicity of several polyacrylate polymers of varying molecular weight has been measured in algae (**Table 4**). According to USEPA TSCA criteria and the test data presented, the polyacrylate polymers are characterized as of moderate to low concern to algae.

Chronic toxicity

The chronic toxicity of p(AA) materials has been assessed. The Rohm and Haas Co. reported an NOEC of 12 mg l^{-1} based on a 21 day life-cycle study in *D. magna*. Contrastingly, an NOEC of 450 mg l^{-1} using a similar study design has been reported. The preceding toxicity endpoints are qualitatively similar, that is, low to moderate toxicity. The discrepancies in the endpoints could be due to the nature of the test water and its influence on the

solubility or physical behavior of the polymer. It has been our experience that *D. magna* are particularly sensitive to physical influences of large molecular weight substances. Toxicity as a function of induction of clumping or agglomeration in cultures is particularly important with regard to invertebrates and algae. Fish can also be affected by the physical nature of the test substance. Nonspecific toxicity could be due to adherence of the test substance in the gill areas of fish. **Table 5** illustrates the chronic toxicity of polycarboxylates to fish.

With regard to aquatic toxicity testing with p(AA) or polycarboxylates in general, USEPA and the Organization for Economic Cooperation and Development (OECD) testing guidelines prescribe test with low DOC in fish and

Table 4 Acute aquatic toxicity of polyacrylates in algae

Polymer tested (molecular weight, Da)	Algal species	Study type/ endpoint	mg l^{-1}
p(AA) 4500	Scenedesmus subspicatus	96 h EC10	180
p(AA) 70 000	Scenedesmus subspicatus	96 h EC10	32
p(AA) 70 000	Scenedesmus subspicatus	96 h EC10	≥200
p(AA) 78 000	Scenedesmus subspicatus	4–14 d EC10	82

Table 5 Chronic aquatic toxicity of polyacrylates in algae

Polymer tested (molecular weight, Da)	Algal species	Study type/ endpoint	mg l^{-1}
p(AA) 78 000	Brachydanio rerio	14 d NOEC	1000
p(AA) 70 000	Brachydanio rerio	14 d NOEC	40
p(AA) 4500	Brachydanio rerio	28 d ELS NOEC	450[a]
p(AA) 70 000	Brachydanio rerio	42 d NOEC	40[a]
p(AA) 4500	Pimephales promelas	28 d ELS NOEC	124[a]

[a]Highest concentration tested.

D. magna studies. Additionally, toxicity testing with algae is generally conducted under low water hardness conditions. There are indications, however, that the testing outcomes based on these experimental designs are conservative and do not reflect the actual toxicity of these materials under environmentally relevant conditions.

It is hypothesized that polycarboxylates may cause toxicity in algae through a mechanism whereby essential divalent cations such as calcium and magnesium are sequestered by the anionic polycarboxylates. Studies have indicated that the addition of calcium in the form of $CaCO_3$ to a total hardness of approximately $150\ mg\ l^{-1}$ mitigates toxicity to algal cultures. Mitigation of polycarboxylate toxicity to fish and invertebrates has also been demonstrated with the addition of DOC at a level comparable to that anticipated in the environment.

Terrestrial organism toxicity

The toxicity of p(AA) in terrestrial organisms is considered low. In the earthworm (*E. foetida*), the LC50 was reported as $>1000\ mg\ kg^{-1}$ for p(AA) 4500. LC0 values in earthworm for the p(AA) 78 000 and p(AA–MA) 70 000 equaled 1000 and $1600\ mg\ kg^{-1}$ soil, respectively.

With regard to terrestrial plants p(AA) 4500 did not inhibit the growth of corn, soybean, wheat, and grass seed up to application rates of $225\ mg\ kg^{-1}$ soil, the highest concentrations tested. In turnip seed, the NOEC for p(AA) 78 000 was $1000\ mg\ kg^{-1}$ soil. In oat seed, the NOEC for p(AA) 70 000 equaled $400\ mg\ kg^{-1}$ soil.

Biodegradable Water-Soluble Polymers

Because of the recalcitrance to degradation of typical plastic materials, research has been directed recently toward development and testing of biodegradable polymers.

Ecotoxicological data regarding biodegradable water-soluble polymers are limited. The degradative capacity of carboxymethyl cellulose (CMC) in a continuous flow activated sludge (CAS) simulation test has been assessed. The results obtained in the CAS test indicated that CMC degraded partially, yielding approximately 50% of the spiked CMC carbon. Additionally, undiluted effluent from the CAS reactor containing degradation intermediates formed by the biodegradation process was shown not to be acutely toxic to the zebra fish (*Brachydanio rerio*) and *D. magna*. Chronic toxicity studies were conducted with the undiluted CAS effluent. The results indicated that the biodegradation intermediates contained within the effluent were not toxic to *Pseudomonas putida*, *D. magna*, and the algae *S. capricornutum* using standard chronic testing protocols. In addition, CMC intermediates produced by a pure culture of CMC-degrading bacteria were also shown not to be toxic; no effects were observed at the highest

concentration tested, $0.5\ g\ l^{-1}$ for *S. capricornutum*, $1.0\ g\ l^{-1}$ for *D. magna*, and $1.0\ g\ l^{-1}$ for the zebra fish.

Synthetic copolyesters containing aromatic constituents have been shown to be susceptible to degradation by microorganisms. The extent of biodegradation of an aliphatic-aromatic copolyester (Ecoflex®) in culture containing bacterium isolated from a compost system was investigated. The polymer composition was terephthalic acid, 22.2 mol.%; adipic acid, 27.8 mol.%; and butanediol, 50 mol.%. The individual strain of the isolated compost microorganisms was *Thermomonospora fusca*. Ecotoxicity studies with the degradation intermediates were performed with *D. magna* and with *P. phosphoreum*, a luminescent bacterium. Following 22 days of degradation more than 99.9% of the polymer had depolymerized. With regard to the degradation of the diacid and diol components, only the monomers of the copolyester (1,4-butanediol, terephthlate, and adipate) could be detected by gas chromatography/mass spectrometry. In ecotoxicological tests with *D. magna* and *P. phosphoreum*, no significant toxicological effects were observed, with the monomeric or oligomeric degradation intermediates.

Charged/Uncharged Polymers

Cationic, Anionic, Nonionic, and Amphoteric Polymers (Including Quarternary Aminoesters)

Water-soluble cationic polymers are used as coagulants and or flocculants in processes that include clarification of drinking water, sludge dewatering, paper manufacturing, and mining, and as coating resins. Water-soluble charged polymers are classified according to their charge potential as cationic, anionic, nonionic, and amphoteric. Cationic polymers contain a positive charge density. Many of the polymers contain tertiary or quaternary nitrogens that provide a net positive charge to the polymer. Anionic polymers are negatively charged. Nonionic polymers are not charged because they do not contain an ionizable moiety. Amphoteric polymers are zwitterionic in nature having both cationic and anionic functional groups. The expression of the charge in amphoteric polymers is a function of the pH of the resident media. In addition to mechanistic or nonspecific toxicity that may be evident in fish, invertebrates, and algae, cationic polymers can exert toxic effects via physical interactions with the negatively charged gill surface of fish. Reduced oxygen transfer ensues with associated adverse effects.

The effects of cationic polymer chemistry, charge density, and molecular weight were assessed in acute and chronic exposure to the rainbow trout (*S. gairdneri*, *O. mykiss*). The cationic polymers that were evaluated consisted of two major classes. The epichlorhydrin/dimethylamine copolymers carrying a quaternary nitrogen on the backbone of the polymer. The second type of cationic copolymer was the acrylamide/acrylate copolymers which carry a quaternary nitrogen on the ester side chain of the polymer. The polyamines evaluated ranged in molecular weight from 10 to 200–250 kDa). The acrylamide/acrylate ester copolymers varied in charge density from 10% to 39%. Acute studies were conducted under static nonrenewal and also under flow-through conditions. Chronic studies were conducted via flow-through exposures. For the acute nonrenewal studies, the LC50 values were highly varied. Acute LC50 values from the nonrenewal studies equaled 592, 271, 779, and 661 μg l^{-1} for the three polyamines and one acrylamide, respectively. The polyamines, that is, the polymers with the quaternary nitrogen on the backbone of the polymer, appeared to be in general more acutely toxic than the acrylamide-based polymers (quaternary nitrogen on the ester side chain of the polymer). Under flow-through conditions, the toxicity appeared to increase versus the nonrenewal studies. Acute LC50 values from the dynamic studies equaled 42.6, 96, 156, and 384 μg l^{-1} for the three polyamines and one acrylamide, respectively. The LC50 ACRs for the dynamic flow-through and chronic studies were low for both the polyamines and polyacrylamide tested, indicating that chronic toxicity LC50 values were not unlike the acute values. The low ACRs indicate therefore that the resulting toxicity was a function of rapid acute effects rather than long-term cumulative effects. A trend suggesting decreases in toxicity with increasing molecular weight was noted. With regard to sublethal effects, the polyamine evaluated for chronic toxicity did not induce adverse effects on growth parameters. In fact, both polyamines induced concentration-related increases in growth parameters. For the acrylamide tested, a significant decrease in body weight of the surviving trout was noted. One can conclude from these studies that cationic charge and the physical bulk of the polymer were the determining factors in the toxicity noted in the nonrenewal system. Flow-through conditions increased the toxicity of the polymers compared to that under static conditions. Polymer molecular weight and toxicity were inversely proportional. In the flow-through systems, the cationic polyamines appeared to be more toxic than the cationic polyacrylamides.

The acute toxicity of a number of cationic polymers was assessed in *D. magna*, fathead minnow (*P. promelas*), gammarids (*Gammarus pseudolimnaeus*), and midges (*Paratanytarsus parthenogeneticus*) using *in vitro* testing methods. Additionally, a microcosm test employing fish or invertebrate species and ten algal species has been conducted. Acute toxicity studies were conducted with *D. magna* and fathead minnow at polyelectrolyte concentrations of 100 mg l^{-1}. If the test concentration of 100 mg l^{-1} proved toxic to either or both test organisms, then the electrolyte was tested using the less sensitive gammarid. Some of the electrolytes were tested using

the midges. The LC50 values for four of the polycations was greater than 100 mg l^{-1} for *D. magna* and/or the fathead minnow. Of the remaining 11 cationic polymers, the LC50 values ranged from 0.09 to 70.7 mg l^{-1} for *D. magna* and from 0.88 to 9.47 mg l^{-1} for fathead minnow. According to USEPA TSCA criteria, the acute toxicity of these polycations ranges from low concern (LC50 > 100 mg l^{-1}) for several to moderate to high concern (LC50 < 100 mg l^{-1} to LC50 < 1.0 mg l^{-1}). *Paratanytarsus parthenogeneticus* LC50 values were less than 100 mg l^{-1} for three of the eight cationic polymers tested (<6.25 to 50 mg l^{-1}). The LC50s for gammarids were 8.1–33.4 mg l^{-1} for seven of 13 polymers tested. In the microcosm studies, algal growth was delayed at the higher cation concentration. It was not evident however that the polymers induced direct toxic effects on the algae and the delayed cell growth was speculatively attributed to potential physical interactions of the algal cells and the polymers. Alterations in species composition in the microcosm were attributed to the polyelectrolytes but grazing activity was not ruled out as the reason for species diversity alterations in treated microcosms.

The acute toxicity of several polyelectrolytes to rainbow trout (*O. mykiss*), lake trout (*Salvelinus namaycush*), a mysid (*Mysis relicta*), a copepod (*Limnocalanus macrurus*), and a cladoceran (*D. magna*) in Lake Superior water has been evaluated. Additionally, a 21 day life-cycle study in *D. magna* was undertaken to examine the effects of the polycationic polymers on reproduction in this invertebrate species. The cationic polyelectrolytes tested were Superfloc® 330 (Calgon Corp.), Calgon M-500, Gendriv 162 (General Mills Chemicals), Magnifloc® 570C (Calgon Corp.), and Magnifloc® 521C. Under static conditions, the 96 h LC50 values for rainbow trout ranged from 2.12 mg l^{-1} for Superfloc® 330 to 218 mg l^{-1} for Gendriv 162. The toxicity characterization is of low to moderate concern according to USEPA TSCA criteria. For lake trout, the 96 h LC50 value for Superfloc® 33 equaled 2.85 mg l^{-1} and for Calgon M-500, 5.70 mg l^{-1}. These data are indicative of moderate toxicity to this species of fish. For *D. magna*, the 48 h LC50 ranged from 0.34 to 345 mg l^{-1}, a wide range, with toxicity characteristics according to TSCA of low to high concern. In a 21 day *D. magna* life-cycle study, Superfloc® 330 and Calgon M-500 impaired reproduction in the invertebrates at lower concentrations, that is, 0.10 and 1.0 mg l^{-1}, respectively, than those permitting survival, that is, 1.10 and 2.85 mg l^{-1}. The data are indicative of some response variation, likely a consequence of charge density. Additionally, the data also indicate that at least for several polyelectrolyte cations the associated toxicity in aquatic organism can be substantial.

Studies have shown that mitigation of the toxicity of cationic polymers can be facilitated via the introduction of anionic polymers and/or organic matter added as foodstuffs to exposed species. Specifically, the toxicity of cationic polymeric material has been reduced via the addition of humic acid. The addition of humic acid to cultures of rainbow trout was shown to reduce the toxicity of cationic polymers up to 75-fold depending upon the concentration of the humic acid in the cultures. In sum, these data indicate that the addition of organics to cultures containing polycationic polymers reduce toxicity. The practical implication of this is that while standard toxicity studies conducted without the addition of organic material such as humic acid allows for the comparison of toxicity across test materials, the addition of organics allows for the evaluation of toxicity under more plausible, environmentally relevant conditions.

The mechanism of polymer toxicity in algal cultures has been hypothesized to be a function of nutrient trace metal sequestration. This hypothesis has been tested using water-accommodated fractions (WAFs) of aqueous mixtures of three multicomponent lubricant additives. The WAFs were utilized because of the insoluble nature of a proportion of the lubricant additives. Resulting toxicity data for *S. capricornutum* generally indicated that the WAFs were very toxic, exhibiting median effective loading concentrations (EL50s) based on cell density increases or growth rates of less than 1 mg l^{-1}. Contrastingly, for *O. mykiss* and *D. magna*, the resulting EL50 values were in excess of 1000 mg l^{-1}. In addition, tests designed to determine whether the lubricant WAFs were algistatic (the concentration that inhibits algal growth without reducing cell levels) or algicidal were included. Results of these studies indicated that the algal toxicity was indirect, resulting from sequestration of essential micronutrients. WAF fortifications in the form of iron or disodium ethylenediaminetetraacetic acid (EDTA) ranging from 200% to 1000% of the standard algal medium concentration mitigated any toxicity noted in unamended cultures. Algal cultures removed from the WAF-containing medium and resuspended in fresh culture medium resumed exponential growth. One can draw several conclusions from these studies: (1) sequestration of micronutrients by charged polymeric materials will likely impart significant toxicity to the exposed organisms, algae being particularly sensitive to logarithmic phase growth reductions due to essential nutrient depletion; and (2) testing of materials using standard testing protocols may overestimate toxicity because the correlation of the limited nutrient supply in standard media and that of the natural dynamic waters is low.

An environmental risk assessment case study has been conducted for a C_{12}–C_{18} monoalkyl quaternary ammonium compound (MAQ). The MAQ is a cationic surfactant that functions in combination with other laundry detergent components. In the case study information regarding the physical and chemical properties of the test material, predicted environmental concentrations and

environmental fate were presented. Additionally, environmental effects data were discussed for the MAQ. The 96 h EC50 values for green and blue-green algae and diatoms ranged from 0.12 to 0.86 mg l^{-1} MAQ. Algistatic concentrations ranged from 0.47 to 0.97 mg l^{-1}. Daphnid 48 h EC50 values averaged 0.06 mg l^{-1} for five tests in laboratory water. The chronic NOEC and LOEC in a 21 day *D. magna* life-cycle study equaled 0.01–0.04 mg l^{-1}. EC50 values for the marine invertebrates, mysid, and pink shrimp equaled 1.3 and 1.8 mg l^{-1}, respectively. The 96 h LC50 for four species of freshwater fish were a function of chain length. The LC50 values equaled 2.8–31.3 mg l^{-1} for MAQs with chain lengths of C_{12}–C_{14} and 0.10–0.24 mg l^{-1} for the MAQs with chain lengths ranging from C_{15} to C_{18}. The measured 28 day chronic NOEC and LOEC in fathead minnow early-life-stage studies equaled 0.46–1.0 mg l^{-1} for C_{12} MAQ and 0.01–0.02 mg l^{-1} for C_{16}–C_{18} MAQs. Clearly these materials have significant toxicity based on laboratory studies. Because these materials are likely to be treated in WWTFs, the toxicity of the material to activated sludge microorganisms was evaluated. The concentration of the MAQ required to cause a 50% reduction in heterotrophic activity was approximately 39 mg l^{-1}.

Acute and chronic toxicity tests were conducted with MAQ in river and lake waters. The rationale was to evaluate the effects of dissolved organics contained with the natural waters in terms of the bioavailability of the polymer. Both the acute LC50 values and the chronic LOEC levels averaged threefold higher in natural surface water for daphnids, the most sensitive species. The LC50 values ranged from 0.1 to 0.5 mg l^{-1} MAQ in seven river and lake water tests (LC50 in laboratory water averaged 0.06 mg l^{-1}). Measured chronic NOEC and LOEC values in four different surface water tests ranged from 0.05 to 0.10 mg l^{-1} MAQ (NOEC and LOEC in laboratory water ranged from 0.01 to 0.04 mg l^{-1}). The results of two river water acute toxicity tests with bluegill and fathead minnows were comparable to the laboratory studies; LC50 values equaled 6.0 mg l^{-1} in river water versus 2.8–31.0 for the same-chain-length MAQ in laboratory water.

Microcosm studies were also conducted where replicate populations of *D. magna*, chironomid midges, and colonized river periphyton were exposed to concentrations of C_{12} MAQ that were anticipated to be lethal to *D. magna*. The microcosms were flow-through systems with natural river water and clean sediment. The organisms were exposed for up to 4 months ensuring the exposure of multiple generations. Based on the results of the study, there were no significant effects on *D. magna* density or biomass at C_{12} MAQ concentrations up to 0.110 mg l^{-1}. The first effect occurred at 0.180 mg l^{-1} in populations that were initially exposed to that test concentration. Populations acclimated to lower concentrations and subsequently exposed to 0.180 mg l^{-1} were not adversely affected. Significant reductions in both pre-exposed and control-reared populations occurred at $0.310 \, mg \, l^{-1}$. The results were attributed to compensatory changes in invertebrate population dynamics where the loss of sensitive individuals was compensated for by increases in reproductive capacity of tolerant populations after multigenerational exposures.

Finally, field studies were conducted in rivers and lakes in good biological condition and receiving quantifiable amounts of WWTF effluents. Natural phytoplankton and zooplankton structural and functional parameters were assessed as well as biodegradation rates. Laboratory-derived EC50 values for green and blue-green algae and diatoms were some 12–23-fold less than in the *in situ* concentration that affected photosynthetic activity or community structure. Biodegradation by pre-exposed microbial communities was rapid and reflected biodegradation of naturally occurring organics. Indigenous fish, macroinvertebrates, and periphyton were much less sensitive to MAQ than was the most sensitive laboratory species *D. magna*. In an effluent-dominated stream, there were no significant adverse effects noted for any of the indigenous communities exposed to a concentration of $0.27 \, mg \, l^{-1}$ MAQ, more than twice that of the acute EC50 for daphnids based on laboratory studies.

Lake trout fry, *Salvelinus namaycush*, were exposed in laboratory experiments to two wastewater treatment polymers, one anionic (MagnaFloc® 156) and one cationic (MagnaFloc® 368; Ciba Specialty Chemical) to determine if these chemicals used in mining operations were toxic to exposed fish. The polymers are added to wastewater to facilitate the settling and removal of suspended particulates. Cationic polymers function primarily as coagulants and adsorb to the surface of negatively charged particles, thus neutralizing electrostatic surface charges. Anionic polymers function primarily as flocculants binding together suspended particles into higher molecular weight aggregates that more readily settle out of solution. Results indicated that the cationic polymer MagnaFloc 368 was substantially more toxic to lake trout fry than the anionic polymer MagnaFloc 156. MagnaFloc 368 had a 96 h LC50 of $2.08 \, mg \, l^{-1}$ whereas the LC50 for MagnaFloc 156 could not be determined. At the highest tested MagnaFloc 156 concentration, $600 \, mg \, l^{-1}$, 5% mortality was observed.

The toxicity seen in these fry were attributed to charge density. The stronger the electrostatic charge of the polymer, the greater its toxicity. Lesser molecular weight polymers are also typically of greater toxicity. The mechanism of toxicity is hypothesized to be that the charged polymers are attracted to and interact with the negative charged gill surfaces of the exposed fish. The toxic effect of cationic polymers in fish is consistent with hypoxia and is evidenced by associated histopathology including increased vascularization, increased lamellar thickness

via cellular proliferation, and decreased lamellar height. The histopathologic findings support the physiologic mechanism of impaired respiratory efficiency and ion regulation at the gill membrane. For anionic polymers, it is hypothesized that these materials sequester important nutrients in the media such as trace metals magnesium and/or iron. Alternatively anionic materials could also influence ion regulation within the gill membrane.

Fluoropolymers

Perfluorooctane sulfonic acid (PFOS) and perfluorooctanoic acid (PFOA) have been identified as ubiquitous environmental contaminants. These materials are not natural products and are purely anthropogenic in origin. The perfluorinated acids (PFAs) in general are a class of anionic fluorinated materials characterized by a perfluoroalkyl chain and a sulfonate or carboxylate solubilizing group. The perfluoroalkyl chain is commonly referred to as a telomer or synonymously as a fluorotelomer. The perfluorinated compounds are used as precursor materials in the synthesis of very high molecular weight fluorinated polymers. The environmental liabilities of the high molecular weight polymers are limited due to their size, that is, molecular size exclusions and general recalcitrance to degradation. Any potential environmental liabilities are a consequence of residual telomers in the formulated end-use products and any degradation of the high molecular weight polymers. The following discusses telomer environmental toxicity.

Table 6 illustrates the acute toxicity of PFOS to fish, invertebrates, and algae. The data indicate that PFOS is practically nontoxic to freshwater algae and aquatic vascular plants, that is, *Lemna gibba*. PFOS exhibits only slight toxicity to invertebrates and is considered 'of moderate concern' to fish according to the USEPA TSCA criteria. **Table 7** suggests that fish are more sensitive to PFOS than invertebrates or algae based on subchronic or chronic exposures.

For PFOA, the majority of the aquatic ecotoxicity studies have been conducted with the ammonium salt (APFO) of the prefluorooctanoic acid. Under environmentally relevant conditions in aqueous environmental compartments, PFOA will exist as the fully ionized component (COO^-). Since a likely route of emission of fluoropolymers will be via WWTF effluent, the toxicity of PFOA to bacteria has been evaluated. The 30 min and 3 h EC50 values for sludge respiratory inhibition studies ranged from >1000 to $>3300 \, mg \, l^{-1}$. For algae, the lowest 96 h EC50 and NOEC values reported for algal assays using *Pseudokirchneriella subcapitata* were 49 and $12.5 \, mg \, l^{-1}$, respectively. Overall, 96 h EC50 values (based on growth rate, cell density, cell counts, and dry weights) ranged from 49 to $>3330 \, mg \, l^{-1}$. NOEC values ranged from 12.5 to $430 \, mg \, l^{-1}$. Based on USEPA TSCA

Table 6 Acute toxicity of PFOS to fish, invertebrates, and algae

Organism	Toxicity endpoint	Time (h)	Concentration[a] (mg l^{-1})
Selenastrum capricornutum	EC50 growth rate	96	126
		72	120
Selenastrum capricornutum	EC50 cell density	96	82
Selenastrum capricornutum	EC50 cell count	96	82
Anabaena flos aqua	EC50 growth rate	96	176
	NOEC growth rate		94
Navicula pelliculosa	EC50 growth rate	96	305
	NOEC growth rate		206
Lemna gibba	IC50	168	108
	NOEC		15.1
Daphnia magna	EC50	48	61
	NOEC		33
Daphnia magna	EC50	48	58
Freshwater mussel	LC50	96	59
	NOEC		20
Fathead minnow	LC50	96	9.5
	NOEC		3.3
Rainbow trout	LC50	96	7.8
Rainbow trout	LC50	96	22

[a]Potassium salt of PFOS (PFOS-K$^+$).

Table 7 Chronic toxicity of PFOS to fish and invertebrates

Organism	Toxicity endpoint	Time (d)	Concentration[a] (mg l^{-1})
Daphnia magna	NOEC	21	12
	Reproduction, survival, growth		
Daphnia magna	EC50 reproduction	21	12
	NOEC reproduction	28	7
	EC50 reproduction	28	11
Fathead minnow	NOEC survival	42	0.30
	NOEC growth	42	0.30
	NOEC hatch	5	> 4.6
Fathead minnow	NOEC	30	1
	Early life stage		
Bluegill sunfish	NOEC mortality	62	> 0.086 < 0.87

[a]Potassium salt of PFOS (PFOS-K$^+$).

criteria, PFOA would be characterized as of low concern to algal species. Daphnid 48 h EC50 values (based on immobilization) ranged from 126 to >1200 mg l^{-1}. The 10 day NOEC for the sediment-dwelling *Chironomus tentans* was shown to be >100 mg l^{-1}. Additionally, in laboratory studies, no effects on *C. tentans* were evident following 10 day exposures to PFOA at concentrations up to 100 mg l^{-1}. Based on these toxicity endpoints, PFOA would be characterized according to USEPA TSCA criteria as of low concern to aquatic invertebrate species. With regard to vertebrate fish species, measured 96 h

LC50 values ranged from 280 to 2470 mg l^{-1}. Based on the LC50 values for fish, PFOA would be characterized as of low concern according to USEPA TSCA criteria.

The available chronic toxicity data include 14 day algal EC50 values of 43 and 73 mg l^{-1} (in addition to the 96 h NOEC values), 21 day daphnid reproduction NOECs ranging from 20 to 22 mg l^{-1}, 35 day mixed zooplankton community LOECs from freshwater microcosm studies ranging from 10 to 70 mg l^{-1} and chronic fish NOECs ranging from 0.3 mg l^{-1} for steroid hormone levels in male fish measured in 39 day microcosm studies to 40 mg l^{-1} based on survival and growth from an 85 day rainbow trout early-life-stage study. The reductions in steroid hormone levels in fish were accompanied by only limited increases in time to first oviposition and limited decreases in overall egg production. Thus, hormonal fluctuations induced by chronic exposures to PFOA have limited, moderate-term consequences on fish reproductive capacity. Some uncertainty does exist however regarding the longer-term consequences of PFOA exposures and reproductive capacity of exposed populations. According to USEPA TSCA criteria, PFOA would be characterized as of low chronic concern to algae and low to moderate chronic concern to invertebrates and fish. Based on the available data, the ecotoxicity of PFOA is considered low to aquatic organisms.

See also: Ecotoxicology: The History and Present Directions.

Further Reading

Biesinger KE and Stokes GN (1986) Effects of synthetic polyelectrolytes on selected aquatic organisms. *Journal of Water Pollution Control Federation* 58: 207–213.

ECETOC (1993) *Joint Assessment of Commodity Chemicals, No. 23: Polycarboxylate Polymers as used in Detergents*. Brussels: ECETOC (ISSN-0773-6339-23).

Goodrich MS, Dulak LH, Friedman MA, and Lech JJ (1991) Acute and long term toxicity of water-soluble cationic polymers to rainbow trout (*Oncorhynchus mykiss*) and modification of toxicity by humic acid. *Environmental Toxicology and Chemistry* 10: 509–515.

Guiney PD, McLaughlin JE, Hamilton JD, and Reinert KH (1997) Dispersion polymers. In: Hamilton JD and Sutcliffe R (eds.) *Ecological Assessment of Polymers. Strategies for Product Stewardship and Regulatory Programs*, pp. 147–165. New York: Van Nostrand Reinhold.

Guiney PD, Woltering DM, and Jop KM (1998) An environmental risk assessment profile of two synthetic polymers. *Environmental Toxicology and Chemistry* 17(10): 2122–2130.

Hall WS and Mirenda RJ (1991) Acute toxicity of wastewater treatment polymers to *Daphnia pulex* and the fathead minnow (*Pimephales promelas*) and the effect of humic acid on polymer toxicity. *Research Journal Water Pollution Control Federation* 63: 895–899.

Hamilton JD, Morici IJ, and Freeman MB (1997) Polycarboxylates and polyacrylate superabsorbants. In: Hamilton JD and Sutcliffe R (eds.) *Ecological Assessment of Polymers. Strategies for Product Stewardship and Regulatory Programs*, pp. 87–102. New York: Van Nostrand Reinhold.

Hamilton JD, Reinert KH, and Freeman MB (1994) Aquatic risk assessment of polymers. *Environmental Science and Technology* 28(4): 187A–192A.

Hamilton JD and Sutcliffe R (eds.) (1997) *Ecological Assessment of Polymers. Strategies for Product Stewardship and Regulatory Programs*. New York: Van Nostrand Reinhold.

Hekster FM, de Voogt P, Pijnenburg J, and Laane RWPM (2002) Perfluoroalkylated substances. Aquatic environmental assessment. *Report RIKZ/2002.043*. University of Amsterdam, Amsterdam.

Hennes EC (1991) *Fate and Effects of Polycarboxylates in the Environment*. Strombeek-Bever: Procter and Gamble.

Lyons LA and Vasconcellos SR (1997) Water treatment polymers. In: Hamilton JD and Sutcliffe R (eds.) *Ecological Assessment of Polymers. Strategies for Product Stewardship and Regulatory Programs*, pp. 113–145. New York: Van Nostrand Reinhold.

Opgenorth H-J (1987) Umweltvertraeglichkeit von Polycarboxylaten. *Tenside Detergents* 24: 366–369.

Schumann H (1990) Elimination von 14Cmarkierten Polyelektrolyten in Biologischen Laborreaktoren. Fortschritt-VDI Berichte, Reihe 15: Umwelttechnik 81. VDI, Duesseldorf, 1 – 190.

Van Ginkel CG and Gayton S (1996) The biodegradability and nontoxicity of carboxymethyl cellulose (DS 0.7) and intermediates. *Environmental Toxicology and Chemistry* 15(3): 270–274.

Ward TJ, Rausina GA, Stonebraker PM, and Robinson WE (2002) Apparent toxicity resulting from the sequestrating of nutrient trace metals during standard *Selenastrum capricornutum* toxicity tests. *Aquatic Toxicology* 60: 1–16.

Woltering DM and Bishop WE (1989) Evaluating the environmental safety of detergent chemicals: A case study of cationic surfactants. In: Pastenbach DJ (ed.) *The Risk Assessment of Environmental Hazards*, pp. 345–389. New York: Wiley.

Uranium

R M Harper, Western Washington University, Bellingham, WA, USA

C Kantar, Mersin University, Mersin, Turkey

Uranium Sources
Redox Chemistry
Aqueous Chemistry

Toxicity
Further Reading

Uranium Sources

Uranium is a naturally occurring radioactive element with an atomic number of 92 and is considered a heavy metal. Uranium makes up *c.* $2-4\,\mathrm{mg\,kg^{-1}}$ of the Earth's crust and can occur as a significant constituent in numerous minerals such as uraninite, pitchblende, carnotite, autunite, torbernite, schoepite, uranophane, and coffinite. Uranium has been mined both for nuclear power and for nuclear weapons and so is present in the ecosphere naturally as a mineral as well as a waste product of the nuclear industry. It can be found at low levels in soils, rocks, and water in nature.

There are 19 different known isotopes of uranium, with half-lives ranging from 1 μs (^{222}U) to 4.47×10^9 years (^{238}U) and atomic masses ranging from 218 to 242 g mol^{-1}. Natural uranium has three isotopes: ^{238}U (half-life 4.47×10^9 years, energy 4.196 MeV), ^{235}U (half-life 7.038×10^8 years, energy 4.598 MeV), and ^{234}U (half-life 2.446×10^5 years, energy 4.777 MeV), with abundances of approximately 99.275%, 0.72%, and 0.054%, respectively. The natural uranium isotopes undergo radioactive decay by releasing alpha (α) particles accompanied by weak gamma (γ) radiation. The decay process terminates with the release of stable, non-radioactive isotopes such as ^{206}Pb and ^{207}Pb.

Uranium in ores can be extracted and chemically converted to uranium dioxide (UO_2) or other chemical forms that are usable in the nuclear industry. Enriched uranium (increased ^{234}U and ^{235}U), used as a fuel in nuclear reactors, typically contains 2–4% ^{235}U, and >90% ^{235}U in nuclear weapons and nuclear submarines. The uranium that remains after the enrichment process is known as depleted uranium (DU), and has less ^{235}U and ^{234}U than the isotopic composition of naturally occurring U. DU is generally used in military armor and armor penetrators due to its high density. It is also used as counterweights on certain wing parts in airplane and helicopter construction, shields for irradiation units in hospitals, and containers for transport of radioactive materials.

Redox Chemistry

The most important property of uranium with respect to its environmental behavior is its oxidation state since solubility, complexation, and sorption behavior differ from one oxidation state to another. Uranium isotopes have a long half-life, and are observed to have coordination chemistry consisting of multiple stable oxidation states and stable solid and aqueous forms within the ecosphere. Uranium can exist in oxidation states ranging from II to VI, although U(IV) and U(VI) are the most dominant oxidation states in natural systems. Under standard environmental (oxidizing) conditions, uranium typically occurs in the hexavalent form (U(VI)). Aqueous U(III) readily oxidizes to U(IV) under most oxidizing conditions. The aqueous uranyl U(V) species (UO_2^+) easily dissociates into U(IV) and U(VI). Under strongly reducing conditions, U(VI) can be reduced to U(IV) (uranous). Microbial activities and natural organic ligands may play a significant role in U(VI) reduction to U(IV).

Aqueous Chemistry

The aqueous chemistry of uranium depends on a number of chemical factors, including redox reactions, hydrolysis, precipitation/dissolution, solution complexation, and sorption as illustrated in **Figure 1**. Uranium is chemically very reactive and readily reacts with both inorganic and

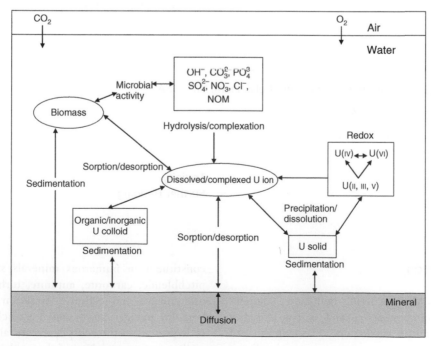

Figure 1 Biological/chemical processes affecting U distribution in natural systems.

organic ligands in the environment. The U^{4+} ion hydrolyzes and forms sparingly soluble hydroxides, phosphates, or oxides (e.g., uraninite ($UO_{2(s)}$)). Due to very low solubility of U(IV) solids, the concentration of U(IV) in water is rather low, ranging from 3 to 30 $\mu g\,l^{-1}$. The most common hydrolytic species of U(IV) include UOH^{3+}, $U(OH)_2^{2+}$, $U(OH)_3^+$, and $U(OH)_4$ (**Table 1**). These hydrolysis species dominate U(IV) speciation under slightly acidic to alkaline conditions. In groundwater systems, however, complexation of U(IV) with inorganic and organic groundwater ligands such as carbonate and natural organic matter may increase its solubility due to the formation of nonsorbing aqueous-phase U(IV)–ligand complexes.

In aqueous systems, U(VI) forms mono- and polynuclear hydrolysis species. The formation of aqueous U(VI) species is highly dependent on water chemistry, particularly the pH and concentrations of total dissolved U(VI) and complexing ligands. In acidic environments, the uranyl ion (UO_2^{2+}) is the dominant species. Under slightly acidic to alkaline conditions (pH > 5), the mononuclear

Table 1 Aqueous-phase U reactions

$U^{4+} + H_2O = UOH^{3+} + H^+$

$U^{4+} + 2H_2O = U(OH)_2^{2+} + 2H^+$

$U^{4+} + 3H_2O = U(OH)_3^+ + 3H^+$

$U^{4+} + 4H_2O = U(OH)_4 + 4H^+$

$U^{4+} + 5H_2O = U(OH)_5^- + 5H^+$

$UO_2^{2+} + H_2O = UO_2OH^+ + H^+$

$UO_2^{2+} + 2H_2O = UO_2(OH)_{2(aq)} + 2H^+$

$UO_2^{2+} + 3H_2O = UO_2(OH)_3^- + 3H^+$

$UO_2^{2+} + 4H_2O = UO_2(OH)_4^{2-} + 4H^+$

$2UO_2^{2+} + H_2O = (UO_2)_2(OH)^{3+} + H^+$

$2UO_2^{2+} + 2H_2O = (UO_2)_2(OH)_2^{2+} + 2H^+$

$3UO_2^{2+} + 4H_2O = (UO_2)_3(OH)_4^{2+} + 4H^+$

$3UO_2^{2+} + 5H_2O = (UO_2)_3(OH)_5^+ + 5H^+$

$3UO_2^{2+} + 7H_2O = (UO_2)_3(OH)_7^- + 7H^+$

$4UO_2^{2+} + 7H_2O = (UO_2)_4(OH)_7^+ + 7H^+$

$UO_2^{2+} + CO_3^{2-} = UO_2CO_{3(aq)}$

$UO_2^{2+} + 2CO_3^{2-} = UO_2(CO_3)_2^{2-}$

$UO_2^{2+} + 3CO_3^{2-} = UO_2(CO_3)_3^{4-}$

$3UO_2^{2+} + 6CO_3^{2-} = (UO_2)_3(CO_3)_6^{6-}$

$3UO_2^{2+} + CO_3^{2-} + 3H_2O = (UO_2)_3(CO_3)(OH)_3^+ + 3H^+$

$2UO_2^{2+} + CO_3^{2-} + 3H_2O = (UO_2)_2(CO_3)(OH)_3^- + 3H^+$

$UO_2^{2+} + PO_4^{3-} = UO_2PO_4^-$

$UO_2^{2+} + PO_4^{3-} + H^+ = UO_2HPO_{4(aq)}$

$UO_2^{2+} + PO_4^{3-} + 2H^+ = UO_2H_2PO_4^+$

$UO_2^{2+} + PO_4^{3-} + 3H^+ = UO_2H_3PO_4^{2+}$

$UO_2^{2+} + 2PO_4^{3-} + 4H^+ = UO_2(H_2PO_4)_{2(aq)}$

$UO_2^{2+} + 2PO_4^{3-} + 5H^+ = UO_2(H_2PO_4)(H_3PO_4)^+$

$UO_2^{2+} + SO_4^{2-} = UO_2SO_{4(aq)}$

$UO_2^{2+} + 2SO_4^{2-} = UO_2(SO_4)_2^{2-}$

$UO_2^{2+} + NO_3^- = UO_2NO_3^+$

$UO_2^{2+} + Cl^- = UO_2Cl^+$

$2UO_2^{2+} + 2Cl^- = (UO_2)_2Cl_2^{2+}$

(UO_2OH^+) and polynuclear species (e.g., $UO_2(OH)_2$, $UO_2(OH)_3^-$) dominate the U(VI) speciation in the absence of complexing ligands such as dissolved carbonate, sulfate, and phosphate (**Table 1**).

U(VI) also forms several monomeric and polynuclear UO_2^{2+}–carbonate complexes in systems containing dissolved carbonate. The formation of such species depends highly on the pH and total dissolved carbonate concentration. These species usually dominate U(VI) speciation in alkaline conditions. While $(UO_2)_2CO_3(OH)_3^-$ is the dominant U complex under slightly alkaline conditions and at high U concentrations, U–carbonate species such as $UO_2(CO_3)_2^{2-}$ and $UO_2(CO_3)_3^{4-}$ dominate U speciation at pH > 8. Under slightly acidic conditions (i.e., pH 6–7), UO_2CO_3 becomes the dominant U complex, especially at high carbonate concentrations. Complexes of U(VI) with inorganic ligands such as sulfate, phosphate, and chloride are important in aqueous systems where concentrations of these ions are high. Such U–ligand complexes usually form under acidic conditions. The affinity of UO_2^{2+} for complexation with inorganic ligands decreases in the order of $CO_3^{2-} > PO_4^{3-} > SO_4^{2-} > NO_3^- > Cl^-$.

In most aquatic systems, species of natural organic matter (NOM) such as fulvic and humic acids constitute an important pool of ligands for complexing metals. The uranyl ion, for example, displays a strong affinity for organic ligands due to its high effective charge. The effective charge of U in UO_2^{2+} is about 3.3. U migration, solubility, and bioavailability are highly affected by the formation of such stable UO_2^{2+}–organic ligand complexes in subsurface soil systems. Uranium also forms highly soluble complexes with low molecular weight organic ligands such as citrate, oxalate, and ethylenediaminetetraacetic acid (EDTA).

Toxicity

Uranium is nonessential for biological processes. Both chemical and radiation hazards can be of concern with radioactive heavy metals such as uranium, although the radiotoxicity of uranium has been shown to be low. All radioisotopes of uranium can result in chemical toxicity. With radiation hazards in general, the intensity of the source is important; for example, radioisotopes with shorter half-lives are generally more toxic. Additionally, the degree of toxicity is related to the type of emission; beta particle emission and gamma rays are generally more toxic than alpha particle emission. For both chemical and radiological hazards, the route of exposure is an important component of toxicity and will modify degree of toxicity.

For chemical toxicity of uranium, the route of exposure and chemical form are of importance. In mammals, toxicity is increased when the route of exposure is inhalation compared to ingestion. With ingestion (through either food or water), the majority of uranium is not absorbed through the gastrointestinal (GI) tract and is, therefore, eliminated via feces. A small mass of the soluble species (such as UCl_4, UO_2F_2, and $UO_2(NO_3)_2 \cdot 6H_2O$) is absorbed into the blood, distributed systemically, and either eliminated via urine or stored in kidneys, bones, and soft tissues. With inhalation exposures, the size of the particle and the solubility of the uranium are important. Large particles will be removed from the respiratory tract by mucociliary action and enter the GI tract. Smaller particles can be deposited in the alveoli where the insoluble forms (such as UF_4, UO_2, and UO_3) can remain for years, resulting in local toxicity to the lungs. The soluble forms can be absorbed, transported systemically, and ultimately excreted via urine, although some are retained in the kidneys and bones. The primary target organ of inhaled soluble uranium species, therefore, is the kidney in mammals, not the lungs.

Naturally occurring isotopes of U decay by alpha particle emission. This does not pose a significant radiological hazard when the exposure occurs outside of the body, as the alpha particles cannot readily penetrate the skin. Ionizing radiation of enriched uranium, however, is greater than depleted or natural uranium, because of the greater abundance of ^{234}U and ^{235}U, which have the shortest half-lives of the environmentally relevant U isotopes. Cancer has not been linked to natural U or DU exposures, but may be linked to bone sarcomas with enriched U exposures in humans. The greatest radiological hazard would be one where the uranium is in close proximity to a cell for a long period of time, such as with an inhalation exposure of an insoluble form of uranium that is small enough to pass through the pulmonary system to the alveoli. A low risk of lung cancer in humans has been identified in this exposure scenario, because additional exposure factors, such as cigarette smoke and radon, have confounded the analysis. For nonhumans, changes in natural populations (i.e., distribution and abundance) have not been demonstrated as a result of radiotoxicity. Of greater radiological risk and concern with a uranium exposure are the radionuclides in the transformation series, such as Rn, Rd, and Th, which emit alpha, beta, and gamma radiation, and have been shown to be carcinogenic.

Additional hazards may be present in aquatic environments where the primary routes of exposure to aquatic organisms include ingestion and dermal/gill absorption. Uranium toxicity in aquatic environments is highly dependent upon the metal speciation, but acute toxicity has been determined to be in the ppb to low ppm (m/v) range for most aquatic organisms. Water chemistry alters toxicity by changing the chemical form and, subsequently, the bioavailability of uranium. An increase in hardness, for example, results in a decrease in toxicity in fish. LC50 values in teleost fish have been shown to increase $\sim$2 orders of magnitude in hard water from the low ppm (m/v) in soft water. The presence of organic ligands, such as NOM, can also reduce toxicity, as the U:ligand complexes are generally less bioavailable and toxic than the free ion species. pH is an additional modifier of U toxicity. The changes in toxicity caused by pH are complex, and can generally occur in two competing ways. First, pH can alter the chemical form of uranium that is present in the environment and, therefore, affects the concentration of toxic forms that are available to interact with exposed organisms. For example, at pH < 5, the predominant uranyl species is the UO_2^{2+} ion, which is bioavailable and, therefore, is likely to be absorbed by cellular membranes. UO_2CO_3 would be a predominant U:carbonate species at pH 6–7. This carbonate species is also considered bioavailable. At higher pH, the predominant uranium carbonate is $UO_2(CO_3)_3^{4-}$, which is not expected to pass through cell membranes. The toxicity of uranium in a carbonate system, therefore, is expected to increase at lower environmentally relevant pHs. Second, pH may directly affect toxicity by changing the cell surface where initial exposure occurs or by competition mechanisms, resulting in changes in membrane permeability and uranium uptake. At decreased pH, for example, the H^+ concentration is increased. It is proposed that the H^+ can compete with uranium for binding sites of a cell and has been shown to lead to a decreased uptake of the toxic forms of uranium (i.e., UO_2^{2+} and UO_2OH^+). The free ionic activity model (FIAM) can be used to explain this relationship between uranium ions and toxicity.

Uranium can bioaccumulate in aquatic organisms. Uranium binds to both proteins and lipids at biologically relevant pHs, and has been found to accumulate both extracellularly and intracellularly in microorganisms, such as bacteria, fungi, and algae and in invertebrates. The accumulated uranium may be stored as granules, reducing the toxicity, or may inhibit enzyme activity, such as ATPase resulting in increased toxicity. Adsorbed uranium on unicellular organisms, such as algae, may be a source of ingested uranium in higher-trophic-level organisms.

See also: Bioavailability; Bioaccumulation; Ecological Risk Assessment; Ecotoxicology: The Focal Topics; Exposure and Exposure Assessment; Ecotoxicology: The History and Present Directions; Radioactivity; Risk Management Safety Factor; Plutonium.

Further Reading

Agency for Toxic Substances and Disease Registry (ATSDR) (1999) *Toxicological Profile for Uranium*. Atlanta, GA: US Department of Health and Human Services, Public Health Service.

Craft ES, Abu-Qare AW, Flaherty MM, *et al.* (2004) Depleted and natural uranium: Chemistry and toxicological effects. *Journal of Toxicology and Environmental Health* 7B: 297–317.

Guillaumont R, Franghanel T, Fuger J, *et al.* (2003) *Update on the Chemical Thermodynamics of Uranium, Neptunium, Plutonium, Americium and Technetium*. New York: Elsevier.

Irwin RJ, Van Mouwerick M, Stevens L, Seese MD, and Basham W (1997) *Environmental Contaminants Encyclopedia – Uranium Entry.* Fort Collins, CO: National Park Service, Water Resources Division, Water Operations Branch.

Krupka KM, Kaplan DI, Whelan G, Serne RJ, and Mattigod SV (1999) *Understanding Variation in Partition Coefficient, K_d, Values. Vol. II: Review of Geochemistry and Available K_d Values for Cadmium, Cesium, Chromium, Lead, Plutonium, Radon, Strontium, Thorium,*

Tritium (^3H), and Uranium. Washington, DC: US Environmental Protection Agency.

Pacific, Northwest National Laboratory (PNNL) (2003) Uranium. In: *Methods and Models of the Hanford Internal Dorimetry Program.* PNNL-MA-860. http://www.pnl.gov/eshs/pub/pnnl860.html (accessed December 2007).

Veterinary Medicines

B H Sørensen, Copenhagen University, Copenhagen, Denmark

Introduction
Fate and Exposure Routes
Effects

Scenarios
Analysis of Drugs in the Environment
Further Reading

Introduction

Veterinary medicines were not discussed in this context before the early 1990s. It was proposed in the European Union (EU) in the beginning of the 1990s to distinguish between drugs with extremely low concentrations in the environment and which therefore would be no threat to the environment on the one side and drugs which could occur in sufficiently high concentrations to cause deterioration of the environment. The latter group of drugs should be examined further by setting up stepwise an environmental risk assessment (ERA). This has been reached and the EU has adopted to require ERA for all new veterinary drugs from 1 January 2004 (EU note for guidance EMEA/CVMP/055/96).

A survey of the current literature on the topic 'drugs in the environment' shows that the primary drugs investigated for environmental impacts are:

1. the antibiotics,
2. the antiparacetic agents,
3. the antinoleptic agents, and
4. the hormones.

Fate and Exposure Routes

Figures 1 and **2** shows the anticipated exposure routes to the environment for different types of drugs. To identify the exposure routes, it is necessary to divide the substances used for human treatment (**Figure 1**) and veterinary treatment (**Figure 2**) even though identical drugs are often used for humans and animals. The latter group can be further subdivided into exposure due to substances used as growth promoters for pigs (antibiotics

as growth promoters were phased out in Denmark on 1 January 2000), as therapeutics in livestock production, cocciodiostatica used for poultry production, into therapeutics for treatment of livestock on fields, for example, antiparacetic agents, or into feed additives in fish farms. An identification of the exposure route is crucial for estimating the corresponding environmental loading because, as for pesticides, it is the dose of drugs and the duration of treatment that gives the environmental loading. The same drug may be used for several applications, for example, therapeutical treatment for pigs and in fish farms, resulting in a different dose and duration of treatment. Combined with different exposure routes through various environmental matrices, the fate of the drug may also vary, resulting in quite different environmental concentrations.

The drugs used by humans will be discharged to the sewer systems together with the urine and feces (**Figure 1**, F1) and enter the sewage treatment plant (STP) (**Figure 1**, F4). The possible fate of drugs in the STP may, as for all other xenobiotics, be divided as:

1. The drug or metabolites of the parent drug are mineralized by microorganisms to carbon dioxide and water, for example, aspirin.

2. The drug or metabolites of the parent drug are more or less persistent in the STP, which implies that depending on the lipophilicity or other binding possibilities, such as ionic bindings, a part of the substance will be retained in the sludge. If the sludge is used as soil conditioner (**Figure 1**, F5), drugs may be dispersed on agricultural fields. Again the fate of the drugs depends on the lipophilicity or some other ability of binding to sludge or soil. Drug molecules often have many functional groups, for example, carboxylic acids, aldehydes, and

Drugs for human treatment

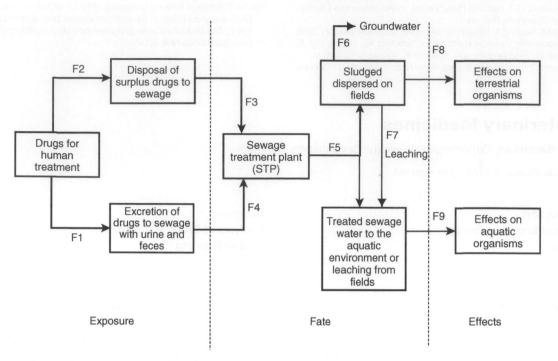

Figure 1 Anticipated exposure routes of drugs for human treatment in the environment.

Drugs for veterinary treatment

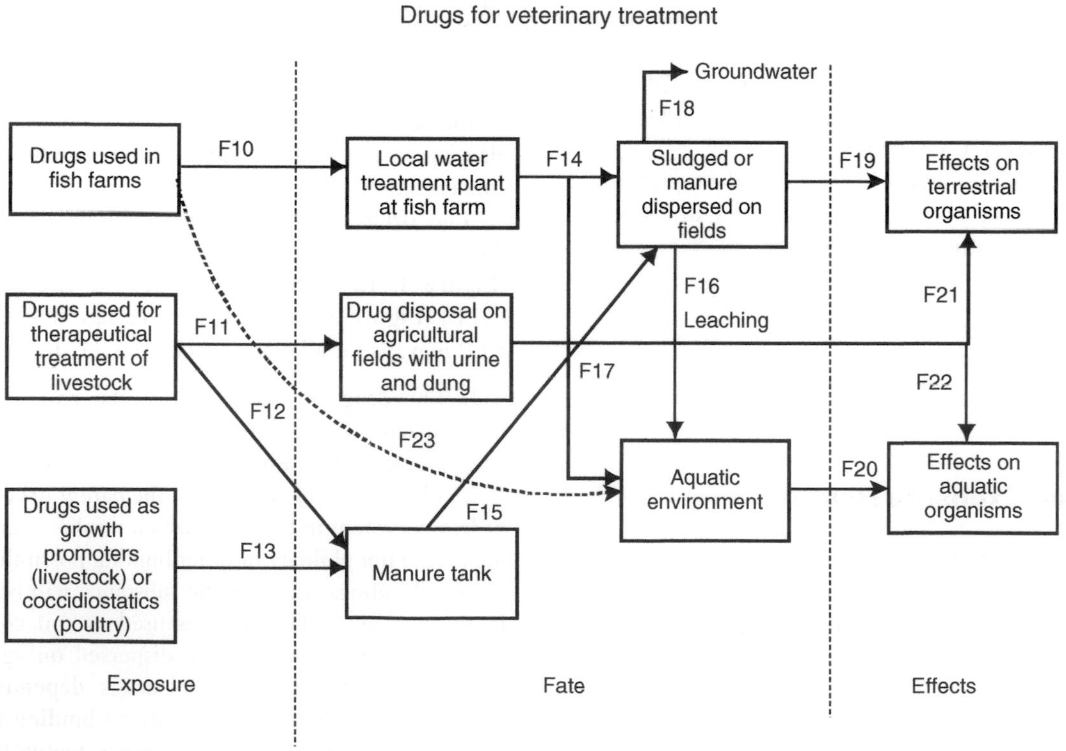

Figure 2 Anticipated exposure routes of drugs for veterinary treatment in the environment.

amines, which make the binding capacities of the molecules to solids dependent on pH or other constituents (e.g., complexation) in the solids matrix. Drugs that are mobile in the soil may be a threat to the groundwater (**Figure 1**, F6) or leach to a nearby stream (**Figure 1**, F7). Depending on the ability of the drug to bind solids, either organisms in the terrestrial ecosystem (**Figure 1**, F8) or aquatic ecosystem (**Figure 1**, F9) may be exposed.

3. The drug or metabolites of the parent drug are persistent and at the same time very polar and nonbinding to solids. The substance will thus neither be retained nor degraded in the STP and therefore easily reach the aquatic environment (**Figure 1**, F5), and affect the aquatic organisms.

An unknown portion of drugs marketed for human treatment ends in the sewer system as surplus medical substances (**Figure 1**, F2). After entering the STP (**Figure 1**, F3), the fate of these drugs will be almost identical with the excreted drugs. The only difference is that the wastewater will not include the drug metabolites often produced by humans before excretion.

Most of the drugs used for veterinary purposes (**Figure 2**, F12 and F13) will end up in manure. Manure is conserved in tank systems for a period of time before being dispersed on fields (**Figure 2**, F15). This conservation period and field immersion standards depend on legislative regulations on national level. As previously explained, the mobility of the drugs or drug metabolites in the soil system predicts if the drug will threaten the groundwater (**Figure 2**, F18), affect the terrestrial organisms (**Figure 2**, F19), or the aquatic organisms (**Figure 2**, F20), due to leaching from fields (**Figure 2**, F16).

If the drugs are applied to grazing animals, drug or drug metabolites will be urinated or defecated directly on the field (**Figure 2**, F11) and principally undergo the same fate as the drugs dispersed with manure. The only difference is that the latter situation may give much higher local environmental loadings of drugs.

Drugs used as a feed additive in fish farming will be discharged directly to the receiving water as parent compound (**Figure 2**, F10 and F17 or F23), because a large portion of the applied medicated feed is not eaten by the fish. A local water treatment plant often treats the water from aquaculture before going to the aquatic environment. The sludge from these treatment plants is also used as soil conditioner, so drugs used in fish farming may also end up on agricultural soil (**Figure 2**, F14), and undergo the same fate as previously described for the growth promoters and therapeutically used drugs.

Generally, the exposure routes are of particular importance for drugs, because the total amount of drugs emitted to the environment is relatively low at least compared with the most widely used chemicals.

However, a local discharge of drugs may cause environmental problems due to concentrations exceeding $10–100\,\mathrm{ng\,l^{-1}}$. It is therefore preposterous that drugs were excluded when ERA for all new chemicals was introduced. Drugs are biologically very active chemicals – they are actually drugs because they have a biological activity that is used to cure or prevent diseases. This implies that drugs often are made to interfere with specific biological systems, for example, receptors or specific enzymes. Probably, it was assumed that they occur in sufficiently low concentrations to be harmless for the environment independent of their biological activity. It may be the case if all drugs discharged to the environment would be diluted in the total environment. But several drugs are discharged locally, for instance antibiotics in manure which is spread on a few hectares of agricultural land, or hospital wastewater discharged via a relatively small treatment plant to a local stream or lake. In addition, the consumption of some (few) drugs is surprisingly high (see **Table 1**, where the annual consumption of some selected drugs in Denmark, representing 5 million inhabitants in a typical high-industrialized country, is shown).

In 1996, 25 different drugs were identified in natural environmental samples. This number has risen to more than 150 in 2006. It seems therefore that we have reached the same situation as known from the pesticides where we find most of all the compounds that we are looking for. We have however only a few analytical methods capable of detecting drugs at environmnetally relevant concentrations. Therefore, we need to ensure the quality of the sampling and analytical procedure whenever we are identifying drugs in natural samples.

Effects

We can classify the possible effects of veterinary medicines into three groups. The first group comprises the normal toxic effect we can observe for all xenobiotics. The impact may be on any level of the biological hierarchy: cells–organs–organisms–population–ecosystems–the ecosphere. Effects are evaluated on line with other xenobiotics.

Antibiotics have a different effect, because bacteria are the target organism of antibiotics. As antibiotics are nature's own weapons for establishment and maintenance of all microbial ecosystems, resistance is a natural part of the regulatory factors in any ecosystem and genes coding for resistance have existed as long as microbes. The increased use of antibiotics during the last five decades has caused a genetic selection of more harmful bacteria. The genetic pool of microorganisms in nature has changed significantly, simply due to our increasing production and consumption of antibiotics. This is a long-term and, to a high extent, irreversible effect. We can for instance

Table 1 Consumption in Denmark (5.2 million inhabitants) of some high-volume medical substances in 2005

Substances	DDD per year (millions)[a]	DDD WHO (g)[b]	Applied weight (t)
Human treatment			
Single substances			
Ibuprofen	27.7	1.2	33.2
Furosemid	91.9	0.040	3.7
Estrogens in comb. with gestoden or desorgestrel	58.3	6.5×10^{-5}	3.8×10^{-3}
Estradiol	24.3	0.002	0.049
Therapeutic groups			
Antibiotics	25.1	1.5	37.7
Analgesic (NSAID-type)	56.6	0.5	28.3
Hypotensiva	41.0	0.010	0.41
Diuretica (loop)	95.3	0.040	3.8
Antiasthmatics	110.5	0.015	1.7
Psychleptics	147.5	0.050	7.4
Veterinary treatment			
Growth promoters (livestock production pigs)[c]			
Single substances			
Carbadox			1.181
Olaquindox			16.213
Avilamycin			1.665
Avoparcin			5.690
Bacitracin			7.910
Flavomycin			0.048
Monesin			5.007
Salinomycin			0.850
Spiramycin			0.507
Tylosin			52.275
Virginiamycin			2.590
Therapeutics applied in livestock production[d]			
Antibiotics			49.687
Therapeutics applied in poultry industrys			
Coccidiostatics			16.165
Antibiotics feed additive			2.491
Fish farms, feed additives[e]			
Single substances			
Oxolinic acid and inoxyl			3.406
Oxytetracycline			0.082
Sulfa-contaning substances (therapeutics)			0.182

[a]Defined Daily Doses in millions (pers. comm. Danish Drug Administration).
[b]Anatomical Therapeutic Chemical (ATC) index, WHO (1995).
[c]Pers. comm. Plantedirektatoriet, Lyngby, Denmark.
[d]Pers. comm. Plantedirektoratet, Lyngby, Denmark.
[e]Pers. comm. Velje Amt, Teknik og Miljø, Tilsyn med dambrug (1995).

reintroduce a fish in a stream if it has been exterminated due to pollution, but we cannot reestablish the same gene pool in a microorganism population. Moreover, it seems that the development of antibiotic resistance is favored by low concentrations of antibiotics. In conclusion, we are confronted with another type of effect that we do not know from other chemicals – it is irreversible and exerted at even very low concentrations.

The discovery of the so-called endocrine disrupters, that is, chemicals which can disturb the normal function of hormones (third group of effects), has added to our suspicion that drugs may also cause environmental damages even if they may be found in low concentrations.

It was found more than 15 years ago that some compounds, as for instance phthalates and bisphenol A, can in some contexts replace estrogen and thereby disturb the hormone balance of an organism. As these effects are observed in the concentration range of a few nanograms per liter, it is obvious that other biologically active compounds may give adverse effects at extremely low concentrations. The adverse effects may be due to disturbances of 'biological signals' or due to cascade effects, which are both known from biochemistry to occur at extremely low concentrations. As hormones are used directly as drugs to cure various diseases, it might not be excluded that this type of effect is also associated with our use of drugs.

Therefore, the question is that do we have the needed test system, on laboratory and field scale, to evaluate the environmental effects of drugs? We think that the answer is no.

Scenarios

During the 1990s, we have observed an increasing interest in the risks and effects of drugs. The full step, to make ERA for all drugs and consider them 'in principle not different' from other chemicals, has not yet been taken, but it is strongly recommended to take this step as soon as possible. To distinguish between drugs and other chemicals when they are discharged to the environment is preposterous. Chemicals used to kill insects in a field or to kill microorganisms in our body may of course cause the same damaging results to the living organisms in our environment.

During the last one or two decades, we have got a better understanding of ecosystems and their functions. We understand today much better why they sometimes are extremely robust and sometimes very sensitive to our impacts. The ecological network which links everything together in an ecosystem can explain why it is sometimes possible to repair damage and why the network sometimes collapses when an indispensable part of the network is impaired. Risk assessment requires that we take all known types of effects of a chemical on all levels of the biological hierarchy from cells to the entire ecosphere in account. It is of course not possible to claim beforehand that a risk assessment is unnecessary. It is important to consider the buffer capacity of the ecosystem toward any change and also especially for antibiotics the possible change of the gene pool.

As emphasized, drugs are not environmentally different from other chemicals. It means that it is the same properties that are of interest in an environmental evaluation of a drug. Water-soluble compounds are more mobile and may therefore contaminate the hydrosphere and threaten the groundwater. Lipophilic compounds are bioaccumulated or accumulated in sediment and sludge. The biodegradability is a key factor, because we cannot tolerate a chemical to accumulate somewhere in the environment for years, because it may cause unexpected adverse effects due to unexpected high long-term concentrations or due to synergistic effects. The bioavailability is equally pertinent for drugs as for other chemicals, as the effect due to the availability is different when the drug is adsorbed, bound more or less firmly to other compounds or dissolved in water. In this context, we should determine not only the properties of the drugs but also of their intermediate metabolites which may sometimes be a cumbersome task.

The predicted environmental concentration (PEC) should always be determined in the systems where it is anticipated that the highest values will be found. This implies that models should be developed for these cases. Consequently, the most pertinent models to be developed for drugs are:

1. antibiotics in manure disposed on agricultural land, including the possibilities to contaminate adjacent streams and the ground water;
2. antibiotics used in aqua culture where it will contaminate the water flowing from the aqua culture to the next aquatic ecosystem;
3. drugs and their metabolites in wastewater discharged to a treatment plant, where they will biodegrade, be discharged with the treated water, or accumulate in the sludge (for the latter case, model (1) can be applied, if the sludge is used as soil conditioner in agriculture).

Regional fate models can hardly be applied in the ERA, although they may be used to compare therapeutically identical drugs. Air pollution models are probably not needed.

The above-mentioned types of models ((1)–(3)) are developed. The gained experience is, however, limited up to now and none of the models has been fully validated, yet. It is important in the coming years to get sufficient data to allow the validation of these models, because it is prerequisite for their more general application.

Analysis of Drugs in the Environment

It is necessary to determine concentrations of drugs in the environment at least in a limited number. Models are able to identify where and when measurable concentrations will occur, but it is absolutely necessary to calibrate and validate the applied models against real data. The actual concentrations are in the nanograms per liter range and often associated with complex matrices (sediment, soil, etc.), that make heavy demands on the analytical work and the preconcentration procedures. Consequently, it is recommended to apply the most demanding quality control of the analytical results, for instance, by use of two or more different analytical methods to determine the same substances.

Another analytical complication is the determination of drug metabolites. Many drugs are partially metabolized in the human body before excretion. The metabolites may also be harmful to the environment and it is therefore necessary to include them in the investigations. All in all, analytical chemistry has in the coming years two challenging tasks: to develop and validate analytical methods that may determine low concentrations of drugs and their metabolites in complex environmental matrices.

See also: Antibiotics in Aquatic and Terrestrial Ecosystems; Biogeochemical Approaches to Environmental Risk Assessment.

Further Reading

Arcand-Hoy LD, Nimrod AC, and Benson WH (1998) Endocrine-modulating substances in the enviornment: Estrogen effects of pharmaceutical products. *International Journal of Toxicology* 17: 139–158.

Baguer AJ, Jensen J, and Krogh PH (2000) Effects of the antibiotics oxytetracycline and tylosine on soil fauna. *Chemosphere* 40(7): 751–757.

Boxall ABA, Oakes D, Ripley P, and Watts CD (2000) The application of predictive models in the environmental risk assessment of ECONOR®. *Chemosphere* 40(7): 775–781.

Carson RL (1962) *Silent Spring*. Boston: Houghton Mifflin.

Eckman MK (1994) Chemicals used by the poultry industry. *Poultry Science* 73: 1429–1432.

European Union (1996) Note for guidance: Environmental risk assessment of veterinary medical products other than GMO-containing and immunological products. EMEA/CVMP/055/96. London: EMEA.

Halling-Sørensen B (2000) Algal toxicity of antibacterial agents used in intensive farming. *Chemosphere* 40(7): 731–739.

Halling-Sørensen B, Jacobsen A-M, Jensen J, *et al.* (2005) Dissipation and effects of chlortetracycline and tylosin in two agricultural soils – A field-scale study in southern Denmark. *Environmental Toxicology and Chemistry* 24(4): 802–810.

Halling-Sørensen B, Nors Nielsen N, Lansky PF, *et al.* (1998) Occurrence, fate and effects of pharmaceuticals in the environment – A review. *Chemophere* 36(2): 357–393.

Holten Lützhøft H-C, Halling-Sørensen B, and Jørgensen SE (1999) Algal toxicity of antibacterial agents applied in Danish fish farming. *Archives of Environmental Contamination and Toxicology* 36: 1–6.

Holten Lützhøft H-C, Vaes WHJ, Freidig AP, Halling-Sørensen B, and Hermens JLM (2000) 1-Octanol/water distribution coefficient of oxolinic acid: Influence of pH in relation to the interaction with dissolved organic carbon. *Chemosphere* 40(7): 711–714.

Ingerslev F, Holten Lützhøft H-C, and Halling-Sørensen B (1999) Humant anvendte lægemidlers vej til miljøet er gennem rensningsanlægget. *Dansk kemi* 80(6–7): 22–25 (in Danish).

Jacobsen P and Berglind L (1988) Persistence of oxytetracycline in sediments from fish farms. *Aquaculture* 70: 365–370.

Jørgensen SE (ed.) (1984) *Development in Environmental Modelling l6: Modelling the Fate and Effects of Toxic Substances in the Environment*, 342pp. Amsterdam: Elsevier.

Jørgensen SE (2002) *Integration of Ecosystem Theories: A Pattern*, 400pp. Dordrecht, The Netherlands: Kluwer.

Jørgensen SE, Holten Lützhøft H-C, and Halling-Sørensen B (1998) Development of a model for environmental risk assessment of growth promoters. *Ecological Modelling* 107: 63–72.

Kümmerer K, Al-Ahmad A, Bertram B, and Wießler M (2000) Biodegradability of antineoplastic compounds in screening tests: Influence of glucosidation and of sterochemistry. *Chemosphore* 40: 767–773.

Kümmerer K, Al-Ahmad A, and Mersch-Sundermann V (2000) Biodegradability of some antibiotics, elimination of the genotoxicity and affection of waste water bacteria in a simple test. *Chemosphere* 40: 701–710.

Kümmerer K, Steger Hartmann T, Baranyai A, and Bürhaus I (1996) Evaluation of the biological degradation of the antieoplastics cyclophosphamide and ifosfamide with the closed bottle test (OECD 301D). *Zeitschrift fur Hygiene* 198: 215–225.

Kümmerer K, Steger Hartmann T, and Meyer M (1997) Biodegradability of the anti-tumor agent ifosfamide and its occurrence in hospital effluents and sewage. *Water Research* 31: 2705–2710.

Loke M-L, Ingerslev F, Halling-Sørensen B, and Tjørnelund J (2000) Primary elimination of tylosin A in manure containing test systems by high performance liquid chomatography. *Chemosphere* 40(7): 759–765.

Migliore L, Civitareale C, Brambilla G, and Di Delupis GD (1997) Toxicity of several important agricultural antibiotics to artemia. *Water Research* 31(7): 1801–1806.

Migliore L, Civitareale S, Cozzolino P, *et al.* (1998) Laboratory models to evaluate phytotoxicity of sulphadimethoxine on terrestrial plants. *Chemosphere* 37(14–15): 2957–2961.

Migliore L, Cozzolino S, and Fiori M (2000) Phytotoxicity and bioaccumulation of flumequine used in intensive aquaculture on the aquatic weed, *Lythrum salicaria* L. *Chemosphere* 40(7): 741–750.

Petersen A, Olsen JE, and Dalsgaard A (1997) Antibiotikaresistente bakterier I det akvatiske miljø. *Dansk Veterinærtidskrift* 80(16): 15–18 (in Danish).

Rabølle M and Spliid NH (2000) Sorption and mobility of metronidazole, olaquindox, oxytetracycline and tylosin in soil. *Chemosphere* 40(7): 715–722.

Raich-Montiu J, Krogh KA, Sporring S, Jönsson JA, and Halling-Sørensen B (in press) Determination of ivermectin and transformation products in environmental waters using hollow-fibre microporous membrane liquid–liquid extraction and liquid chromatography–mass spectrometry/mass spectrometry.

Richardson ML and Bowron JM (1985) The fate of pharmaceutical chemicals in the aquatic environment – A review. *Journal of Pharmacy and Pharmacology* 37: 1–12.

Schneider J (1994) Problems related to the usage of veterinary drugs in aquaculture – A review. *Quimica Analitica* 13(supplement 1): S34–S42.

Shore LS, Gurevitz M, and Shemesh M (1993) Estrogen as an environmental pollutant. *Bulletin of Environmental Contamination and Toxicology* 51: 361–366.

Shore LS, Shemesh M, and Cohen R (1988) The role of oestradiol and oestrone in chicken manure silage in hyperoestrogenisms I cattle. *Australian Veterinary Journal* 65: 67.

Sommer C (1992) *Environmental Consequences of Ivermectin Usage*. Ph.D. Thesis, Institute of Population Biology, University of Copenhagen, Denmark (in English).

Sommer C and Overgaard Nielsen B (1992) Larvae of the dung beetle *Onthophagus gazella* F. (Col., *Scarabaeidae*) exposed to lethal and sublethal ivermectin concentrations. *Journal of Applied Entomology* 114: 502–509.

Sommer C, Steffansen B, Overgaard Nielsen B, *et al.* (1992) Ivermectin excreted in cattle dung after subcutaneous injection or pour-on treatment: Concentration and impact on dung fauna. *Bulletin of Entomological Research* 82: 257–264.

Stuer Lauridsen F, Birkved M, Hansen L, Holten Lützhøft H-C, and Halling-Sørensen B (2000) Environmental risk assessment of human pharmaceuticals in Denmark after normal therapeutical use. *Chemosphere* 40(7): 783–793.

Tabak HH, Bloomhuff RN, and Bunch RL (1981) Steriod hormones as water pollutants. Part II: Studies on the persistence and stability of natural urinary and synthetic ovulation-inhibiting hormones in treated and untreated waste water. *Developments in Industrial Microbiology* 22: 457–519.

Wollenberger L, Halling-Sørensen B, and Kusk KO (2000) Acute and chronic toxicity of veterinary antibiotics to *Daphnia magna*. *Chemosphere* 40(7): 723–730.

Wratten SD and Forbes AB (1996) Environmental assessment of veterinary avermectins in temperate pastoral ecosystems. *Annals of Applied Biology* 128: 329–348.

APPENDICES

APPENDICES

Appendix 1: The periodic table of ecotoxicology

Periodic Table

1A	2A		3B	4B	5B	6B	7B	8B			1B	2B	3A	4A	5A	6A	7A	8A
1 **H** Hydrogen																		2 **He** Helium
3 **Li** Lithium	4 **Be** Beryllium												5 **B** Boron	6 **C** Carbon	7 **N** Nitrogen	8 **O** Oxygen	9 **F** Fluorine	10 **Ne** Neon
11 **Na** Sodium	12 **Mg** Magnesium												13 **Al** Aluminium	14 **Si** Silicon	15 **P** Phosphorus	16 **S** Sulfur	17 **Cl** Chlorine	18 **Ar** Argon
19 **K** Potassium	20 **Ca** Calcium	21 **Sc** Scandium	22 **Ti** Titanium	23 **V** Vanadium	24 **Cr** Chromium	25 **Mn** Manganese	26 **Fe** Iron	27 **Co** Cobalt	28 **Ni** Nickel	29 **Cu** Copper	30 **Zn** Zinc	31 **Ga** Gallium	32 **Ge** Germanium	33 **As** Arsenic	34 **Se** Selenium	35 **Br** Bromine	36 **Kr** Krypton
37 **Rb** Rubidium	38 **Sr** Strontium	39 **Y** Yttrium	40 **Zr** Zirconium	41 **Nb** Niobium	42 **Mo** Molybdenum	43 **Tc** Technetium	44 **Ru** Ruthenium	45 **Rh** Rhodium	46 **Pd** Palladium	47 **Ag** Silver	48 **Cd** Cadmium	49 **In** Indium	50 **Sn** Tin	51 **Sb** Antimony	52 **Te** Tellurium	53 **I** Iodine	54 **Xe** Xenon
55 **Cs** Cesium	56 **Ba** Barium	57-71 Lanthanides	72 **Hf** Hafnium	73 **Ta** Tantalum	74 **W** Tungsten	75 **Re** Rhenium	76 **Os** Osmium	77 **Ir** Iridium	78 **Pt** Platinum	79 **Au** Gold	80 **Hg** Mercury	81 **Tl** Thallium	82 **Pb** Lead	83 **Bi** Bismuth	84 **Po** Polonium	85 **At** Astatine	86 **Rn** Radon
87 **Fr** Francium	88 **Ra** Radium	89-103 Actinides	104 **Rf** Rutherfordium	105 **Db** Dubnium	106 **Sg** Seaborgium	107 **Bh** Bohrium	108 **Hs** Hassium	109 **Mt** Meitnerium	110 **Ds** Darmstadtium	111 **Rg** Roentgenium	112 **Uub** Ununbium	113 **Uut** Ununtrium	114 **Uuq** Ununquadium	115 **Uup** Ununpentium	116 **Uuh** Ununhexium	117 **Uus** Ununseptium	118 **Uup** Ununoctium

Lanthanides	57 **La** Lanthanum	58 **Ce** Cerium	59 **Pr** Praseodymium	60 **Nd** Neodymium	61 **Pm** Promethium	62 **Sm** Samarium	63 **Eu** Europium	64 **Gd** Gadolinium	65 **Tb** Terbium	66 **Dy** Dysprosium	67 **Ho** Holmium	68 **Er** Erbium	69 **Tm** Thulium	70 **Yb** Ytterbium	71 **Lu** Lutetium
Actinides	89 **Ac** Actinium	90 **Th** Thorium	91 **Pa** Protactinium	92 **U** Uranium	93 **Np** Neptunium	94 **Pu** Plutonium	95 **Am** Americium	96 **Cm** Curium	97 **Bk** Berkelium	98 **Cf** Californium	99 **Es** Einsteinium	100 **Fm** Fermium	101 **Md** Mendelevium	102 **No** Nobelium	103 **Lr** Lawrencium

Green: these elements are present in biological material. They participate in all ecological processes and cycle in ecological networks in all ecosystems. Blue: these elements are trace elements and are in small concentrations often present in biological material. These elements may in some context have ecotoxicological effects, particularly at higher concentrations. Red: These elements have ecotoxicological effects. It is important to know the environmental concentration and the effects. Po, Rn, Ra, U, Np, Pu, and Th have radioactive effects. Yellow: These elements are of minor interest ecology due to their low concentrations in biological material and low effects to ecological processes.

Appendix 2

This appendix gives an overview of the combinations of chemical components and their effects found in Parts B and C of this book. Chronic and acute toxicity cover a wide range of effects, and this cannot be covered by an overview table. It cannot be excluded that the combination of a chemical compound and its effect not included in the table and the book may play a (minor) role in ecotoxicological management applied in practice. For some compounds it is indicated in brackets that low biodegradability plays a role. The brackets indicate that low biodegradability is a property found in many of the chemical components of the group but it is not necessarily valid for all the components in the group. The brackets [] are used for radioactivity to underline that it is not a biological decomposition.

The matrix includes a few compounds that are not covered by a chapter in Part C, but the combinations of the chemical component and effect(s) are mentioned elsewhere in the book.

1. Acute Toxicity
2. Chronic Toxicity
3. Bioaccumulation
4. Low Biodegradability
5. Biomagnification
6. Effects as Endocrine Disruptors
7. Reproductive Toxicity
8. Mutagenesis
9. Teratogenesis
10. Carcinogenic

Appendix 2

	1	2	3	4	5	6	7	8	9	10
Antifouling chemicals		x								
Antibiotics	x	x								
Benzene	x	x	(x)	x						x
Copper	x		x						x	
Crude oil, oil, gasoline, and petrol	x	x	x					x		
Dioxin (see also POP)	x	x	x	x	x	x		x		x
Endocrine disruptors	x					x		x		
Halogenated hydrocarbons	x	x	x	x	x		x	x	x	x
Lead	x	x	x							
Nitrogen	x	x								
Persistent organic chemicals	x	x	x	x	x	x	x			
Phenols	x	x	x	(x)	x					
Pheromones	x						x			
Phthalates	x	x	x	(x)	x	x				x
Plutonium	x	x	x	[x]				x		x
Polychlorinated biphenyls	x	x	x	x	x	x	x		x	x
Polycyclic aromatic hydrocarbons	x	x	x	x	x			x	x	x
Radioactivity	x	x	x							x
Synthetic polymers	x	x	x	(x)						
Uranium	x	x	x							x
Veterinary medicines	x	x	x	(x)	x	x				
Bisphenol A	x	x	x	x	x	x				
Cadmium	x	x	x					x	x	x
UV light	x	x					x	x	x	x
Atrazine	x	x	x	x	x				x	x

INDEX

Printed and bound by CPI Group (UK) Ltd, Croydon, CR0 4YY

03/10/2024

01040328-0019